Manufacturing F

with introduction to

ERP, SCM, and CRM

About the Author

Professor Khalid Sheikh graduated in Mechanical Engineering from Government College of Engineering, Amravati (Maharashtra, India). Afterwards he did his post-graduation in Production Engineering from VJTI Mumbai. Subsequently he worked with:

- Nelco Limited, Mumbai;
- Western Regional Instrumentation Center (WRIC) of University of Bombay;
- Industrial Design Center of Indian Institute of Technology (IIT), Bombay;
- K J Somaiya College of Engineering, Mumbai; and
- National Institute of Industrial Engineering (NITIE), Mumbai.

It was while working at NITIE that he realized the need for a comprehensive text on MRP II and authored this book. Currently he is working as *Associate Professor* at S P Jain Institute of Management & Research (SPJIMR), Mumbai.

His areas of interest include:

- Concurrent Engineering,
- Japanese Manufacturing Systems,
- Enterprise Resource Planning (ERP),
- Supply Chain Management (SCM),
- Customer Relationship Management (CRM), and
- The Extended e-Enterprise.

Manufacturing Resource Planning (MRP II)

with introduction to

ERP, SCM, and CRM

KHALID SHEIKH

Associate Professor

S P Jain Institute of Management & Research (SPJIMR)

Mumbai

McGraw Hill

Boston Burr Ridge, IL Dubuque, IA Madison, WI New York San Francisco St. Louis
Bangkok Bogotá Caracas Kuala Lumpur Lisbon London Madrid Mexico City
Milan Montreal New Delhi Santiago Seoul Singapore Sydney Taipei Toronto

McGraw-Hill Higher Education

*A Division of The **McGraw-Hill** Companies*

Manufacturing Resource Planning (MRP II)
with introduction to ERP, SCM and CRM
International Edition 2002

10 09 08 07 06 05 04 03 02
20 09 08 07 06 05 04 03
CTF SLP

ISBN 0-07-042060-2

When ordering this title, use ISBN 0-07-120399-0

Printed in Singapore

Dedication and Acknowledgement

At a personal level, I dedicate this book to my parents

Abdul Wahid Sheikh and *Fasiha Sheikh*

who have been the support and source of strength for me throughout my life.

As a learner of MRP II, I dedicate this book to

Joseph Orlicky

who developed MRP I,

Oliver Wight

who was instrumental in extending the concept of MRP I into MRP II,

George Plossl

whose writings have had the greatest influence in shaping my own thinking and understanding of the subject, and

all the members of APICS

who freely shared their invaluable knowledge of the subject through their publications, which alone made it possible for a novice like me to compile this book.

Finally, I acknowledge my indebtedness to

National Institute of Industrial Engineering (NITIE),

Mumbai-400 087, India; and its

Advanced Manufacturing Center of Excellence (AMCE)

where I started learning the A, B, C and 1, 2 of MRP II in June 1997 and finished the first draft of this book in November 1999.

Preface

I found that the entire MRP literature consisted of 26 items—good, bad, and indifferent—all of which were either articles, excerpts, special reports, or trade-press "testimonials". Tutorial material on MRP basics was lacking entirely. The job of "getting it all together" remained to be done. Someone had to write a book on "MRP from A to Z" and I concluded I may have to be the one. So I wrote this book.

JOSEPH ORLICKY
in the preface of his book
Material Requirements Planning (1975)

Whenever I tried to think of something to write in the preface to this compilation, Joseph Orlicky's words, which he wrote in the preface of his book, would invariably come to my mind—so I have quoted it right at the beginning as I have reproduced so many things from his masterly original work. The one and the only similarity between what Orlicky wrote and what I should be writing is that I also genuinely felt that "getting it all together" remains to be done. Apart from this minor similarity, there are only differences—major differences!

Firstly, Orlicky endeavored to get all the MRP I knowledge together, whereas what I have tried to get together is related to MRP II—a concept much more comprehensive than MRP I. Secondly, there was not much of literature on MRP I available for Orlicky to draw upon. In my case, there was already an enormous amount of MRP II literature available—albeit in bits and pieces. What was lacking was a single, comprehensive and coherent source of MRP II knowledge that covered all the aspects of MRP II. Thirdly, when Joseph Orlicky said, with the humility that is a characteristic of the truly great people, that "getting it all together" was what was required to be done and he wrote the book to that end, he was not telling the truth for the simple reason that Orlicky himself was the one who had developed the MRP I concept. What he got together was his own contribution to the science of production and operations management. However, when I say "getting it all together, is what I have attempted here", it is simply a statement of the truth!

This book is more of a compilation of MRP II knowledge from the authorities on the subject. My original contribution apart from the compilation, arrangement, and editing of the vast amount of MRP II knowledge, has been a presentation of the MRP II concepts in an orderly and easy-to-understand format. Wherever possible, I have tried to present the concepts through block diagrams and tables that present the information in the most compact format. However, in the process of making this small contribution, I might have contributed enormous amounts of mistakes which, of course, are all mine! I hope the enlightened readers would take the trouble to inform me about these mistakes so that they can be corrected in future editions. The readers are welcome to contact me directly through e-mail (khalid_sheikh@hotmail.com) or by post.

The intended audience for the book is anybody who has interest of any kind in the manufacturing module of any ERP package. ERP students, implementers, trainers, users, application programmers all will find it useful in acquiring the knowledge of what exactly happens inside the manufacturing module. It can also serve as a supplementary study material for APICS CPIM examinations. The knowledge of MRP II concepts is a necessary prerequisite for understanding the working of any ERP package; but it is not sufficient. A thorough understanding of the processes built in the particular ERP package through its manuals and/or online help is a must. About 70 to 80% of the concepts described here might be similar to those adopted in the package. However there might be significant differences as each individual ERP vendor tries to introduce as many unique features as possible to make it a unique selling point for his package, if not for anything else.

The book is organized such that it would first provide an overview of MRP II and then give details of each of the components of the MRP II system. However, a detailed knowledge of MRP I is a prerequisite to understanding the MRP II system, and hence a detailed study of MRP I should be done right in the beginning. Vollmann, Berry, and Whybark—the authors of one of the finest texts on manufacturing planning and control systems—have made a similar observation in the preface of their book *Manufacturing Planning and Control Systems*: "We have found that starting with MRP (I) record processing has been pedagogically sound. We find that understanding time-phased records is helpful for understanding the entire manufacturing planning and control (MRP II) system." Putting MRP I right at the beginning, however, disrupts the order in which the various MRP II functions should be understood. Hence, the book makes a compromise by including detailed study of MRP I right in the overview itself so that the entire MRP I process is understood in the beginning without disturbing the order.

The book is intended to be a single comprehensive sourcebook for all the MRP II-related knowledge. Except for aggregate planning, which has been intentionally kept outside the scope of the book, all relevant topics have been included. However, people whose interests are limited to acquiring knowledge related to ERP package implementation only can skip Chapters Three and Four on strategic and business planning, and sales and operations planning respectively. Chapter One provides most of the background knowledge related to manufacturing. Those already familiar with manufacturing engineering can skip parts of this chapter. Everyone should, however, go through the topics on manufacturing environments, particularly assemble-to-order demand response strategy and the related topics like planning bills of material and product configurator, which are essential for a thorough understanding of the manufacturing module of a sophisticated ERP package.

While the manuscript was going through the review and production process, it was strongly felt that chapters giving comprehensive introduction to the emerging SCM and CRM systems should be added at the end of the book. Some additions were also made to the chapter on ERP. It is hoped that, with these additions, the book will be even more useful to the students and professionals working in the area of ERP.

The book is based on APICS body of knowledge and has freely borrowed from the authors associated with APICS. Care has been taken to acknowledge all the sources through references, yet some references might have been unintentionally omitted, which is sincerely regretted. It is hoped that these authors would be magnanimous enough to forgive such omissions. It is earnestly hoped that this book would facilitate the understanding of the MRP II concepts and would promote, in its own small way, the use of MRP II in the manufacturing industries of all types and sizes realizing the vision of the founders of APICS!

KHALID SHEIKH

Address for Correspondence:
PROF. KHALID SHEIKH
S P Jain Institute of Management & Research (SPJIMR)
Munshi Nagar, Dadabhai Road
Andheri (West), Mumbai-400 058, India
(e-mail: khalid_sheikh@hotmail.com)

Contents

Preface *vii*

Chapter 1

Industrial and Manufacturing Typologies and Their Characteristics 1

1.1 *Production and Manufacturing 1*
1.2 *Three Modes of Production 2*
1.3 *Classification of Basic Industries 5*
1.4 *Classification of Production Industries 13*
1.5 *Classification of Production Based on the Production Volume 14*
1.6 *Manufacturing Environments 26*
1.7 *Evolution of Manufacturing 48*
1.8 *Product Life Cycle and Product Positioning Strategy 49*
References 55

Chapter 2

An Overview of Manufacturing Resource Planning (MRP II) Including Detailed Study of Material Requirements Planning (MRP I) 57

2.1 *Three Kinds of Flows Supporting Effective Manufacturing 57*
2.2 *Principles of Effective Manufacturing 60*
2.3 *What is Manufacturing Resource Planning? 63*
2.4 *A Framework for the MRP II System 65*
2.5 *Universal Manufacturing Equation 67*
2.6 *Elements of MRP II 68*
2.7 *Strategic and Business Planning 69*
2.8 *Resource Requirements Planning (RRP) 70*
2.9 *Demand Management 71*
2.10 *Product Configurator 73*
2.11 *Sales and Operations Planning (S&OP) 73*

2.12 *Master Production Scheduling (MPS) and Final Assembly Scheduling (FAS)* 75
2.13 *Independent and Dependent Demand* 77
2.14 *Rough-Cut Capacity Planning (RCCP)* 81
2.15 *Engineering Change Control* 83
2.16 *Material Requirements Planning (MRP I)* 87
2.17 *Capacity Requirements Planning (CRP)* 175
2.18 *Plant and Supplier Scheduling* 179
2.19 *Closed-Loop Approach* 188
2.20 *Problems Associated with MRP II* 189
2.21 *Benefits of and Prospects for MRP II* 193
2.22 *MRP II and JIT* 197
2.23 *Role of MRP II in Process Manufacturing* 201
References 203

Chapter 3

Strategic and Business Planning 205

3.1 *Strategic Planning* 205
3.2 *Intended and Adaptive Strategies* 206
3.3 *Need for Devising Company Strategies* 207
3.4 *Making Decisions in an Organization* 208
3.5 *The Five Tasks of Strategic Management* 211
3.6 *Characteristics of Strategy-Making Process* 212
3.7 *Strategy-Making as a Group Task* 213
3.8 *Strategic Vision and Mission* 213
3.9 *Corporate Objectives* 215
3.10 *Corporate Strategy* 218
3.11 *Competitive Strategy* 227
3.12 *Functional Strategies* 232
3.13 *Manufacturing Strategy* 233
3.14 *Manufacturing Tactics* 235
3.15 *Developing a Manufacturing Strategy* 235
3.16 *Necessity of Involving Manufacturing in the Corporate Strategy Debate* 237
3.17 *Strategic Integration—Linking Manufacturing to Corporate Planning and Marketing* 239
3.18 *Manufacturing Strategy Development Process Developed at Hewlett-Packard* 244
3.19 *Business Planning—Management Levers* 245
3.20 *Make or Buy Decisions—Vertical Integration* 245
3.21 *Capacity Decisions* 246
3.22 *Facilities Decisions* 246
3.23 *Process and Technology Decisions* 248
3.24 *Manufacturing Infrastructure Decisions* 252
3.25 *Human Resources Decisions* 253
3.26 *New Product Introduction* 253
3.27 *Supply Chain Relationships* 253
References 253

Chapter 4

Sales and Operations Planning (S&OP) 255

4.1 *Introduction 255*
4.2 *Objectives of Sales and Operations Planning 258*
4.3 *Factors that Affect S&OP 259*
4.4 *Issues to be Addressed by S&OP 261*
4.5 *S&OP and MPS 262*
4.6 *Key Management Considerations to Make S&OP a Success 263*
4.7 *The S&OP Process 263*
4.8 *Integration Issues 273*
4.9 *S&OP Measurements 274*
4.10 *Benefits of S&OP 275*
4.11 *Summary 275*
References 276

Chapter 5

Demand Management 277

5.1 *Introduction 277*
5.2 *Fundamental Requirements of Demand Management 280*
5.3 *Functions of Demand Management 281*
5.4 *Linkage of Demand Management and Production Planning 283*
5.5 *Interactions between Demand Management and MPS 283*
5.6 *The Spare Parts Issue 287*
5.7 *Objectives of Demand Management 287*
5.8 *Forecasting System 288*
5.9 *The Strategic Role of Forecasting in Supply Chain Management and TQM 289*
5.10 *Classification of Forecasting Methods 291*
5.11 *Projective Methods (Time Series Methods) 295*
5.12 *Seasonal Adjustments 299*
5.13 *Error Measurements 299*
5.14 *Qualitative, Subjective, or Judgmental Forecasting 302*
5.15 *Use of Manager's Views to Update Forecasts 305*
5.16 *Aggregating and Disaggregating Demand Forecast 307*
5.17 *The Principles Related to Forecast Accuracy 309*
5.18 *Book Date and Quantity versus Demand/Shipment Date and Quantity 310*
5.19 *Order Entry System 311*
5.20 *Available-to-Promise Calculation 312*
5.21 *Methods for Analyzing Customer Order: Abnormal Demand or Extra Demand 314*
5.22 *Forecast Consumption Logic 315*
References 326

R

Chapter 6

Master Production Scheduling (MPS) 328

6.1 *Introduction 328*
6.2 *The Need for Master Scheduling 329*
6.3 *The Objectives of Master Scheduling 330*
6.4 *Functions of MPS 330*
6.5 *Management and the MPS 331*
6.6 *MPS: The Communication Link Between Strategic and Tactical Planning 332*
6.7 *Master Scheduling: An Iterative Process 333*
6.8 *Items that Are to Be Included in The MPS 334*
6.9 *VAT Analysis 335*
6.10 *Certain Other Items that are Likely to be Master Scheduled: Multilevel Master Scheduling 337*
6.11 *Planning Horizon and Time Periods (Buckets) of MPS 338*
6.12 *The Frequency of MPS Revision (Updating/Maintenance) 340*
6.13 *Major Inputs to Master Scheduling Process 341*
6.14 *Factors that Go into Developing an MPS 341*
6.15 *Order Management 345*
6.16 *Fallacy of Fences 349*
6.17 *Providing Safety And Hedges 350*
6.18 *Master Schedule, or Master Production Schedule Report 354*
6.19 *Projected Available Balance (PAB) 355*
6.20 *Available To Promise (ATP) 357*
6.21 *An Example of MPS Report 361*
6.22 *Master Scheduling in Make-to-Order Environment 364*
6.23 *MPS Family Totals Per Month 364*
6.24 *Mixed-Model Master Scheduling 365*
6.25 *Exception Messages 366*
6.26 *The Challenge of Managing Change and MPS Decisions 368*
6.27 *MPS Decisions 371*
6.28 *Summary of Master Scheduling Activities 373*
6.29 *Role of the Master Scheduler 374*
6.30 *Characteristics of a Master Scheduler 376*
6.31 *Master Planning 376*
6.32 *Symptoms of a Mismanaged Master Production Schedule 377*
6.33 *Measures of Performance 377*
6.34 *Computer Assistance in Master Scheduling 378*
6.35 *Master Production Scheduling in Process Industries 379*
6.36 *Final Assembly Schedule (FAS) or Finishing Schedule 380*
6.37 *The Mechanics of FAS 383*
6.38 *Benefits of FAS 384*
References 385

R

Chapter 7

X

Bill of Material (BOM) 387

7.1 *Introduction 387*
7.2 *Bill of Material Structuring 390*
7.3 *Single-Stage and Multiple-Stage Bills of Material 392*
7.4 *Types of Bills of Material 393*
7.5 *Single-Level Bill of Material 393*
7.6 *Multilevel or Indented Bill of Material 394*
7.7 *Planning Bills of Material 396*
7.8 *Manufacturing Bills of Material (M-Bills) 404*
7.9 *Phantom, Blowthrough, or Transient Bill of Material 405*
7.10 *Summarized Bill of Materials (Parts List) 410*
7.11 *Matrix Bill of Material or the Cross-Classification Chart* 411
7.12 *Costed Bill of Material 412*
7.13 *Functions of Bills of Material 412*
7.14 *Where-Used Bills of Material 414*
7.15 *Maintenance of the Bill of Material 416*
7.16 *Item Identification 417*
7.17 *Part Numbering Prerequisites and Principles 419*
References 422

Chapter 8

Capacity Management 424

8.1 *Capacity 424*
8.2 *Capacity Expansion Strategies 425*
8.3 *Load 426*
8.4 *Capacity Planning Process 428*
8.5 *Measuring Capacity 432*
8.6 *Determining Capacity Available 432*
8.7 *Resource Requirements Planning 436*
8.8 *Rough-Cut Capacity Planning (RCCP) 437*
8.9 *Capacity Requirements Planning (CRP) 442*
8.10 *Inputs to CRP 443*
8.11 *Scheduling Strategies 450*
8.12 *Infinite versus Finite Loading 451*
8.13 *CRP Outputs 452*
8.14 *Role of the Capacity Planner 454*
8.15 *Role of the Manufacturing Supervisor 455*
8.16 *Production Smoothing 456*
8.17 *Advantages and Limitations of CRP 459*
8.18 *Capacity Planning in Repetitive Manufacturing 460*
References 461

Chapter 9

Shop Floor Control 462

9.1 *Shop Floor Control* 462
9.2 *Factory Coordination* 465
9.3 *Production Environment Design (PED)* 466
9.4 *The Control Task within Factory Coordination* 468
9.5 *Building Blocks Associated with the Factory Coordination Control Task* 468
9.6 *Overview of Production Activity Control (PAC)* 469
9.7 *Data Requirements* 471
9.8 *Planning Files* 471
9.9 *Control Files* 473
9.10 *Order Preparation* 473
9.11 *Scheduling* 474
9.12 *Input-Output Control* 482
9.13 *Plant Scheduling* 488
9.14 *Kanban* 489
9.15 *Production Reporting and Status* 490
9.16 *Measurements for Control* 490
9.17 *Order Disposition* 491
9.18 *Importance of Shop-Floor Control* 492
References 493

R

Chapter 10

An Introduction to Enterprise Resource Planning: Architecture, Implementation, and Cases 494

10.1 *Introduction* 494
10.2 *IT as Enabler* 495
10.3 *Need for ERP* 497
10.4 *Architecture of SAP R/3 ERP Solution* 499
10.5 *Architecture of Baan ERP Solutions* 503
10.6 *Selecting the Right ERP Package* 512
10.7 *ERP Implementation* 517
10.8 *Benefits of ERP* 521
10.9 *Future of ERP* 523
10.10 *ERP Implementation Case Studies* 524
References 535

Chapter 11

Supply Chain Management 538

11.1 *Supply Chain (Logistics Network)* 538
11.2 *Characteristics of Supply Chains* 539

11.3 *Supply Chain Management 540*
11.4 *Objectives of a Supply Chain Management System 541*
11.5 *Supply Chain Management and Logistics Management 542*
11.6 *Supply Chain Management and Enterprise Resource Planning 543*
11.7 *Difficulties and Issues in Supply Chain Integration 544*
11.8 *Effective Supply Chain Management: The Next Frontier 545*
11.9 *Functions of IT Systems in SCM 545*
11.10 *Supply Chain Performance Measures 546*
11.11 *Improving the Supply Chain Structure 547*
11.12 *Improving the Supply Chain Infrastructure 548*
11.13 *Process Optimization 549*
11.14 *Supply Chain Application Modules 549*
11.15 *Market Leaders 551*
References 551

Chapter 12

Customer Relationship Management and the Extended Enterprise 552

12.1 *Introduction 552*
12.2 *CRM Objectives 553*
12.3 *Need for CRM 555*
12.4 *Effect of IT and Internet on Supplier-Customer Relationship 556*
12.5 *What CRM Can Do for Various Users? 557*
12.6 *The CRM Process 558*
12.7 *CRM Process Competencies 559*
12.8 *Three Key Enablers for Implementing a CRM Strategy 561*
12.9 *CRM Benefits 562*
12.10 *CRM Market Leaders 563*
12.11 *How CRM Can Support Various User Roles? 563*
12.12 *The Extended Enterprise 568*
12.13 *Enterprise IT Solutions: "Best-of-Breed" versus "Wall-to-Wall" 569*
References 571

Appendix 1

Related Web Sites 572

Index 575

Chapter 1

Industrial and Manufacturing Typologies and Their Characteristics

1.1 PRODUCTION AND MANUFACTURING

Production is the making of something new, either tangible (products) or intangible (services that disappear in the very act of their creation). Thus, production is the process that converts inputs into goods or services. Today even intangible 'ideas' might be included under the heading of production.

Production is one of the most basic and important functions of human activities in modern industrial societies and is now viewed as a cultural activity (Hitomi, 1996). According to *The Shorter Oxford English Dictionary of Historical Principles* (1973), the English term 'production' appeared in 1483, stemming from the Latin word *producere* which means to 'lead forward'.

Manufacturing is the production of tangible goods (products); it is a basic historical process, which has lasted for several thousand years. Manufacturing matters, in that case is a basic means of human existence, it creates the wealth of nations, and contributes towards human happiness and world peace (Hitomi, 1996).

Manufacturing (and in a narrow sense production) is understood to be 'the transformation of raw materials into products by a series of energy applications, each of which affects well defined changes in the physical or chemical characteristics of the materials' (Danø, 1966). A place where manufacturing is done is a 'factory' or a 'workshop'. Firms concerned with manufacturing are 'manufacturing firms' or 'manufacturers'.

Meaning of Manufacturing and Production

The original meaning of manufacture was to make things by hand. However, the present meaning has quite widened. *Manufacturing* is 'the conversion of a design into a finished product', and 'production' has a narrower sense, namely the physical act of making the product (Young and Mayer, 1984).

In 1983 CIRP (International Conference on Production Research) defined manufacturing as 'a series of interrelated activities and operations involving the design, materials selection, planning, manufacturing production, quality assurance, management and marketing of the products of the manufacturing industries'. Note how this definition includes the inextricable linking of design into the manufacturing system. Manufacturing is now seen as a process integrating the activities of three participants: suppliers of materials and services, manufacturing plants, and customers. Thus 'manufacturing' should now be regarded as a series of productive activities that encompass planning, design, procurement, production, inventory management, marketing, distribution, sales, management, and final disposal/recycling.

1.2 THREE MODES OF PRODUCTION

According to *Encyclopedia Americana* (1965), the mode of production has changed with time in the following three ways: extraction (primary), construction and manufacturing (secondary), and services (tertiary).

Primary Industries (Extraction) These industries are involved in the pre-industrial extraction of materials from their natural source. It includes basic productive activities, such as agriculture, hunting, fishing, mining, quarrying, forestry and their like, which totally rely on nature as the sole source of wealth.

Secondary Industries (Construction, and Manufacturing, which Includes Refinement, Conversion, Fabrication and Assembly) In the late eighteenth century and early nineteenth century some economists included 'manufacture' as an element in the creation of wealth, introducing the concept of '*vendibility*: production for the market'. The term 'manufacture' originally appeared in 1622, stemming from the Latin *manu factum*, which means 'made by hand'. Production in this sense puts special emphasis on making things which are tangible (products). This category is now known as secondary industry, which includes manufacturing, construction, and public utility generation.

Tertiary Industries (Services) Towards the later part of the nineteenth century, the concept of *utility* came to be used by the marginal utility economists. Economically, utility is an index expressing the degree of satisfaction of a human want. This term originally appeared in 1440, stemming from the Latin *utilis* meaning 'useful'. With the introduction of this concept, the meaning of production was widened:

Production is a creation of utility. The production function (the supply side) creates utility by providing form, time, and place utilities for the produced goods. Manufacturing provides the form utility, while physical distribution (a service function) provides the time and place utilities.

Hence, the term production further includes intangible products called 'services', that is, accompanying transportation, sales, trade, and other service activities. Service industries provide a service that involves the expenditure of labor and/or the use of equipment to perform a task without a physical product delivered to the customer. Now this category is known as tertiary industry.

Goods are tangible things that we can carry away with us, whereas services are intangible and perishable and are consumed in the process of their creation. It is often easier to describe the making of goods than the provision of a service. You can probably imagine a car being built on an assembly line, but how is an insurance policy developed? Table 1.1 describes

Table 1.1

Differences between Services and Goods

(Adapted from Waters, 1996)

Sr. No.	*Services*	*Goods*
1.	The output is intangible.	The output is tangible.
2.	Services cannot be kept in stock.	Goods can be kept in stock until needed.
3.	There is simultaneous production and consumption.	There is a delay between production and use.
4.	Services are more likely to be unique and cannot be mass produced.	Goods might be unique, similar, or identical; mass production is possible.
5.	There is a lot of interaction between the server and the customer.	There is little or no interaction between the manufacturer and the customer.
6.	Mostly, customers participate in the service.	Customers do not participate in the manufacturing.
7.	Facilities are located close to the customers.	Factories are located away from the customers.
8.	Services are labor intensive	Manufacturing is largely automated
9.	Quality of service is difficult to measure	Quality of the goods is much easier to measure
10.	Quality depends largely on the server	Quality does not depend on one person
11.	It is difficult to measure the output	The output can usually be measured or counted

some of the differences between services and goods. These might not always be true, but they do give some useful guidelines.

Service organizations are not all similar. There is a range of size from an individual server to the United Nations; and a range of activities from saving souls in churches to destroying them in armies (Waters, 1996). It might appear that service organizations and goods manufacturing organizations are completely separate; but in reality, all producers provide a combination of goods and services. Consider, for example, a restaurant; it provides a service, but at the same time, it makes meals. A company that makes washing machines is clearly a manufacturer, but it also gives an after-sales service. So there is not really a clear distinction between goods and service organizations. It would be better to describe a spectrum of organizations. At one end of the spectrum are the organizations that mainly supply goods, at the other end of the spectrum are organizations that mainly provide services. Between the two extremes of this spectrum are the organizations that provide a combination of goods and services (Waters, 1996).

Distribution (also called Physical Distribution) One of the most important service functions, is defined by APICS Dictionary as the function that involves the activities associated with the movement of material, usually finished products or service parts, from the manufacturer to the customer. These activities encompass the functions of transportation, warehousing, inventory control, material handling, order administration, site and location analysis, industrial packaging, data processing, and the communications network necessary for effective management. It includes all activities related to physical distribution, as well as the return of goods to the manufacturer. In many cases, this movement is made through one or more levels of field warehouses (Cox III and Blackstone Jr, 1998). Table 1.2 compares manufacturing and distribution system.

Table 1.2

Comparison of Manufacturing and Distribution Systems

	Manufacturing	*Distribution*
Items in Inventory.	Inputs to manufacturing system: raw materials, components, and work-in-process.	Outputs of manufacturing systems: end items, service parts and repair kits.
Source of Demand.	Master production Schedule (MPS).	Forecasted and actual customer orders.
Types of Inventory Systems.	MRP I and/or Just in Time (JIT).	Statistical Order Point System or Distribution Requirements Planning (DRP I).

Other services provide transformations for individuals, organizations, and equipment and structures, such as:

- *For Individuals*: psychological, esthetical, physiological, and educational transformations
- *For Organizations*: informational, organizational, and technological transformations
- *For Equipment and Structures*: functional and esthetical transformations

We often judge the wealth of a nation by the amount of goods it produces. However, in most countries this manufacturing output is only a small part of the gross domestic product (GDP). This is one effect of economic development, which can be considered in three stages. In the first stage, the economic activity is concentrated in the preindustrial primary industries, such as agriculture, fishing, mining, etc. In the second stage, the emphasis is on the secondary industries of manufacturing and construction. In the third stage, as prosperity continues to grow, economic activity moves to the post-industrial service sector of government, education, health, retailing, catering, etc. Most countries still have economies that are in the first stage. India, China, Brazil, South Africa and other developing countries rely on agriculture, but are developing by encouraging manufacturing. Much of Europe, North America, Japan and some other developed countries are in the third, post-industrial stage. These have about 5% of workforce in primary industries, 20% in manufacturing, and the remaining 75% in services (Waters, 1996).

1.3 CLASSIFICATION OF BASIC INDUSTRIES

There is a wide variety of basic industries, including not only manufacturing but all others as well. The analysis of industrial typology brings into focus the critical parameters of each type by comparing their different requirements and thus helps achieve a better understanding of the mechanics of the underlying methods. We can classify basic industries into various categories as shown in Fig. 1.1.

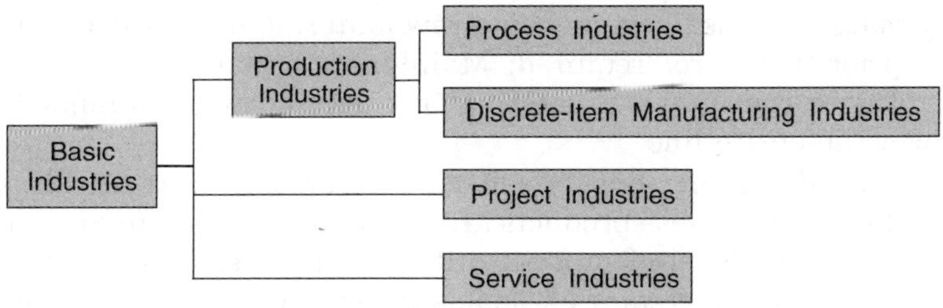

Fig. 1.1 *Types of Industries*

1.3.1 Production Industries

These are the industries engaged in the production of goods (finished products) that have value in the marketplace. These industries are further classified into the following two categories.

Process Industries (Flow-Production or Continuous-Process Production Industries)

Process manufacturing is defined by APICS Dictionary as *the production that adds value by mixing, separating, forming, and/or performing chemical reactions. It may be done in either batch or continuous mode* (Cox III and Blackstone Jr, 1998).

These industries manufacture highly standardized non-discrete products in extremely large volumes using a continuous process. Product that flows in a *continuous* stream falls into this manufacturing category. Process manufacturing is characterized by wet or dry product flow measurable by weight or volume, small work-in-process, and short lead times. A feature of this type of production is that the components or ingredients cannot be readily identified in a finished product and the product cannot be disassembled. Further, the end product is not naturally divisible, hence an additional step such as bottling or canning may be needed to get the product into discrete units. Products are typically bulk commodities, some of which may be packaged to order. In some cases, these products have become so standardized that there are virtually no real differences between the products of different firms. Steel, ores, cloth, pharmaceuticals, gases, chemical products, petroleum products, rubber, cement, paints, commodity ceramics, and early stages of food and paper processing are all examples of process production.

Process industries develop extensive long-range resource requirement plans. Because the process industries are more capital intensive than the fabrication and assembly (discrete-item) industries, Capital budgeting decisions have a greater impact on financial performance in the process industries. Plant capacity and location, long-range plans for materials, manpower, energy, and waste disposal are all important in the process industries. Continuous process scheduling will be capacity-driven with the only material consideration being raw material and without any shop floor or priority control required. Maintenance planning is especially important with continuous processing, as a breakdown will immediately shut down the entire line.

Flow manufacturers tend to sell high volumes of a relatively small number of products. These products are often quite similar to competitor's products. This low level of product differentiation leads to a marketing strategy that places a greater emphasis on product availability and price in flow manufacturing than in job manufacturing. High product availability is achieved by producing products *to stock* and storing finished goods in warehouses in principal demand centers. Flow manufacturing products have relatively few design changes, low unit values, and relatively high transportation costs. Costs are minimized by using high-volume, dedicated equipment arranged in *process trains* or *production lines* (Taylor and Bolander, 1997).

In this type of industry, the production process generally follows a specific and fixed sequence of operations (routing). An obvious attribute of these processes is that the production equipment is utilized for a relatively small group of similar products. The facility is typically a maze of pipes, conveyors, tanks, valves, and bins. The layout follows the processing stages of the product, and the output rate is controlled through equipment capacity and flow and mixture rates. The nature of the plant lends itself to the use of massive automation. Combining this attribute with the continuous-flow characteristic allowed investment in automated equipment that realized benefits of mass production (such as low variable operating costs and low unit costs) well before such advantages were seen in other categories of production. Labor input in this type of industry is one of policing (monitoring and maintaining) rather than being highly skilled as an integral input to the overall process. Process computers and electronic control devices were first applied in these industries, and today many continuous cycle factories run virtually unmanned. A principal weakness of flow manufacturing is its lack of flexibility.

Batch Continuous Production Note, however, that plants with fixed routings may employ either batch operations, continuous operations, or a mixture of batch and continuous operations. Using a continuous process on an intermittent basis constitutes *batch continuous production*. Thus, materials may move through flow manufacturing lines continuously or in batches; however, families of similar products are often produced on the same production line. Thus, continuous-flow operations often schedule blocks of time for different products on the same production line. This gives rise to *production lots*, which are very similar to production batches. Dependence on setup considerations for specific operations in the process routing may call for maintaining inventories at differing stages of the process and will result in larger work-in-process (WIP) inventories while still being produced in a relatively short lead-time. Scheduling considerations for batch process manufacturing does call for additional planning and control (Toomey, 1996). There is often no reason in flow manufacturing operations that all units in a particular lot must complete an operation before the lot can begin processing at the next operation in the process train. This flexibility facilitates job overlapping, which, in turn, leads to low work-in-process inventories and fast throughput times (Taylor and Bolander, 1997).

Jobshop Process Industries Another segment of the process industries is *job shop manufacturers*. Jobshops are characterized by *variable product routings*. For example, a speciality chemical plant may have a variety of batch reactors, mixers, and other process equipment. Each speciality chemical product has its own particular requirements met by a routing specific to the product. The planning and scheduling needs of such process-industry job-shops are best met by the manufacturing resource planning (MRP II) systems (Taylor and Bolander, 1997).

Table 1.3 compares process and discrete item (fabrication and assembly) industries. Some of the characteristics of the process industries are:

1. Variability in raw material quality
2. Variability in formulas (recipes or BOMs)
3. Product yields may vary widely
4. May have large demands for intermediate products
5. Co-product demand must be balanced
6. Product or raw materials may have a shelf life

The process industries often obtain their raw materials from mining or agricultural industries. These raw materials have natural variations in quality, which leads to variations in formulas. For example, variations in the acidity, concentration, and so on, of the raw material may cause variations in the proportion of ingredients or components required to make a finished product.

Discrete-Item Manufacturing Industries

Discrete manufacturing is characterized by discrete, countable products and component parts. A feature of this type of production process is that it produces a single component or unit of product each time an operation is performed. Machining, fabrication, and assembly operations are typical. Discrete-part products can be disassembled and reassembled. This type of manufacturing entails the production of discrete units of production—components and assemblies—in various quantities.

Difficulties in Categorizing Manufacturing Processes

Steel and aluminum extrusion or rolling operations may produce various lengths of a product at relatively slow rates and be classified as 'discrete.' Alternatively, they may produce long lengths of product at very high rates and be classified as 'process.' Similarly, mass production of discrete items like fasteners and injection-molded parts can be better controlled as continuous flow processes. Thus, according to Plossl (1997), categorizing manufacturing processes is at best indefinite and can be misleading, particularly when selecting optimal planning and control strategies and techniques.

Table 1.3

Differences Between Process and Discrete-Item Industries

Criteria	*Process Industries*	*Discrete Industries*
Relationship with the Market		
Product Type.	Commodity.	Custom or standard.
Product Assortment.	Narrow.	Broad.
Demand per Product.	Generally very high.	Low, medium, high.
Cost per Product.	Low.	High.
Order Winning Criteria (OWC).	Price; delivery guarantee.	Speed of delivery; product features.
Transporting Costs.	High.	Low.
New Products.	Few.	Many.
The Production Process		
Bills of Material.	Shallow (fewer levels).	Deep (multiple levels).
Routings.	Fixed.	Variable.
Layout.	By product.	By function or by product groups.
Flexibility.	Low.	High.
Production Equipment.	Mostly specialized, highly automated involving programmable logic controllers (PLCs), process computers, and other electronic control devices; fixed path material handling equipment.	Mostly universal either general purpose or with flexible automation involving computer numerical control (CNC) technology; also dedicated equipment such as transfer lines and special purpose machines; variable path material handling equipment.
Equipment Failure Affects.	Plant.	Machine.
Labor Intensity and Skill.	Low; operating and monitoring skills required to operate the equipment.	High; skilled craftsmen required to build the product.

(Contd.)

Criteria	*Process Industries*	*Discrete Industries*
Capital Intensity and Overheads.	High.	Low.
Capacity Definition/Lead Time to Increase Capacity.	Easy capacity definition, but involves longer lead times to commission.	Difficult capacity definition, but involves shorter lead times to commission.
Changeover Times.	High.	Low.
Work in Process.	Low, WIP is rarely stored.	High, WIP is often stored.
Volumes.	High.	Low.
Quality		
Environmental Demands.	Yes.	Hardly.
Danger.	Sometimes.	Almost never.
Quality Measurement.	Sometimes long.	Short.
Planning and Control		
Production.	Mostly to-stock.	To-stock and to-order.
Long-Term Planning.	Capacity.	Product design.
Short-Term Planning.	Utilization of capacity.	Utilization of personnel.
Starting Point of Planning.	Availability of capacity.	Availability of materials.
Material Flow.	Divergent + Convergent.	Convergent.
Yield Variability.	Sometimes high.	Mostly low.
'Explosion' via.	Formulas or recipes.	Bill of material (BOM).
By- and Co-Products.	Sometimes.	Never.
Delay in Receipt of Purchased Parts Affects.	Plant.	Customer Order.
Lot Tracing.	Mostly necessary.	Mostly not necessary.

1.3.2 Project Industries

The key identifying characteristic of a project industry is that the materials, tools, and personnel are brought to the location where the product is to be fabricated. A project is defined by APICS Dictionary as *an endeavor with a specific objective to be met within the prescribed time and cost limitations and that has been assigned for definition or execution* (Cox III and Blackstone Jr, 1998). In the context of engineering industries, a project is a one-at-a-time major task such as construction, shipbuilding, erection and commissioning of plants, and design of a new product.

The project industries are characterized by the *construction-site type of an organization.* This type of manufacturing organization is based on a single product, generally of large size, which keeps it busy for months or even years. Each product may thus be considered a prototype, because its structure is modified in each case to the requirements of the customer (Sartori, 1988). It should be noted here that aircraft, defense equipment, and space vehicle manufacturing industries are not generally classified as project industries even though they have similar characteristics. These industries differ from project industries in two aspects: first, the fabrication and assembling operations are performed inside a factory using a fixed site type of process organization, and secondly, the same site is used for making several products. Hence, these industries are generally classified as *discrete-item job-shop industries with fixed-site process organization.* In this case, fixed-site production is actually used in conjunction with other processes since, after the product reaches a certain size, it is often more practical to keep it stationary and move the necessary components to its location.

The tools provided by information technology find limited application in the management of a construction site, as its problems are intrinsically different from those of other industries. Inventory is mainly concentrated in the product itself, and materials always flow in the same direction. Inventory and order management functions are not required for an application that merely involves the control of supplies that are in any case all accumulated in the product being constructed. Reducing inventory simply means shortening the delivery time and is equivalent of using equipment to the maximum of its capacity. The three objectives of manufacturing—namely, minimizing delivery time, minimizing inventory investment, and maximizing equipment use, usually considered conflicting—coincide in this case and deprive MRP II of a good deal of its relevance.

Undoubtedly, careful planning of supplies is still useful to avoid having to take delivery of all components required at the start of the process. Consignments should instead be scheduled according to actual use. An adequate level of precision in this respect, however, requires reliable lead times, and this can become a problem due to the diversity of the successive 'prototypes.' Many SFC functions also loose all significance because it is difficult to apply the concept of work-order progress to operations that often consist merely of putting the piece in place (Sartori, 1988).

The *Program Evaluation and Review Technique (PERT)* and the *Critical-Path Method (CPM)* are typical of the scheduling techniques developed for project execution. These approaches have been found more effective than other tools for controlling the key variable, 'time'. Automation also has been little used in construction sites because of the difficulty of defining the individual operations in an abstract way and of automating the process as a sequence of patterns intelligible to a robot.

1.3.3 Service and Other Industries

Here, we include those industries that do not produce goods, but provide certain services. (Industries which do not fit into any particular category are also included here). Their operations might, in fact, include a number of different categories. Table 1.4 lists a few examples of each of these types of industries.

Table 1.4

Examples of Basic Industries

(Adapted from Groover, 1987)

Process (Flow-Production) Industries	*Discrete-Item Manufacturing Industries*	*Project Industries*	*Service and Other Industries*
Beverages, Cement, Chemicals, Cosmetics, Drugs, Food Processing (Canned Foods, Dairy Products, Meats, Packaged Foods, Oil), Glass, Metals (Aluminium, Copper, Steel, etc.), Paper and Cardboard Petroleum Products, Plastics, Rubber, Soaps, Textiles, Tires.	Aerospace, Automotive (Cars, Trucks, Buses), Clothing (Garments, Shoes), Equipment and Machinery (Agricultural, Construction, Electrical and Electronics, Household Appliances, Industrial, Machine Tools, Office Equipment, Computers, Railroad Equipment, Steam-Generating, etc.).	Construction (Civil), Erection and Commissioning of Plants and Factories Shipbuilding	Advertising, Financial (Banks, Investment Companies, etc.), Hotel/Motel, Insurance, Mass Media (Publishing, Radio, TV) Motion Pictures, Natural Resources (Coal, Forest, Oil, etc.), Restaurant, Retail (Food, Departmental Stores, etc.), Transportation (Trucking, Railroad, Shipping, Airlines, etc.), Utilities (Electric Power, Natural Gas, Telecommunications).

1.4 CLASSIFICATION OF PRODUCTION INDUSTRIES

Another way of classifying production industries is to place a company into one of the following three categories:

1. Basic Producer (Raw Material Extractor and Refiner)
2. Converter (Material Processor)
3. Fabricator and Assembler

The three types form a connecting chain in the transformation of natural resources and basic raw materials into goods for the consuming public, as shown in Fig. 1.2.

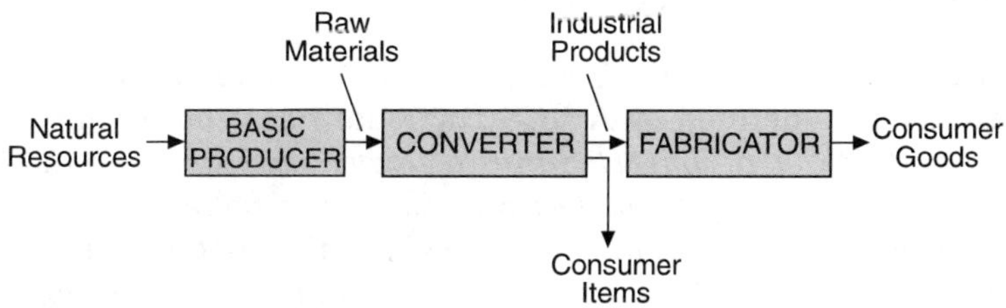

Fig. 1.2 *Transformation Chain*

1.4.1 Basic Producers (Raw Material Extractors and Refiners)

The basic producers take the natural resources and transform these into the raw materials used by the other manufacturing firms. It involves obtaining materials from their natural resources and refining them. Refinement describes the processes primarily concerned with changing the chemical properties of materials although physical properties may also be changed. A typical example is a steel producer that transforms iron ore into steel ingots; others are those making wood pulp, glass, and rubber. These processes usually require heavy capital investment in highly specialized and automated processing and material-handling equipment. Production tends to be either continuous flow or large batch process. In general, refinement processes offer little or no flexibility because of the specialized nature of the equipment. Quantitative techniques such as linear programming have proved useful in assisting management in selecting the proper mix of raw materials to achieve an output with optimal cost, chemical properties, and demand fulfillment. Network planning and control techniques also have proved very useful in scheduling tear-down, cleanup, and restart of refining processes.

1.4.2 Converters (Material Processors)

The converters take the output of the basic producer and transform these raw materials into various industrial products and some consumer items. They represent the intermediate link in the chain. For example, the steel ingot is converted into bar stock, tubing, plate, or sheet metal. Chemical firms transform petroleum products into plastics for molding. Paper mills convert wood pulp into paper. Soap and dyes are other examples. A distinguishing characteristic of the converter is that its products are uncomplicated in physical form. The products are not assembled items. The production processes used to make the products may be complex but the products themselves are not.

1.4.3 Fabricators and Assemblers

The third category of manufacturing firms is the fabricator and assembler. A fabricator is defined by APICS Dictionary as *a manufacturer that turns the products of a converter into a larger variety of products. For example, a fabricator may turn steel rods into nuts, bolts, and twist drills or may turn paper into bags and boxes.* Fabrication is defined by APICS Dictionary as *manufacturing operations for components, as opposed to assembly operations* (Cox III and Blackstone Jr, 1998). These firms fabricate and assemble final products. The bar stock and sheet metal are transformed into machined engine components and automobile body panels. The plastics are molded into various shapes. Then these parts are assembled into final products, such as trucks, automobiles, appliances, garments, and machine tools. Fabricators include both the firms that produce the components and those that assemble the components into consumer goods.

Complicating Factors in this Classification There are several complicating factors in this classification. Some firms posses a high degree of *vertical integration,* which means that their operations include all three categories. The major oil firms are examples of vertical integration. They convert natural resources into finished petroleum products and then market these products directly to the consumer. Another complicating factor is that some companies—the conglomerates—are in so many different types of business that it is difficult to classify them. Some of their operations are in the basic producer category; others are converters; and still other lines of business fall into the fabricator category.

1.5 CLASSIFICATION OF PRODUCTION BASED ON THE PRODUCTION VOLUME

Another way of classifying production activity is according to the quantity of product made. In this classification, there are four types of production:

1. Job Shop Production
2. Batch or Intermittent Production
3. Mass Production (Flow or Repetitive)
4. Cellular Production

This classification is associated with both process and discrete-item production. In process industries, some chemicals are produced in batches (batch production), whereas others are produced by continuous-flow processes (mass production).

1.5.1 Job Shop Production

The distinguishing feature of job shop production is *low volume*. The manufacturing lot sizes are small, often one of a kind, hence, a more definitive term for this category could be *single job lot*. Job shop production is engineered- and/or made-to-order and is commonly used to meet specific customer orders, and there is a great variety in the type of work the plant must do. In view of the small lots and the large variety in the type of work, the manufacturing process is intrinsically variable and cannot be optimized once and for all, and no special purpose machines or tooling can be economically justified. Therefore, the production equipment must be flexible and general-purpose to allow for this variety of jobs that require different sequences through the production equipment. In addition, the skill level of job shop workers must be relatively high so that they can perform a range of different work assignments. Besides working with general-purpose machines is a very lengthy and often error-prone process.

Examples of products manufactured in a job shop include space vehicles, aircraft, machine-tools, special tools, jigs, fixtures, custom clothing, special machinery or equipment to perform specialized and specific tasks, models and prototypes of future products for design proving, components and/or assemblies to provide replacement for parts in existing machinery, etc. Large turbo-generators, large engines, boilers, processing equipment, special electronic equipment, material handling machines, and many other manufacturing activities are in the job shop production category. However, it should be noted that construction work and shipbuilding are *not* normally identified with the job shop category, even though the quantities are in the appropriate range. Although these two activities involve the transformation of raw materials into finished products; the work is not performed in a factory.

Functional or process layout, that is the spatial and administrative organization of similar equipment by functions such as milling, drilling, turning, forging, and assembly, is characteristic of a job shop. However, job shops might also utilize *fixed-position layout,* in which the fixed position refers to the product. For example, while assembling a large aircraft, because of the large size and weight of the product, the product remains in one location and the material, tools, equipment, and personnel are brought to the location where the product is being fabricated. Most machines are general-purpose types. Materials move from one work center to another in small discrete batches, and there is very little overlapping or splitting. The routing of each component through the plant is unique. Inventory is carried only as raw material, work in process, and support materials (Plossl, 1997).

Because of the large number of different products and demands for those products, scheduling and routing problems are enormous. Owing to the uncertainty of an optimal plan and schedule in variety production, executing production activities in the job shop is complex. Machine breakdown, absence of workers, manufacture of inferior products, etc. occur on many occasions. Consequently, production activities must rely upon past experience and intuition rather than upon an exact plan/schedule based upon optimum decision-making. Production control is also complex and liable to change. As a result, automation has at best been restricted to individual components of the job shop (sometimes referred to as *islands of automation*), and there have been few attempts to automate the entire facility by integrating the islands of automation. Following are the key characteristics of job shop type of production:

1. Investment in automation is for general-purpose process technology rather product specific investment.
2. Many different products are run throughout the plant and material handling has to be modified and adjusted to suit many different product types.
3. Detailed planning will evolve around sequencing requirements for each product, capacities for each work center and order priorities; because of this, scheduling is relatively complicated in comparison to repetitive line manufacture.

Clearly the efficient management of a job shop is a difficult task, since every output must be treated differently. In addition, the resources available for processing are limited. Furthermore, not only is it management's task to ensure the performance of the proper functions of each output, where considerations of quality and deadlines may vary, but management must also be sure that the available resources (staff, equipment, materials, supplies, capital) are being efficiently utilized. Often there is a difficult trade-off between efficiency and flexibility of operations. Job-based processes tend to emphasize flexibility over efficiency.

A job shop has planning and scheduling needs that are well matched by the tools provided by manufacturing information systems, as its activities require both detailed operational plans and frequent revisions of purchase and production activities. Volume and mix decisions are frozen only a short time before their implementation is scheduled to take place. Till that time they can be freely rescheduled in terms of quantity and timing. It is nevertheless important to prepare work plans in advance so that any choices made, go in the right direction. At the operational stage, inventory and order management and shopfloor control, can make the fullest use of all their functions to maintain under control a situation which develops its own momentum.

Advantages of the Job Shop

1. Ability to Produce a Wide Variety of Outputs at Reasonable Cost In job shops, general purpose equipment is used, and this is in greater demand and is usually available from many suppliers at a lower price than special purpose equipment. In addition, used equipment is more likely to be available, further reducing the necessary investment. There is a larger base of experience with general-purpose equipment; therefore, problems with installation and maintenance are more predictable, and replacement parts are more widely available. Finally, since general-purpose equipment is easier to modify or use elsewhere, and disposal is much easier, the expense of obsolescence is minimized.

2. Certain Resources may be Centralized or Segregated if required Since job shops use functional layout of the equipment, resources for a function requiring special staff, materials, or facilities (such as, heat treatment, painting, or audio-visual equipment) may be centralized at the location of that function, and the organization can thus save expense through high utilization rates. Distracting or dangerous equipment, supplies, or activities may also be segregated from other operations in facilities that are soundproof, airtight, explosion-proof, and so forth.

3. Responsibility and Pride in One's Work, Morale, and Learning Increases With more highly skilled work involving constantly varying jobs, responsibility and pride in one's work are increased, and boredom is reduced. Other advantages to the staff are that concentrations of experience and expertise are available and morale increases when people with similar skills work together in centralized locations. Because all workers who perform similar activities are grouped together, each worker has the opportunity to learn from others and the workers can easily collaborate to solve difficult problems.

4. Output-Based Incentives can be Set Up Because pace of the work is not dictated by a moving 'line', incentive arrangements can be set up.

5. The Entire Set of Operations do not Stop Due to Breakdown of One Function Other functional areas can continue operating, at least until *in-process* inventory is depleted. Also, other general-purpose resources can usually substitute for the non functioning resource: one machine for another, one staff member for another, one material for another.

Disadvantages of the Job Shop

1. Higher Per-Unit Cost General-purpose equipment in the job shops is usually slower than special-purpose equipment, resulting in higher variable (per unit) cost. In addition, cost of direct labor for the experienced staff, necessary to operate general-purpose equipment further increases unit costs.

2. High Raw Material and In-Process Inventory Not only do many types of raw materials, parts, and supplies have to be kept for the wide variety of outputs anticipated, but in-process inventories become very large.

3. High Cost of Material-Handling Equipment and Greater Space Requirements Because there are so many inventory items that must travel between operating departments in order to be processed, the cost of handling material is also typically high. Since job routings are not identical, inexpensive fixed material-handling mechanisms like conveyor belts cannot be used. Instead, larger and more costly equipment, such as fork-lift trucks are used, therefore corridors and aisles must be large enough to accommodate it.

4. Difficult Managerial Control Because the output varies in terms of function, processing, quality, and timing, the managerial tasks of routing, scheduling, cost accounting, and such, become nearly impossible when demand for the output is high. Expediters must track down lost jobs and reorder priorities. In addition to watching the progress of individual jobs, management must continually strive to achieve the proper balance of materials, staff, and equipment. Otherwise, imbalance of work load would cause some of the highly expensive resources to sit idle due to lack of work whereas some other resources would have overload and thus become bottlenecks adding to the uncertainties of cost and time.

1.5.2 Intermittent (Job-lot or Batch) Production

Intermittent production is defined by APICS Dictionary as *a form of manufacturing in which the jobs pass through the functional departments in lots, and each lot may have a different routing* (Cox III and Blackstone Jr, 1998). This category involves the manufacture of medium-sized lots of the same item or product. The lots may be produced only once, or they may be produced at regular intervals. The purpose of batch production is often to satisfy continuous customer demand for an item. However, the plant is capable of a production rate that exceeds the demand rate. Therefore, the shop produces to build up an inventory of the item. Then it changes over to other orders. When the stock of the first item becomes depleted, production is repeated to build up the inventory again. The intermittent production caters to make-to-stock, assemble-to-order, and also to make-to-order manufacturing environments.

The manufacturing equipment used in batch production is general-purpose but designed for higher rates of production. For example, turret lathes capable of holding several cutting tools are used rather than engine lathes. The machine tools used in batch manufacture are often combined with specially designed jigs and fixtures, which increase the output rate. Batch production plants include machine shops, casting foundries, plastic molding factories, and press-working shops. Some types of chemical plants are also in this category. Most production equipment is arranged in functional layout, although trends now are toward more work cells and flow-centers. There is consistency in the routings among jobs, and materials are moved between work centers in discrete batches. Job splitting and/or overlapping are common. Batch quantities can vary from a few to many thousands of units. Inventories are carried at all levels of the product's bill of material, and work-in-process is usually large (Plossl, 1997). Automation, especially for lower volumes of batch manufacturing tends to be general purpose rather than dedicated to a particular product whose volume does not demand product-specific investment in automation.

Examples of items manufactured in batch-type shops include industrial equipment, furniture, textbooks, and component parts for many assembled consumer products (household appliances, lawn mowers, etc.). Job shop and batch production is a very common feature in industry. It has been estimated that in the United States, perhaps as much as 75% of all parts manufacturing is in lot sizes of 50 pieces or less. Machine tool work, especially capstan and turret lathes, presswork, forging and casting processes, some glass manufacturing, and chemical processes very often operate on a batch basis.

Imprecise information due to the dynamic situation of orders, a variety of production processes, frequent changes in ordered products, etc., make optimum production planning and scheduling, as well as cost estimation for jobs difficult. Scheduling is, therefore, complicated and has to be completely reviewed on a regular, ongoing basis. For the same reason, operators have to be able to perform a number of functions. Greatest benefits can be achieved in batch production by reducing setup times. Setup time is the time needed to perform the work required to change a specific machine, work center, or assembly line from making the last good piece of Item A to making the first good piece of item B.

1.5.3 Mass Production (Flow or Repetitive)

The manufacture of non-discrete products using a continuous process is called *flow* production, whereas the manufacture of discrete parts or assemblies using a continuous process is called *repetitive* production. This is the continuous specialized manufacture of identical products. Common to both flow and repetitive processes is the need to work with large volumes to attain the necessary economies of scale and the tendency to concentrate resources on a few finished products belonging to the same technological family.

Mass production of discrete items, also called *repetitive* manufacturing, was advanced in the United States in the 19th century by employing the *Principle of Interchangeability,* which originated in France in the 18th century. With this principle, standardized parts are produced within a specified tolerance; a product can be made up by assembling a set of parts selected randomly.

Mass production and mass consumption formed a sort of production culture. Giant corporations quickened mass production; a prime example was Ford's model T car of which 15 million were produced during the period 1908–27. The establishment of conveyor systems in 1913 advanced productivity, reduced the cost/price of a car, and raised the wage rate.

It should be noted that flow production or continuous-process production discussed earlier, also comes under the category of mass production. In fact, mass production of discrete items attempts to emulate the characteristics of continuous-flow production for discrete products that are produced in high volumes with relatively small variations in product. As a result, mass production has historically realized many benefits from mechanization and automation. Not only is the equipment dedicated to one product, but the entire plant is often designed for the exclusive purpose of producing the particular product. In a sense, the production skill has been transferred from the operator to the machine. Consequently, the skill level of labor in a mass production plant tends to be lower than in a batch plant or shop.

Repetitive Production

When volumes exceed certain limits, a fixed assignment of resources, which would otherwise be risky, may be considered justified. Each piece of machinery is dedicated to a well-defined phase of the process, closely linked to the adjacent ones, and performs at high speed the operations that are always the same, thus justifying the name given to this category. Specialized departments disappear, and their place is taken by production lines, at least as far as assembly operations are concerned. A direct consequence of this is a great reduction in the product range. The goal of repetitive manufacturing is to achieve a process flow for fabrication and assembly operations similar to process manufacturing, and it is one of the essential ingredients of just-in-time (JIT) manufacturing. Example of the repetitive manufacturing is automobile assembly.

Continuous type of material-handling equipment, such as conveyors (having low per unit operating costs) can be used because the operations are standardized and, typically, all outputs follow the same path from one operation to the next. This standardization of treatment provides for a known, fixed throughput time, giving managers easier control of the system and more reliable delivery dates. The flow shop is easier to manage for other reasons as well: routing, scheduling, and control are facilitated because each output does not have to be individually monitored and controlled. Standardization of operations means that less skilled workers can be used and each manager's span of control can increase (Shafer and Meredith, 1998).

The control techniques used in this kind of environment puts particular emphasis on volume-related variables. Formulating a well-chosen final schedule is thus a task of major importance involving all the basic master scheduling functions, with particular emphasis on the resource requirements planning. This module replaces Capacity Requirements Planning (CRP), as it identifies the optimum work-load profile for the various lines, without making reference to the work-order concept, which is in practice irrelevant to the monitoring of shop floor operations in a repetitive manufacturing plant. The Shop Floor Control (SFC) system is used merely to collect the data on the quantities produced, if necessary, and to relate the production to the day or week concerned.

Characteristics of Repetitive (or Modified-Flow) Production

1. Very high production rates (the same job runs steadily for a few hours or days).
2. Very high demand rates for the product.
3. Dedicated lines, specialized equipment or facilities configured linearly, with the exception of a few feeder departments providing components used on several lines or certain service departments such as heat-treatment shops that provide services to several lines.
4. Very similar routings (process plans) and rates of flow.
5. Shorter lead times.
6. More synchronized work centers requiring less scheduling and reporting controls.
7. Material movement between operations in small batches, approximating continuous flow.
8. Inventories carried in finished goods, purchased raw materials and components; low work-in-process.

Examples of repetitive products are cosmetics, hand tools, small household appliances, and fasteners. The manufacturing environment is make-to-stock or assemble-to-order.

Mixed-Model Repetitive Flow These processes are also used to manufacture two or more models. However, the changeover time between models is minimal (frequently zero), and the different models are intermixed on the same line. The following characteristics typify a mixed-model repetitive flow line (Fogarty *et al.* 1991):

1. The equipment is general-purpose in order to facilitate manufacture of several models.
2. Workers capable of performing many different tasks on one line or moving to different lines in the same plant.
3. Setup times are very short. The ideal setup time is small enough to accommodate run lengths of one unit, switching models after every item.
4. The line produces at the market rate of sales. Workers are added

and subtracted and manufacturing tasks are reassigned as needed to support the market rate on all models.

Flow Production

It refers to the production or processing of fluids, wastes, powders, basic metals, and other bulk items. While complete continuity of flow is rare, a 'pipeline-type' system, where each section of the plant processes a certain quantity of materials and passes it on to the next section when it is ready to receive, is instead frequent. In this case, the characteristics of the flow are similar to a repetitive system, but the units of measure are always weight, length, or volume units.

Characteristics of Flow Production

1. Very high production rates specified in units per time period, varying run lengths—from short periods to continuous operations.
2. Very high demand rates for the products.
3. Facilities or lines of flow that are completely dedicated to the manufacture of a particular group of products.
4. All equipment linearly configured, with the exception of material preparation (blending, mixing, and weighing) and packaging.
5. High degree of automation.
6. Identical routings for each product processed in a specific plant or production unit.
7. Highly reliable, synchronized flow of material at specified rate through production lines.
8. Production control, often called flow control, is usually done by varying the rate of flow; however, there are processes in which the rate can be varied little if at all.
9. To guard against any possibility of continuation, a thorough cleaning of the equipment which has been used for producing one particular substance is absolutely essential before it can be used for producing another substance. This requires a significant amount of time and effort.
10. Only raw materials and finished products carried in inventory; very small work-in-process (usually just enough to fill the process units).

There are different ways to vary the production rate in a continuous-process (flow) production, such as:

- The line equipment may be operated for more or fewer hours.
- The line may be operated with more or fewer personnel.
- The line may be operated intermittently, with personnel doing other work.
- The line may be operated in two or more stages by pulling the

partially completed product from the line and accumulating as in-process inventory. After this inventory is built up to a certain point, the first phase of the process may be shut down and the second phase operated by feeding in the in-process inventory. The same personnel would be used in both processes.

Repetitive production is, sometimes, further classified into the following two categories:

1. Quantity Production
2. Continuous Production

Quantity Production This involves the mass production of single parts on fairly standard machine tools such as punch presses, injection molding machines, and automatic screw machines. These standard machines have been adapted to the production of the particular part by means of special tools designed for the part in question, such as die sets, molds, and form-cutting tools etc. The production equipment is devoted full time to satisfy a very large demand rate for the item. Examples of items in quantity production include components for assembled products that have high demand rates (components for automobiles, some household appliances, light bulbs, typewriters, etc.), hardware items (such as screws, nuts, and nails), and many plastic molded products (Groover, 1987).

Continuous Production This is the other category of mass production. The terminology is actually borrowed from the continuous-process type of production. The idea is that here the items are made to 'flow' through a sequence of operations by material handling devices (conveyors, moving belts, transfer devices, etc.). Examples of flow production include automated transfer machines for the production of complex discrete parts, and manual assembly lines for the assembly of complex products.

In repetitive production, full advantage should be taken of repetitive operations in the design of production of auxiliary aids, such as special tools, fixtures, positioners, feeders and material handling systems, inspection devices, and weighing and packing equipment. Not only is the equipment dedicated to one product, but the entire plant is often designed for the exclusive purpose of producing the particular product.

In continuous production of discrete items, the demand rate and the production rate are approximately equal. The equipment is special-purpose rather than general-purpose. The investment in machines and specialized tooling is high. The very high volume justifies the use of special purpose-machines and transfer lines to decrease the cost of production substantially. Also these ensure that a very high degree of accuracy can be achieved with these systems. As the production skill has been transferred from the operator to the machine, the skill level of labor in a repetitive production plant tends to be lower than in a batch plant or a job shop. However, these manufacturing facilities, once designed and fabricated, are very inflexible and can only be used for a single product.

Further, the lead-time taken from the product design stage to the setting up of the manufacturing facility is very large (Groover, 1987).

Advantages of Repetitive/Flow Production

1. *Low Per-Unit Cost* Owing to specialized high-volume equipment, bulk purchasing of raw materials, lower labor rates, efficient utilization facility, low in-process inventories, and simplified managerial control.

2. *Lower Wages* Because the machines are specialized, operators can be less skilled, and therefore lower wages can be paid. In addition, fewer supervisors are needed, further reducing costs.

3. *Maximum Utilization of Space* Since the flow shop is generally continuous, with materials handling often built into the system itself, the operations can be designed to perform compactly and efficiently with narrow aisles, thereby making maximum use of space.

4. *Simplified Managerial Control* Because a standardized product is produced, inventory control and purchasing decisions are routine. Equipment is automated and process is standardized.

Disadvantages of Repetitive/Flow Production

1. *Changes in Product Design and Production Rate are difficult to make* Not only is the variety of output difficult to obtain; even changes in the rate of output are hard to make. Changing the output may require using overtime, laying off workers, adding additional shifts, or temporarily closing the plant. Also, because the equipment is so specialized, minor changes in the design of the product require substantial changes in the equipment.

2. *Boredom and Absenteeism among the Labor Force; Absence of Output-Based Incentives* Since the equipment performs the skilled tasks, there is no challenge for the workers. The constant, unending, repetitive nature of the manufacturing line can also dehumanize the workers. Since the rate of work flow is generally set (*paced*) by the line speed, incentive pay and other output-based incentives are not possible.

3. *Shortage of Supplies or Breakdown of Even a Single Machine or Conveyor can Halt the Entire Line* Such occurrences are prohibitively expensive.

4. *Pace of the Line is Decided by the Slowest Element* Since all machines and labor must work at the same repetitive pace in order to coordinate operations, the work loads along the entire line are generally balanced to the pace of the slowest element.

5. *Large Support Staff and Large Stock of Raw Materials are Required to Keep the Line Running Smoothly.*

6. *Complex and Costly Initial Setup and Maintenance* Simplicity in ongoing operations is achieved at the cost of complexity in the initial setup. The planning, design, and installation of the typically complicated,

special-purpose, high-volume equipment are mammoth tasks. The equipment is costly not only to setup originally but also to maintain and service. Furthermore, such special-purpose equipment is very susceptible to obsolescence and is difficult to dispose off or modify for other purposes (Shafer and Meredith, 1998).

1.5.4 Cellular Production

The cellular form is based on group technology, which seeks to achieve efficiency by exploiting similarities inherent in parts. In production, groups of parts that have similar processing requirements are grouped into a part family. After the parts are divided into families, a cell is created that includes the human skills and all the equipment required to produce a part family. Since the outputs are all similar, the equipment can be setup in one pattern to produce the entire family and does not need to be set up again for another type of output (as is necessary in a job shop). In cellular manufacturing, a team is completely responsible for conducting the work within each cell. The team members usually schedule and inspect the work themselves, once they know when it is due. Occasionally, work must be taken outside a cell for a special treatment or process that is unavailable within the cell, but these operations are minimized whenever possible. Cellular production combines the advantages of the job shop and repetitive shop to obtain the high variety possible with the job form and the reduced costs and short response times available with the repetitive form.

Mini Plant Another form of cellular production is called a mini-plant. Here, the cell not only does the manufacturing but also has its own industrial engineer, quality manager, accountant, marketing representative, and sales person, and almost all other support services that a regular plant has. Only far-removed services, such as R&D and human resources, are not dedicated to the mini-plant. The entire facility of the firm is thus broken down into a number of mini-plants, each with its own general manager, production workers, and support services so that it can operate as an independent profit center.

Advantages of Cellular Production

1. Reduced Machine Setup Times, Improved Productivity and Flexibility In cellular production, machine setup times are minimized because each cell processes only parts that have similar (or identical) setup and processing requirements. Reduced setup times result in improved productivity, smaller batch sizes, higher flexibility, and reduced lead times.

2. Less In-Process Inventory and Space Requirements.

3. More Accurate Forecasting Shorter lead times facilitate more accurate forecasting as the period for which the forecast is to be made becomes shorter. Shorter lead times may also provide a competitive advantage.

4. *Simplified Shop Floor Control, Reduced Material Handling, and Reduced Defects due to Increased Accountability, Responsibility, and Autonomy of Workers* In a cell, all the workers are totally responsible for completing every job. Thus, the effect is to enrich the work, provide challenges, encourage communication and teamwork, meet due dates, and maintain quality.

5. *Minimal Cost Required to move to Cellular Production* Although some cells may be highly automated, with expensive special-purpose equipment, it is not necessary to invest any additional capital in order to adopt the cellular form. It requires only the movement of equipment and labor into cells. Or with even less trouble, the firm can simply designate certain pieces of equipment as dedicated to a single part family, without actually relocating them. These are termed as *virtual cell* or *logical cell*, since the equipment is not physically adjoining but is still reserved for production of only one part family. There is, however, some loss of efficiency.

Disadvantages of Cellular Production

1. *Breakdown of a Single Machine or Conveyor can Halt the Entire Cell* If a piece of equipment should breakdown, it can stop the entire production in the cell. However, there may be machines in other cells that can be temporarily used to get a job out.

2. *Breaking up the Cell and Redistributing the Equipment may be Necessary* If demand for a family dries up, it may be necessary to break up that cell and redistribute the equipment, or reform the families. In the short run, the labor can be assigned to whatever cell needs it.

3. *Cellular Manufacturing is Neither as Efficient as Repetitive Manufacturing nor as Flexible as Job Shop* Volumes are too small in cellular production to allow the purchase of the high-volume, efficient equipment that flow shops use. The cellular form also does not allow for the extent of customization usually found in job shops.

4. *Fostering of Specialized Knowledge is Reduced* The fostering of specialized knowledge associated with various operational activities is reduced because the workers who perform these activities are spread out and therefore have limited opportunities to collaborate (Shafer and Meredith, 1998).

1.6 MANUFACTURING ENVIRONMENTS (Product Positioning Strategies or Demand Response Strategies)

Discrete manufacturing can be classified into the following four manufacturing environment categories based on the level of direct customer contact in production planning and scheduling:

1. Make to Stock (MTS)
2. Assemble to Order (ATO)
3. Make to Order (MTO)
4. Engineer to Order (ETO)

The above typology can also be referred to as product-positioning strategies classified on the basis of the type of inventory an organization chooses to maintain, or demand response strategies.

1.6.1 Make to Stock (MTS)

In the MTS demand response strategy, the producer stocks the finished product in inventory for immediate shipment. MTS emphasizes immediate delivery of good quality, reasonably priced, off-the-shelf, standard items. Rather than starting with the customer, the cycle starts with the producer specifying the product, acquiring the raw materials, and producing it for stock. In this environment, a customer is not willing to tolerate a delay in receiving the product. Hence, management is required to maintain a stock of finished goods. Often the stock of finished goods held is quite large due to the need to provide a variety of sizes, colors, and features. Thus, in MTS environment, products are made and put into inventory before an order is received from a customer. If the price and quality are acceptable, the customer orders the product. Assuming the product is indeed on hand, the producer ships immediately from stock. Off-the-rack clothing, frozen food, and bicycles are examples of this kind of manufacturing.

MTS implies the manufacture of products based on a well-known and predictable demand pattern. Operation management focuses entirely on replenishment of inventory; actual customer orders cannot be identified in the production process. At any particular time, the actual level of production may bear little correlation to the level of actual orders being received.

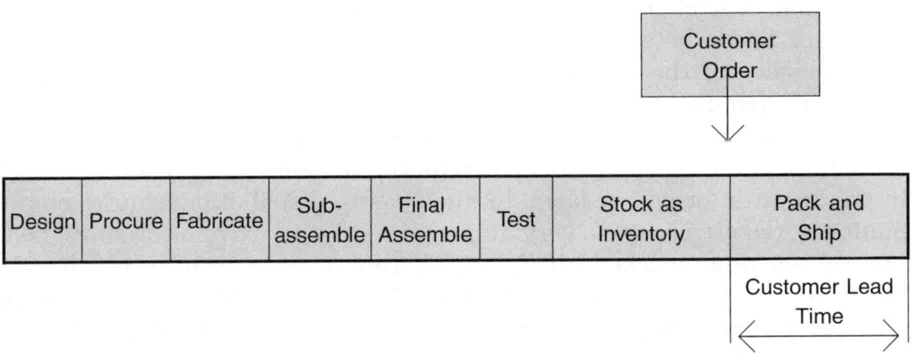

Fig. 1.3 *Make-to-Stock Manufacturing Strategy and Lead-Time*

In this environment the interface with the customer is rather distant. The production volume of each sales unit tends to be high and customer delivery time, normally short, is determined by the availability of finished goods inventory. The finished goods inventory acts as a buffer against uncertain demand and stockouts.

The MTS system has the advantage of having short delivery time, but inventory costs in the factory are large and customers are unable to express preferences as to the design of the product. The MTS environment also assumes reasonably long and predictable product life cycles (McMahon and Browne, 1993).

Generally, firms make to stock when:

1. Demand is fairly constant and predictable.
2. There are few product options.
3. Delivery time demanded by the marketplace are much shorter than the time needed to make the product.
4. Product has a long shelf life.

The information needed to make a production plan is as follows:

1. Forecast by period for the planning horizon.
2. Opening inventory.
3. Desired ending inventory.
4. Any past-due customer orders. These are orders that are late for delivery and are sometimes called *back orders.*

$$\text{Total Production Required} = \text{Total Forecast} + \text{Back Orders} + \text{Ending Inventory} - \text{Opening Inventory}$$

In MTS environment, the customer is identified only after production has been completed, often through the channels of a more or less complex distribution chain. The actual manufacturing function is therefore 'blind,' in the sense that it does not perceive market demand directly, but only through the abstract medium of forecasts. The point where sales and production control merge is the finished goods inventory. Keeping the inventory levels between the minimum and maximum values, ensures the efficiency of the distribution operation (Waters, 1996). Thus, MTS items are generally mass consumed items, and pass through multiple channels before reaching the end user. As most of the data about customers is unknown, it is the distribution network that has to be used to create an information base. In such cases, distribution requirements planning (DRP) becomes very important. In the extreme case of fast moving consumer goods (FMCG), DRP becomes the heart of system which links various stocking and sales points. Demand is also calculated from these outlets and aggregated for production purposes (Mishra, 1998).

In most MTS cases, production is less complex, product recipe or product structure is stable and the rate of production is fairly fixed. Thus, Capacity Requirements Planning (CRP) and Shop Floor Control (SFC) are not likely to be critical. The most important aspect of Enterprise Resource Planning (ERP) in such environments is to create a totally integrated distribution and production system which keeps a stable flow of products at the point of consumption and is responsive to any change. System designers have to decide how closely it should represent the actual distribution network. Since there is no direct contact with the end user, the distribution system has to act as ears and eyes for the organization and support accurate forecasting and trend analysis (Mishra, 1998). Engineering becomes involved in MTS manufacturing environment only if and when engineering changes are made through engineering change orders (ECO) and when new products are introduced. When an engineering change is to be introduced, product engineers design the new product and provide revised or new drawings, BOMs, etc.; whereas manufacturing engineers design the new or modified process and communicate it to the MRP II system through routings.

Among the main risks inherent in this approach is the possible inaccuracy of sales forecasts. Errors can arise from the neglect of seasonal factors or from unexpected phenomena affecting demand trends. Their results are, on the one hand, the saturation of stocks of the unwanted product and, on the other, an insufficient production of other goods that have been deprived of production resources and therefore cannot be shipped in time, although in great demand. On the other hand, the ready availability of finished goods allows MTS manufacturers to reach a wide range of buyers and stay in a market where features and prices of competing products are similar. Television sets, motor cycles, and personal computers are just a few examples of sectors in which the consumer decides to buy, often on impulse, on the basis of the choice that the retailer can offer at that particular moment.

1.6.2 Assemble-to-Order (ATO) (Includes Finished- or Packaged-to-Order)

Where several options exist, such as in computers and automobiles, and where the customer is not willing to wait until the product is made, manufacturers produce and stock standard component parts. When manufacturers receive an order from a customer, they assemble the component parts selected by the customer. Since the components are stocked, the firm needs only time to assemble before delivering to the customer. It is a system which uses the same core assemblies for product and has the capability to vary other components of the final assembly. This demand response strategy is used by companies having modular products with a number of final products having common modules. ATO type of an organization attempts to make a final product, which is configured from certain standard modules as per the requirements and choice of the customer, and yet keeps the cost and the customer lead-time low. Thus, ATO attempts to combine the product variety/customization of engineer- or make-to-order environments with the low cost and short lead-time of the MTS environment. The most important characteristic of ATO type of manufacturing environment is that the product design is modular.

Modular Design

A modular system is composed of units (or modules) that are designed independently but still function as an integrated whole. A module is a self-contained component having standard interfaces to other components or sub-assemblies in the system. Modularity implies the product can be built as a series of building blocks where all the product variants can be created with sets of modules. The essence of modular concept is to design, develop and produce those parts which can be combined in the maximum possible number of ways. The term modular production was used to indicate a concept of producing a variety of products to meet the market needs by optimally combining standard parts.

A Product Variant (or Exclusive Choice) In modular configurations, it is that part of the product which, while it must be present, may be specified to the customer's choice from the options offered. For example, every automobile must have a transmission system. It might, however, be manual or automatic depending upon the choice of the customer.

Add-ons, Attachments, Accessories (or Additive Choices) These are the additional features that are not mandatory. For example, an air-conditioner or a factory-fitted stereo system for an automobile, or a multimedia system for a personal computer.

Modular products have many variants and add-ons for customers to select from and thus provide a large number of possible option combinations in specific models. As a principle of production, modularity has a long history. Manufacturers have been using it for a century or more because it has always been easier to make complicated products by dividing the manufacturing process into modules or cells. Carmakers, for example, routinely manufacture the components of an automobile at different sites and then bring them together for final assembly. They can do so because they have precisely and completely specified the design of each part (Baldwin and Clark, 1997).

We have now entered a period of great advances in modularity. Breakthroughs in material science and other fields have made it easier to obtain the deep product knowledge necessary to specify the design rules. For example, engineers now understand how metal reacts under force well enough to ensure modular coherence in body design and metal-forming processes for cars and big appliances. Improvements in computing, of course, have dramatically decreased the cost of capturing, processing, and storing that knowledge, reducing the cost of designing and testing different modules as well.

Through the widespread adoption of modular designs, the computer industry has dramatically increased its rate of innovation. Indeed, it is modularity, more than speedy processing and communication or any other technology, that is responsible for the heightened pace of change that managers in the computer industry now face. Strategies based on modularity are the best way to deal with that change (Baldwin and Clark, 1997).

Developing Modular Design (Making Modular Sub-Assemblies)

Technique for Achieving Modularity Designers achieve modularity by partitioning the design information into the following two categories:

1. Visible Design Rules (or Visible Information)
2. Hidden Design Parameters (or Hidden Information)

Modularity is beneficial only if the partitioning is precise, unambiguous, and complete.

1. The Visible Design Rules These are decisions that affect subsequent design decisions. The visible design rules establish and enforce the visible overall design rules that determine how the different modules of the machine would work together. Ideally, the visible design rules are established early in a design process and communicated broadly to those involved. The different teams have to adhere absolutely to these rules. However, each team has full control over the hidden elements of design in its module, that is, those elements that have no effect on other modules.

The module designers are free to try out a wide range of approaches as long as they obey the design rules ensuring that the modules would fit together. The more experiments and the more flexibility each designer has to develop and test the experimental modules, the faster the industry is able to arrive at improved versions. Visible design rules fall into three categories (Baldwin and Clark, 1997).

1. *An Architecture* Which specifies what modules will be part of the system and what their functions will be.
2. *Interfaces* These describe in detail how the modules will interact, including how they will fit together, connect, and communicate.
3. *Standards* These are for testing a module's conformity to the design rules (can Module X function in the system?) and for measuring one module's performance relative to another (how good is Module X compared to Module Y?).

2. The Hidden Design Parameters These are design decisions that do not affect the design beyond the local module. Hidden elements can be chosen late and changed often and do not have to be communicated to anyone beyond the module design team (Baldwin and Clark, 1997).

Advantages of Modular Design

- Modularity is a strategy for organizing complex products and processes efficiently.

- Modularity in use allows consumers to mix and match elements to come up with a final product that suits their tastes and needs (Baldwin and Clark, 1997).
- Modular design offers the ability to standardize diversity because it allows a product to be customized by using different combinations of standard components (Redford and Chal, 1994).
- Modular design resists obsolescence and shortens the redesign cycle. A new generation of product can usually utilize most of the old modules (Redford and Chal, 1994).
- Modularity in use can spur innovation in design: the manufacturers can independently experiment with new products and concepts, and find ready consumer acceptance as long as their modules fit the standard dimensions (Baldwin and Clark, 1997).
- Costs are reduced and ease of service and repair are enhanced because a defective module can be quickly replaced by a good one (Redford and Chal, 1994).
- Most importantly, modular design simplifies final assembly because there are fewer parts to assemble and each module can be fully checked prior to installation.
- Modularity has enabled companies to handle the increasingly complex technology. By breaking up a product into sub-systems, or *modules,* designers, producers, and users have gained enormous flexibility. Different companies can take responsibility for separate modules and be confident that a reliable product will arise from their collective efforts (Baldwin and Clark, 1997).

Drawbacks of Modular Design

1. *Adds Cost and Complexity* Modular design can add cost and complexity because of extra fitting and interconnections required. Therefore, modular design should not be used unless its advantages are needed.
2. *Difficult to Design* Modular systems are much more difficult to design than comparable interconnected systems. The designers of modular systems must know a great deal about the inner workings of the overall product or process in order to develop the visible design rules necessary to make the modules function as a whole. They have to specify those rules in advance. And while designs at the modular level are proceeding independently, it may seem that all is going well; problems with incomplete or imperfect modularization tend to appear only when the modules come together and work poorly as an integrated whole (Baldwin and Clark, 1997).

An example of a modular product is the farm tractor family shown in Table 1.5. Customers obviously like to have as many variants and

Table 1.5

Farm Tractor Family
(Adapted from Plossl, 1994)

Sr. No.	*Function*	*Options*	*Total No. of Options*	*Modules to be Master-Scheduled*	
1.	Wheel Arrangement	1. Four-Wheel Construction 2. Three-Wheel Construction—Single front wheel 3. Three-Wheel Construction—Double Front Wheel	3	Basic Tractor Wheel Arrangement	1 3
2.	Fuel	1. Gasoline 2. Diesel 3. LPG	3	Engine	6
3.	Horsepower	1. 56 hp 2. 68 hp	2		
4.	Trans-mission	1. Stick Shaft 2. Automatic	2	Transmission	2
5.	Steering	1. Mechanical 2. Power	2	Steering	2
6.	Rear Platform	1. Regular 2. Low	2	Rear Platform	2
7.	Axies	1. Standard 2. High-Clearance	2	Axies	2
8.	Hitch	1. Mechanical 2. Hydraulic	2	Hitch	2
9.	Power Takeoff	1. With Type A 2. With Type B 3. Without Type A or B	3	Power Takeoff	3
10.	Radiator Shutters	1. With 2. Without	2	Radiator Shutters	1
11.	Operator Cab	1. With 2. Without	2	Operator Cab	1
		Total No. of Modules to be Master-Scheduled			25

Total No. of Possible Unique Configurations

$= 3 \times 3 \times 2 \times 2 \times 2 \times 2 \times 2 \times 2 \times 3 \times 2 \times 2$

$= 6912$

add-ons as possible to get a more or less 'customized product.' Hence, marketing would want the product designers to offer large number of variants and add-ons for each product family. On the other hand, greater product variety, which invariably results in reduced production volumes, is the last thing the manufacturing people want. Manufacturing organizations want low product variety and high volumes so that advantages of specialized mass production, such as reduced number of setups, specialized equipment requiring lesser operator skill, and simplified production planning and control, can be availed of. Thus there is a conflict between marketing and manufacturing: marketing wants more and more number of possible unique configurations while manufacturing wants less and less number of modules to be master-scheduled for production. The beauty of modular construction, as can be seen from the farm tractor example, is that the addition of modules increases the number of possible unique configurations, the desirable part, in geometric progression whereas the number of modules to be master-scheduled, the undesirable part, increases in arithmetic progression.

As we can see, if we manufacture 25 modules for customers to select from, there could be 6912 tractors with no two identical tractors; each of the 6912 tractors would have a unique configuration of optional features. If, on the other hand, we were to manufacture the tractor in the conventional way, we would have, in any case, manufactured the basic tractor and 11 modules, which could be assembled into only a single tractor configuration. *Thus, by increasing the number of modules—the undesirable part—from 12 to only 25 the number of possible configurations—the desirable part—has increased from 1 to 6912.*

Now suppose we convert one more feature in the common parts of the tractor to a modular design giving one more option to the customer by adding just one module, it would result in $6912 \times 2 = 13{,}824$ possible configurations with 27 modules to be master-scheduled.

Following additional features are required for manufacturing planning, order entry, and final assembly of ATO products:

1. The Planning, Super, or Generic Bill of Material
2. The Product Configurator

The Planning, Super Modular, or Generic Bill of Material

Specific details of options desired by customers must be provided for customer order entry, pricing, and promising deliveries. These details will define the specific modules required to build the products customers want. These details are needed also for final assembly scheduling, production, and shipping in the execution phase of manufacturing. In the planning phase, however, it should be obvious that attempts to forecast demand for a multitude of specific combinations of many options are doomed to fail. For example, in case of tractor family, forecasting the demand, developing the MPS, and setting up bills of material for all the possible 6912 configurations, some of which customers may never order, is neither practical nor needed.

The starting point for manufacturing planning in ATO is again the forecast, since we must manufacture and keep ready all the standard modules in inventory before the customer comes into picture. Forecasting how many of each of the possible end configurations of the product family would be sold in a particular time period is impossible. For example, in case of farm tractor family it is impossible for anybody to forecast how many of each of the 6912 tractor models would be sold in a particular month. Even visualizing what exactly these 6912 configurations are is virtually impossible. If the manufacturer is to produce 500 of this tractor family per month, how many of which of the 6912 models should make up the 500? The problem is not eliminated even at higher volumes (Plossl, 1994).

The solution to the above problem is to forget, in the planning phase, the end product configurations, and instead, forecast and plan each of their top-level components (that is, the modules) separately. In the farm tractor family example, if the production volume for a particular month is 500 units; schedule 500 sets of components common to all configurations (models). This is facilitated by grouping all these components in a modular bill of material for a *pseudo item* called *basic tractor*. We call this item as 'pseudo' item because the components included in this bill of material are not really assembled into a single sub-assembly; rather they would be fitted at various places, in various sub-assemblies, in the final tractor assembly. Thus, this bill of material and the item 'basic tractor' is not real; it is an artificial grouping of components to facilitate the planning.

Next, there are two options for the feature called transmission: stick shaft and automatic. We might be having the historic data as to how many customers have gone for the stick shaft option and how many have gone for automatic option in the past. We can apply these fractions to the forecast volume of the product family to get the volume of each of the options. Thus, if the past demand for stick shaft option is 75% and 25% for automatic option, then (500×0.75 =) 375 units of stick shaft transmission module and (500×0.25 =) 125 units of automatic transmission modules is a reasonable schedule. To take care of the possible fluctuation in the percentages, each type of transmission modules can be 'overplanned' based on past variations in actual orders. This overplanning might be in the form of increased fractions, say, instead of applying the fractions 0.75 and 0.25, we can apply 0.8 and 0.35 respectively to the two modules to get extra modules (in this case the total of the fractions would obviously be more than 100%). Alternatively, overplanning might be in the form of inflating the MPS quantities by adding certain quantities. Say we add 50 to the master-scheduled quantity of 375 units of stick shaft option making it 425, and add 25 units to the master scheduled quantity of 125 units of automatic transmission option making it 150. In case of new products where no historical data is available, the fractions are forecasted based on the judgement of the people in marketing and production. This type of planning is facilitated by the planning bill of material shown in Fig. 1.4. It is also called *super modular* bill of material or *generic* bill of material.

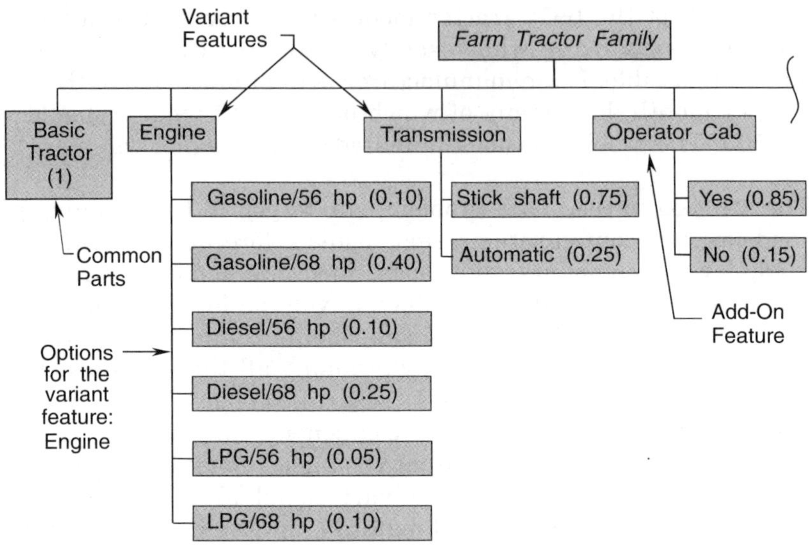

Fig. 1.4 *Planning, Super Modular, or Generic Bill of Material with no Options 'Overplanning' (Only a Part of the Bill is Shown)*

Planning bills are basically used to simplify and improve master scheduling. As shown in Fig. 1.4, these bills clearly show what variants or add-ons are to be master-scheduled and what is the relative volume of each of the modules relative to the forecast volume of the total family. Once the aggregate forecast for the product family is available, further planning can be done using this bill.

The Product Configurator

(a) *Menu-Type Order-Entry System* Since the end product is not predefined, there is nothing like a predefined traditional catalog of end product part numbers in ATO. Hence the first thing needed for customer order entry is a means through which the end product can be defined by the customer by selecting modules of his or her choice from those offered by the manufacturer. To accomplish this what is required is a system that will display the available options for each variant or add-on feature in the form of a menu. The information about all the available options is included in the super modular bill of material. The order entry mechanism should prompt for the entry of the type of product, show which variants are required and which add-ons are available. This task is accomplished with the help of a piece of software called *product configurator.* Thus, the first function of the product configurator is to take the super modular bill as input and create a menu type of order entry system. The end product is typically configured by the customer by clicking the modules of his choice.

(b) *Configuration Validation* Many times, there are certain modules in the product family, which are not compatible with each other. For example, in case of personal computers, video monitor is a module for which monochrome and color monitors of various screen sizes are the possible options. A corresponding module is the video card for which again color and monochrome drivers are the options. In this case, color monitor and monochrome video card are not compatible. Since not all variants/options necessarily work with all other variants/options, the system should eliminate the choices that are not valid based on the choices made earlier. The product configurator ensures that incompatible modules are not selected. This is typically achieved by blocking out the incompatible modules from the menus based on the selections already made; or alternatively, a message can be displayed when two incompatible modules are selected. Thus, the second function of the configurator is to use the product constraints to monitor the product specifications as submitted by the customer so that incompatible modules are not included in any configuration. Similarly, if selection of one module mandates selection of some other module then the configurator must be programmed to ensure that both the modules are included in the configura-

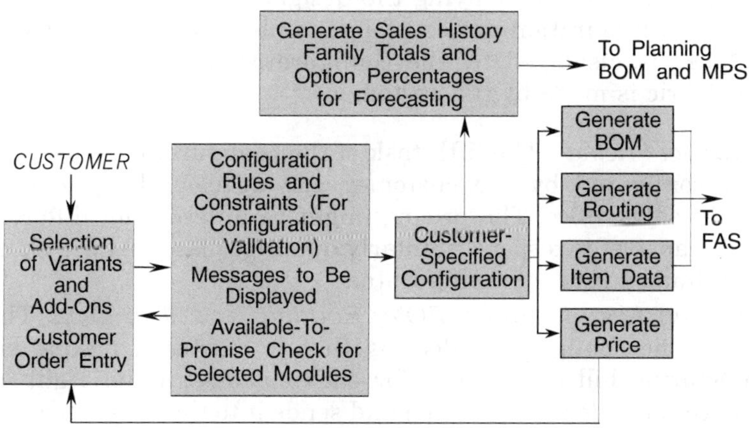

Fig. 1.5 *The Product Configurator*

tion. Thus, through the configurator the customers are elevated to a level of technical abstraction, where they do not need to know which components fit together and which don't, or how the components fit together. These pieces of information are programmed into the configurator. The workings of the *Product Configurator* is shown in Fig. 1.5.

(c) *ATP Check for Common Parts and Each of the Selected Modules* To make the delivery promise for a particular customer order, availability of all the customer-selected modules has to be checked. The order can be promised only when all the customer-selected modules are available in the required time period. The availability data of modules in inventory is made available to the sales department through the on-line 'available-to-promise (ATP)' figures. Consequently, the order entry mechanism should check the available to promise for the type of product and the available to promise for each module as shown in Fig. 1.6. If not all of the customer-selected modules are available by the requested date, then the order should be flagged for human review and approval. This is the third function of the configurator.

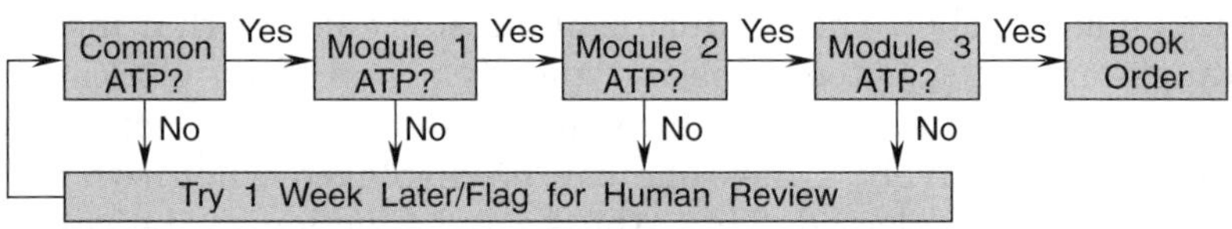

Fig. 1.6 *Available-to-Promise Logic with Modular BOM Architecture*

(d) *Displaying Designated Messages* The fourth function that the configurator performs is displaying the designated messages on the screen during the configuration process; for example, 'leather fabric will cost an extra 30%,' is displayed by the configurator when a selection having leather fabric is made by the customer.

(e) *Product Pricing* The fifth task of the configurator is to cost the end product configured by the customer and develop the pricing for the particular sales office. The pricing might be different for different sales offices because of taxes, etc. Similarly quantity discounts etc. should also be programmed into the configurator.

(f) *Creating Manufacturing BOM, Routing, and Item Data* The sixth task that the configurator does is that it creates a unique end-item manufacturing bill of material for each customer order right at order entry once the order is confirmed and sends it to the shop where the final assembly is to be done. In this bill, the option descriptions are converted into part and bill-of-material numbers by the configurator. A routing to final-assemble, finish, and/or package the product is also created by the configurator. In the absence of formal instructions, the finishing operations are usually performed based on the knowledge of the shopfloor personnel resulting in production errors and loss of uniformity and quality, especially when staff turnover occurs. A configuration system that produces detailed instructions for product assembly can greatly reduce all these problems. All the information about the various items (item data) required for the particular order is also generated by the configurator.

(g) *Providing Historical Data for Forecasting of Family Totals and Option Fractions* Finally, as all the orders are entered through the product cofigurator, it has all the historical information about the total number of end products ordered in various time periods and also the fractions of orders asking for particular options for each of the features. Using this information the configurator can help in forecasting for the aggregate as well as fractions for preparing the super modular bill of material for master-scheduling.

To summarize, the primary purpose of the configurator is rapid development of a product structure through the inputs of a customer without requiring the detailed technical knowledge usually necessary to build a BOM and routing. It is a tool used to create manufacturing BOMs, routings, and prices based on user-selected options and pre-defined rules (constraints). The required knowledge base (BOM, routing, and configuration rules) resides in the configurator rather than with a particular individual. Thus, configurator facilitates creation of a user-independent knowledge base for product structuring. The sales volume of the ATO product should be enough to justify investment of time and effort in creating and loading the required knowledge base. For example, if the ATO item is going to be built only a few times, then it may not be worth the time and effort to develop the features, options, and constraints required to make the product configurator work.

Following are the main benefits of using a product configurator:

1. Rapid development of product structure using a user-independent data base.
2. Order cycle-time reduction.
3. Integration of customer specifications directly into the planning system.
4. Reduction in the training time and effort for effective order entry.
5. Reduction in the number of technical people in presales and sales, as the customers can configure the product themselves without any assistance from sales people.

Non-Programming Environment A configuration system should enable users to enter new product models or customize existing models without complicated programming. Programmers, who often have little product knowledge, cannot be relied upon to maintain and update the configuration system. Asking them to do so can cause delays, inaccuracies, and cost overruns and defeat the very purpose of the system. Those who know the product best should be able to define the variables, rules, and documentation (Virmani, 1999).

Web-Enabled Product Configuration and Order Booking Many manufacturers now offer online quotes for customer products. Customers can visit a company's web site; configure a product to their own specifications; and receive information back about the product's validity, price for the particular configuration and particular delivery terms, and estimated delivery date. The configuration system generating the information on the requested product would then enable the customer to place an unassisted order via the Internet. This frees sales staff from handling routine quotes so they can concentrate on more challenging tasks (Virmani, 1999). Thus, through the Internet, customers from all over the world are able to configure complex products directly at the site of the manufacturer, without needing the required technical background. Further, the price is dynamically returned to the user almost instantaneously completing the web-based presales process.

Assemble-to-Order Demand Response Strategy

In the ATO demand response strategy, all sub-assemblies or modules are available in inventory. When a customer orders a product, the producer quickly assembles the modules and ships the final product. Thus, these companies can respond to customer demand much more effectively by forecasting and stocking the modules, and then assembling the final product upon receipt of the customer order. The planning of standard modules is done by master production scheduling (MPS), which is driven by the forecast. The planning of the end items is done by final assembly scheduling (FAS) based on the actual customer orders as shown in Fig. 1.7.

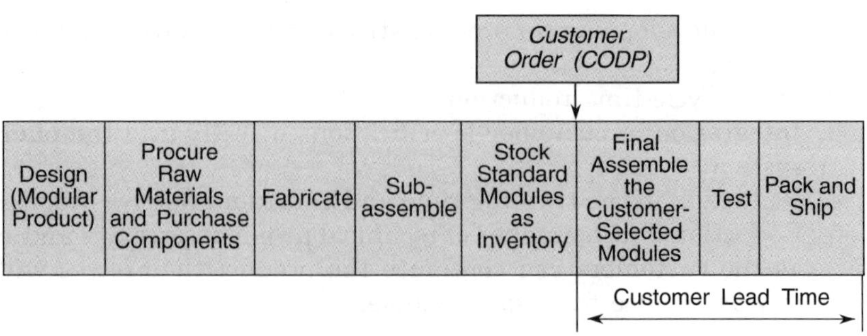

Fig. 1.7 *Assemble-to-Order Manufacturing Strategy and Lead Time*

Customer Order Decoupling Point (CODP) It is the point at which production changes from forecast-driven standard production to backlog-driven customized (that is, customer-specific) production. The CODP is the point within the product structure that a product takes on a customer-specific configuration. By establishing this point strategically within the product structure and within the total lead-time, manufacturer can achieve both a high level of customization and also a reasonable lead time for which the customer is ready to wait. The needs of the forecast-driven (push) systems are well addressed by MRP II type systems while the backlog-driven (pull type) systems require a project management system. However, these two systems must work together, particularly in an ATO type of environment, and their point of interface is the CODP. CODP varies depending upon the nature of the product and the current position of the product in its life cycle.

A manufacturing environment working on ATO strategy primarily has contact with the customer only at a sales level. Customer involvement in the design of the product is limited to selecting the standard modules (variants and add-ons). The delivery time is medium to low and customer delivery time is based on the availability of major sub-assemblies (modules). Demand uncertainty is handled by overplanning components and sub-assemblies. Assembly only takes place on receipt of a customer order and buffers of modules or options may exist. The product routing in the factory is typically fixed. No final product inventory buffer exists and the customer has limited input into the design of the product.

The following information is needed to make a production plan for MTO products:

1. Forecasts for family total and option fractions by period for the planning horizon.
2. Opening backlog of FAS customer orders for each module.
3. Desired ending FAS backlog for each module
4. Desired Change in inventory levels for each module.

Total production in a period
= {Sales plan for each module (that is, family forecast × option fraction for the module) + Planned reduction in backlog of the module + Planned increase in the module inventory}
= {Sales plan + (Opening FAS backlog of the module – Ending FAS back log of the module) + (Ending inventory of the module – Opening inventory of the module)}

1.6.3 Make to Order (MTO)

In this environment, manufacturers wait until an order is received from a customer before starting to make the goods. It describes a manufacturing facility in which the final product is usually made from items already designed but may include some custom designed components as well. It has many of the base components available along with the engineering designs, but the product is not actually completely specified. The order cycle begins when the customer specifies the product that he or she wants. The producer can assist the customer to prepare the specifications. Sometimes the design might be provided by the customer himself. The configuration of the product is likely to change from the initial specification during the course of processing. Generally speaking, successive versions of the same object produced for different customers are not identical, and this lack of interchangeability complicates document and inventory control.

Thus, in MTO demand response strategy, only the product and component designs and some standard raw materials and components are held in inventory as shown in Fig. 1.8. Examples of this kind of manufacture are custom-tailored clothing, machinery, or any product made to customer specifications. Very expensive items are usually made to order.

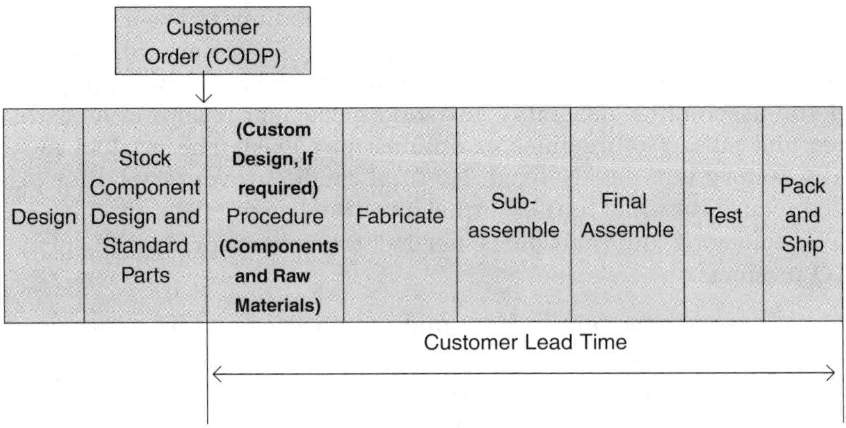

Fig. 1.8 *Make to Order Manufacturing Strategy and Lead-Time*

Generally, firms make to order when:

1. Goods are produced to customer specifications.
2. The customer is willing to wait while the order is being made.
3. The product is expensive to make and store.
4. Several product options are offered.

Interaction with the client is extensive, normally involving sales and engineering, while the delivery time ranges from medium to large. The producer quotes a price and delivery time based on the customer's request. The quotation process itself is less costly and complex compared to the engineer to order system described next. The customer and producer frequently discuss alternatives to reduce cost, reduce time to ship, and/or meet the customer's actual needs more closely. Although the products are standard products that have been made before, the processing activities are tailored to each individual customer order. Promise for completion of orders is based on the available capacities in manufacturing and engineering.

In the MTO environment, as customer order changes, production schedule changes significantly from one period to another. It is entirely possible that some types of products are not made during a period at all. Production in such environments is very dynamic. At any time, several jobs are being worked upon and their priority keeps changing. In the MTO environment, organizations invest in capacity—they build large capacities in anticipation. Capacity is not really expected to be fully utilized. However, loads on various machines keep changing and hence capacity requirements planning (CRP) and shop floor control (SFC) become critical. Distribution is not very complex as fewer products flow directly to customer locations according to a predetermined schedule (Mishra, 1998).

1.6.4 Engineer to Order (ETO)

In this demand response strategy, nothing is inventoried in the producer's system, not even the design. These products have not been made before, at least not by this company. Generally, the customer defines the specifications in terms of functionality and performance levels and then the manufacturer designs the product as per the requirements. In other words, the customer's specifications require unique engineering design or significant customization. Usually, the customer is highly involved in the product design. The customer usually asks for a quotation of cost and time from the producer. The quotation process itself can be complex and costly. When the customer places order, the producer first develops the design for required product, which can involve considerable time and expense, receives customer approval of the design, and then orders the needed material. Upon receipt of the material, the producer fabricates components, assembles the product, and ships it to the customer. Thus, ETO is the extreme case where not only manufacturing but engineering is also customer-specific as depicted in Fig. 1.9.

Characteristics Common to MTO and ETO Environments

ETO is an extension of the MTO system with the engineering design of the product based on the customer requirements and specifications. The same characteristics apply as in case of the MTO, but clearly, customer interaction with the product supplier is even greater. True one-of-a-kind products are engineered to order (McMahon and Browne, 1993).

It is usually the technologically advanced sectors that are organized along these lines. The aeronautical, defense, and machine-tool industries all produce goods with a high unit value, which can be neither stand-

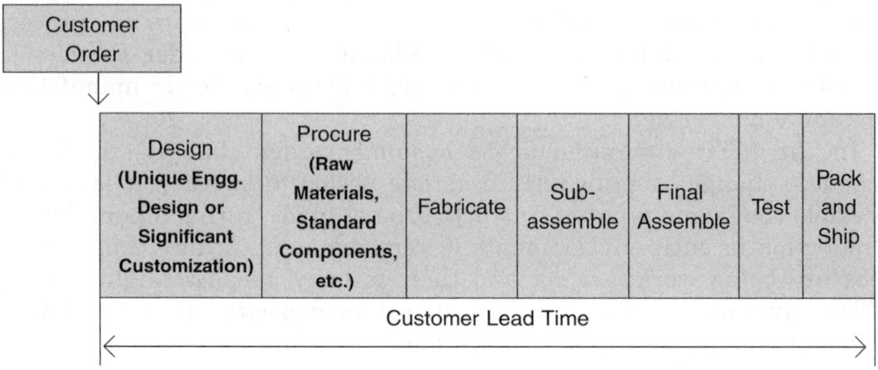

Fig. 1.9 *Engineer to Order Manufacturing Strategy and Lead-Time*

ardized nor kept in stock. Actually, each customer needs them for a different application and buys them on the basis of a full-fledged long-term plan. For these customers, delivery time, even measured in months, is no problem.

For the manufacturer, a large component of risk is removed because all resources used in the factory have a precise economic justification, and inventory carrying charges are also reduced. While it is difficult to achieve economies of scale in purchasing or to set up manufacturing operations in the most effective way, these additional costs can be passed on to the customer and are swallowed up in the high unit price. The main issues for firms that manufacture 'to order' are in the commercial and the engineering areas. First of all they have to get the order in competition with rival suppliers—this phase often takes the form of a tender—and then they have to find the technically best solution to make the special parts.

New products are the rule, not the exception, and the firm's catalog exists only in a partial form as a list of products made in the past. A single order is sufficient to cover the costs of research and development. Investment planning is therefore less crucial than a fast design process allowing accurate cost forecasts. To offer a competitive price, it is necessary to evaluate all the direct costs and ensure an acceptable profit margin. Long-term investments are often made in acquiring and increasing the firm's expertise in a specific sector and thus gaining a competitive edge. A manufacturer of earth-moving machines can, for example, convert the drive and transmission components from mechanical to hydraulic. A missile manufacturer can likewise decide to introduce on a large scale either analog or digital control devices. However, if subsequent developments in the industry demonstrate that wrong choices have been made, the company can find itself in a serious economic plight (Sartori, 1988).

Table 1.6

Comparison of MTS, ATO, and MTO/ETO Environments
[Adapted from Wemmerlov (1984), and Sari (1997)]

Aspect	*Make to Stock (MTS)*	*Assembled to Order (ATO)*	*Make-/Engineer to Order (M/ETO)*
Interface between Mfg. and Customer.	Low/Distant.	Primarily at sales level.	Engineering and sales level.
Delivery Time.	Short.	Medium.	Long.
Production Volume of Each Sales Unit.	High.	Medium.	Low.
Product Range.	Low.	Medium/high.	High
Basis of Production Planning and Control.	Forecasts of sales unit (end item).	Option Forecasts for master scheduling, and backlog for final assembly scheduling.	Backlog. Customer Intelligence Forecasts*.
Order Promising (based on . . .).	Availablity of finished goods inventory.	Availability of standard modules and common parts.	Availability of manufacturing and engineering capacity.
Handling of Demand Uncertainty.	Safety stocks of sales units.	Overplanning of standard modules and common parts.	Little uncertainty exists about materials For uncertainty in capacity requirements, excess capacity is usually planned.
Master Scheduling Unit.	End Items/Sales units.	Standard Modules—Variants, Add-ons, Attachments, Accessories, and Intermediates; Common Parts.	Customer Orders; Management authorizations of long lead time components and speculative items**; Standard raw materials and components;

(*contd.*)

Aspect	*Make to Stock (MTS)*	*Assembled to Order (ATO)*	*Make-/Engineer to Order (M/ETO)*
Items Held in Inventory.	Finished Products (Very high inventory carrying costs).	Major modules/Sub-assemblies, Common Parts (Lower inventory carrying costs for the total variety).	Product Designs and Standard Raw Materials (Lowest inventory carrying costs).
Final Assembly Schedule.	Close correspondence to the master schedule.	Determined by customer orders received by order entry.	Covers most of the assembly operations.
Bill of Material Structuring.	Standard BOMs for each sales item.	Planning BOMs.	Unique BOMs for each order.
Critical/Important MRP II/ERP Modules.	Distribution Resource Planning.	Product Configurator, MPS, MRP I, and FAS.	FAS, CRP, SFC.

* Intelligence forecasts are based on the information collected from the 'listening posts,' which a manufacturer might maintain near potential customers.

** Many times management authorizes the shop to start manufacturing certain items having long lead-times without receiving any orders for them as a way to reduce the lead time of executing anticipated orders. This might also provide a competitive advantage to the manufacturer as in the event of a customer needing that product this manufacturer would be able to provide the product in a shorter lead time compared to the competitors. However, here the risks involved are very high and hence such items are also termed as speculative items. Thus, the production of speculative items commences not as a response to a customer order but through an authorization given by a senior manager to the production function to go ahead. If and when the customer order for the speculative item appears it naturally replaces the authorization.

Project Execution

Project execution is the classical ETO type of manufacturing. Project operations are of large scale and finite duration; also they are non-repetitive, consisting of multiple, and often simultaneous, tasks that are highly interdependent. However, the primary characteristics of the tasks are their limited duration and their immobility during processing. Generally, staff, materials, and equipment are brought to the output and located in a nearby staging area until needed; (*staging* is the term used to describe the act of pulling material for an order from inventory before the material is required; this is often done to identify shortages of components and raw materials). Projects have particularly limited lives. Resources are brought together for the duration of the project; some are consumed, and others, such as equipment and personnel, are deployed to other uses at the conclusion of the project (Shafer and Meredith, 1998).

Often the manufacturer is required to deliver the product on a turn key basis. For example, an oil exploration client may ask an engineering company to supply an oil production unit, complete with pipelines, platforms, metering device and refining unit. While the number of such orders are few and a company may get no more than one order a year, yet each order can fetch large amount of money.

Customer interaction is very high in ETO and technical complications are immense. Because delivery schedules are short, it is often necessary to carry out both design and manufacturing activities in parallel. In such cases, the entire project is divided into major sub-activities, with the basic engineering being done first. These sub-activities are planned on the network, using techniques like Critical Path Method (CPM). This method helps to identify critical processes. It allocates time and resources to every sub-activity, and each sub-activity is further divided into tasks and sub-tasks. This kind of network planning replaces classical MRP I and CRP to calculate manufacturing and fabrication schedules. Those schedules are integrated with various shop floors, fabrication units and even vendor's operations (Mishra, 1998).

Traditionally, only project management techniques have been used in ETO environments, but now there is an increasing trend to use an ERP solution integrated with networking tools. The critical issues in such environments are as follows:

1. *Shortening of Project Completion Time and Proper Sequencing and Interconnecting of Various Activities* The project activities include basic design, cost and time estimation, detailed design, manufacturing/erecting, commissioning and testing. How to shorten the total lead-time and how to sequence and interconnect all the activities and sub-activities together are the key issues.

2. *Accurate Estimation and Control of Cost and Completion* Time The estimation of the cost and completion time is a key activity as it is done

right in the beginning during the bidding stage when the details have not even been worked out. Similarly, monitoring and controlling the actual cost and time against the estimated is another critical activity.

3. Vendor Management The size and complexity of the task mandates that a large number of component parts and equipment should be purchased from outside suppliers. Similarly, a number of tasks have to be sub-contracted. This makes vendor development and management one of the key activities.

4. Risk Management Project execution usually involves coordination of activities performed by several different groups. The tasks and activities might be spread at multiple locations. Further huge amounts of money are tied up in the project. All this makes risk management the most vital issue in project execution.

Table 1.6 compares and contrasts the various categories of manufacturing system. In reality, very few firms belong specifically to a particular category. Many firms would be classified as hybrids of the above. A firm may be hybrid of MTS and ATO. This implies that it holds assembled products, for which there is a steady demand, in stock, but also has a facility whereby products can be configured according to customer needs.

1.7 EVOLUTION OF MANUFACTURING

Following are the major determinants of the product positioning strategy:

(i) The manufacturing lead time
(ii) The time a customer is willing to wait for product delivery
(iii) The degree of customization desired by the customer

If the time a customer is willing to wait for product delivery is less than either manufacturing lead time or assembly lead time, an organization must maintain an inventory of finished goods for immediate delivery or lose the business to competitors who have suitable items already available (Fogarty *et al.* 1991). If the manufacturer is able to reduce the manufacturing or assembly lead time to the extent that the customer is willing to wait that much for the product delivery, the manufacturer can then change to ATO or MTO product positioning strategy leading to a

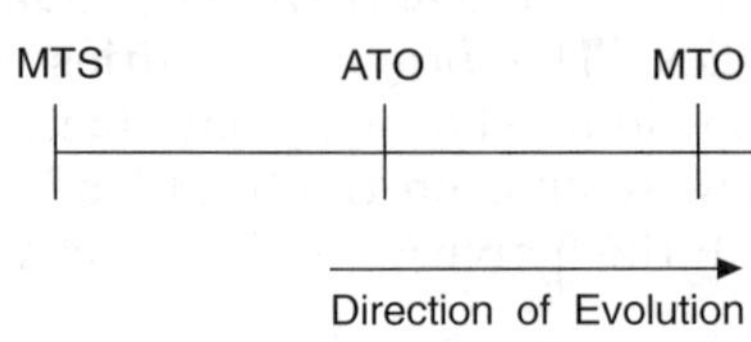

Fig. 1.10 *Evolution of Manufacturing*

drastic reduction in inventory investment with a simultaneous increase in product variety offered. Thus, a challenge for any MTS manufacturer is to reduce manufacturing lead-time to move to an ATO or MTO strategy as shown in Fig. 1.10.

It is not unusual for an organization to have different strategies for different product lines. A company may even have two strategies for one product. For example, automobile companies are primarily make-to-stock but do have some assemble-to-order business (Fogarty *et al.* 1991). In some cases, components and sub-assemblies may be used on different end products and so require different strategies than the end products.

Today, market pressures are forcing companies previously involved in mass production to develop more flexible batch-production oriented systems. This is particularly true of the automotive industry and of manufacturers of consumer goods. Increasingly, it seems customers demand a greater variety of products and are unwilling to accept mass-produced products. Customization to consumer needs is the emerging trend. Over the past twenty years or so manufacturing has moved along the continuum in a direction from 'make-to-stock' to 'engineer-to-order' as customers increasingly demand customized products.

The availability of computer-based information systems (such as MRP I and II), modeling and design support tools [such as computer-aided design (CAD) and computer-aided process planning (CAPP) systems] and computer controlled manufacturing technology [such as computer numerical control (CNC) machine tools and robots] has certainly facilitated this drift away from MTS systems and towards customer-driven manufacturing system.

1.8 PRODUCT LIFE-CYCLE AND PRODUCT POSITIONING STRATEGY

Almost all products have a limited life span. Experience shows that demand for most products follows a standard life cycle. The life cycle of a product consists of five phases, differentiated by the sales volume: introduction, growth, maturity, decline and withdrawal.

1. Introduction A new product appears on the market and demand is low while people learn about it, try it and see if they like it. The introduction stage is dedicated to setting up the production process and carrying out intensive sales promotion. Here, the key factor is innovation and the objective is to conquer new market segments. This requires major investments with limited profits. To avoid loosing opportunities opened up purely by the novelty value of its offerings, the firm must be able to satisfy market needs as soon as they appear; this makes delivery times the most important variable. To ensure the greatest promptness in deliveries to customers, considerable stocks of the product are often built before it is announced.

2. Growth During this stage, the market has already accepted the product and is demanding ever-greater quantities. Promotional expenses do not decrease, but are spread out over a larger volume that is able to absorb them better. Profitability thus improves, in part as a consequence of the beneficial effects of the learning curve in the factory, and partly because the scale of production can be set at the most appropriate level. Further, investments are therefore required in view of future profits that are becoming increasingly easier to forecast. The variable requiring the best management at this stage is existing production capacity, while plant expansions are being completed.

3. Maturity During the maturity phase most potential customers know about the product and are buying it in steady numbers, hence the market stabilizes at a constant level, and the volume of production does not experience any great upsurges. If anything, it follows the periodical trends of demand, whose characteristics are by now well known. The business is in the best position to obtain the maximum profit from its product and thus reap the benefits of the investments previously made. The variables to keep under control are inventory levels, delivery times, and use of equipment, all equally important to allow great quantities to be produced at reduced costs. These factors must be managed paying particular attention to the seasonal factors that are important in many market sectors.

4. Decline Sales fall as customers start buying new, alternative products that become available. During the decline of the sales, inventory control becomes the dominant issue in order to avoid accumulating a product in ever-decreasing demand. There is no way to fully utilize a production capacity sized for the former levels of demand, but fortunately by then the facility is already depreciated. On the other hand, the abundance of equipment makes it easy to meet sometimes stringent delivery due dates. Investments are reduced to a minimum and are limited to ordinary maintenance and replacement operations. The efficiency of the process is still high, but suffers from the negative impact of unused residual capacity. To extend the life of the product as long as possible with positive results, the firm must above all reduce costs (Sartori, 1988).

5. Withdrawal Demand declines to the point where it is no longer worth making the product (for example, black and white television sets, telegrams, and three-wheel cars).

The exact shape of the life-cycle changes from product to product, but the most important variable is its length. Each edition of a newspaper has a life-cycle of a few hours; clothing fashions and fad computer games have life-cycles of months or even weeks; consumer durables like washing machines have a life-cycle of five or 10 years; some basic commodities like soap and coffee remain in mature stage for decades. Figure 1.11 illustrates a typical product life-cycle curve and Table 1.7 compares the various stages of the product life-cycle.

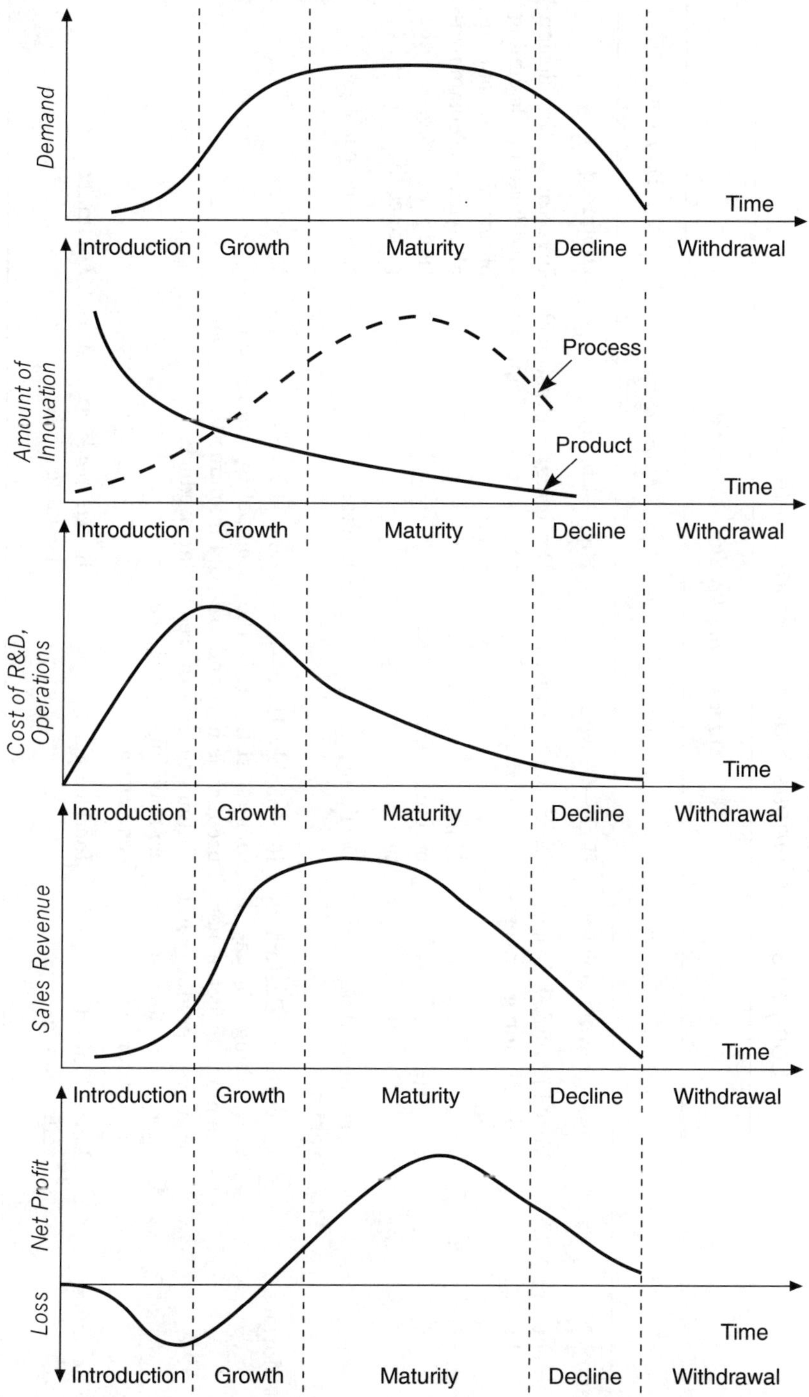

Fig. 1.11 *Product Life-Cycle Characteristics*

Table 1.7

Comparison of Various Stages of Product Life-Cycle

Criteria	*Stages in the Product-life Cycle*			
	Introduction	*Growth*	*Maturity*	*Decline*
Demand	Low—growing slowly	High—Growing	High—Steady.	Declining.
Investment in R&D, Mfg. Facilities, etc.	Very high R&D, design, and planning investments.	Declining investment in product. More investment in manufacturing facilities.	Declining investment in process improvement.	Investment in design modification, designing of additional or optional features, customization to extend the product life.
Unit Costs.	Very high due to low volumes.	Declining.	Lowest.	Low—increasing.
Profit Margins.	Very High—Customers are willing to pay a premium to get a new or novel product and there are few or no competitors.	Reasonably High—customers still view the product as new and are willing to pay a reasonably high price, no competitors.	Low—by this time the competitors have started manufacturing similar products.	Very Low.
Sales Revenue.	Low—limited by low sells volume.	Increasing.	Reach peak and start declining.	Declining.

(contd.)

Criteria	Stages in the Product-life Cycle			
	Introduction	*Growth*	*Maturity*	*Decline*
Net Profit or Loss.	There is a net loss as fixed costs are yet to be recovered.	Product starts to make an overall profit as the fixed cost are recovered by then.	Increasing profits till the time competitors introduce similar products, from there the profits decline.	Declining profits due to reduced utilization of capacity.
Strategic Focus.	Expand market awareness.	Market penetration.	Defend share.	Exit profitably.
Success Factor.	Innovation, promotion	Quantity.	Quantity and costs.	Costs.
Operations Focus.	Research and Development. Problem solving (designing out any initial problems).	Meeting due dates and specifications; design and quality improvements; Improvements in production process—introduction of mass production. Improvements in procurement, distribution and marketing.	Further improvements in production process—mass production and increased use of automation. Full utilization of capacity. Cost reduction and improved productivity.	Design modifications, offering additional features, and greater customization to beat the competition, repackaging.
Product Positioning.	Usually, ETO followed by MTO.	MTO → MTS	MTS	MTS → ATO/MTO

Unfortunately, there are no real guidelines about the length of a life cycle. In practice, most organizations keep a range of products at different stages in their life-cycles. This gives long-term stability, with new products being phased-in while older ones are declining and being withdrawn. As a result, overall production is smoothed rather than having wide fluctuations (Waters, 1996).

A manufacturer must be able to mix and change the methods of manufacture to best suit the needs of the moment. A manufacturing planning and control system must allow the customer order decoupling point (CODP) to change dependent upon the product, its position in its life cycle, or market condition.

The diverse orientations characterizing the various stages of the life-cycle shows how inconvenient it can be to make products at different stages of maturity in the same plant. A production unit manager cannot promote costly innovations for some products while simultaneously, starving certain other products of any investment for development. Such responsibilities must be separated and assigned to different people in different places. This is one of the reasons that encourage business to make use of subcontractors: the more mature technologies are handled using external resources, while the more recent ones, still being developed, remain in house (Sartori, 1988).

1.8.1 Ways of Extending the Life of a Product

Beyond the maturity stage when profits decline, improved production methods, experience and higher productivity can offset the decline, but profits will inevitably fall. At some stage demand and profit fall to an unacceptable level and the organization withdraws the product. However, there are several ways of extending the life of a declining product; some of which are described below:

- Increasing advertising and market support.
- Finding new uses for the product and hence new markets.
- Modifying the product to make it appear new or different—by redesign, additional/optional features, customization, etc.
- Changing the packaging with new sizes, different emphasis, and so on.
- Selling the product in new geographical areas.

These changes are certainly worth trying as it is usually much cheaper to extend the life of an existing product using these methods than to introduce a new one. The disadvantage of these changes is that they are usually short-term and only really give cover until a new product is available (Waters, 1996).

References

Ashayeri, J, A Teelen, and W Selen, 'Computer-Integrated Manufacturing in Chemical Industry', *Production and Inventory Management Journal (P&IMJ);* Vol. 37, No. 1, First Quarter 1996; pp 52–57.

Baldwin, Carliss Y and Kim B Clark, 'Managing in an Age of Modularity', *Harvard Business Review,* Vol. 75, No. 5, September–October 1997, pp 84–93.

Cox III, James F, and John H Blackstone Jr, *APICS Dictionary* (ninth edition), American Production and Inventory Control Society, Falls Church, V A, 1998.

Danø, S, *Industrial Production Model,* Springer, Vienna, 1966.

Fogarty, Donald W, John H Blackstone Jr, and Thomas R Hoffmann, *Production and Inventory Management* (second edition), South-Western Publishing Co., Cincinnati, Ohio, 1991.

Groover, M P, *Automation, Production Systems and Computer-Integrated Manufacturing,* Prentice Hall Inc., Englewood Cliffs, New Jersey, 1987.

Hitomi, Katsuno, *Manufacturing Systems Engineering* (second edition), Taylor & Francis Ltd, London, 1996.

McMahon, Chris and Jimmie Browne, *CADCAM: From Principles to Practice.,* Addison-Wesley Publishing Company, Wokingham, England, 1993.

Mishra, Anupam, 'Manufacturing Planning Based on Type of Customer Contact', *Express Computer,* May 18, 1998

Plossl, George, *Orlicky's Material Requirements Planning* (second edition), McGraw-Hill, Inc, New York, 1994.

Plossl, George (Ed) 'The Production and Inventory Control Environment', in *Production and Inventory Control Handbook* (third edition), McGraw-Hill, Inc, New York, 1997.

Redford, Alan H and Jan Chal, *Design for Assembly: Principles and Practice,* McGraw-Hill Book Company, London, 1994.

Sari, F John, 'Master Production Schedule', in *Production and Inventory Control Handbook* (third edition), McGraw-Hill, Inc, New York, 1997.

Sartori, Luca G, Manufacturing Information Systems, Addison-Wesley Publishing Co., Wokingham, England, 1988.

Shafer, Scott M and Jack R Meredith, Operations Management: A Process Approach with Spreadsheets, John Wiley & Sons, Inc., New York, 1998.

Taylor, Sam G and Steven F Bolander (Eds), 'Production Planning and Scheduling for Process Flow Manufacturers', in *Production and Inventory Control Handbook* (third edition), McGraw-Hill, Inc, New York, 1997.

Virmani, Vijay, 'Custom-Build a Model for Efficiency', in *APICS: The Performance Advantage,* May 1999, Vol. 12, No. 5, pp 34–36, 1999.

Waters, Donald, *Operations Management: Producing Goods and Services,* Addison-Wesley Publishing Co, Harlow, England, 1996.

Wemmerlov, U, 'Assemble to Order Manufacturing: Implications for Materials Management', *Journal of Operations Management,* 4(4), pp 347–368, 1984.

Young, R E and R Mayer, "The Information Dilemma: To Conceptualize Manufacturing as Information Process", *Industrial Engineering,* 16(9), 1984.

Chapter 2

An Overview of Manufacturing Resource Planning (MRP II) Including Detailed Study of Material Requirements Planning (MRP I)

The Japanese industrial miracle was not accomplished through sophisticated computer systems. They concentrated on the fundamentals of flawless execution of production tasks supporting a coherent strategy and sound planning, based on two beliefs:

(i) inventory is an evil, and
(ii) perfection of the process is possible.

Computer-based systems built around the proper techniques are necessary but not sufficient. Greater sophistication in the system provides only more opportunities for underqualified people to self-destruct at higher speeds. Skill in operating businesses with systems determines success.
Effective planning systems make tight control possible, but people make it happen.

George W Plossl (1997)

2.1 THREE KINDS OF FLOWS SUPPORTING EFFECTIVE MANUFACTURING

For efficient manufacturing activities, unification and harmonization of the following three flows are vital:

1. Flow of Materials Conversion of raw materials into products, that is, technical production.

2. Flow of Information Planning and control of production.

3. Flow of Cost Cost continues to flow *into* and gets accumulated in the product right from the time the product is conceived through production, distribution, sale, right up to the end of its warranty period. Figure 2.1 depicts the above written.

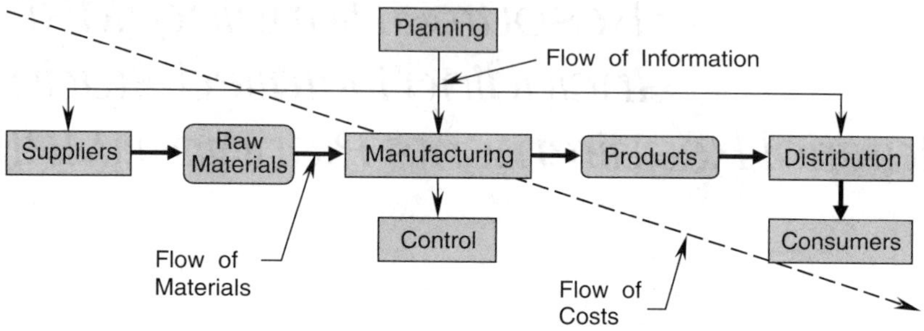

Fig. 2.1 *Three Kinds of Flows Supporting Effective Manufacturing*

Flow of Materials Production of goods is basically, the utilization of resources of production (manpower, materials, machines, money, and information) in the manufacture of the finished product. This input-output system is referred to as the 'flow of materials' (technical production). From a wide viewpoint, this flow constitutes a serial chain of functions: procurement, production, distribution, inventory and sales, as depicted in Fig. 2.1. Throughout this process, raw materials are procured from outside suppliers, processed and assembled at workshops and stored in warehouses as inventories, and finished products are delivered to consumers as commodities throughout the distribution stages. Also involved will be the means of physically handling materials through and between these functions.

Flow of Information Planning and Control Planning is selection of the future course of actions. It consists of selecting measurable objectives and deciding how to achieve them. Without plans there is no basis for action (execution) and no basis for evaluating the results achieved and correcting them (control). According to the plans (schedule) developed, production executes practical activities to make products in a workshop. Thus, execution is the carrying out (performance) of plans. Control is the measuring of actual results and comparing it with the desired results and either correcting the performance of the activities in order to make sure that the management objectives and plans are being accomplished or revising the objectives to make them realistic and hence achievable. The

Manufacturing Planning and Control (MPC) system provides information to efficiently manage the flow of materials, effectively utilize people and equipment, coordinate internal activities with those of suppliers and communicate with customers about market requirements. The MPC system does not make decisions or manage the operations—managers perform those activities. The system provides the support for them to do so wisely (Vollmann *et al.* 1997).

Planning, execution, and control are iterative processes that should occur continuously. Initiation of control does not require that plans actually be executed. What is required is only that their results be simulated and evaluated. Thus, at times it is difficult to identify an activity as uniquely planning or uniquely control. However, describing planning, execution, and control separately leads to a better understanding of these activities (Fogarty *et al.* 1991).

Flow of Costs Through the production processes, raw materials are converted into finished products with value added. This added-value process is the 'flow of value' (economical production). It is more specifically called the 'flow of costs' as depicted in Fig. 2.1. Through the production processes, costs occur and are accumulated with successive production activities (Hitomi, 1996).

The 'flow of materials' is a basic indispensable function in manufacturing. This flow is accompanied by the 'flow of costs' as the raw materials are converted into products with high value. The driving force of this function is 'the flow of information.' The flow of materials, which generates the flow of costs proceeds according to the instructions issued from the flow of information based upon market needs. This synergistic action realizes high quality, low cost, and just-in-time (or quick response) delivery. Realization of this ideal production mode is now observed in automated factories or those concerned with Computer-Integrated Manufacturing (CIM).

The essence of manufacturing according to Plossl (1994) is

- (a) flow of materials from suppliers, through plants to customers, and
- (b) flow of information to all parties about
  - (i) what was planned,
  - (ii) what has happened, and
  - (iii) what should happen next.

This is true regardless of what is made, how and when it is made, and who makes it where. The first law of manufacturing according to Plossl (1994) is:

> *All benefits will be directly proportional to the speed of flow of materials and information.*

This is a universal law applying to every type of manufacturing. Difficulties in controlling manufacturing will decrease and planning will become more effective as material and information flows speed up. The best use of resources comes from eliminating problems that interrupt or slow down these flows.

2.2 PRINCIPLES OF EFFECTIVE MANUFACTURING

1. Production Problems must and can be Eliminated, they cannot be Covered up Successfully with Cushions of Inventory and Time Any manufacturing problem can be minimized, most can be eliminated. Providing inventory and time cushions wastes inventory, makes planning less valid, and actually aggravates the problems. Even worse, it diverts attention from the real solution—eliminating the problems (Plossl, 1997).

2. Inventory is more of a Liability than an Asset, having Real Value only when it is Flowing through Operations or used to Support them Money invested in inventory, although listed on balance sheets as assets, is, in reality, wasted money that could be better spent in product or process development. Excessive work-in-process not only adds no value to the product and holds money captive, it increases manufacturing lead time and reduces shop floor flexibility while requiring additional storage space (Toomey, 1996).

In Japan, inventory is seen as an evil in itself. It is a resource sitting idle, wasting money. But more important, *inventory tends to hide problems.* In the traditional plant, inventories are used to buffer operations so that problems at one stage don't have an impact at the next stage. However, inventories also hide problems, such as defective parts, until the inventory is needed and then is found to be defective. For example, in a plant with lots of work-in-process inventory, a worker who discovers a batch of defective parts can simply put them aside and work on something else. By the time, the worker returns to the defective batch, if ever, so much time has elapsed since the batch was processed upstream that the cause of the problem is unlikely to be discovered and corrected to prevent recurrence. In contrast, in an environment where there is little or no buffer inventory, a worker who discovers a defective batch has no choice but to work on the defective batch. Furthermore, the worker is in a good position to notify upstream operations of the problem so that they can correct it and ensure that it does not occur in the future (Shafer and Meredith, 1998).

3. Tolerating some Downtime while Striving to Eliminate their Causes is better than Preventing Idle time by Manufacturing Items not Required Immediately The tangible costs of idle people or machines are easily calculated, but the true total costs of avoiding them are not. Among the costs incurred in avoiding idle time by manufacturing items not required immediately are excess inventory, commitment of flexible resources to the wrong items, poor reaction to customer's needs, and less valid planning data.

4. More Frequent and Precise Replanning, such as Computing Daily Rather than Weekly does not make it More Accurate Computer systems can easily handle myriad revisions, but plants and people cannot. The real problem arising of this myth is the neglect of efforts to make a more valid plan, requiring short horizons and quick execution.

5. Replanning is Admitting Failure, it is no Substitute for Sound Execution

6. Plans Impossible to Execute are Worse than Useless This principle stresses the importance of realistic planning. Prior information and material schedules cannot be valid, however desirable, if resources are or will be inadequate; such plans are impossible to execute. Planning can be done by computer-based systems with minimum human involvement, but successful execution depends on people taking the proper actions promptly.

7. Lead times cannot only be Monitored and Adjusted but can also be Controlled A little known (and difficult to accept) fact is that the *lead times will be what you say they are*! Setting planned lead times reasonably lower than current average actual value will shut MRP I off from recommending the release of new orders for the period that lead times were cut. Work centers will work on the orders already in work-in-process and the queues will shrink, making actual lead times shorter. Increasing planned lead times triggers a vicious circle called the 'lead time syndrome'. For example, if production lead times are increased by say a week, MRP I will immediately signal the need to release an extra week's work to the shop. If this recommendation is followed and output is not increased, these jobs will join the queues as Work-in-Process (WIP). Thus, WIP will increase by one week and the lead time will actually increase by one week. If the planning system responds to this second increase in lead time by releasing orders even earlier, the control system enters a vicious spiral. Thus, actual lead times will get longer and more erratic, and more orders will be late. *Lead times are actually self-fulfilling prophecies.* Similarly, if a supplier sees that its backlog of unshipped customer orders exceeds its ability to deliver all of them within quoted lead times, he is often tempted to inform customers that lead times now are longer. When these increased lead times are inserted into a customer's MRP I program, it signals the additional orders to go to the supplier, who thinks that 'business is picking up and good times lie ahead.' In fact, the actual load on his facilities is larger now, and quoted lead times are again too short. The only way out of this lead time syndrome is to increase output or decrease input. When there is a downturn in the economy, the opposite effect occurs. An announcement of a decrease in lead time means that customers delay placing orders, with the result that lead times fall even further.

8. Reducing Setup Time is Worth the Effort Cutting setup time not only shortens the working elements of lead time but also makes possible dramatic reductions in the dominant element—*queue time*—by smoothing and speeding up materials flow through all operations.

In the traditional firm, long lead times are often thought to allow more time to make decisions and get work performed. But in the JIT firm, short lead times mean easier, more accurate forecasting and planning. Moreover, a way to capitalize on the increasing strategic importance of fast response to the customer is to minimize all the lead times. If lead times are reduced, there is less time for things to go awry, to get lost, or to be changed. For example, it is not uncommon for an order placed two months ago to be changed every three weeks until it is delivered. However, if the delivery time is one week, the customers get exactly what they need and can delay ordering until the week before they need it—when they know best what they will actually need (Shafer and Meredith, 1997).

9. Planning Defines Resources Needed to Make What is Planned, Execution Applies Available Resources to Make What Customers Want Now
This principle highlights the difference between planning and execution. It is a fallacy trying to do both with the same system programs. The planning data, called *soft data,* are very different, they look precise but often are changed. Execution data need to be precise; they are *hard data* (Plossl, 1994).

10. Only Resources Requiring Long Periods for Actions Should be Planned far Ahead, Detailed Plans Should Cover only very Short Horizons. Do not Commit any Flexible Resources to a Specific use Until the Latest Possible Moment Plans cover future occurrences, they undoubtedly will change, and the amount and effects of changes will be more severe as planning horizons lengthen. The later specific allocations of resources are made, the fewer the changes necessary, the less severe their effects, and the less waste of resources (Plossl, 1994).

11. Time is the Most Precious Resource Employed in the Manufacturing Process Time as a resource has following unique characteristics (Plossl, 1994):

- Equal amounts are available to everyone.
- Time moves relentlessly; it cannot be slowed, stored, extended, reused, or recycled.
- Wealth can buy no more of it.
- Wasting time causes irretrievable loss.

This principle demands intolerance of delays and time cushions, concentration on speeding up all needed activities, and eliminating un-needed ones. Lost time is the worst waste.

12. There is One Manufacturing Planning and Control System Framework Common to all Types of Manufacturing This principle is true regardless of products made, materials used, processes employed, and customers served (Plossl, 1997). However, the importance of each element along with the techniques employed can differ among various types of businesses. It derives inevitably from the common logic, the similarity of needs of major parties, the nature of the resources employed, and the identical types of data involved.

13. Well-run Operations do not Require Complex Systems The bulk of complexity in production and inventory control systems arises while correcting data errors, or while trying to cope with unexpected upsets or surprises. This principle focuses attention on eliminating problems, not coping with their effects (Plossl, 1997).

14. All Employees Need Continuous Education Manufacturing, needs to constantly change to meet market needs, utilize latest technologies, comply with government regulations, and respond to management goals and policies. This requires that all employees undergo continuous education to improve their understanding and skills (Plossl, 1997).

A Word of Caution About Safety Stocks

Safety Stocks (and Safety Lead Times) Pollute Credibility It results in workers being asked to complete something sooner than it is needed. At times this does not cause a problem. But whenever a number of jobs are competing for scarce resources, working on one ahead of time may prevent working on another that is needed. Furthermore, it is hard to justify a factory working overtime simply to make components or products earlier than required. Similarly, it is hard to explain, paying suppliers a premium or paying for airfreight shipments only to have the materials arrive sooner than required. To have credibility in the plans, 'drop-dead dates,' which contain no fat, no lies, and no padding, should be the goal. Where safety stock is judged to be necessary, two good rules are: *Use as little as practical, and make it visible.* When users can see the gap between due dates and need dates, they can make better decisions leading to better performance (Goddard, 1997).

2.3 WHAT IS MANUFACTURING RESOURCE PLANNING?

Manufacturing Resource Planning (MRP II) is a method for the effective planning of all resources of a manufacturing company. It is made up of a variety of interlinked functions, as shown in Fig. 2.2 such as:

1. Strategic and Business Planning.
2. Demand Management.
3. Sales and Operations Planning (S&OP, also called Production/Aggregate Planning).
4. Master Production Scheduling (MPS) with Rough-Cut Capacity Planning.

5. Material Requirements Planning (MRP I).
6. Capacity Requirements Planning (CRP) and Vendor Requirements Planning (VRP).
7. The Execution Support System for Capacity and Material [Shop Floor Control (SFC), and Purchase Planning and Control].

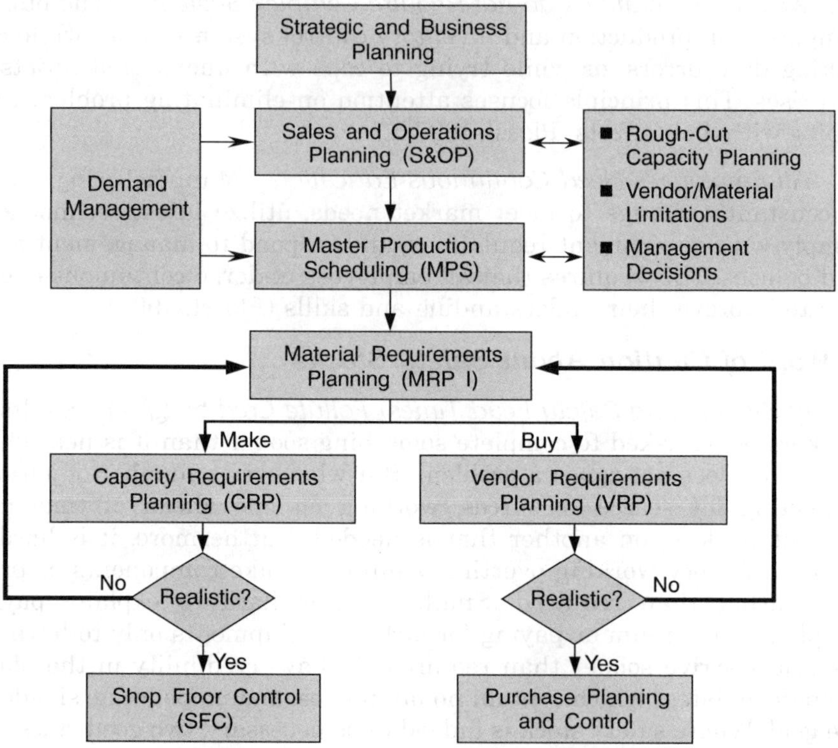

Fig. 2.2 *MRP II Framework*

- Ideally an MRP II system addresses
  - (i) operational planning in sales units, and
  - (ii) financial planning in monetary units (for example in rupees or dollars etc.).
- Further, an ideal MRP II system has a simulation capacity to answer 'what if' questions.
- Output from various MRP II functions is usually integrated with financial reports such as the business plan, purchase commitment report, annual budget, inventory projections in monetary units, etc.
- MRP II is also often integrated with a company's warehousing and distribution processes.

MRP II is an approach to managerial planning, execution, and control of productive activity. It integrates in a closed loop or feedback manner, the forecasting of demand, production planning, production scheduling, production activity control, and purchase planning and control. MRP II software can relate many information needs of all the functions and departments, including purchasing, accounting, payroll, distribution, marketing, engineering, production activity control, and general managerial planning, measurement, evaluation, and control.

MRP II is an explicit and formal manufacturing information system that integrates marketing, finance, and operations. It coordinates the sales and manufacturing plans to assure their consistency. It converts resource requirements, such as facilities, equipment, personnel, and material, into financial requirements and converts production outputs into monetary terms. It evaluates the organization's ability to execute the plan financially and also evaluates the financial merit of the plan in terms of such measures as profit, return on investment, and return on assets (Fogarty *et al.* 1991).

2.4 A FRAMEWORK FOR THE MRP II SYSTEM

Following is an overview of an MRP II architecture, which can be divided into three hierarchical planning levels as shown in Fig. 2.3. Planning horizons for different levels are given in Table 2.1.

Table 2.1

The Three Hierarchical Planning Levels and their Planning Horizons

Sr. No.	*Planning Level*	*Planning Horizon*
1.	Strategic planning	One to five years
2.	Tactical planning	One month to three years
3.	Operational level	Quasi-real time to one week

This MRP II architecture shown in Fig. 2.3 reflects a situation where a factory has been subdivided (in so far as possible) into a series of Group Technology (GT) based cells, where each cell is responsible for a family of products, assemblies or components and is managed by a Production Activity Control (PAC) system. The factory coordination module ensures that the individual cells interact to meet an overall production plan.

In any firm, Manufacturing Resource Planning (MRP II) encompasses three distinct activities or phases:

1. Master Planning (Strategic Planning/Front End) The first activity is the establishment of the overall direction for the firm, with the resulting management plan stated in manufacturing terms, such as end items or product options. The manufacturing plan must be consistent with the

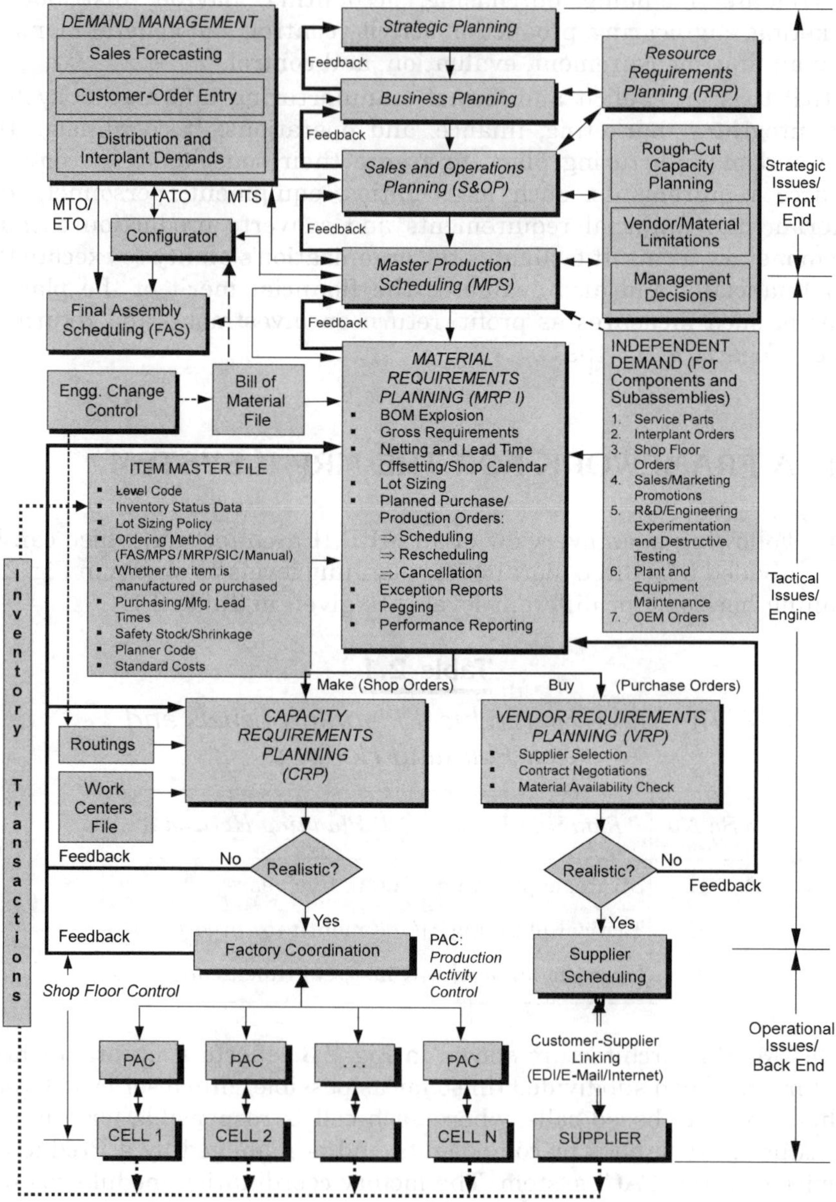

Fig. 2.3 *Information Flow in MRP II*

company's direction and the plans for other departments of the firm. Thus, master planning is the set of activities and systems for overall direction setting it creates overall plans stated in production terms.

2. Detailed Planning of Material Flows and Capacity (Tactical Planning/Engine) The second MRP II activity is the detailed planning of material and capacity needs to support the overall production plans.

3. Execution of the Plans (Operations Planning/Back End) The third and final MRP II activity is the execution of the material and capacity plans in terms of detailed shop scheduling and purchasing actions. *Shop floor control systems* establish priorities for all shop orders at each work center so that the orders can be properly scheduled. *Purchasing systems* provide detailed planning information for *vendor scheduling*. This information relates to existing purchase orders as well as to planned purchase orders.

2.5 UNIVERSAL MANUFACTURING EQUATION

The major task of the Manufacturing Planning and Control (MPC) system in a manufacturing company is to find answers to the following five questions. Oliver Wight identified the last four questions as 'the universal manufacturing equation.' All manufacturing companies regardless of their size, types of processes, or complexity of products must answer them. Table 2.2 describes the five questions and how they would be answered in a manufacturing company.

Table 2.2

Universal Manufacturing Equation

The Questions	*The Answers Come from*
1. What are we going to sell?	Demand Management. It consolidates demands for an item from all the sources, which might include: 1. Sales forecast. 2. Anticipated replenishment of distribution centers. 3. Backlog (customer orders booked but not yet shipped). 4. Quotes entered but not yet approved. 5. Interplant orders. 6. Service parts requirements.
2. What do we need to make? (What is to be made? How many, and when are they needed?)	1. *The Sales and Operations Planning (S&OP)*, which establishes rates of production for families of products, answers this question at the aggregate level. 2. *The Master Production Schedule (MPS)*, which is the anticipated manufacturing schedule, answers this question at the detail level.

(*Contd.*)

The Questions	*The Answers Come from*
3. What resources (material, equipment and manpower) are required to make it?	1. *The Bills of Material (BoMs), Formulas or Recipes.* It says what raw materials, components, and/or ingredients are needed to make the product. (Detailed engineering specifications and drawings are to be provided by the engineering group whenever required) 2. *The Routing.* It indicates what kind of capacity — manpower and/or equipment (machinery and tooling — is needed.
4. Which of these resources do we already have, and which others will be available in time?	1. *The inventory records* showing what materials are on hand, what's allocated, and what's scheduled to be received. 2. *The planned or demonstrated capacities* for equipment and manpower.
5. What more do we have to get and when?	*The company's detailed schedules* for both material and capacity. It tells us what we need, and also when we need it.

The universal manufacturing equation offers a structured approach to describing what Production and Inventory Control (PIC) is all about. PIC is immersed in all five questions, receiving answers for some and supplying answers to others (Goddard, 1997). In today's modern factories, MRP II is the standard for Management Information Systems (MIS) and an important component of Computer-Integrated Manufacturing (CIM).

2.6 ELEMENTS OF MRP II

Following sections describe the various elements that comprise a manufacturing resource planning (MRP II) system. These sections briefly describe the various functions performed in an MRP II system to give a broad overview of the information flow in an MRP II system in the order in which these information flows actually occur. A detailed knowledge of MRP I is a pre-requisite to understanding the MRP II system, and hence a detailed study of MRP I should be done right in the beginning. Vollmann, Berry, and Whybark—the authors of one of the finest texts on manufacturing planning and control systems—have made a similar observation in the preface of their book: 'We have found that starting with MRP (I) record processing has been pedagogically sound. We find that understanding time-phased records is helpful for understanding the entire MPC system (Vollmann *et al.* 1997).' Putting MRP I right at the beginning, however, disrupts the order in which the various MRP II functions should be understood. Here, a compromise is made by including detailed study of MRP I right in the overview itself so that the entire MRP I process is understood in the beginning without disturbing the order.

2.7 STRATEGIC AND BUSINESS PLANNING

Strategic Planning It forms the basic input to the MRP II from the top management and covers long time horizons (typically three to five years). Strategic plans include a statement of charter of business, a brief summary of the type of business it is and will become. They also include strategies for the marketplace entered, types of products, and manufacturing practices.

Business Planning Business plan is a statement of objectives for all aspects of the business at an aggregate level. It has a long time horizon (typically 12 to 18 months, which is short compared to the time horizon of strategic planning) and focuses on product families to meet strategic goals. Business planning provides the plans that are necessary to drive the sales, manufacturing, and financial activities of an organization. In general, the business planning process has more far-reaching strategic implications and is more financially oriented than the Sales and Operations Planning (S&OP), which is next in the planning hierarchy. Through business planning process, the top management sets boundaries (budgets) on the monthly production plan to be developed in S&OP. The business plan defines:

1. The product and market strategies such as:
   (i) market share,
   (ii) new product planning,
   (iii) analyses of competition,
   (iv) distribution strategies,
   (v) profit planning, and
   (vi) product positioning to meet customer service objectives.
2. The markets to be addressed (target demographic and geographic areas).
3. Annual sales plan or forecast for the overall business, or particular market segments, or other high level groupings of existing products and also of new projected products and services. (It states what the company expects to sell annually in monetary units, for example in dollars or rupees).
4. Make or buy decisions.
5. Process decisions (choosing a particular process from the available alternatives for a manufacturing plant).
6. Capacity decisions (equipment, factory floor space, and manpower) and facilities decisions (assignment of specific products, customers, and markets to individual facilities, plant location and layout):

scenarios for new facilities, plant and equipment purchases, and retirement of old items.

7. Human resource decisions: design of the organization and reward system, plans for all levels of personnel and their training and development.
8. Capital and investment planning (such as, amount of money to be invested in R&D, new facilities, new technologies and new processes).
9. Manufacturing infrastructure decisions to formulate organization and policies by which the work is accomplished, such as:
   (i) structure of the manufacturing organization,
   (ii) production and inventory control system (planned inventory levels and turnover objectives),
   (iii) cost and quality control system,
   (iv) workforce management policies, and
   (v) consideration of new management approaches and systems (that is, empowered teams, just in time, or enhanced information systems).

Business Forecasting Business as well as strategic planning is based on business forecasting, which evaluates political, economic, demographic, technological, and competitive factors that will affect the demand for a firm's products. Business forecasts are aggregated into large product families. Top management is responsible for this activity. It is not unusual to have a long-range planning task force reporting directly to the Chief Executive Officer (CEO) and also to employ external forecasting consultants (Fogarty *et al.* 1991).

- Business planning decisions are explicit commitments to an organizational direction. It is usually difficult in the short run to change them. The wisdom of marketing decisions substantially influences organizational growth and prosperity (Fogarty *et al.* 1991).
- Production plans and budgets for non-production related areas are developed from the business plan.

2.8 RESOURCE REQUIREMENTS PLANNING (RRP) (Capacity Planning Conducted at the Business Plan Level)

Resource Requirements Planning (RRP) develops resource acquisition plans. It has been defined by APICS Dictionary as *the process of establishing, measuring, and adjusting limits or levels of long-range capacity. Resource planning is normally based on the production plan but may be driven by higher level plans (such as, the business plan) beyond the time horizon for the production plan. It addresses those resources that take long periods of time to acquire. Resource planning decisions always require top management approval* (Cox III and Blackstone Jr, 1998).

Resource planning is the most highly aggregated and longest-range capacity planning decision. It typically involves converting monthly, quarterly, or even annual data from the extended production plan into aggregate capacity requirements such as gross labor-hours, floor space, machine-hours, and the like. The units of measure might be constant monetary (rupee/dollar) output rates, man-hours, head counts, machine-hours, key facility hours, tons of output, or some other output measure. The need is to plan capacity at least in aggregate terms, for a horizon at least as long as it takes to make major changes. The time horizon for RRP should cover at least the lead time for acquisition of resources and extend over the economic life of the assets.

Resource requirements plans might include strategic plans for building a new plant or expanding an existing plant, long-term contracts with suppliers for major raw materials, labor contracts, energy supply contracts, purchase and implementation of new information systems, and a disposal plan for waste materials. These strategic decisions must be closely coordinated with strategic marketing and financial plans and have a major impact on the long-range success of the firm (Taylor and Bolander, 1997).

Decisions made at this top level of the planning system hierarchy constrain the lower-level production plans and operating schedules by specifying upper limits on the available plant capacity. Resource planning is directly related to production planning, since, in the short-term, the resources available provide a set of constraints to production planning. In the longer run, the resource requirements can be planned based on extended production plans.

2.9 DEMAND MANAGEMENT

It is the function of recognizing and managing all of the demands for products to ensure that the master scheduler is aware of them. Demand management encompasses the areas of:

1. Advertising, promotions, pricing, adding and deleting products.
2. Forecasting anticipated business.
3. *Customer order servicing,* which involves the following activities:
    (a) receiving and responding to the customer inquiries, and
    (b) interacting with the master scheduler on availability of products.
4. Order delivery date promising.
5. Customer order entry (entering new orders and processing changes to existing ones).
6. Distribution Requirements Planning (DRP), and interplant orders. (DRP may be thought of as an upward expansion of the MRP I

logic from the MPS through the distribution network to the ultimate customers.)
7. Other customer-contact-related activities.

Demand management also concerns other sources of demand for manufacturing capacity, such as:

1. Service parts demands.
2. Intra-company requirements (including inventory changes, promotion requirements, quality assurance needs, R&D, engineering testing, pilot project requirements, etc.).
3. Pipeline/transportation inventory stocking.

Pipeline/Transportation Inventory Stocking It is the inventory to fill the transportation network and the distribution system including the flow through intermediate stocking points. It is the materials and products in transit between locations in a distribution system between plants or to customers. The function of this inventory is to relocate inventory to the point of need. The value of the transportation inventory can be considerable for many companies. The flow time through the pipeline has a major effect on the amount of inventory required in pipeline. Time factors involve order transmission, order processing, shipping, transportation, receiving, stocking, review time, etc. Other factors to be considered are distance and mode of transportation. Transportation inventory is calculated primarily as a function of the in-transit time between locations. Using faster modes of transportation normally reduces the in-transit lead time but could also increase distribution costs. Therefore, the level of transportation inventory should reflect the costs of transportation. Management should evaluate the total costs related to the trade-off between inventory-carrying costs and transportation costs when deciding how best to manage this inventory function.

Following are the objectives of demand management:

1. To develop the most reasonable projection of future requirements.
2. To update this projection of future requirements when change is warranted.
3. To avoid trivial changes to S&OP and MPS by properly managing the different demand streams.
4. To identify any meaningful differences in the marketplace (such as trends) at the earliest possible moment so that effective action can be taken.

In addition to forecasting the total demand, it is also necessary to manage the consumption (transformation) of the forecast by actual customer orders. That is, one must deal with order entry, customer-order promising, and order backlog as these actual orders replace forecast information. Output of the demand management module is one of the most important considerations in business planning, and also in sales and operations planning. Similarly, demand management module interfaces directly with the Master Production Scheduling (MPS) module in case of Make To Stock (MTS) production environment.

2.10 PRODUCT CONFIGURATOR

In case of Assemble To Order (ATO) production, demand management module does not interface directly with the MPS but through the *product configurator.* This is because in case of ATO, the individual standard modules are master-scheduled rather than the end products. The end products in this case are actually a very large number of uniquely configured products which are assembled from a limited number of variants and options which along with the common parts, form the standard modules to be master scheduled. If we can assemble or finish to order, we only have to create a BOM for the end items we sell. We don't have to create, store, and maintain a bill for every possible configuration of end items.

Product configurator interfaces with the customer order entry in the demand management and its principal function is to convert the uniquely configured customer orders into the standard modules to be master scheduled. Rather than a traditional catalog of part numbers, the configurator presents a 'menu' of options to select from the customer. Further, if there are certain modules that are not compatible with certain other modules in the configured system, then the configurator should also provide constraints for configuring a product. The configurator also displays any messages designated to be displayed to assist the customer in configuring the product. The configurator also creates a unique manufacturing BOM for each customer order right at order entry. Configurator software converts the description of options for a product into part and BOM numbers. All items coded and required for the option are configured into an end-item BOM unique for a customer order. In addition, the configurator will develop the product price for each sales office, and display to the customer. On confirmation of the order it creates the routing to build it and sends it to the final assembly scheduling along with BOM and other item data. Finally, it helps in generating sales history for the forecasting of anticipated sales volume and options fractions for creating the planning BOM. This is a great solution for assemble-to-order factories.

2.11 SALES AND OPERATIONS PLANNING (S&OP)

The best way to determine what to make, how many and when to make it is by using a process in which all departments meet regularly to asses future market demand and determine whether the factory or plant can meet this demand and if so, how. This process is called Sales and Operations Planning (S&OP). The business plan is stated in dollars whereas the plans that people will ultimately be accountable for meeting are stated in terms of units or hours. The first step in converting financial plans into operational plans happens in the S&OP. S&OP evaluates anticipated sales and planned production in terms of product families (five to ten families in most companies). One output from this process is the production plan, which is typically stated in terms of rates of output by product family—for example, 200,000 small electronic motors per month, 400,000 gallons of silicone fluid per week, or three steam turbines per quarter. S&OP is the responsibility of the general manager or the president.

S&OP is the ability to state future sales and operations plan by product families. It is an aggregate planning process aimed at keeping aggregate balance between supply and demand, and making shipments to meet customers' needs with a minimum of time, energy, and effort. Managers have a month-to-month forecast of total demand for all outputs (combined) for the year or so. They are expected to capitalize on this demand by supplying whatever portion of it will maximize long-run profitability. That is, not all the demand need be Satisfied if the attempt to fill that demand will result in lower overall profits. But a loss of market share might result, which, in turn, may reduce long-run profitability.

Sales and operations planning is defined by APICS Dictionary as *a process that provides management the ability to strategically direct its businesses to achieve competitive advantage on a continuous basis by integrating customer-focused marketing plans for new and existing products with the management of the supply chain*. The process brings together all the plans for the business (sales, marketing, development, manufacturing, sourcing, and financial) into one integrated set of plans. It is performed at least once a month and is reviewed by the management at an aggregate (product family) level. The process must reconcile all supply, demand, and new product plans at both detail and aggregate level and tie to the business plan. It is the definitive statement of the company's plans for the near to intermediate term covering a horizon sufficient to plan for resources and to support the annual business planning process. Executed properly, the S&OP process links the strategic plans for the business with its execution and reviews performance measures for continuous improvement.

S&OP focuses on managing rates of manufacture to meet targeted inventory and/or backlog levels. Make-to-stock businesses target finished goods inventory levels to maintain desired customer service levels, while make-to-order businesses target backlog levels (usually expressed as the desired customer lead times) to maintain desired customer service levels (Sari, 1997). S&OP provides the framework within which the master production schedule (MPS) is developed.

S&OP brings people together. It is a formal mechanism to ensure coordination among various business functions. The executive who has the profit-and-loss responsibility for the business—that is, the divisional president or general manager—must preside over the formal S&OP meeting and should never delegate it to lower-level managers. S&OP is attended by the vice presidents (managers) of sales and marketing, product development, finance, engineering, materials, manufacturing, human resources, etc., as each of these functions has an impact or can be impacted by shifts in plans on what will be sold and what needs to be made. As a result of S&OP process, all areas of the company take ownership of that final unit number to be manufactured and the information gets disseminated formally and quickly. S&OP is the right place for making decisions about finished goods, inventories, backlogs, sales, and production. S&OP should review trends, identify opportunities, and approve changes in plans to support the overall strategic objectives. This is because the right people are available to make the best decisions, the planning process is structured, management's expectations are known, consensus is developed, and commitments are made and communicated.

Following are the activities performed in the formal S&OP process (Sari, 1997):

1. Sales forecasts are reviewed and revised (usually based on the review of planned and actual sales in the recent past).
2. Current inventory and/or backlog positions are documented.
3. Demonstrated capacities are documented.
4. Production rates are revised within the constraints of both material and capacity availability.
5. Projected inventory and/or backlog positions are calculated and compared to targets. Financial projections are developed and reviewed.
6. Contingency plans are developed based on reasonable variations of sales and manufacturing plans.
7. Alternatives are presented to the top management for discussion, review, and approval.

It should be noted that substantial pre-meeting communication and preparation takes place before the formal S&OP process. Middle management discusses and prepares alternatives for review and agreement.

2.12 MASTER PRODUCTION SCHEDULING (MPS) AND FINAL ASSEMBLY SCHEDULING (FAS)

Master production scheduling develops the data (schedules) showing *which products* should be produced, *how many*, and *when*. That is, MPS is the *statement of production by item, date, and quantity for as far out as the planning system extends*. These schedules are constrained by the strategic business and production plans and drive all detailed operating plans. MPS also takes into consideration capacity, material, and vendor limitations along with management decisions (policies and guidelines) to make sure that the schedules developed are valid and realistic. MPS is the disaggregated version of the production plan and must sum up to it.

'End item planning' (in master production scheduling) and 'customer order entry' (in demand management) are the activities that involve detailed trade-offs between marketing and manufacturing. Master production scheduling is a challenging process that involves constant analysis and reaction in an attempt to balance the demands of the marketplace with the resources of the company. When demands of the marketplace are satisfied and the manufacturing resources are fully utilized at the same time, the MPS has achieved the best balance. Another trade-off involved in the master-scheduling process is the one between stability and flexibility. On one hand, it is desirable to provide enough *stability* so that manufacturing can reasonably be held responsible for meeting plans. On the other hand, it is worthwhile to provide sufficient *flexibility* so the company can respond competitively to actual customer needs.

Master production scheduling sits between the sales forecasts and Material Requirements Planning (MRP I). Its prime purpose is to exert control over changes to the plan so that changing customer requirements are met as far as possible, whilst keeping disruptions, both on the factory floor and in MRP I, to a level which can be accommodated (Kenworthy, 1997).

There is rarely a perfect or optimal MPS—one that represents the best balance of responsiveness to customers and resource utilization. And although there is no 'right' answer, some choices are clearly better than others, and some are so close that, when you weigh the pros and cons, you can only flip a coin in the end. The outcome is many times the least-worst choice. For these reasons, master scheduling is both frustrating and highly rewarding (Landvater, 1997).

Final assembly schedule, according to the APICS Dictionary, *is a schedule of end items to finish the product for specific customers' orders in a make-to-order or assemble-to-order environment*. It is also referred to as the *finishing schedule* because it may involve operations other than just the final assembly; also, it may not involve assembly, but simply final mixing, cutting, packaging, etc. The FAS is prepared after receipt of customer order as constrained by the availability of material and capacity, and it schedules the operations required to complete the product from the level where it is stocked (or master scheduled) to the end-item level.

In make-to-stock environment, the MPS and FAS generally coincide, sometimes, however, there might be an MPS for end items (bulk) and a separate FAS to schedule packaging variations of the bulk. In Assemble-To-Order (ATO) manufacturing environment, FAS is connected to the MPS through product configurator. In make-to-order and engineer-to-order environments the planning of the customized (non-standard) items is done using network planning tools, such as PERT and CPM. If the customer order involves certain standard items which are planned by MPS and/or MRP I, then those standard requirements are planned using MPS and/or MRP I. FAS *is the tool for releasing assembly orders and the issuing of component pick lists*. In some situations, the assembly process may include sub-assembly operations, finishing, painting, and/or even fabrication not included in the MRP I system. These additional activities would be listed in the routing file and must be covered in the lead-time allowance.

2.13 INDEPENDENT AND DEPENDENT DEMAND (Distribution and Manufacturing Inventories)

Independent demand items are *distribution inventories* comprising of final or finished products that are not a function of, or dependent upon, the demand of any other item or upon internal production activity. Independent demand is usually external and, thus, is beyond the direct control of the organization. Cars are an example of an independent demand item, since, in general, there is no product that creates demand for a car. The demand for wheels used in the car is dependent on the production of cars (Russell and Taylor III, 1998). Independent demand appears to be random—that is, caused by chance events and must be forecast.

Generally speaking, independent demand items are sold, or at least shipped out of the factory, rather than used in making another product in the same factory. There is no feasible way to precisely determine how many of these items will be demanded during any given time period, because demand is typically somewhat random. Therefore, forecasting plays an important role in stocking decisions for independent demand items. Independent demand can range from smooth to very sporadic or 'lumpy,' depending on the industry and the customer consumption patterns. Most independent demand items must be in stock virtually all the time, and usually require safety stock.

Dependent demand is when the demand for a particular item depends on the demand for another item. Dependent demand items are *manufacturing inventories,* which typically include sub-assemblies, component parts, or raw materials, which are used internally in the process of producing a final product. Table 2.3 compares the various characteristics of dependent and independent demand. Since the dependent demand is directly related to the demand for higher-level assemblies or products, it can be calculated by multiplying the forecast quantity of the end item by the usage quantity of the component in the end item. MRP I is designed to do this calculation.

Table 2.3

Comparison of Independent and Dependent Demand

Sr. No.	*Independent Demand (Distribution Inventories)*	*Dependent Demand (Manufacturing Inventories)*
1.	The purpose of distribution inventories is to supply quickly the items the customers demand. Distribution inventory is customer oriented. These inventories are the end result of manufacturing and are created from manufacturing inventories.	The purpose of manufacturing inventories is to satisfy production requirements. Manufacturing inventory is product oriented.
2.	Demand is external, based on market place needs. Further, the demand is usually difficult to forecast, tends to be erratic, and is typically made up of small demands originating from many separate sources.	Demand is internal based on production schedules—MPS or FAS. Since the same component might be used in several end items, demand in any period will typically consist of a limited number of individual demands for multiple quantities.
3.	Demand is random and relatively continuous. Independent demand can range from smooth to very sporadic or 'lumpy,' depending on the industry and the customer consumption patterns.	Dependent demand can also range from smooth to very sporadic depending on how components are used in higher-level assemblies; but usually it tends to be lumpy and discontinuous.
4.	Demand must be forecast. Uncertainty exists at the item level. Many unit demands originating from separate sources combine to create the total demand.	Component and raw material requirements are geared to the quantities of the parent items in the MPS. Hence, demand need not be forecast; it can be calculated, based on the MPS and company's inventory procurement policies. Uncertainty exists only at the MPS level.
5.	Most independent demand items must be in stock virtually all the time. Service levels are defined, and safety stock is used to attain the target service level to ensure a certain reliability of product delivery.	Dependent demand items need only to be stocked just prior to the time they will be needed in the production process. The relative predictability of dependent demand item usage means that there is little or no need for safety stock to ensure a 100% service level. Defectives, however, are taken care of by shrinkage or yield factor.

(*Contd.*)

Sr. No.	*Independent Demand (Distribution Inventories)*	*Dependent Demand (Manufacturing Inventories)*
6.	The inventory investment level is determined by the trade-off between shipping, warehousing, and carrying costs of the inventory investment and added sales revenue realized through improved availability to customers. Higher inventory levels add the cost but also improve customer service levels.	The inventory investment level is determined by manufacturing considerations such as setup times and costs. Work-in-process inventory may constitute a significant portion of the investment. The level of WIP inventory is primarily a function of planned manufacturing lead times. No trade-off is involved as Inventory that exceeds the minimum required brings no extra revenue.
7.	*Independent Demand Inventory Items:* 1. Retail. 2. Wholesale finished goods. 3. Manufactured finished goods. 4. Service and replacement parts, service kits. 5. Transportation inventory. 6. Maintenance, repair, and operating (MRO) supplies.*	Dependent Demand Inventory Items: 1. Assemblies. 2. Sub-assemblies. 3. Fabricated components. 4. Sub-assemblies and components in process. 5. Purchased components. 6. Raw materials.

*MRO supplies are the items used in support of general operations and (internal) maintenance such as maintenance supplies, spare parts (required for the internal maintenance of the manufacturing plants and other facilities), and consumables used in the manufacturing processes and supporting operations. It includes items such as adhesives and solvents consumed in manufacturing, machine lubricants, spare parts for production equipment and other facilities, and general supplies used throughout the firm.

MRP I differs from statistical inventory control, in that, it considers intra-company demands as dependent on the demand for finished goods. This is quite obvious in principle but surprisingly quite difficult to organize in practice. The concept of dependent demand is often called the *fundamental principle of MRP I.* It provides the way to remove uncertainty from the requirement calculations.

The production schedule (MPS) for the end item creates the demand for the components/sub-assemblies. Once the master scheduler has authorized production of a particular quantity, the people in manufacturing do not need to know whether or not there is a customer demand for that end item. Because we can calculate the exact quantity of the components/raw materials needed, and the exact date when they will be needed, we can answer all the three basic inventory management questions (what, when, and how many) without requiring a forecast. And because our precise calculations are much more accurate than any forecast, we do not need to forecast dependent consumption of components, sub-assemblies, or raw materials.

An item can have both a dependent and independent demand, a service or replacement part, for example, has both. It should be noted that the MPS is for planning to satisfy the independent demand for end items or standard modules, which is based on the forecast. The dependent demand for components and sub-assemblies, which is derived from the forecasted independent demand, is handled by MRP I. An exception is made in case of forecasted independent demand for components and/or sub-assemblies (such as spare parts), which are mostly handled by the MRP I, but may sometimes be planned in the MPS if they are critical in terms of material and capacity requirements.

In Fig. 2.3 the block captioned 'Independent Demand (for Components and Subassemblies)' lists the various sources of independent demand for components and sub-assemblies. This block is connected with a continuous arrow to the MRP I implying that this demand is *usually* addressed by the MRP I logic. However, it is also connected to the MPS with a dotted arrow implying that *sometimes* this demand is addressed by the MPS, although it is a component requirement. The reason might be that it requires significant amounts of capacity at critical work centers or it is critical from the purchase or financial point of view and hence, should be addressed in the MPS itself so that it is subjected to RCCP, material and vendor limitations, and management decisions before the MPS is finalized.

Although useful concepts in terms of understanding demand, the distinction between independent and dependent demand is becoming increasingly arbitrary as more and more companies recognize the advantages of sharing information between supplier and customer in so called *supplier partnerships.* In the simplest case, the manufacturer makes available to the supplier the output from his MRP I system covering planned purchase order releases, and this serves as demand for the supplier planning. In effect, what used to be independent demand has become very close to being dependent. Similar effects can be achieved by linking a distribution planning system to the manufacturing planning system (Kenworthy, 1997). When MRP I is integrated with distribution requirements planning, a DRP I system has the effect of moving independent demand out to the distribution points nearest to the customers and creating dependent demand within the supply chain back to the master schedule and to the MRP I. With DRP I and MRP I integrated, there may be little or no independent demand directly impacting the MPS or MRP I (Rice and Hatch, 1997). What used to be independent demand for the manufacturer now becomes dependent on the distribution plans.

Horizontal and Vertical Dependencies In addition to being dependent regarding the type of demand placed on them, manufacturing inventories also exhibit dependencies in other significant ways.

1. *Vertical (Parent-Component) Dependency* The obvious dependence between the parent assembly and its components is called *vertical dependency.*
2. *Horizontal Dependency (between Components of the Same Parent)* Not only do components and sub-assemblies depend on their parent assemblies (vertical dependency), they also depend on one another (horizontal dependency). For example, if three assemblies and six components are required to assemble a final end product, all nine must be available at the same time to perform the final assembly. And if the final assembly will not begin for another week, then all the other sub-assemblies and components are not needed for another week either, and hence they can be *de-expedited* so that capacity can be utilized for items needing expediting. Thus, the need (or priority) for each item is dependent on each of the others.

2.14 ROUGH-CUT CAPACITY PLANNING (RCCP)

To give the production plan developed in the S&OP process a reality check and ensure that it can be achieved, people use an abbreviated form of capacity planning called Rough-Cut Capacity Planning. RCCP provides a reasonability test of capacity needed to accomplish production plan and/or the MPS at intermediate levels of detail. RCCP is rather *rough and ready* but *simple and quick* to use. It is typically used to look from a few months to several years ahead. Master schedulers need a quick way of assessing the feasibility of any proposed changes to the plan and RCCP was developed for this purpose. RCCP techniques vary among companies, but most involve two essential ingredients: identified critical resources and load profiles.

Critical Resources These are company specific. They normally include bottleneck factory processes, both equipment and labor, constraining supplier capacities, design or engineering capacities, quality assurance or test facilities, and so on. The critical few resources truly dictate overall throughput rather than the large number of specific capacities. In most companies, 10 or 20 critical resources are evaluated before the plan is taken to the next level of detail. If the plan is not achievable, it is changed before proceeding.

Load Profiles These identify the loads placed on critical resources as production planning families of products or specific MPS items are planned. They are developed by summarizing the detailed routings used to prepare the various components of a product and by analyzing cost data or labor reporting, or they may simply be estimates prepared by knowledgeable individuals.

The capacity requirements planning (CRP) at this level might be done only for a 'representative' item for the family group at overall family rates of manufacturing and not for each item in the group, provided, such a representative item with average capacity requirements exists. Otherwise, a more precise RCCP analysis is made by evaluating item-specific plans as stated in the MPS. In addition, only approximate methods, which are simpler, faster, and better, are used. Hence, it is called 'rough-cut' capacity planning. Through experience, companies develop a proper level of RCCP precision appropriate to their needs. The production plan or master production schedule (MPS) is the primary information source for rough-cut capacity planning. The capacity requirements of a particular MPS can be estimated by any of the following techniques:

1. Capacity Planning using Overall Factors (CPOF).
2. Capacity Bills (Bills of Resources).
3. Resource Profiles.

These techniques provide information with which to modify the resource levels or material plan in the medium range to ensure efficient execution of the MPS.

1. Capacity Planning Using Overall Factors (CPOF) In this method, the MPS items and quantities are multiplied by the *total time* required to build each item to provide the *total* number of hours on all the work centers to produce the schedule. Historical work center percentages are then applied to the total number of hours to provide an estimate of the hours per individual work center to support the master schedule. This technique eliminates the need for engineered time standards. CPOF is the simplest of all the four techniques and is based only on accounting data.

2. Capacity Bills or Bills of Resources It is a listing of the required capacity and key resources needed to manufacture one unit of a selected item or family. RCCP uses these bills to calculate the approximate capacity requirements of the MPS. Resource planning may use a form of this bill.

3. Resource Profiles 'Resource profiles' is, perhaps, the most commonly used technique in RCCP. Here, the top-level products are extended by a summarized resource profile for each product. The resource profile also gives information on the timing of the capacity requirements. The resource profile identifies:

(a) the key resources required,
(b) the number of hours of the key resources required to make the product, and
(c) approximately when those resources are required relative to the completion of the product.

Thus, after multiplying the MPS quantity with the number of hours of a particular resource we can add these hours to the correct time bucket in which these hours would be required relative to the delivery time of the item. Both the CPOF and capacity bills calculate the resource requirements and show them in the same time period in which the MPS item is scheduled for delivery; that is, there is no time phasing. Consideration of time phasing is accomplished with the resource profile approach, a refinement to the capacity bills. Each key resource is time phased through the use of a resource profile.

- Rough-cut capacity plan can be calculated from either the production plan or the MPS. The production plan would be used when the items in an S&OP family are similar (that is when they use about the same amount of key resources and hence one resource profile will work for the entire family). If, however, the number of hours of key resources required for the different items of the S&OP family are significantly different, then a resource profile is needed for each MPS item.
- Most companies with flow-type manufacturing use RCCP *only*. This includes most process manufacturers and Just-in-Time (JIT) companies with extensive use of kanban to control the movement of material.
- Almost all companies start with RCCP. Generally, this is because they implement MRP II in stages with the first stage consisting of RCCP, MPS, and MRP I modules.
- RCCP tools are designed to permit a rapid analysis of how reasonable the production plan and/or MPS is in light of demonstrated capacities. Highly interactive, computer–assisted RCCP is very desirable. 'What-if' questions abound in most manufacturing companies, and RCCP is a major 'what-if' evaluation tool. It is used whenever changes of consequence are proposed in production plans and/or the MPS (Sari, 1997).
- If a company finds that RCCP is sufficient to meet the needs of the business, it has no reason to implement detailed CRP. However, if RCCP doesn't show a capacity shortage problem but capacity bottlenecks appear on the manufacturing floor, then CRP is needed.

2.15 ENGINEERING CHANGE CONTROL

Products change and evolve, getting better, which means we are better serving our customers' needs. We correct the rough edges, streamline our production, do a better job. Design department communicates these changes through Engineering Change Notes (ECNs). Engineering changes, if not properly planned and controlled, can result in high costs owing to:

(i) scrapping obsolete product, raw materials, and components,
(ii) excessive tooling cost (tooling for old and new designs),
(iii) production inefficiency particularly during changeover, and
(iv) late shipments (Oden *et al.* 1993).

Nevertheless, maintaining market position usually requires ongoing product improvement, and the pace of change continues to increase. The challenge is how to implement engineering changes efficiently with the least disruption throughout the company. Most companies have an engineering change committee that:

(i) reviews the change,
(ii) approves or disapproves the change, and
(iii) decides when approved change will go into effect.

The central role Bills of Material (BOMs) play in MRP I programs makes it essential that engineering changes be handled correctly. Such changes are introduced into BOMs by two methods:

1. Through Effectivity Dates

This method is by far the most common method of implementing an engineering change and it is also easier to use. Effectivity dates are entered into BOM process file data for individual items affected, together with references to the engineering change numbers initiating them. MRP I uses BOM as a framework for planning and replanning; the *timing of changes* affects the ways MRP I handles requirements, netting, and order scheduling. An *expiry date* can also be specified for each engineering change. This field is optional and its absence indicates that the change is valid for an indefinite period. As soon as the decision is made, the engineering change coordinator enters the effectivity date on which the old part will be made non-effective and the new part added. The MRP II system will execute the change automatically. When MRP I explodes the planned shop order, it checks the effective dates in the BOM, and creates gross requirements for *only* those components that are *effective* on the planned shop order *start date*.

2. Through Serial Number, Lot Number, or Work Order Number

The second method of implementation involves using something other than a date to control the effectivity of the change. Serial number, lot number, and work order number have been used. Serial number effectivity is considerably more complex than date effectivity. First, the specific sub-assembly where the change occurs must be identified, according to the end-product serial number into which it will eventually be assembled. The net result of serial number effectivity is that the identity of the serial number of end product must be carried all the way down the entire product structure. The MPS must contain both product-identifying model number and serial number. Products below the cut off serial number in the MPS will use the existing bill, and products *at* and *above* the cut off serial number will use the changed bill. Once the serial or lot number arrives, the new design goes into effect.

In regard to implementation, engineering changes fall into following two broad categories:

1. Mandatory or Immediate

These result from:

(a) product functional failures, and operating problems that did not show up until the product had been out in the field for a while,
(b) potential harm or injury to users which might lead to safety or liability problems,
(c) unavailability of materials, and
(d) new or changed government laws or regulations.

These engineering changes must be implemented immediately. The management will immediately halt production and implement the change, without regard to cost, delay in production, disruption of other operations, or any other factors. Perhaps, even the products already in customers' hands will be recalled. Costs and internal problems are irrelevant.

Implementing mandatory changes is simple but very expensive, erase the old parts and replace them with the new, then deal with the consequences. Old stock could be:

1. Scrapped,
2. Reworked to the new specifications,
3. Used in other products, and
4. Retained for service or repair use only.

2. Discretionary, Optional, or Deferrable (Phase-ins and Phase-outs)

These include the great bulk of changes resulting from:

(a) efforts to improve product performance and safety,
(b) customer preferences and competitor's innovations,
(c) new/improved materials and processes,
(d) cost reductions, and
(e) improving operations.

Management has considerable leeway in deciding *how* and *when* to implement this type of changes. Timing is usually specified as '*Soon as possible,*' '*When current material gone,*' '*Minimum cost,*' '*On X date,*' or '*With Y serial/lot number.*' These should be manually converted into *effectivity dates* before they can be input to BOM processor software.

Four factors make controlling engineering changes difficult:

1. *Design Engineering* works with and transmits data via parts lists, thus manufacturing has to work with part lists that have been modified by others for their use.
2. The timing of introducing changes into planning BOM is critical, and many factors deserve consideration.
3. The sheer volume of changes can be daunting.
4. The effects on all BOM levels can be difficult to trace.

Further, changes rarely come uniformly, floods follow immediately after new or radically revised product designs have been released. Here, groups of engineering changes for periodic release, called *block changes,* have significant benefits over individual releases. The benefits of block changes include:

(i) more stable BOM,
(ii) more accurate maintenance at lower costs,
(iii) minimal disturbance of operations,
(iv) simpler service parts support,
(v) reduced obsolescence, and
(vi) better planning.

Factors to be Considered in Determining the Timing of a Change (Oden *et al.* 1993)

1. Availability of new items required to implement the change: raw materials, tooling, equipment, etc.
2. Need for recalling or retrofitting products in use.
3. Time for minimizing cost, including obsolete inventory, tooling, equipment alterations.
4. Time for depletion of existing product and component inventories.
5. Need to retain supply of current parts for customer service.
6. Time of changes to other products.
7. Time of model changeover or start of a new contract.
8. Impact on our plant and supplier's plant capacity.
9. Importance of change to competitive position.
10. Need for documentation of changes.
11. Time for approval by regulatory body.

Engineering Change Affects Only the Individual Parent-Component Relationship For example, if item X is being phased out of parent A on a particular date, entering this change into the BOM only affects that particular usage of item X. It can still be used in many other places. The action does *not* render item X obsolete. In fact, item X can be phased back into the bill of material for parent A at a later time (Oden *et al.* 1993).

2.16 MATERIAL REQUIREMENTS PLANNING (MRP I)

The material requirements planning (MRP I) is a computerized inventory control and production planning system. It is responsible for scheduling the production of all items beneath the end item level. It recommends the release of work orders and purchase orders, and issues rescheduling notices when necessary.

(Russell and Taylor III, 1998)

- The material plan, which is based on the MPS, describes *what* materials you need, and *when* you need it. MRP I determines the period-by-period (time-phased) plans for all component parts and raw materials required to produce all the products in the MPS. This means taking a time-phased set of MPS requirements and producing a resultant time-phased set of component part and raw material requirements. Thus, MPS and MRP I are the elements used for the planning and replanning process to maintain proper due dates on required materials. This function of determining *what* materials are required and *when* is called *priority planning*.
- Managerial objective of MRP I is to provide 'the right materials/parts at the right time to the right place' to meet the schedules for completed products. This goes out as far as the planning horizon. MRP I is the way to make the MPS happen. To do this MRP I provides *formal* plans for each part number, whether raw material, component, or finished good. Accomplishing these plans without excess inventory, overtime, labor, or other resources is also important (Vollmann *et al.* 1997).
- MRP I translates the top-level plans into individual schedules for the thousands of items that have to be manufactured or purchased in the typical company. These schedules are concrete enough that people can go out, make them happen, and be accountable for them (Landvater, 1997).
- In technical terminology, MRP I is a *deterministic planning model.* That is, the results of each step in calculation are determined directly from the prior step without considering probability distributions for any of the variables. A probabilistic model would take into account such considerations. That approach is used where variables have a wide range of possible values. The deterministic MRP I model is useful and sufficient for planning day-to-day manufacturing activities (Rice and Hatch, 1997).
- The various functions performed by MRP I make for a set of repetitive clerical tasks, which lends itself to being performed by a computer.

- All items listed in a BOM (purchased, fabricated, or sub-assembled) *can* be controlled through the MRP I system. Notice the use of the word 'can.' Some items such as hardware or packaging may be more easily controlled through alternative methods such as two-bin or reorder point (or manual systems) but would still be listed in the BOM for purposes of product costing and understanding. Software systems allow the user to code such items as 'non-MRP' just as end items are coded as 'MPS' rather than MRP I (Toomey, 1996). For each item an order control system, such as FAS, MPS, MRP I, SIC, or manual, is specified through a field, sometimes called 'order policy code' in the item master file. This code determines which system would plan the production/purchase orders for that item.

The planning of detailed component and raw material needs to support the manufacturing plan can be quite complex. The basic MRP I calculations only involve simple arithmetic—addition, subtraction, and an occasional multiplication, but it is tedious and repetitive and best left to the computer in practical applications. This is because for many firms, MRP I involves the creation of detailed plans for tens or even hundreds of thousands of individual parts for a year or more into the future, and changes happen everyday. Further, as more additional features, such as lot-sizing, safety stock, and scrap factors are introduced, even simple calculations become somewhat complex. It was the development of high-speed digital computers that finally allowed people to carry out the calculations in a reasonable period of time (several hours) and keep their plans up to date. Today MRP I is still a simple set of calculations used to create and maintain valid schedules that show what items are required and when they are needed. It answers the questions: 'What do we need, and when do we need it?'

2.16.1 Development of Material Requirements Planning

Much of the credit for introducing MRP I and educating industry about its benefits goes to three individuals: Joseph Orlicky, George Plossl, and Oliver Wight, and to a professional society they endorsed, known as the American Production and Inventory Control Society (APICS). It was the advent of cheap computing power in the 1950s and 1960s that led to a rapid growth in the number of companies using MRP I. After accounting and payroll systems, MRP I was one of the earliest computer applications to become widespread (Childe, 1997). Unlike the experience in accounting, in which formal and consistent manual procedures were available to be computerized, manufacturing management was characterized by largely *ad hoc* informal procedures that were different for each company.

MRP I was initially called *'processing the bill of materials,'* and BOMP (bill of material processing) software packages were developed. The early developers were manufacturers of highly complex products, such as heavy construction and agricultural equipment. These products had many thousands of parts, sub-assemblies, and assemblies. The products had many steps or levels of manufacturing, necessitating multilevel BOMs and typically many features, options, and accessories. A large variety of service parts had to be stocked and supplied at many levels of the BOMs. Often these products were part of a product family that shared a variety of assemblies, parts, and raw materials. People responsible for scheduling such products were in need of a managerial tool that would simplify and structure the process. It was in 1961 that Joseph Orlicky developed MRP for one such company called the JI Case Company. The JI Case Company manufactured a variety of heavy equipment for the construction industry, such as bulldozers, graders, other earth-moving equipment, farm tractors and a variety of agricultural implements. Following are some of the milestones in the development of MRP I/II and APICS (Noujaim, 1997).

1956: Some 26 production and inventory control professionals met in Cleveland, Ohio (USA) and formed the American Production and Inventory Control Society (APICS).

1961: Joseph Orlicky developed MRP I at the JI Case Company.

1970: APICS published the first edition of Production and Inventory *Control Handbook* (The latest edition, the third, of this handbook has been published in 1997). APICS also decided to establish the *Curriculum and Certification Council* to develop the criteria for testing an individual's knowledge in Production and Inventory Control (PIC) and to award a certification of professionalism in that field.

1970s: APICS launched an 'MRP Crusade' to publicize MRP I, to educate practitioners, and to promote the development and implementation of MRP I (Rice and Hatch, 1997).

1972: The first examinations were given for the granting of the designation *Certified in Production and Inventory Management* (CPIM).

1973: The first examinations were given for the granting of the designation *Certified Fellow in Production and Inventory Management* (CFPIM). (CFPIM recognizes those who actively educate others in the industry.)

1974: Oliver Wight published *Production and Inventory Management in the Computer Age* in which he explained the shortcomings of informal systems and the rationale for developing formal, computerized materials and distribution requirements planning and control. He also outlined how MRP I may be expanded into a generalized logistics and resource management system, which he christened as Manufacturing Resource Planning or MRP II (Rice and Hatch, 1997).

1975: Orlicky published his now-classic book *MRP — Material Requirements Planning* in which he laid down the formalized computational logic for dependent-demand materials requirements planning. The book was revised by Plossl and published in 1994 under the title *Orlicky's Material Requirements Planning.*

1991: APICS took certification to the next level by unveiling the *Certified in Integrated Resource Management* (CIRM) program. The CIRM was designed to help transform manufacturing professionals from functional specialists into organizational leaders by giving them knowledge of 13 business functions in a manufacturing organization and how they interact.

1995: APICS introduced *Computer-Based Testing* (CBT) for its CPIM and CIRM programs.

By 1997, APICS had more than 70,000 members, and more than 52,000 CPIM, more than 2000 CFPIM, and over 1500 CIRM designations had been awarded (Noujaim, 1997).

MRP I offered companies the ability to order goods according to actual requirements, rather than simply replenishing standard stock levels (such as through statistical inventory control, SIC). Initially, MRP I was used mainly for purchased items, but the extension of its use to include in-house manufactured items followed, especially in those factories where there was irregular demand and a wide range of items, such as in the aerospace industry (Childe, 1997).

2.16.2 Objectives and Applicability of MRP I

1. Determine Requirements to Support MPS

The main objective of MRP I is to determine what components are needed to meet the MPS and, based on lead time, to calculate the periods when the components must be available. It must determine the following:

(a) What to order (b) How much to order

(c) When to order (d) When to schedule delivery

MRP I has the ability to suggest ordering, at the right time, the right items, in the right quantities, for delivery on the right date. It also provides future data on an item-by-item basis. MRP I frequently is criticized for ignoring capacity constraints. MRP I is designed to answer the question 'What should be produced to support MPS?' rather than to answer the question 'What can be produced with existing capacity?' It assumes that capacity constraints were considered when MPS were developed. MRP I implicitly trusts the MPS that is driving it. *Invalid MPS guarantee invalid MRP I outputs.* Of course, invalid MRP I outputs can be caused also by bad data in BOM, inventory, lead-time, and activity-reporting files. The value of MRP I depends on the quality of data it uses (Plossl, 1994).

2. Maintaining the Lowest Possible Inventory

One objective of MRP I is to maintain the lowest possible level of inventory. MRP I does this by determining *when* component items are needed and scheduling them to be ready at that time, *no earlier and no later.* Thus in MRP I, raw materials are not procured nor components machined simply because they 'may' be required. The theoretically ideal way of running an MRP I system is with zero safety stock and zero safety lead time. Further in MRP I, *inventory control is proactive, not reactive* (Plossl, 1994). The system anticipates inventory requirements and plans for it in advance.

In the days before computers, it was necessary to maintain extensive manual systems and to have large inventories and long lead times. These were needed as a cushion due to the lack of accurate, up-to-date information and the inability to perform the necessary calculations quickly. Somehow, someone in the organization figured out what was required sooner, or very often, later than needed. *Get it early and get lots of it* was a good rule then (Arnold, 1998).

The operating philosophy of MRP I is that materials should be *expedited* when their lack would delay the overall production schedule, and *deexpedited* when a schedule change postpones their need. To this end, MRP I logic will always plan inventory to the lowest possible amount, unless instructed otherwise by *order modifiers* like safety stock and lot-sizes.

3. Scheduling the Production

MRP I was the first inventory system to recognize that inventories of raw materials, components, and finished goods may need to be handled differently. Firstly, the demand for component parts being a dependent demand does not have to be forecasted, it can simply be derived from the forecasted independent demand for the finished product. Secondly, with component items, demand does not necessarily occur on a continuous basis. Because of the batch sizes and lead times involved in manufacturing and assembly processes the demand is lumpy or discrete, not continuous. Using an inventory system such as EOQ for component items would result in inventory being held that we know we won't need until a later date. The excess inventory takes up space, soaks up funds, and requires the additional resources for counting, sorting, and moving. MRP I relates the quantities and timings of component items to the quantities and timings of the parent items.

In the process of planning the inventory levels for the raw materials, components, and finished goods, the MRP I system also planned:

(a) purchasing activities (for raw materials and purchased components),
(b) manufacturing activities (for component parts and assemblies), and
(c) delivery schedules (for finished products).

Thus, the MRP I system was more than an inventory control system, it became a production scheduling system as well.

4. Keeping Schedules Valid and Up-to-date

One of the few certainties in a manufacturing environment is that things rarely go as planned—orders arrive late, customers change orders, components get used up, suppliers are late with delivery, machines break down, workers are absent, excessive scrap occurs, designs are changed, and so on. In this ever-changing world, a material requirements plan must be able to recognize priorities to keep plans current. It must be able to add and delete, expedite, delay, and change orders. With its computerized database, MRP I is able to keep track of the relationship of job orders so that if a delay in one aspect of production is unavoidable, other related activities can be rescheduled, too. MRP I planning is change-sensitive and reactive having the ability to keep schedules valid and up-to-date.

5. Especially Useful in Complex and Uncertain Manufacturing Environments

Industries that manufacture complex products, requiring the coordination of component production, find MRP I especially useful. MRP I was designed for planning products that are complex enough to have more than two levels in the BOMs. A complex product may have hundreds of component parts, dozens of assemblies, and several levels of assembly. MRP I tries to ensure that multiple components of an assembly are ready at the right time so that they can be assembled together. MRP I may not be suitable for products that involve only a single assembly operation of a collection of staged parts (that is, parts pulled from inventory before they are required) that have no intermediate sub-assemblies, and no raw materials other than those used directly in the final assembly, and when there are no after-market service parts that require planning. The reason is that in these situations, no product structures need to be exploded and netted, and other computer software approaches may be adequate and simpler. Even in these situations, however, the short-term scheduling features of MRP II for materials, labor, and capacity might prove useful.

The advantages of MRP I are more evident when the manufacturing environment is complex and uncertain. Manufacturing environments in which customer orders are erratic, each job takes a different path through the system, lead times are uncertain, and due dates vary, need an information system such as MRP I to keep track of the different jobs and coordinate their schedules. These are really the characteristics of *batch,* or *job shop* processes. Although MRP I is currently available for continuous and repetitive manufacturing, it was designed primarily for systems that produce goods in batches. Finally, MRP I systems are also very useful in ATO types of industries.

When MRP I was introduced, revolutionary changes occurred and new premises were established. Orthodox practices and techniques were challenged. The existing inventory-control literature, indeed an entire school of thought, had to be re-examined. It became evident that the basic tenet of old inventory-control theory—that low inventory investment and high customer service were incompatible—was mistaken, successful users of MRP I programs reduced inventories and improved delivery service at the same time (Plossl, 1994).

There are a number of other functions in the MRP II system that are dependent on the material plan. These include Capacity Requirements Planning (CRP) and the execution of material and capacity plans. In addition, MRP I provides much of the detailed information used to integrate the financial and operating systems. Therefore, MRP I forms a vital central link in the whole closed-loop process. Following sections describe the functions that the MRP I performs.

2.16.3 Exploding through the Bill of Material Using Low-level Codes

MRP I determines quantities of raw materials, base components, fabricated items, purchased components and assemblies, intermediates (solutions, mixtures, formulations), bulk production, etc. needed to support the MPS, and *when* they should be scheduled. This process is referred to as *'BOM Explosion.'* It is the stage in the MRP I process, where the end-item requirements listed in the MPS are broken down, using the BOM, into *time-phased* requirements at all levels of sub-assembly. It uses the BOM process file of each MPS item to find out the quantities and timing of raw materials, components etc. required to produce the end item. The BOM is, in turn, kept updated by the 'engineering change control' module which provides the updated BOMs, engineering change numbers, and also their 'effectivity' and 'expiry' dates.

Each component shown on the BOM is planned for by the MRP I system. For convenience, it is assumed that each component will go into inventory and be accounted for. Whether the components actually go into a physical inventory or not is not important. However, it is important to realize that planning and control take place for each component on the bill. Raw material may go through several operations before it is processed and ready for assembly, or there may be several assembly operations between components and parent. These operations are planned and controlled by Production Activity Control (PAC) in the shop floor control module, not MRP I.

Low-Level Coding

The process of BOM explosion gives the required quantity of each component for the total quantity of *all* the end items scheduled in the MPS. It is quite possible that one particular component might be used at different levels in the same end item and also at different levels in different end items. If a component is on more than one level in a BOM, it is necessary to make sure that all gross requirements for that component have been recorded before netting takes place. If we process the time-phased record for this common part before all of its exploded requirements have been accumulated, the computations must be redone. Thus, the process of requirements explosion of a particular component should continue until all the end items in which the component is used, have been considered. Further, for each component the explosion process should accumulate the requirements down to the *lowest* level in any BOM of MPS items in which the particular end item appears. If another end item or parent at a lower level item exploded more gross requirements into the component later, MRP I would have to replan that component, dramatically increasing processing time. Further, the explosion process for a particular item should stop as soon as the lowest-level in which the component occurs has been scanned. Only then, the calculations for gross requirements and netting should be performed.

The MRP I technique used to make sure that the process of computing gross requirements of a particular component is done down to the lowest level at which the component appears in any of the BOMs of *all* the end items in the MPS that contain the particular component is called 'low-level coding.' Low-level code is a number that identifies the lowest level in *any* BOM at which a particular component appears. LLC prevents MRP I from calculating net requirements for a given item until all of that item's gross requirements have been computed. Net requirements for a given component are not calculated until all the requirements have been calculated down to that level for all the end items in which that component appears. Thus, every item included in the MRP I plan must include a number that represents the level at which its planning record should be balanced. This level number must obviously be the lowest level at which that item appears among *all* product structures that contain that item.

By convention, the top final-assembly level is denoted as level 0. All immediate component sub-assembly/part numbers of the final assembly are at the level 1. Thus, the higher the level number, the lower in the product structure the part is used. Consequently the last level number (code) assigned to a part indicates the lowest level of usage and is the low-level code retained for that part. Low-level codes are normally calculated and maintained automatically by the BOM processor program, which also posts it to the parts master record in the item master file. Low-level codes need to be updated whenever new BOMs are introduced or some old BOMs are deleted from the BOM file. A session to update low-level codes is invoked to update the LLCs. Low-level codes are determined by starting at the lowest level of a BOM and, working up, recording the first level at which it occurs as the low-level code for the item. If a part occurs on a higher level, its existence has already been recorded. The low-level code for a part used exclusively for service would be zero. If it is used as a component part as well, the low-level code would be determined the same way as for any other part.

Once low-level codes are established, MRP I record processing proceeds from one level code to the next in all BOMs of the end items in the MPS of a particular period, starting at level zero and going till the low-level code and *no further*. The computer scans the level zero of all the BOMs of the MPS items and creates a spreadsheet for each of the item occurring at this level, then it copies the MPS quantities or independent demand figures for these items into the respective spreadsheets. Then it scans level one of all the BOMs of the MPS items and places the requirements (that is, MPS quantity of the parent item × usage factor) into the MRP I spreadsheets of these items as gross requirements in the appropriate time bucket. The procedure is repeated for level two and so on until the BOM level designated by the low-level code has been scanned. Then the record is balanced. Within a level, the MRP I record processing is typically done in part number sequence. Cost roll-up programs start at the lowest level in a BOM and 'roll' costs upward; the LLC facilitates this process.

There is no logical requirement for low-level codes in MRP I, their absence will not prevent the program from arriving at correct results. Use of low-level codes yields higher computer-processing efficiency, which usually outweighs the cost of developing and maintaining the coding. Practically every BOM processor software program does this automatically (Plossl, 1994).

2.16.4 Gross Requirements

The planned order release of the parent item in a period becomes (with the appropriate multiplier) the basis for the requirement of the components in the same period. In order to determine the net requirements for common-usage items correctly, total exploded requirements stemming from all parent-item planned orders must first be determined.

Total exploded requirement of the component in a period
= Sum, for all end items in the MPS in which the component is used, of (Planned order release quantity of the parent item in the same period × usage quantity of the component from the BOM of the parent)

Most companies make more than one product and often use the same components in many of their products. The material requirements planning system gathers the planned order releases from all the parents and creates a schedule of gross requirements for the components. The explosion process is illustrated in Fig. 2.5. In this example, the MPS consists of two items A and B. The product structures for A and B are as shown in Fig. 2.4. Figure 2.5 illustrates how and when the orders for various

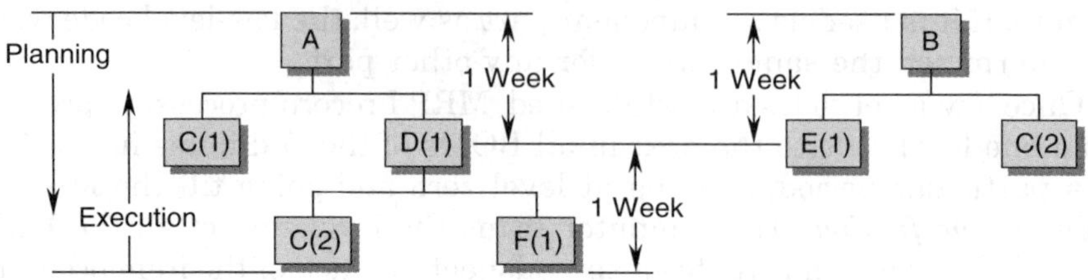

Fig. 2.4 *Bills of Material for MPS Items A and B*

items in the product structure are released for 10 units of the MPS item A. It should be noted that although the planning and scheduling process starts from the top of the product structure, the actual order release process starts at the bottom of the product structure. Whereas, the scheduling and order-release for the end item is done through MPS, the

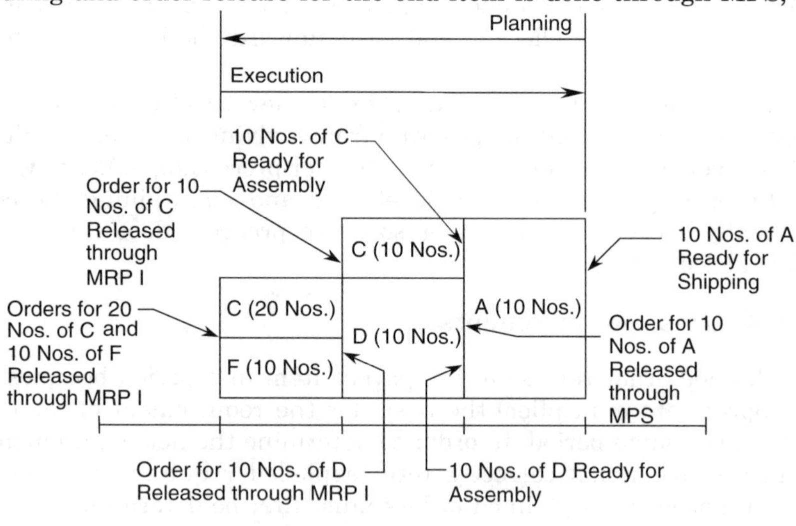

Fig. 2.5 *Order Release Process for 10 Units of the End Item A*

scheduling and releasing of orders for the sub-assemblies and components is done through MRP I. The planning for the entire product structure is done using backward scheduling from the shipping date for the end item. Order releases for all the sub-assemblies and components are offset by their lead times from the time when they should be available for the assembly process. First, the order for purchase of the required quantities of items C and F, which are at the lowest level in the product structure are released. The order for sub-assembly D is released after a period equal to the lead time for C and F, so that C and F would be available for assembly when the order for their parent assembly D is released. The process is continued till the time the order for the required quantity of end item A is released through MPS. The order-release process shown in Fig. 2.5 is just to illustrate the order-release process for ten units of A, and is not related to the explosion process shown in Fig. 2.6(a).

Figure 2.6(a) illustrates what happens in a multi-product, multi-level explosion process. Part C is a component of both A and B. Items A and B are MPS items. Lead time for all the items is 1 week. The computer accumulates the requirements level-by-level, starting with level 0, and proceeding sequentially to levels 1, 2, and so on. Within a level, the

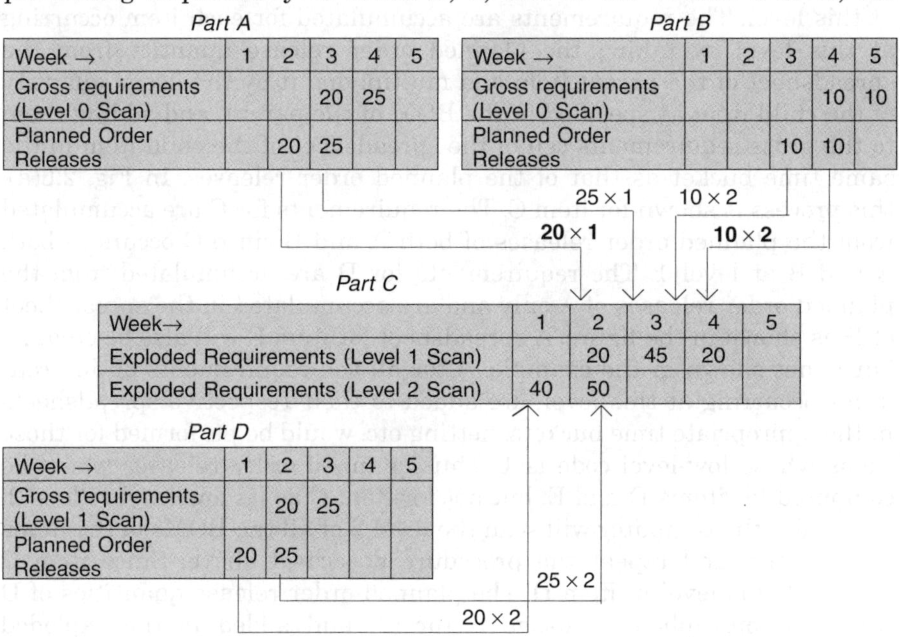

Part A

Week →	1	2	3	4	5
Gross requirements (Level 0 Scan)			20	25	
Planned Order Releases		20	25		

Part B

Week →	1	2	3	4	5
Gross requirements (Level 0 Scan)				10	10
Planned Order Releases			10	10	

Part C

Week→	1	2	3	4	5
Exploded Requirements (Level 1 Scan)		20	45	20	
Exploded Requirements (Level 2 Scan)	40	50			

Part D

Week →	1	2	3	4	5
Gross requirements (Level 1 Scan)		20	25		
Planned Order Releases	20	25			

Fig. 2.6(a) *Multiproduct and Multilevel Bill of Material Explosion Process*

scanning is in the item-number order. First, level 0 is scanned which has only two items in the MPS, items A and B. The requirements of these items are entered in the respective MRP I spreadsheets for these items. In case these items are MPS items, gross requirements are the same as MPS quantities and are copied from the MPS. In case these items are sub-assemblies, their requirements are calculated by multiplying the planned order release quantities of their parent items by the usage quantity of these items in the BOM of their parent items.

After all the requirements of these items are entered, further processing of this MRP I tableau, that is netting, lot-sizing, and lead-time offsetting, would be done only if all of item's requirements down to the level, identified by low-level code, have been accumulated. Since in this example, the low-level codes of both A and B are 0, the MRP I procedure would now be performed and planned order releases would be generated. To simplify, the example shows the case where no netting or lot-sizing is required for any of the items (that is, there are no on-hand balances of any of the items and lot-for-lot policy is used). Hence, planned order releases are simply the gross requirement quantities offset by their lead times, which have been assumed to be one week for all the items.

Secondly, the computer scans the level 1 of all the BOMs of the items in the MPS and accumulates the requirements for all the items occurring at this level. The requirements are accumulated for each item occurring at this level by taking the planned order release quantity from the spreadsheet of the parent item and multiplying it by the usage quantity of the child item as specified in the BOM of the parent, and adding them to the gross requirements cell of the spreadsheet of the child item in the same time bucket as that of the planned order releases. In Fig. 2.6(a), this process is shown for item C. The requirements for C are accumulated from the planned order releases of both A and B since C occurs in both A and B at level 1. The requirements for D are accumulated from the planned order releases of A only and are accumulated in the spreadsheet of D as shown in the figure. A spreadsheet for item E will also be created, but is not shown in the example. After all the requirements of different items occurring at this level, are added to their respective spreadsheets in the appropriate time buckets, netting etc. would be performed for those items whose low-level code is 1. Thus, planned order releases would be computed for items D and E, but not for item C as its low-level code is 2.

Thirdly, the computer will scan the level 2 of all the BOMs of the items in the MPS and repeat the procedure described above. Since item C occurs at this level in item D, The planned order release quantities of D would be multiplied by usage factor (2) and added to the exploded requirements row of C in the same time buckets as shown in Fig. 2.6 (a). Since the low-level code for C is 2, the BOM-explosion process for C can be terminated here. The total exploded requirement for C in this example are 40, (20 + 50 =) 70, 45 and 20 for periods 1 to 4 respectively.

Fourthly, there might be additional (independent) demand for components from other sources such as for service parts, interplant orders, shop floor orders, sales and marketing promotions, R&D/engineering experimentation and destructive testing, plant/equipment maintenance, and OEM orders (that is, for other original equipment manufacturers who might be using these components in their own products). These *inde-*

pendent demands are added to the *dependent (exploded)* demand to find out the *'gross requirements'* of each component.

Gross Requirements in a period
= Total exploded requirements for the item in the period
+ Total independent demand for the period

For proper handling with MRP I, the forecasted independent demand values for the components and sub-assemblies are placed in the gross requirements row of the component *before* the MRP I explosion process begins. Then, the MRP I logic adds calculated dependent (exploded) demands (if any) to the independent demands, so that, the total gross demand for the component is properly derived.

The total exploded requirements computed are added to these independent demand figures in the respective time buckets. The addition of these exploded (dependent) and independent demands gives the gross requirements for C

After the gross requirements for C are computed, the netting, lot-sizing, and lead-time offsetting process is performed for item C and its planned order releases are generated as shown in Fig. 2.6(b). To simplify the processing, lot-for-lot policy is used and it is assumed that no inventory of C is available in any period so that there is no need to apply lot-sizing and netting processes.

Part C (Purchased Item, Lead Time 1 Week, L4L, No inventory of C)

Week →	1	2	3	4	5
Independent Demand (Placed in the spreadsheet before the scanning process starts)	10	05	15	10	15
Total Exploded Requirements (Added as a result of scanning process)	40	70	45	20	
Gross Requirements	50	75	60	30	15
Scheduled Receipts	50*				
Net Requirements		75	60	30	15
Planned Order Receipts		75**	60	30	15
Planned Order Releases	75**	60	30	15	

*The order for the 50 units must have been released at the beginning of the last week since the lead time is one week. This should be received at the beginning of this week. The table shows the state at the beginning of the week when the scheduled receipt has *not* yet been received. As soon as these 50 units are received this scheduled receipt would be wiped out and the on-hand balance would be increased by 50 units.

**The order for 75 units should be released at the beginning of this week so that it will be received at the beginning of the next week. As soon as the planner releases this order, both the planned order release of 75 units in period 1 and planned order receipt of 75 units in period 2 would be wiped out and a scheduled receipt of 75 units would be shown in period 2.

Fig. 2.6(b) *Simplified MRP I Spreadsheet for Item C*

2.16.5 Netting

From these gross quantities the amounts available in the inventory and the amounts expected (on or before the time of need) through scheduled receipts (released orders) is deducted to find out the net requirements of each component. This is called the *'netting'* process, which determines the need to manufacture or buy parts with an emphasis on timing. Net requirement is the projected shortage for a particular period unless we take action. A negative net requirement, when the sum of quantities on hand and on order exceeds gross requirements, is considered to be zero. The logic of the net requirements computation is given below:

Net Requirements = Gross Requirements + Time-phased Allocation in the Period – (Projected Available from the previous period + Scheduled Receipts)

2.16.6 Lot-Sizing

Lot-sizing is a specification of the order quantity, that is the amount of material to produce or purchase at one time. All lot-size choices, whether intuitive or scientific, are based on *the principle of economy of scale.* It is usually less costly to purchase (and transport) or produce a bunch of material at once than to order it in small quantities. If orders for large quantities are specified, there will be fewer orders placed. For purchasing, this means that quantity discounts and transportation efficiencies may be realized. For manufacturing, large lot-sizes mean long runs of the same work with less time lost on setup of equipment or changeover of a production line. On the other hand, larger lot-sizes result in more inventory, and inventory is expensive to hold. Thus, the basic trade-off usually involves elimination of one or more setups at the expense of carrying inventory longer.

Thus, lot-sizing means producing or purchasing a quantity that may exceed the immediate need due to economies of scale, or due to constraints inherent in the production process. Policies regarding lot-size quantities and how they will be determined must be provided as an input to the MRP I process. The planned order quantity is calculated using an order policy for an item. Considerations on lot-sizing apply equally well to master production scheduling and to MRP I. All lot-sizing techniques that have been conceived and investigated are based, in one way or another, on the principle of offsetting various costs involved in the ordering or manufacturing processes. The costs associated with lot-sizing are listed in Table 2.4.

Table 2.4

The Costs Associated with Lot-Sizing

For Purchase Orders	*For Manufacturing Orders*	*For Carrying the Inventory*
1. Supplier selection, follow-up, and other contacts. 2. Preparation and handling of authorization document (preparing requisitions, purchase orders, receiving documents for purchased materials). 3 Receiving, inspection, and handling. 4. Accounts payable.	1. Preparation of production order paperwork (preparing requisitions, and shop packets for manufactured items). 2. Setup of equipment or assembly line changeover. 3. Inspection, scrap, and re-work that results from start-up after a new setup. 4. Tracking and reporting on work orders in the plant.	1. Storage facility costs—cost of space, equipment, heat, light, and people. 2. Deterioration from long-term storage and handling. 3. Counting, transporting, handling, and record keeping. 4. Risk of obsolescence due to market, engineering changes, competitor's product innovations, or spoilage. 5. Insurance and taxes. 6. Risk of loss due to pilferage. 7. Cost of capital invested in inventory or opportunity cost—the alternative investments that could be made with the money tied up in inventory.

An order policy is really just a set of rules that are used to determine the order quantity. Seldom will a single lot-size rule be appropriate for all levels of end products, assemblies, and components. There are also substantial cash flow and cost consequences associated with lot-size decisions. Therefore, the formulation of lot-size rules must be done with great care. Order policies are exclusive of one another, one order policy is specified for each item and an item cannot have several order policies at the same time. Commonly used order policies generally fall into the seven categories described in Table 2.5.

Discrete and Non-Discrete (or Demand-Rate) Lot-Sizes

Discrete Lot-Size It is an order quantity that represents an integer number of periods of demand. Thus, discrete lot-sizing is defined as any lot-sizing method that does *not* result in 'remnants' of inventory carried over from one period to another that are insufficient to cover all of the requirements of the next period. In other words, discrete lot-sizes satisfy—fully and exactly—the requirements of one or more periods, hence, the inventory carried over from one period to another would be either exactly equal to the requirements of the next one or more periods or no inventories would be carried over.

Non-Discrete or Demand-Rate Lot-Sizes These lot-sizes would carry over 'remnant' inventories from one period to another which are not enough to fully satisfy the requirements of the next period.

Static and Dynamic Lot-Sizes

- A static lot-size method would not recalculate the order quantity during each replanning cycle. Once computed, static lot-size quantity continues unchanged over the planned-order horizon. Only the fixed order quantity (FOQ) is always static. The EOQ technique will also be static if its usage component is not recalculated during a replanning cycle.
- A dynamic lot-size is recalculated every time an MRP I record is balanced based on the current set of requirements. Thus, a planned order release may be for one quantity during one planning cycle but, because of changes in requirements, may result in a different planned quantity in a subsequent cycle. This only applies to planned orders. A scheduled order that has already been released would not be altered. The EOQ technique would also qualify as dynamic if usage component is recalculated using the current planned requirements during each replanning cycle. All other lot-sizing policies except FOQ are dynamic.
- It is only with the static FOQ method that an actual lot-size quantity would be specified in the MRP I data. For the dynamic methods, the MRP I record would contain a code designating the *rule* to be used, such as L4L, POQ, etc.

Significance of Lot-Sizing

- A change to a lot-sizing technique that, overall, results in larger lot-sizes than before increases cash-flow requirements. Even with a single lot-sizing method, a change in the cost assumptions, such as carrying cost per year or setup cost, could have a significant impact. Hence, changes to lot-size methods or cost assumptions should be made with great care and, if possible, with some amount of simulation before the final decision is made.
- Dynamic lot-sizes pose a special problem. Based on a planned order for a parent assembly in some future period, orders for components farther down the structure, may have already been released. If a change in requirements resulted in a new lot-size quantity for the parent order, even if in the same time period, it would adversely affect the open orders of the components. It is also possible that the

time period of the parent order might also change, in addition to the quantity, requiring expediting or deexpediting of the component scheduled orders. This phenomenon sometimes exhibits itself in what is called 'nervousness' of the MRP I system, especially if lot-for-lot ordering is employed. Even a very small change in requirements (perhaps because of slightly more or less scrap than anticipated) could affect parent planned orders and therefore component scheduled receipts.

Lot-Size Adjustments

The planned order quantity determined by any of the lot-sizing techniques is subject to certain practical constraints, some of which are described below:

1. Minimums and Maximums A lot-size calculated by any technique should be adjusted if the quantity to be ordered is impractical. The minimum, for example, might be the smallest order quantity a supplier will accept. Maximums frequently are imposed by management policies. Minimums and maximums usually are stated in quantities — 'no less than 50 and no more than 500 units'—but the limits may be stated in period coverage—'no less than 4 weeks and no more than 13 weeks,' or 'not to exceed one year's supply.'

2. Scrap Allowance A *scrap allowance,* or *shrinkage factor,* is a quantity added to the computed lot-size to compensate for anticipated scrap or other losses in process, and to ensure that the required quantity of 'good' pieces is received. This is important only in instances of discrete lot-sizing, because the order quantity exactly equals net requirements in an integral number of periods (no 'remnants'). The scrap allowance will vary from item to item, being based usually on past experience with scrap, and is at best a guess. The proper way to handle scrap allowances in the time-phased MRP I record is to add them to (include them in) the planned-order quantities. When the planned orders are released, the full quantity, including the scrap allowance, should be considered to be on order. This quantity would then be reduced as and if scrap losses were posted to the record. (Application of scrap and shrinkage is covered in detail in a later section.)

3. Multiples Another constraint on lot-sizing is the requirement that a given item be ordered in multiples of some number. This may be dictated by process equipment considerations such as the number of pieces constituting a full load, packaging lots, and capacity of plating or heat-treating equipment. The calculated sizes, in these cases, is increased to the nearest multiple specified.

4. Raw Material Cutting Factors Raw material cutting and wastage due to residual odd-sized material often require adjustments to lot-sizes.

The lot-sizing technique, unaware of the form in which the item's raw material comes, may set an odd quantity or understate the total raw material quantity. For example, if a metal bar can be used to machine seven pieces of a turned component, a lot-size of 25 pieces would result in a remnant of bar stock, which would be difficult to keep track of and use up somehow later. A better alternative would be to produce 28 pieces.

- In cases, where more than one type of adjustment is to be made to the order quantity for a given item, the several adjustments should be made consecutively in a logical sequence.
- Minimums and maximums, scrap allowances, multiples, and raw material cutting factors may result in excess inventory. This excess, however, is temporary. It will be applied by MRP I against later gross requirements and will not accumulate.

The planned order releases become much more lumpy and discontinuous because of the lot-sizing, particularly when the lot-sizes are large. This lumpiness increases as the orders are exploded down through the product structure. Since the capacity to meet the MPS is determined in Rough-Cut Capacity Planning (RCCP) based on the MPS, the redistribution of orders resulting from lot-sizing during MPS process may result in an unbalanced workload even though the overall capacity is sufficient. Extensive lot-sizing in MRP I makes the load balancing operations in Capacity Requirements Planning (CRP) much more difficult.

- There is no one 'best' lot-sizing technique for a given manufacturing environment, for one class of items, or even for a single specific item. Experience with MRP I has shown that lot-for-lot approach gives good results, where setup costs are small, where these costs are high, as in the fabrication of components, least total cost, part period balancing, or period order quantity can provide satisfactory results (Plossl, 1994).

Good rules of thumb for a first cut at selecting lot-sizing techniques, which help minimize nervousness at very little potential cost, according to Plossl (1994), are as follows:

1. Use fixed order quantity at the MPS level. FOQ procedure passes along only order timing changes (and not changes in order quantity), thus, limiting changes to timing only. Secondly, nervousness can be damped by using firm planned orders.
2. At intermediate BOM levels, use lot for lot where setup costs are low, or use period order quantity or least total cost where setup costs are significant.
3. At bottom BOM levels (purchased items), use period order quantity.

Lot-Sizing in MRP I

In principle, MRP I is a lot-for-lot planning system that, if executed properly, results in minimum inventory. The principles of MRP I are consistent with just-in-time (JIT) and zero inventories (ZI), emphasizing minimal lot-sizes and minimal safety stocks. MRP I planning can be facilitated by production activity execution systems such as the Japanese *Kanban method* (Rice and Hatch, 1997).

- Most MRP I software packages include lot-sizing policy options. These options are useful in simplifying the ordering process for small, inexpensive category C items such as floor stock fasteners. (The designation of 'category C' is from the ABC Pareto classification system.)
- Users of the lot-sizing algorithms should recognize that the lot-sizing calculations do not consider major costs such as work center congestion, expediting, queues of work-in-process inventory, customer service, and other factors that in total may be much more than setup and holding costs. People are still busy devising methods for multilevel lot-sizing within MRP I based on the oversimplified costs of replenishment, stock-out, and holding costs, because the most important costs are difficult to measure and vary by company and situation. The manager should generally place little faith in lot-sizing formulas, especially those included in MRP I software that users are accepting on faith (Rice and Hatch, 1997).
- Another problem with lot-sizing algorithms in MRP I is that optimal lot-sizes for parent items will result in non-optimal lot-sizes for components and vice versa. Lot-sizing at one level invariably results in increased gross requirements at the next level of the product structure. So, larger lot-sizing near the end item level of the bill of materials cascades down through all levels. Thus, it turns out that *lot-for-lot is better than we might expect in actual practice, particularly at the intermediate levels in the BOMs.* This is especially the case, when a product structure has many levels, and the cascading effect becomes greatly magnified. This cascading effect can be mitigated to some extent for components and raw materials that are very common. When this is the case, again lot-sizing may be appropriate. As a consequence, many firms employ lot-sizing primarily at the end item and basic component levels, while intermediate sub-assemblies are planned on a lot-for-lot basis (Vollmann *et al.* 1997).

Work with simulation models has shown that small order sizes have advantages other than in reducing finished goods stocks. They are easier to schedule effectively, assist flow through plant, reduce lead times and increase the flexibility of response to changes in demand and there is evidence that EOQ formula, which ignores these additional benefits, overestimates the optimum size.

In conclusion, one should be skeptical about the lot-sizes that may be computed by the MRP I software. Management should have clear reasons for departing from lot-for-lot materials planning. A wiser philosophy is to pursue a just-in-time (JIT) policy, which tries to drive down setup costs and manage stock-out costs and so drives lot-sizes toward-lot for-lot, and to a lot-size of one, consistent with the basic MRP I output. MRP I by itself does neither recommend nor discourage use of any lot-sizing policy, it simply accepts the lot-sizing policy specified in the item master file and computes planned orders. If we were to enter a very high value of lot-size for a parent item—say 5000 for item A, MRP I would simply assume that we know what we were doing, accept the change, and respond by ordering enough components to make 5000 of part A. According to Oden and others (1993), *MRP I is not designed to ask the very important question: 'Are you out of your mind?'*

2.16.7 Manufacturing Calendar (or Shop Calendar)

Our Gregorian calendar serves satisfactorily for most social purposes, but has serious deficiencies for inventory planning and production scheduling. For example,

(i) it does not employ a decimal base,
(ii) months have an uneven number of days (28, 29, 30, or 31),
(iii) the pattern of holidays is irregular, and
(iv) workdays are not distinguished from others.

To avoid these problems in both manual and computer-based planning systems, special manufacturing/shop calendars are used (Plossl, 1994). A manufacturing or shop calendar consecutively numbers just those days of the year when production is actually planned, it excludes weekly off days (such as Saturdays and Sundays), holidays, periods of plant shutdown, etc. This makes it much simpler for both humans and computers to figure due dates, lead-time offsets, and other time-related calculations (John, 1993). Typically, weeks are given two-digit designations (00 to 99, or more popularly 01 to 52 to coincide with the calendar year) and working days are given three-digit designations (000 to 999).

It should be noted that with such an arrangement, the week designation numbers have no relation with the day numbers. Week numbers are only a means to conveniently identify different weeks. The former covers a 100-week 'year' and the latter a 1000-day *scheduling* 'year.' Such calendars are programmed into computers and can be distributed on cards or placards to people needing them or, better, translated by computers into calendar days before being shown to people. Figure 2.7 shows the popular 1000-day shop calendar, which counts only the working days.

Shop calendars make scheduling arithmetic straightforward. They allow points in time to be identified and lead times to be offset by means of simple addition or subtraction. With a 1000-day manufacturing calendar, the order release day can be computed using the formula:

Order Release Day = Delivery Day – Lead Time in Days

If release day has a negative value, add 1000 to it.

Time-phased data in computer-based planning and control systems are derived from an internal computer shop calendar. Better systems have internal programs to translate from shop to Gregorian calendars and back, avoiding human errors (Plossl, 1994).

Actual Month	*M-Week*	*MON*	*TUE*	*WED*	*THU*	*FRI*	*SAT*	*SUN*
APRIL 1999	14	977 29	978 30	979 31	980 1	Good Friday 2	3	4
	15	981 5	982 6	983 7	984 8	985 9	10	11
	16	986 12	987 13	988 14	989 15	990 16	17	18
	17	991 19	992 20	993 21	994 22	995 23	24	25
MAY 1999	18	996 26	997 27	998 28	999 29	000 30	1	2
	19	001 3	002 4	003 5	004 6	005 7	8	9
	20	006 10	007 11	008 12	009 13	010 14	15	16
	21	011 17	012 18	013 19	014 20	015 21	22	23
JUNE 1999	22	016 24	017 25	018 26	019 27	020 28	29	30
	23	021 31	022 1	023 2	024 3	025 4	5	6
	24	026 7	027 8	028 9	029 10	030 11	12	13
	25	VACATION WEEK (Subject to Change) 14	15	16	17	18	19	20
	26	031 21	032 22	033 23	034 24	035 25	26	27

The calendar shows the second quarter (that is, week 14 to 26) of the calendar year 1999. The company follows the 4-4-5 rule for representing the months in a quarter, that is, each quarter is of 13 weeks; the first two months of a quarter are of 4 weeks each, whereas the third month of any quarter has 5 weeks. Note that the last week of May is actually shown as the first week of June since it is the ninth week of the quarter and hence, according to 4-4-5 rule, should fall in the third month of the quarter.

Fig. 2.7 *Manufacturing Calendar (also called, 1000-Day, M-day, Planning, Production, or Shop Calendar)*

In addition to simplifying the arithmetic, manufacturing calendars express lead times in terms of work days rather than the usual weeks. Where there are many lead times that are really days or hours rather than weeks, this helps keep down the cumulative lead time (which determines the planning horizon) and provides much more realistic planning data with more accurate placement of planned orders and component gross requirements.

In addition to identifying the working and non-working days, most of the advanced MRP I programs further classify the working days into various 'Day Types.' For each day type, how much percentage of the day length (hours by day) is available for the work center can be specified. This percentage is used to determine the end date. Similarly for each day type, how much percentage of the capacity is available for the workday can also be specified. For example, if normally 10 resource units are available then a value of 70% for this field implies only seven resource units are available on all the days that are designated by this day type.

2.16.8 Lead Time Off-setting and Planned Orders

When the net requirements are converted to lot-sizes and backed off over the lead-time they become planned orders ready to be released to the vendors or the factory. MRP I covers net requirements with planned orders for future release. The principal value of planned orders for an item is providing advance data for a rigorous determination of its component requirements. They provide 'visibility' into the future, and also data for projecting required resources.

The net requirement for a part or assembly is offset by its lead-time to determine when it should be ordered. Because MRP I provides future recognition of need for an item, lead-time in MRP I is expressed as the elapsed time from release of an order until its receipt. MRP I requires a planned lead-time value to offset the planned receipt of an item, using backward scheduling, to properly place the planned order release in the correct time period.

Backward scheduling, which schedules to start each item as late as possible, has several obvious advantages. It will reduce work-in-process, postpone commitment of raw materials to specific products, and minimize storage time of completed components, and more importantly, it minimizes the time between the beginning of the expenditure of money for getting the product ready and recovering of the price from the customer reducing the time for which the money is tied up to the minimum, and thus improves the cash flow. Implementing the backward scheduling approach, however, requires a system that must have accurate BOM data and lead time estimates, some way to ensure all component parts are started at the right times, and some means of tracking components and sub-assemblies to make sure they are all completed according to plans.

In fact, *the combination of back schedules and gross to net explosion is the heart of MRP I* (Vollmann *et al.* 1997).

It is generally conceded, that this lead-time estimate does not require the same accuracy as on-hand balances and scheduled receipts. In fact, actual lead-times are extremely volatile so that the question of accuracy is something of a misnomer. However, there is often a temptation to intentionally overstate lead-times to be on the safe side. When lead times are overstated consistently at each level of the product structure, they are additive, increasing the cumulative lead time for production of the product (and consequently the planning horizon) and thereby, contributing to uncertainty and the likelihood of subsequent MPS changes with their accompanying costs and disruptions. Planned lead times should therefore be reasonable even when they cannot be made totally accurate.

The timing of the component gross requirements is linked directly to time-phased parent planned-order due dates. Planned-order quantities (lot-sizes) are determined by ordering policies described in Table 2.5. The order quantities usually include additional quantities of the item to take care of shrinkage, scrap, or yield. The net requirement of an item is generally divided by a yield factor (less than one) to get the required quantity.

2.16.9 Incorporating Scrap, Shrinkage, or Yield

Occasionally, there may be a known loss of material or parts due to the production process. These are the defective items coming out of the production processes that cannot be reworked and hence are irretrievably lost, however, they can be recycled. It should be noted, that here, we are *not* addressing the scrap which comes out as chips from the machining processes or as runners and risers from the molding/casting processes. It is the task of process planners to select the appropriate processes and raw stock or workpiece, such that, the scrap of this type is eliminated or minimized. The solution to the problem of defectives rejected by quality control is to increase the quantity of material started through the process, in order to have enough good items at the end. This adjustment may be referred to alternatively as scrap, shrinkage, or yield, depending on the industry, and must be accounted for in the MRP I plan. It is desirable to be able to specify a shrinkage against each component so that MRP I will order a correspondingly greater quantity of that component (Kenworhty, 1997).

Following are some of the definitions of related terms from APICS Dictionary:

1. *Scrap* It is the material outside of specifications and possessing characteristics, which make rework impractical.
2. *Fixed Scrap* It is the component quantity lost in setting up and/or the startup of the equipment. In the case of components lost in setting up equipment, the quantity lost is independent of the order size. An ideal system would permit the specification of both a *fixed*

Table 2.5

COMMONLY USED LOT-SIZING POLICIES

Sr. No.	*Lot-sizing Policy*	*Technique*	*Used/Suitable for*
1.	Lot-for-Lot Ordering (L4L), Discrete Ordering, or 'As Needed'.	Ordering only what is necessary. [L4L method is consistent with the just-in-time (JIT) philosophy of eliminating all unnecessary inventory.]	1. Expensive and/or bulky items, perishable items, or items that require little or no setup or ordering cost. 2. Items with highly discontinuous demand (for example service parts). 3. When marginal costs of setup is zero or very low due to availability of excess capacity and fixed labor costs; and the carrying costs are very high.
2.	Ordering Using a Fixed Lot-Size, or Fixed Order Quantity (FOQ).	Make the planned orders equal to the fixed order quantity (FOQ), which is often based on intuition; if the unsatisfied requirement exceeds the FOQ then the planned order is a multiple of the FOQ.	1. Where there is a particular price break or break-even amount the company wants to achieve. 2. Where some physical characteristic of the manufacturing process requires a fixed lot size. 3. Items that are manufactured or purchased in lots; and, in general, for items with conditions not recognized by lot-sizing algorithm.
3.	Economic Order Quantity (EOQ) or Least Total Cost (LTC).	Ordering the quantity that best balances the number of orders placed over time, and the size of the orders to yield the lowest total cost. EOQ calculation assumes that the demand is the same for each week.	Items that are replenished in batches and whose usage rates are low compared to batch size.

(*contd.*)

Sr. No.	*Lot-sizing Policy*	*Technique*	*Used / Suitable for*
4.	Dynamic Lot-Sizing.	A type of EOQ calculation which, unlike the usual EOQ, takes into account lumpy demand. It is done to optimize the planned order quantity.	Items whose demand is not the same for each week.
5.	Fixed Period Requirement (FPR), or Periods of Supply (POS).	Equivalent to the old rule of ordering 'X months' supply,' except that it determines the supply not by forecasting but by adding up discrete future planned net requirements. FPR, unlike FOQ, keeps the ordering interval constant but varies the order quantities.	No stock residues at the end of the fixed period. It can place orders that are unrealistically large or small unless order modifiers like minimum quantity and maximum quantity are incorporated.
6.	Period Order Quantity (POQ), or Economic Time Cycle.	Gross requirements over a specified period are lumped together into a single planned order as in FPR above but the ordering interval is computed using the logic of EOQ.	Necessary for certain bucketless (daily planning period) systems where the lot-for-lot ordering rule may create several orders per week for each unsatisfied requirement.
7.	Specialized Ordering Rules.	Create a single planned order for a quantity of one. The due date is the date of the earliest unsatisfied requirement.	For one-time activities, such as new product/software development, tool design, etc.

quantity and a *proportional* quantity lost, but few systems are this sophisticated and it is usually necessary to base the proportional loss on a typical order size. A practical problem with the application of fixed scrap is that it occurs every time a machine is setup and *not* necessarily every time an order is released. It is quite possible that a new order is released to a work center on which the last item processed was the same item, and hence, no setup is required and, thus, no fixed scrap occurs. Thus, to decide whether or not to apply the fixed scrap, the system should keep track of the last item processed on the workcenter which is extremely difficult if not impossible.

3. *Scrap Rate or Scrap Factor* It is the percentage factor in the product structure used to increase gross requirements to account for anticipated loss within the manufacturing of a particular product.
4. *Shrinkage* It is the reduction in actual quantities of items in stock, in process, or in transit. The loss may be caused by scrap, theft, deterioration, evaporation, etc.
5. *Shrinkage Factor or Shrinkage Rate* A percentage factor used to compensate for the expected loss during the manufacturing cycle of an item. This factor differs from the scrap factor, in that, it affects all components of the item, where the scrap factor relates to only one component's usage.
6. *Yield* It is the ratio of usable output from a process to its input. It is a percentage used to adjust planning data in order to meet actual requirements.

The scrap or shrinkage factor is the inverse of a yield percentage. Whereas, the yield percentage indicates the quantity of 'good' material expected from a process, the scrap factor indicates the quantity expected to be lost, or 'bad.' Following are the formulas used for adjusting the planned quantity:

$$\text{Quantity Planned} = \text{Quantity Needed} \div \text{Yield Percentage}$$
$$= \text{Quantity Needed} \div (1 - \text{Shrinkage Factor})$$

Both, the yield percentage and shrinkage factor are determined based on previous experience and analysis of the production process.

Two Generally Used Methods for Handling Scrap, Shrinkage, or Yield in MRP I

1. Applying Shrinkage/Yield during BOM Explosion Process One method is to carry the adjustment factor as part of the BOM data and apply it to the usage quantity during the explosion process. In effect, adjusting the gross requirements upward to account for the expected loss. One advantage of using this method is that different applications of a product or part could be given different loss (scrap) estimates when appropriate. Thus,

$$\text{Exploded requirements including scrap} = \text{Actual exploded requirements} \div (1 - \text{Scrap Factor})$$

2. Applying Shrinkage/Yield to the Planned Order Receipts The second method would include the scrap, shrinkage, or yield factors as part of the item master data in the item master file, along with the inventory status data. In this case, the percentages would be applied when balancing the MRP I record by applying it to the planned order receipt in order to inflate planned order release quantities to take care of shrinkage. Thus,

$$\text{Planned Order Release} = \text{Planned Order Receipt} \div (1 - \text{Shrinkage Factor})$$

It is a fairly straightforward procedure to use any lot-sizing procedure to determine the lot-size and then adjust the result to take into account the scrap allowance. One issue that arises is whether the quantity shown on the shop paper (and as a scheduled receipt) should be the *starting* quantity (that includes the expected scrap) or the *expected finished* quantity (excluding the anticipated scrap). Practice suggests using the starting quantity. To effect good control over open orders, the *total* amount, including the scrap/shrinkage allowance, should be shown on the shop order, and the scheduled receipt should be reduced as actual yield losses occur during production. This requires, however, that each actual occurrence of scrap be transacted and reflected in updated plans. The procedures for issuing scrap tickets when scrap occurs and procedures for ensuring good parts are accurately counted into inventory must be in place. We have to tightly control the transactions as even an error of as little as one piece in the MRP I records can cause severe problems (Vollmann *et al.* 1997).

Figure 2.8 illustrates the concepts of shrinkage and scrap factors and how they are applied. Since, we are not concerned with the items scrapped at the vendor's shop we ignore it and do not include it while calculating the planned order release quantity. It is assumed that the vendor, the shop, or the work center, would make sure that the quantity equal to the planned order release is dispatched.

- When the scrap is more than the scrap allowance provided, the deficiency can be rectified by increasing the size of scheduled receipts, or creating new planned orders to meet requirements.
- When scheduled receipt has less scrap than the allowance provided, the system will simply accept the excess quantity. The next MRP I run might reduce the quantity of or suggest deexpediting the next scheduled delivery.

From the MRP I user's point of view, the most important thing to note is that the scrap/shrinkage factor stored in the item master file (s_i in the above example) would be applied to the aggregate quantity of a particular item, and, in effect, would inflate the planned order release quantity. On

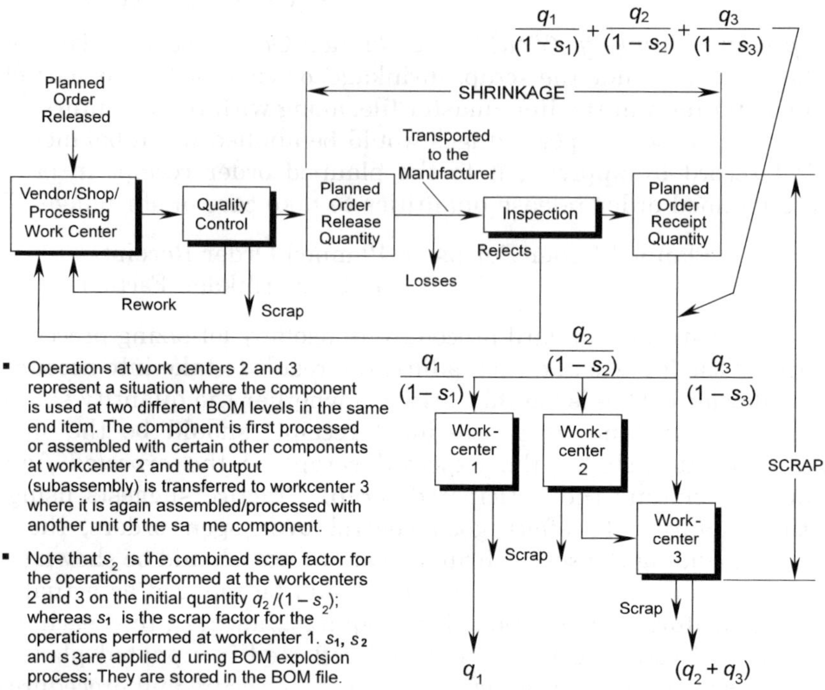

- Operations at work centers 2 and 3 represent a situation where the component is used at two different BOM levels in the same end item. The component is first processed or assembled with certain other components at workcenter 2 and the output (subassembly) is transferred to workcenter 3 where it is again assembled/processed with another unit of the sa me component.
- Note that s_2 is the combined scrap factor for the operations performed at the workcenters 2 and 3 on the initial quantity $q_2/(1-s_2)$; whereas s_1 is the scrap factor for the operations performed at workcenter 1. s_1, s_2 and $s_3$are applied d uring BOM explosion process; They are stored in the BOM file.
- S_t is the shrinkage factor applied to the planned order receipt quantity irrespective of where it is assembled. It accounts for the losses during transportation and receipt, and storage. S_t is stored in the item master file.

Actual exploded requirements $= q_1 + q_2 + q_3$

Planned order receipt quantity = Exploded requirements including anticipated scrap *(Assuming that no netting is needed and that lot-for-lot policy is used)*

$$= \frac{q_1}{(1-s_1)} + \frac{q_2}{(1-s_2)} + \frac{q_3}{(1-s_3)}$$

Planned order release quantity = {1/(1 – st)} ´ Planned Order receipt quantity

$$= \frac{1}{(1-s_t)}\left\{\frac{q_1}{(1-s_1)} + \frac{q_2}{(1-s_2)} + \frac{q_3}{(1-s_3)}\right\}$$

Fig. 2.8 *Scrap and Shrinkage*

the other hand, the scrap factor(s) stored in the BOM file along with the usage quantity would be applied to individual applications of the same item in different parent items or at different levels. In effect, the requirements would be calculated by multiplying the parent item quantity by [usage quantity in the BOM ÷ (1 – s)].

2.16.10 MRP I as an Input-Output Process

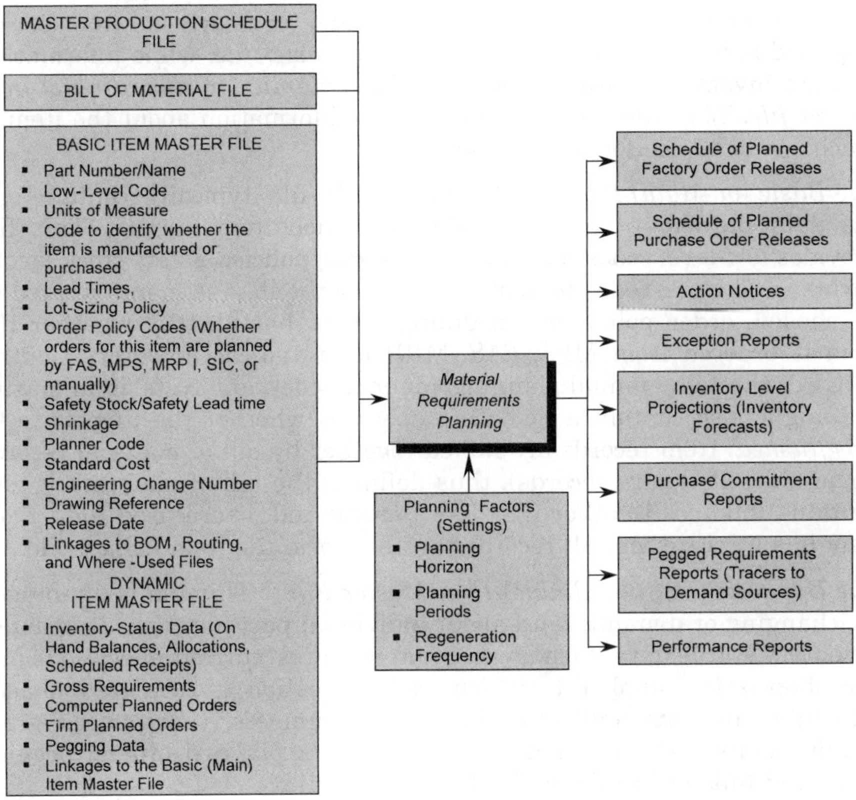

Fig. 2.9 *MRP I Inputs and Outputs*

2.16.11 MRP I Inputs

The MRP I scheduling calculations are ordinarily done using the electronic computer and an appropriate MRP I software package. In addition to the software, certain files are required on the computer. These files contain data that most manufacturing companies maintain whether they use MRP I or not as shown in the above Fig. 2.9. Following are the files required for performing MRP I computations:

1. Master Production Schedule File.

2. Item Master File It contains the 'master' record for each item. This file is keyed by item or part number and contains all relevant data about each manufactured and purchased item. In the days before computers, the only practical way to communicate the heritage or information about an item was to bury intelligence in the part number with significant digits. Today's information technology offers a better alternative. The descriptive intelligence about the item can easily be captured in the item master database. Sometimes the item information is divided into two files. The information that remains the same (or nearly so) from period to period is found in *the basic (main) item master file,* while information on item inventory status is found in the *dynamic (or subordinate) item master file.* With the item master file, the information about the item is much easier to use for search and sorting.

The Basic (or Main) Item Master File This file typically contains all the data needed to completely describe each item number. This file provides low-level codes, lead times, lot-sizing policies, safety stock levels, shrinkage factors, codes to identify whether the item is manufactured or purchased, order policy or scheduling codes that identify whether the item is ordered from MPS, FAS, MRP I, Statistical Inventory Control (SIC), or through manual planning, planner codes, and costs. Item master file might also contain a field that specifies whether the item is to be *backflushed.* Item records are linked together by bill of material records (or product structure records), thus defining the bill of material. It also contains linkages to other files like routing and where used files. Each item has a record and all records together form the item master file.

The Dynamic (or Subordinate) Item Master File This file is often used for changing or dynamic data about individual part numbers. It includes inventory status data, such as, on-hand balances, current allocations and the shop order number to which each allocation is tied, time-phased scheduled receipts and associated order numbers, time-phased gross requirements, computer planned orders, firm planned orders, pegging data, and linkages to the main item master file.

3. Bill of Material File It is the file that stores the BOMs for all the manufactured items; from this file the needed parent-component relationships and the quantities of the component items required are available for each parent item. The bill of material file is typically established on a single-level basis, with each part number linked only to the part numbers of the immediate components required to produce it, that is, the linkages are to one level farther down in the product structure only. By successively linking the part numbers, a full BOM for each item can be developed from the individual single-level linkings. Data elements held in this file usually include component part numbers required to make each individual item, number of each required, units of measure, engineering change numbers, effectivity and expiry dates, active/inactive coding, and where used information.

Description of Some of the Fields in the Main and Subordinate Item Master Files:

1. On-Hand Inventory Balance It represents the total quantity of goods that are physically present in finished goods or component storerooms or warehouses.

- Maintaining a computer record of on-hand inventory is a function of inventory management and is a prerequisite for MRP I.
- Existing inventory data not only must be present but also accurate. Accurate records are an absolute requirement for sound MRP I processing or the entire priority (material-planning) system will breakdown in the execution of the resultant plans.
- A campaign for inventory accuracy should always precede any attempt to implement an MRP I system (John, 1993).

2. Allocated Stock or Uncashed Requisitions (Used Only if the MRP I Grid does not Include a Time-Phased 'Allocations' Row) Allocated quantities are materials reserved within MRP I systems. Of the total amount of inventory on hand, it is frequently the case that certain reservations have been placed on some of it for one reason or another. Allocations mainly consist of components and sub-assemblies reserved for specific shop orders that have been *released,* but whose component parts have not been removed from inventory. Such reservations, or allocations, of on-hand inventory must be recorded as part of the inventory status data so that MRP I will not attempt to plan coverage of future requirements using inventory that is not really available.

Physical, Free, and Allocated Stock Physical stock is physically present in the stores and can be seen and counted. The physical stock increases when materials are received and put away, and decreases when materials are issued. The quantity out of the physical stock that is reserved for some reason is the *allocated stock* and the rest is the *free stock* that is actually available for planning. The relationship between the three is as follows:

$$\text{Physical Stock} = \text{Free Stock} + \text{Allocated Stock}$$

One of the primary examples of stock allocation has to do with the release of factory orders. Once the paperwork is prepared for a factory order, the records are changed to indicate that the order is no longer just planned but has now been released. However, it might be several days or even longer, before someone gets around to withdrawing the materials necessary to complete the factory order from the stockroom. During that period, the material is still physically on hand but must not be considered by MRP I in planned coverage of future needs. When the material is actually withdrawn, both the on-hand and allocated balances are reduced accordingly (John, 1993). It should be noted, that this data field is used only if the MRP I grid does *not* include a time-phased *allocations* row, which is described later.

In the MRP I process, allocated stock is subtracted from the on-hand inventory balance before the record balancing process begins. That is, on hand is reduced to reflect just that amount of inventory that is available for planning purposes to cover gross requirements. This process can be shown by the following formula:

Beginning Inventory Available for Planning = Beginning On-Hand – Allocated Quantity

Then the formula for calculating the projected on-hand row of the planning grid becomes:

Projected On-Hand = On Hand at the Beginning of Period Available for Planning+ Schedule Receipts – Gross Requirements;

or

Projected On Hand = Beginning On Hand + Scheduled Receipt – (Gross Requirements + Allocations)

Note, that the allocated quantity is subtracted from the on-hand inventory only once for the beginning inventory but not in subsequent periods (this does not apply to time-phased allocations described later). The formula for the 'Projected Available Balance (PAB)' row would also be modified to reflect on-hand available at the beginning of a period rather than simply the on-hand balance. It should be obvious from the above formula that the physical inventory present in the store would be larger than the projected on hand, that is,

The physical inventory actually present in the store in a particular period = Projected on-hand from the MRP I records + Allocations

Uses of Allocations The MRP I technique of allocating portions of on-hand inventory balances to 'earmark' or 'reserve' components for specific parent orders has other important uses:

(a) To Get the Correct On-hand Figures for Verifying the Accuracy of Inventory Records Because material planners need to know true available balances available for future planning, 'withdrawals' are posted immediately to computer records as allocations and available inventory quantities are reduced. Some time may elapse, however, before components are withdrawn physically, during which record balances (available for planning) and true stores (on-hand) quantities will be different. Allocations allow cycle counters to get the true stores figures.

(b) Identifying Present or Potential Shortages Allocating the quantity of each component needed to build a customer's order will reveal shortages when the allocation produces a negative quantity available. When a planned order release quantity appears in the current bucket (called *action bucket*) MRP I generates action message to materials planners recommending 'release the order.' Planners review these recommendations and decide whether or not to execute them. One of the considerations in deciding this is the availability of required components

and raw materials. Allocating the required quantities gives the clear picture of gross shortages helping the planners in deciding which orders should be executed. Planners may also choose to change the order quantities if amounts of available raw materials and components are not enough to make the quantity planned, reducing the quantity rather than delaying order release if this is preferable.

(c) Providing Audit Trails on Issues from Stores Allocated balances also provide an audit trail to trace whether or not authorized issues have been made from the physical inventory. Balances remaining in allocations fields beyond a set time period are reported for checking.

3. Safety Stock and Safety Lead-Time

Safety Stock It is normally employed to buffer against the uncertainties of demand and therefore should be utilized for independent demand of end products that are planned with the MPS. Since requirements for assemblies and components developed by MRP I are calculated, rather than forecasted, there is typically little need for safety stock for component parts—that is, intermediates or raw materials. Safety stock is best planned at MPS level, also, so that as it is exploded into component orders through the MRP I process, it forms matched sets.

Safety Stock on Dependent-Demand (MRP I) Items As a general rule, safety stock should not be carried on dependent-demand items and planned with MRP I. There are occasional exceptions such as:

(a) When items are subject to difficult supply, unpredictable yield due to quality or process problems, loss, theft, or similar causes of undependable availability. Similarly potential spoilage or unreliable suppliers might lead to inclusion of some safety stock for components and sub-assemblies. Such problems should be resolved by management action so that safety stocks can be eliminated. Eliminating safety stock for dependent-demand items is a major advantage of using MRP I, which reduces inventory investment (Rice and Hatch, 1997).

(b) Another exception is where an intermediate product is used not only in production of finished products but also to satisfy independent demand (such as for spare parts). In such cases, some safety stock would be called for to buffer against uncertainties of the independent demand of the service part component of the total requirements.

Safety Lead-Time Another way of providing safety is to provide safety lead-time to compensate for uncertainty regarding the receipt or completion of orders. Safety lead time is the adding of extra lead time to the planned lead time to allow for late delivery of either purchased or manufactured items. The amount of safety lead time should be based on the item's history of variation (lateness). This simply means scheduling a receipt one or two periods sooner than it is really needed. *Safety lead time is more practical than safety stock for dependent demand items,* in that the time-phasing adjustment is not in conflict but is still consistent with planned order quantity requirements. If a supply order is late, a safety stock of 200 will not help if the net requirement is 320. Safety lead time should be maintained as a separate number, and although a part of the planned lead time, it should not be built into the supplier or shop order due dates (Toomey, 1996).

Safety lead times, however, have two major disadvantages as describe below:

1. *Safety lead times put false due dates on orders destroying their credibility* When people discover that the dates are false (and they *will* find out!), the credibility of the system will be destroyed. A supplier who expends considerable extra effort to ensure that we receive a shipment on time, and then discovers that we didn't need it for another week or two, will remember that experience for a long, long time.
2. *Safety lead times increase WIP inventories, if used for shop orders* This increases queues at the various work centers. This action is totally antithetical to our attempts to reduce lead times (Oden *et al.* 1993).

In spite of the theory, we cannot recommend that a company mislead its suppliers, because excellent supplier-customer relationships require a foundation of trust. Thus, we recommend using safety stock rather than safety lead time for all buffering including raw materials (Oden *et al.* 1993).

Safety stock can be handled in two different ways by MRP I described as follows:

(a) In the first method the safety stock is treated as a different form of allocated stock and is subtracted from the beginning on-hand balance in the same way that allocated stock is processed. That is, on-hand figures in the MRP I spreadsheet do not include the safety stock. The argument in favor of this approach is that safety stock should always be kept on-hand and is not really meant to be available for planning. It should, therefore, be subtracted from the on-hand balance.

(b) In the second method, which is assumed in the APICS certification (CPIM) examination, the safety stock is *not* subtracted from the beginning on-hand balance that is available for the MRP I process, but the rule for determining net requirements is modified. That is, instead of indicating a net requirement when the on-hand balance becomes negative, the net requirement is indicated by the amount by which the on-hand balance falls below the safety stock level, even though some stock may be present (John, 1993).

Experience with MRP I has shown that when safety stocks and safety lead times are eliminated, cost savings and profit opportunities occur that are much more important than trading off holding versus stock-out costs. These savings include the benefits of less facility congestion, reduced expediting, improved customer and supplier coordination, improved response flexibility, improved quality control, reduced investment in working capital, and other benefits that are not accounted for in theoretical formulas. Because MRP I provides the visibility to know how much stock is actually needed, it is better practice to re-estimate demand frequently, maintain a realistic MPS, focus on data accuracy, execute the MRP I schedules effectively, and depend less on safety stocks. The idea is that effective planning diminishes forecasting errors and thus the need for safety stocks (Rice and Hatch, 1997).

4. *Lead-Time* It is the elapsed time from the release of an order until its receipt. In MRP I, lead time of an item is the time taken to move to that item from one level below in the BOM. It should be noted, that this is not the cumulative lead time. MRP I requires a planned lead-time value to offset the planned *receipt* of an item, using backward scheduling, to properly place the planned order *release* in the correct time period. In practice, many people refer to the estimated lead times as 'plugged' lead times.

5. *Lot-Size* Lot-sizing, as described earlier, means producing or purchasing a quantity that may exceed the immediate need due to economies of scale (such as quantity discounts in purchase or economic batch sizes in manufacturing) or due to constraints inherent in the production process. Policies regarding lot-size quantities and how they will be determined must be provided as an input to the MRP I process (John, 1993). Normally, an order will be placed in *multiples* of this quantity, but it can also represent a *minimum* or *maximum* order quantity or the type of lot-sizing technique (Russell and Taylor III, 1998).

6. *Low-Level Code* It is the lowest level at which the item appears in *all* the product structures.

7. *Item* The name or number that identifies the item being scheduled.

8. *Order Policy or Scheduling Code* This code specifies which system has responsibility for scheduling the item: FAS, MPS, MRP I, Order Point System, or Manual.

9. *Planner/User Code (or Planner Number)* This indicates which planners or users will review the item in their reports. The user codes enable reports to be tailored to the needs of the various departments, while the data is taken from the common files. Items are included or excluded depending upon the destination of the report.

10. *Whether the Item is to be Backflushed* In JIT, a common practice is to stock material at its point of use, therefore eliminating the cost of

the picking process. The inventory accounting in such systems is normally performed when a unit of the finished product is completed. When the product is completed, all components on the single-level BOM are deducted from inventory. This procedure is referred to as *post-deduct inventory transaction processing, explode-to-deduct*, or, more commonly *backflushing.*

Post-Deduct Inventory Transaction Processing or Backflushing It is defined by APICS Dictionary as a method of inventory book-keeping where the book (computer) inventory of components is *automatically reduced by the computer only after completion of activity on the components' upper level parent item,* based on what should have been used as specified in the BOM and allocation records. This approach has the disadvantage of a built-in differential between the book record and what is physically in stock (Cox III and Blackstone Jr, 1998).

This is in marked contrast to classic MRP I, where inventory activity is triggered by shop orders, and inventory accounting for components is performed at the time of stock requisition and issue before production begins. *Pre-Deduct Inventory Transaction Processing* is defined by APICS Dictionary as a method of inventory book-keeping where the book (computer) inventory of components is *reduced before issue,* at the time a scheduled receipt for their parents or assemblies is created via a bill-of-material explosion. This approach has the disadvantage of a built-in differential between the book record and what is physically in stock (Cox III and Blackstone, 1998). However, most MRP I systems now support post-deduct and point-of-use in the BOM module. To gain greater visibility of work-in-process, you can establish *deduct points* at the intermediate workstations, as well as at the end of the production line. A *deduct list* defines a list of components consumed at each deduct point, or pay point (Oden *et al.* 1993). Following are some more terms used in connection with backflushing.

Synchroflushing Usually, it takes several days for an end item to be reported as finished, from the time the order for it is released. However, the process of issuing or using the required components for production begins right from the day of release of the order. Thus, when using backflushing, the issued items will be deducted from their point-of-use inventory record several days after the actual issue or consumption of the components and raw materials. Thus, although the parts have been issued and/or consumed, the record shows they are still there in inventory. The solution to this discrepancy is to relieve the inventory of an item as soon as the particular operation in which that particular item is used is reported complete. To accomplish this, an additional field is added to the BOM record—operation used on. Production is reported after each operation. The inventory is relieved only for the component tied in the BOM record for the corresponding operation number. The inventory record, then, is in sync with the material movement. This backflushing method is sometimes called synchroflushing.

Single-Level Backflush A form of backflush that reduces inventory of only the parts used in the next level down in an assembly or sub-assembly (Cox III and Blackstone Jr, 1998).

Superflush A technique to relieve all components down to the lowest level using the complete bill of material, based on the count of finished units produced or transferred to finished goods inventory (Cox III and Blackstone Jr, 1998).

2.16.12 Planning Factors

1. Planning Horizon

The planning horizon indicates the number of future periods for which plans are made. The planning horizon selected for MRP I development is generally the same as that chosen for the MPS. However, it must be at least as long as the longest cumulative lead time among all products to be produced. Otherwise, the MRP I system is not able to release planned orders of items at the lower level at the correct time.

2. Length of Time Buckets

The period of time covered by each time bucket in the MRP I tableau must be established. Most often, this is one week. At times it might be longer for very long lead-time production or shorter where the manufacturing environment is very dynamic with frequent replanning, such as in just-in-time (JIT) situation. It is also possible to mix bucket lengths, for example, the first few days might be having daily time buckets, the next few might be weekly and further in the planning horizon, it might be monthly or quarterly time buckets.

The selection of a size for planning bucket is more of a convenience for people who must interpret and work with time-phased data than it is processing requirement, the size of lead times involved might be a consideration. Computer systems are certainly capable of dealing with specific dates for planning purposes. In some MRP I implementations, the time-phased grid has been done away entirely with planned quantities simply listed by date. This is sometimes referred to as 'bucketless' MRP I. However, regardless of how output is displayed, the basic logic of the MRP I process remains the same (John, 1993).

- A timing convention is that the current time is the beginning of the first period.
- An item is considered available (and hence, must be available) at the *beginning* of the time bucket in which it is required. This means plans must be so made that any replenishment order will be in inventory at the beginning of the period, in which the gross requirement for that order occurs.
- The immediate or most current period is called the *action bucket.* A

quantity in the action bucket means, that some action is needed now to avoid a future problem.

3. Replanning Frequency

How often should the MRP I application be performed? This question is answered primarily by the nature of the production environment and somewhat by the size of the time bucket selected. A dynamic environment, where changes are frequent or processes are unstable, calls for more frequent replanning than more stable environments.

MRP I can operate in what is referred to as a *regenerative* mode or a *net-change* mode. With regeneration, MRP I can be performed by a command to regenerate. It can also be set to be performed automatically at regular time intervals. Usually, the start of replanning cycle is set to coincide with the end of the time buckets used. That is, if the time buckets represent one week, then MRP I would be run weekly. In a net change mode, the replanning process is automatically triggered whenever a transaction which necessitates replanning is entered into the system.

2.16.13 MRP I Outputs

Following are the three principal outputs of the MRP I module:

1. Schedule of Planned Purchase Order Releases

This output schedules the quantity and time period in which purchasing should plan to issue a purchase authorization.

2. Schedule of Planned Factory Order Releases

This output schedules the quantity and time period in which production control should plan to issue a work order authorization. The schedule of planned factory order releases becomes a primary input for Capacity Requirements Planning (CRP).

- These two reports constitute the *Primary Report* or the *MRP Report.*
- In many cases, requirements change daily. A computer-based material requirements planning system automatically recalculates the requirements for sub-assemblies and components and recreates planned order releases to meet the shift in demand.
- The material planner works with three types of orders: planned, firm planned, and released.

Computer Planned Orders Computer planned orders are automatically scheduled and controlled by the computer, that is, it is created in the computer and will be deleted and recalculated whenever the system is regenerated. As gross requirements, projected available inventory, and scheduled receipts change, the computer recalculates the timing and quantities of planned order releases. The MRP I program recommends to the planner the release of an order, when the order enters the action bucket but does not release the order (Arnold, 1998).

Firm Planned Orders (FPOs) The computer planned orders are merely recommendations from the MRP II system. These orders will only be released *by the planner* when the correct time arrives. As soon as these orders are released to the shop or vendor they are re-classified as 'released orders' or 'scheduled receipts.' However, it is possible to alter manually the dates and quantities of the planned orders in advance before releasing them. Although the logic of MRP I is straightforward and can be implemented very quickly using a computer system; there are times, however, when a human planner needs to override the normal MRP I logic for one reason or another. This is where firm planned orders (FPOs) come in. The use of FPO provides a mechanism for human planners to override the replanning logic of MRP I computer programs, by freezing planned orders in terms of quantity, release date, and/or the due date. This is sometimes necessary to account for special situations that are temporary in nature, and do not call for permanent revision of input data or planning parameters.

The FPOs can, for example, be used to override the normal lead times or calculated net requirements when the production manager is aware of an unusual situation. FPOs are used to effect *one-time* deviations from the standard *item master data* or *system logic* without modifying the item master data, such as to compress lead times (expediting), to change standard order quantities, to change planned order due dates for specific orders. If a supplier is going to shut a plant for vacation, the planner will move the planned orders, which MRP I plans for the vacation shutdown period, to an earlier period, and then make them firm so that MRP I will not reschedule them back to the actual date of need. Similarly, if more rapid delivery is required to meet a special need (such as, using air freight), lead time can be reduced *for that one order.* If a work center will be overloaded for a week or two, some orders can be moved earlier or later (Oden *et al.* 1993). Other uses of FPO are to make one-time substitutions of standard routing or standard BOM with alternate routing or alternate BOM.

Daily changes in orders and production impact the system and cause the computer to revise the plans many times per day. This is called *system nervousness,* and it is a role of the planner to insulate other users from being confused by the dynamic, 'nervous' nature of the system. Firm plan orders provide a mechanism by which the plan is locked by the planner, and daily variables are not allowed to alter the agreed upon plan (Rice and Hatch, 1997). Thus, FPOs are also used to overcome system nervousness. Firming the planned order of a parent order will stabilize all lower-level components, hence, the greatest degree of stabilization is achieved by firming at the MPS level.

It should, however, be kept in mind that FPO command immobilizes that order in the schedule and prevents MRP I from putting another planned order into the 'frozen' time bucket, this may result in an increased net requirement not being fully covered. This capability, therefore, should be used judiciously for one specific planned order only, not for an item's whole planned-order release schedule (Plossl, 1994). FPOs include a heavy administrative cost. Because the computer is prevented from performing its normal MRP I plan for this order, a human must basically perform the entire MRP I review process for this part. Thus, each FPO adds a direct planning burden on a planner. If a planner is too busy to adequately review FPOs, the plan for the item in question, and perhaps all of its components, will become invalid (Oden *et al.* 1993). It should, however, be noted that although a firm order cannot be changed automatically by the computer, the computer will still recommend changes based on MRP I logic through action and exception messages. Often the changing requirements can be covered through safety-stock absorption or lead-time compression, rather than by jerking the system around (Toomey, 1996).

Releasing or Opening the Order (Order Launching) It is the process of releasing or opening orders to the shop (manufacturing orders) or to vendors (purchase orders). Since the objective of the MRP I is to have material available when it is needed and *not before,* orders for material should not be released until the planned order release date arrives. The MRP I logic is to provide a picture of the materials and capacity requirements of the planning horizon but to await specific action *by the (human) scheduler up to the last possible moment.* Hence, the releasing or launching of orders is recommended by MRP I through an action message suggesting planned order release and, therefore, conversion to a *scheduled receipt,* only when the planned order release reaches the current time period (called the *action bucket*).

- *Order release is authorization to make or buy* Releasing an order means that authorization is given to purchasing to buy the necessary material or to manufacturing to make the component. It should be noted that the MRP I system does not consider capacity nor does it use safety stock in its recommendation.
- *Check component availability before releasing a manufacturing order* Releasing of a manufacturing order requires several steps. Before a manufacturing order is released, component availability must be checked. The computer program checks the component inventory records to be sure that enough material is available. If sufficient quantities of each component are available, the shop order can be created. If the order is created, then the system *allocates* the necessary quantities to the particular shop order. If the material is not available, the computer program will advise the planner of the shortage through the action/exception report. The planner will review the situation and will decide a possible course of action. One such possibility might be to release a partial order. It should be

noted, that planners *can* release the order even if the system shows a material shortage.

- *Perform a manual capacity check* Planners should also perform a *manual* capacity check to ensure that capacity is available, prior to releasing the manufacturing order. Presumably, we have already performed CRP, which indicated that the necessary capacity will probably be available under normal conditions. It would be nice if we could do a quick capacity check using *finite loading* to ensure that capacity is really available. However, most CRP systems do not have this capability. All we can do is manually determine if there are any abnormal conditions that would change the available capacity, such as machines down or undergoing maintenance, operators out, or a large backlog. There is little point in releasing an order if capacity is not available; the new order will only increase the backlog. Once a planner releases an order to the shop, the planner has lost much control over that order. If an order will languish in queue, the planner is better advised to hold it in production control, until the backlog situation has been resolved (Oden *et al.* 1993).
- *Order release session is invoked* When the planner decides to release a shop order, typically an order release session is invoked and the order to be released is reviewed. The standard single-level bill of materials is read automatically from the database and displayed as the picking list for the released manufacturing orders. The standard routing is also read from the database if routings are used. The dates and quantities are taken from MRP I. Any of these data can be changed for this particular order if the planner decides to do so.
- *Planned order receipt is converted into a scheduled receipt by the system* As soon as the authorization to purchase or manufacture—that is, the order—is released, the planned order receipt is canceled, and a scheduled receipt is created in its place.
- *Computer allocates the required components and capacity (hours required at various work centers) to that order* When a manufacturing order is released, the computer will *allocate* the required quantities of a parent's components to that order. This does not mean the components are withdrawn from inventory, but, that the projected *available* quantity is reduced. The allocated quantity of components is still in inventory but they are not available for other orders. They will stay in inventory until withdrawn for use (Arnold, 1998). Similarly, the estimated hours required at various work centers to execute that order are allocated in the corresponding work center files.
- *Manufacturing order is made available to all authorized personnel* When the order is actually released, the hard-copy *manufacturing (or shop) order* [or just a small card with a turnaround (setup)

number in a 'paperless' system] is printed for distribution to the shop personnel. The on-line display of the open shop order would be available for all authorized personnel. Labor and materials may now be charged against the order until it is closed. The order would become a part of the prioritized departmental dispatch lists, also available on-line, after a shop re-scheduling program is run (usually with MRP I) (Rice and Hatch, 1997).

- *Picking List is created* Similarly, a *picking list* (also called, *picking ticket, disbursement list, material list, stores issue order, or stores requisition*) is typically created. Picking list is defined by APICS Dictionary as *a document that lists the material to be picked* (that is, to be withdrawn from stock) *for manufacturing or shipping orders.* The picking list facilitates the staging of materials support, for the production of the work order item. (*Staging* is the act of collecting and readying of materials well before the start of the operation, the purpose is to assure that all the items are available and there are no shortages or problems of any kind). The pick list calls for a specified amount of some item number to be removed from some inventory location, on some shop order, to be delivered to a particular department or location. When the picking list has been satisfied (inventory moved), the allocation is removed and on-hand balance is reduced accordingly. For each released order, a *pick date* is also defined, which is the start date of picking components for a production order. On or before this date, the system produces a list of orders to be picked, pick lists, tags, and turnaround (setup) cards.

- *Physical on-hand quantity should match what the records indicate* Availability checking, allocation, and physical stock picking are a type of double entry book-keeping. The result is, that the quantity physically on hand should match what the records indicate is available plus what is allocated. If they don't match, corrective action must be taken. The resulting accuracy facilitates inventory counting and other procedures for maintaining data integrity (Vollmann *et al.* 1997).

- *Drawings, bills of material, route sheets, move tickets, labor tickets, etc. are generated* With the shop order release, related items such as drawings, bills of material, route sheets, move tickets, labor tickets, and so forth are generated. Which specific items to be included are dependent on the execution system. The definitions of the terms *move ticket* and *labor ticket* are reproduced below from APICS Dictionary:

  Move Ticket It is a document used in dispatching to authorize or record movement of a job from one work center to another. It may also be used to report other information, such as the actual quantity or the material storage location.

  Labor ticket (also called labor claim or labor voucher) This is a factory worker's report, listing the jobs the employee has worked on, the number of pieces, the number of hours, etc., and often the amount of money to which the employee is entitled. A labor claim is usually made on a labor chit or time ticket.

- *Confirmation with the supplier and updating of data.* In case of purchase orders, the planner views the requirements screen, heeds the action messages and any late-breaking information, and then releases, changes, cancels, or expedites quantities with suppliers accordingly. When a purchase order is released, purchasing confirms the details with the supplier, and enters the additional data and changed data into the system. With just-in-time (JIT) purchasing, the quantities are 'releases' against long-term supplier agreements (or blanket orders) as opposed to complete purchase orders. Also, the order policy parameters in MRP I are adjusted to reflect the smaller quantities and more frequent deliveries associated with JIT purchasing.
- *Order review and release process normally takes only a few seconds for each item* For an experienced planner, working with well-designed MRP I software, the entire process of order review and release may take only a few seconds for each item if the release follows the system recommendation for dates and quantity and uses the standard BOMs and routing. MRP I user find, this is the case with the majority of items. If there are special problems such as shortages or late orders, additional time is required to analyze the situation and determine a course of action (Rice and Hatch, 1997).
- *Items can be reviewed in a variety of sequences* Software packages offer a variety of sequences in which items can be reviewed. Among those are item number order, description in alphabetic order, commodity type, supplier, purchase agreement number, shop, department and work center, and type and date of action messages (Rice and Hatch, 1997).
- *Order release process can be fully automated* With up-to-date MRP I software, the entire order release process can be automated to any degree as decided upon by the firm. Some companies choose to set the system to automatically release manufacturing and purchase orders on the dates calculated by MRP I. This should, however, be done only as an exception when the company is absolutely certain that the conditions are normal and the orders recommended by the MRP I won't need any reviewing. Review by an analyst prior to release of the order is and should be the normal practice.

Released Orders or Scheduled Receipts When an order is released it becomes an *open order* and is shown as a *scheduled receipt*. These are orders placed on manufacturing or on a vendor and represent a commitment to make or buy. For an order in a factory, necessary materials are committed, and work-center capacity allocated to that order. For purchased parts, similar commitments are made to the vendor. The scheduled receipts row shows the quantities ordered and when they are expected to be completed and available. Scheduled receipts on the MRP I record are open orders on the factory or a vendor and are the responsibility of purchasing and of PAC.

Open Order Files An entire set of files is maintained to support the scheduled receipts (open purchase and shop orders) in the MRP I system. For the purchase orders, we need open purchase orders, open quotations, a vendor master file, vendor performance data, alternate sources, and price/quantity information. The set of records needed to be maintained for shop orders in the factory includes data files describing open orders, routings, work centers, employees, shifts, tooling, and labor/performance reporting.

Closing the Order Finally, when the production is complete, all the issues have been recorded and all the product has been received into stock, the order is 'closed.' In case of purchase orders, when the goods are received into inventory and thus are available for use, the order is closed out, and a transaction is processed to increase the on-hand inventory and eliminate the scheduled receipt. With an integrated system there should be a single receipt transaction, which simultaneously increases the stock record and decreases the quantity on order (scheduled receipt).

3. Action Notices (or Rescheduling Messages)

If the manufacturing process is under control and the material requirements planning system is working properly, the system will work according to plan. However, sometimes there are problems that need the attention of the planner. A good MRP I system generates action notices and exception messages to advise the planner when some event needs attention. These are used in MRP I system 'to separate the vital few from the trivial many.' Notes or messages to a planner are produced indicating when to place orders, or when to expedite/deexpedite orders.

Transactions continually modify inventory status data, which, in turn, provides clues to action. The MRP I software nets, offsets, and explodes requirements and creates planned order releases. It keeps priorities current for all planned orders according to changes in gross requirements for the part. But it does not issue purchase or manufacturing orders or reschedule open orders. However, it does produce action messages suggesting that the planner should act, and what kind of action might be appropriate. Often, planners elect to look at items by exception only. That is, on a given day, they routinely review only those items that have action messages in the order review horizon. Of course, any item can be called up any time if necessary. Action messages are typically expedite, release, change, and cancel. Following are some examples of situations that will generate action/exception messages:

(a) Components for which planned orders are in the action bucket and which should be considered for release.
(b) Open (released) orders for which the timing or quantity of scheduled receipts does not satisfy the plan. Perhaps a scheduled receipt is timed to arrive too early or late, and its due date should be revised.
(c) Situations in which the standard lead times will result in late delivery of a zero-level part. This situation might call for expediting to reduce the standard lead times.

MRP I plans to fill the projected shortages (net requirements) by using existing orders before it suggests any new ones. In this respect, MRP I is designed to conserve the resources of the company (and help controllers sleep well at night). It prints an action message, which suggests that the material planner should reschedule the due date of scheduled receipts to the time period, where the net requirements exist. Thus, if scheduled receipts are in the wrong period, MRP I *suggests* rescheduling them to the period of need, where they will prevent PAB from falling below safety stock level. *MRP I continues its calculations, assuming that the planner will indeed reschedule the scheduled receipts.* In practice, however, planners actually reschedule far too few scheduled receipts, which causes the MRP I plans to be unrealistic. When MRP I has no more scheduled receipts available to use, it turns to its last available tool—the planned order (Oden *et al.* 1993). A planned order receipt is the net requirement rounded to the lot-size multiple.

Examples of Action Messages

Following action messages greatly assist master schedulers and material planners:

1. Reschedule out (Deexpedite) Whenever there is a scheduled receipt arriving earlier than needed to satisfy demands plus safety stock, the action message would recommend changing the due date later, to where it is needed. This avoids users from working on the wrong priorities, building unnecessary inventory, and consuming scarce resources needed for more urgent items.

2. Reschedule in (Expedite) When the need date precedes the current scheduled due date, the action message would recommend rescheduling the order earlier to avoid the potential shortage.

3. Cancel Order If none of the scheduled receipt is required, the recommended action would be to cancel it. Even if this is not possible because it has been started, preventing any additional costs being added is the next best possibility.

4. Release Order (Time to Order) Ordering material prematurely is undesirable because conditions may change. On the other hand, authorizing new orders in less than average lead time puts undesired pressure on the factory and purchasing to compress average lead times. Consequently, the order should be released on the date that we arrive at by moving backwards from the need date through the number of days equal to the average lead-time. This is the case when a planned order appears in the most immediate time period (action bucket).

5. Release Order with Insufficient Lead-Time.

6. Review Order Past Due Whenever a scheduled receipt should have been received prior to today, a warning message should be given to the planner. In a similar manner, if the time to order is late, that is, it should have been ordered earlier than today to give the factory or the purchasing average lead time, the planner needs to be aware of this potential problem.

7. Prepare to Release The time required to prepare paperwork (purchase and stores requisitions, shop order packets, etc.) and to check availability of tooling, test equipment, process specifications, and other resources can be a significant element of manufacturing lead time. MRP I can be programmed to signal the time to begin such actions before the actual release of the order using its forward visibility of planned orders. This element can then be eliminated from planned lead times, making them much shorter. This action message, however, is rarely used.

The action notices involve only the timing of orders (and *not* the quantities); changing quantities once work has begun on order is limited to splitting a lot into two or more sublots with different schedules. It is usually difficult and costly, and may even be impossible, to increase the quantity of an open purchase or work order. It is undesirable to split work orders because the additional setups aggravate capacity problems, which often are the causes of people's desires to split orders (Plossl, 1994). These rescheduling messages are required only in case of firm planned orders and released (open) orders, which cannot be rescheduled by MRP I. Computer-planned orders (CPOs), which have neither been made 'firm' nor released by the planner, are automatically wiped out and recomputed by the MRP I logic during its replanning run, and thus, are made to align properly with the net requirements, requiring no planner action.

During a replanning run, changed gross requirements may necessitate recomputation of projected on-hand data; these give clues to the action required by the planner. When net requirements can be covered by rescheduling of open (released) orders, MRP I will not generate a new computer-planned order but will recommend rescheduling of the nearest open order. MRP I logic creates a planned receipt in the same bucket where there is a net requirement; these planned receipts eventually become scheduled receipts when corresponding planned orders are released. Thus, every scheduled receipt should normally have a corresponding net requirement in the same bucket. This fact is used by MRP I logic to detect a mis-alignment after a replanning run. Following are the two tests that MRP I uses for identifying open-order misalignments:

1. Is there an open order scheduled in a period later than a net requirement?
2. Is there an open order scheduled in a period with no net requirement (the gross requirement is equal to or less than the actual on-hand quantity at the end of the preceding period)?

MRP I carries out these two simple tests whenever netting is recomputed. If a test is positive, the system generates the appropriate rescheduling message (Plossl, 1994), and presents it in the format as shown in Fig. 2.10.

Planner:				Run Date:		
Item Code/No.	Description (Item Name)	Order No.	Quantity	Action	On/From Date	To Date

Fig. 2.10 *MRP I Action Message Report*

- If an item does not need attention, it is not included in the report.
- Before deciding what action, if any, to take with respect to an item on the action report, the planner will normally also review the MRP I primary report and the pegging report for the item.
- Some MRP I systems can organize these action messages according to a priority system. For example, orders to be canceled may be listed in declining monetary values. Order past due may be listed in declining days past due.

4. Exception Reports or Messages

Exception Reports These reports point out various exceptional conditions encountered by the MRP I software *during a replanning run* that call for review and corrective action by management. Exception reports are generated to report errors, incongruities, and out-of-limit situations. Following are a few examples of exception messages:

- Order quantity too small.
- Past-due gross requirement included in the current period.
- Order quantity too large.
- Due date of open order is outside planning horizon.

- Order confirmed.
- Date of gross requirement beyond the planning horizon.
- Inventory below safety stock.
- Planned order offset into a past period but placed in current period. (This condition means an order should have been released in the past. Since it wasn't, lead times through the various levels must be compressed to meet the end item schedule.)
- Confirmation late.
- Excessive scrap.
- Inventory above maximum inventory.
- Allocated on-hand quantity exceeds current quantity on hand. (This condition is analogous to overdrawing a bank account and indicates that unless more inventory is received soon, the firm won't be able to honor all picking lists issued, and there will be a material shortage in the factory.)

Exception Messages In addition to such exception reports, individual exception messages can be generated by the system at the time when certain transactions are not accepted by the system. These messages list the reasons for rejection of transactions by the system. Some typical exception messages are (Plossl, 1994):

1. Part number is non-existent.
2. Transaction code is non-existent.
3. Part number is incorrect.
4. Actual receipt exceeds quantity of scheduled receipt by x per cent.
5. Quantity of reported scrap exceeds quantity on-hand.
6. Quantity of disbursement exceeds quantity on-hand.
7. Order being released exceeds planned quantity.
8. Number of digits entered in a field exceeds size of the field.

Transaction Messages The planner must tell the MRP I software of all actions taken that will influence the MRP I records. For example, when the planner releases an order, or a scheduled receipt is received, or when any change to the data occurs, the MRP I program must be told. Otherwise, the records will be inaccurate, and the plan will become unworkable.

5. Inventory Level Projections (Inventory Forecasts)

Projected on-hand balances in future time periods in component displays are expected inventory balances and are the best possible budget projection of storeroom inventories *if planning is sound and execution is effective.* Open (released) and planned order data provide bases for calculating work-in-process inventory. Finished product inventory projections are shown in time-phased order point displays where this technique is used. Inventory capital investment projections are easily made from MRP I data giving projected on-hand quantities of each item multiplied by its standard cost, summarized for all items by time period over the planning horizon. Prior to MRP I, budgeting inventories was independent of operating plans, being based on desired improvements in turnover rates, and was inflexible. Flexible budgets can be linked directly to operations using MRP I data, and can be revised to reflect significant changes (Plossl, 1994).

6. Purchase Commitment Reports

Open (released) and planned future purchase orders scheduled in time periods can be converted into purchase commitment reports for each supplier. The prices of such orders added in each time period for all suppliers represent the total cash the customer will need to pay in that period for these contracted deliveries (Plossl, 1994).

7-8. Pegged Requirements and Performance Reports

(These are described in Sections 2.16.17 and 2.17.19 respectively)

2.16.14 Typical MRP I Spreadsheet

			Action Bucket							
On-Hand: Safety-Stock: Lead Time: Lot-Size: Low Level Code:		Periods								
		Past Due	1	2	3	4	5	6	7	8
Item	Gross Requirements									
	Scheduled Receipts									
	Allocations									
	Projected On-Hand									
	Projected Available									
	Net Requirements									
	Planned Order Receipts									
	Planned Order Releases									

Fig. 2.11 *Typical MRP I Spreadsheet or MRP I Matrix*
(Adapted from John, 1993)

The MRP I spreadsheet is a universal representation of the status and plans for any single item (part number) whether raw material, component part, or finished good, displaying the information fields as shown in

Fig. 2.11. In this type of spreadsheets, the requirements, inventory status data, planned order releases etc. are displayed in 'bucketed' horizontal arrays divided into time periods (time buckets) of days, weeks, or months. Modern MRP II systems allow the time buckets of different sizes in a single planning horizon, for example, the buckets in the immediate future could be daily to facilitate detailed planning, this could be followed by weekly time buckets, whereas time buckets in the distant future could be monthly periods to give the aggregated figures giving overall view without giving too many details. For distant future, the plans showed should not be so detailed that overall patterns get obscured. The most immediate or current bucket is called the *action bucket.*

Another method of time phasing used in MRP II systems is by using date/quantity data in which the requirements, inventory status data, and planned order releases are displayed by specific dates in vertical 'bucketless' arrays (described later as 'Bucketless MRP I Systems'). The specific method of time phasing employed will determine the way internal arithmetic is carried out by the system. However, the method of *displaying* time-phased data (on paper or on CRT screen) can be selected independently. Most MRP I programs are designed with requirements and inventory data entered, stored, and internally processed by date/quantity (that is in bucketless format) but displayed in time-bucket format. The reason is that the bucketless format, although more efficient from computer processing point of view, lacks the ability to 'make the logic obvious' for easy user understanding and action.

Rolling Schedule

The entire planning process is performed on the basis of a rolling schedule. That is, when you perform MRP I next week, what is currently period one will become past due; and what is currently past due will disappear—it would have rolled into the past. What is now period two will become period one, and likewise period eight will become period seven. Period nine, which is not visible this week because it is beyond the planning horizon, will have rolled into the schedule next week as period eight.

Past-Due

It is the past-due time bucket (period). If an order appears in the past-due time bucket, the schedule is infeasible and an error message will be generated. Projected on-hand entries in this column represent beginning inventory.

2.16.15 Bucketless MRP I Systems

MRP I data is not always displayed in the spreadsheet format shown in Fig. 2.11. Also, it is not the way the data is stored in the computer files or displayed in reports. A bucketless MRP I system specifies the exact release and due dates for each requirement, scheduled receipt, and planned order. Storing requirements data by date is often more efficient from the standpoint of the computer resource. However, the computer can still display the same requirements data in bucket format whenever required. One of the reasons for this is that there are often many blank buckets where no requirements or orders are present. Bucketless MRP I systems are often used for storing requirements data in computers and to facilitate the use of planned lead times that are less than a whole period in length. This is especially useful when lead times need to be expressed in days, rather than in weeks. Some manufacturing orders, and even some purchases, can be accomplished in very short times, but artificially long lead times caused by the use of weekly buckets can result in an unrealistically long cumulative lead time. Figure 2.12 shows a typical bucketless MRP I report, which is also known as 'vertical display' or 'list method.'

PART NUMBER:.................................... PART NAME:..

Lot-Size:
Lead Time:
Beginning On-Hand:

Low Level Code:
Safety Stock:
Allocations:

Date	Week	M-Day	Reference (Work or Purchase Order No.)	Parent (Pegging)	Gross Requirements	Scheduled Receipts	Projected Available	Planned Order Releases

Fig. 2.12 *Sample Bucketless, Vertical, or List MRP I Display*

Bucketless MRP I systems are a better way to use computer. The approach allows better maintenance of lead-time offsets and provides more precise time-phased information. The approach is consistent with state-of-the-art software, and many firms now use bucketless systems. These systems are run on a daily basis (regeneration or net change) utilizing bucketless concepts. With bucketless systems, the planning cycle itself is bucketless, that is, plans are revised as necessary, not on a periodic schedule, and the entire execution cycle is also shortened (Vollmann *et al.* 1997). A comparison between the horizontal display as shown in Fig. 2.11 and the vertical display as given in Fig. 2.12, has been listed in Table 2.6.

Table 2.6

Comparison of Horizontal and Vertical MRP I Displays

Horizontal Display	*Vertical Display*
▪ Intuitive, easy to understand. ▪ Condenses considerable information in a small space. ▪ Has been the industry standard.	Can be far too long, owing to the detail, especially for common fasteners and raw materials.
Longer time buckets hide the problems, create ambiguity about requirements and need dates.	Uses dates to show expected stock transactions, can spot impending shortages inside a week.
Not easy to see any details which would tell the causes of the requirements.	All pegging details are available in the same report.

2.16.16 Fields Used in the MRP I Spreadsheet

The conventions used to define and understand when an activity is taking place within the time period are as follows:

1. All scheduled and planned receipts are due at the *beginning* of the period.
2. The anticipated usage, that is the gross requirements, may be used at any time *during* the period.
3. The projected inventory is as of the *end* of the period.
4. All planned releases are released at the *beginning* of the period.
5. Lead times are in multiples of the time period.

1. Gross Requirements

It is the statement of the anticipated total demand, or usage, for an item by time period. For an end item, this quantity is obtained from the MPS. For a lower-level item, it is derived from the *planned order releases* of its parent. The gross requirements are time-phased, which means they are stated on a unique period-by-period basis rather than aggregated or averaged. This method of presentation allows for special orders, seasonality, and periods of no anticipated usage to be explicitly taken into account.

- Gross requirements may come from more than one source if the component is used in common, on more than one parent. The explosion logic must, therefore, calculate the requirements of the components up to the lowest level (given by the low-level code) in *all* the parent items in which the component is used and add them all to the corresponding periods in the component gross requirements row.

- Gross requirements include independent demands, if any, in case of components or sub-assemblies. For proper handling with MRP I, the forecasted independent demand values for the components and sub-assemblies are placed in the gross requirements row of the component *before* the BOM explosion process begins. Then the MRP I logic adds calculated dependent demands (if any) to the independent demands so that the total demand for the component is properly derived.

2. Scheduled Receipts (Open Orders)

It is the quantity of material that is already ordered (from work orders or purchase orders) and is due at the *beginning* of the period in which it is shown. It is scheduled to be received either physically into inventory or otherwise completed (even though the item may never enter a stockroom but go directly to the next production process). Once a planned order is released, it becomes a scheduled receipt. The scheduled receipt is shown in the time bucket at the *beginning* of which it is to be received, that is, the items received as scheduled receipt would be available to satisfy the gross requirements anytime during the period.

3. Allocations (Time-Phased Allocations)

In certain circumstances, especially with long lead-time production, it may be desirable to allocate on-hand quantities during periods farther out on the planning horizon rather than just at the beginning (by the 'allocated stock' field in the 'inventory status data'). This might be the case, for example, when special customer orders are entered in advance of normal lead time, such as a customer order for a service part component, or when material needed for a released order will not be withdrawn from inventory until several periods into the production process.

Time-phased allocations require another row in the MRP I grid so that allocated quantities can be spread across the horizon as needed. This, however, eliminates the need for the 'allocated stock' field in the 'inventory status data.' The balancing formula would subtract the allocated quantity from the quantity available for planning at the beginning of the appropriate period rather than just from the beginning inventory. In effect, it acts as an additional requirement. Figure 2.13 shows an MRP I spreadsheet with an allocations row.

4. Projected On-Hand

The projected on hand row does not reflect *planned receipts,* only *scheduled receipts* are included.

Projected on Hand = On hand at the beginning of period + Scheduled Receipts – (Gross Requirements + Allocations)

A negative projected on hand in a particular period predicts a shortage

On-Hand: 500 Safety Stock: 10 Lead Time: 3 Lot-Size: 800 Low Level Code: 1		Action Bucket / Periods								
		Past Due	1	2	3	4	5	6	7	8
Item	Gross Requirements		190	10	590	190	590	190	590	10
	Scheduled Receipts				800					
	Allocations		20			5		25		
1	Projected On-Hand	500	290	280	490	295	–295	–510	–1100	–1110
	Projected Available		290	280	490	295	505	290	500	490
	Net Requirements						295		300	
	Planned Order Receipts						800		800	
	Planned Order Releases			800		800				

Fig. 2.13 *MRP I Spreadsheet Showing Time-Phased Allocations* (Adapted from John, 1993)

of so many units of the particular item in that period, and that incidentally becomes the net requirement in that period. In the absence of lot-sizing, the net requirement will be shown as a positive value that corresponds to the *additional negativity* of the projected on-hand in the same period. That net requirement needs to be covered by a planned order receipt in the same period for which a planned order release should be entered in a period, which is offset from the current period by the lead time. In the absence of the lot-sizing, the planned receipt and the corresponding planned order releases will be identical to the net requirement shown for the same period. In case some lot-sizing policy is used, the net quantity would be rounded off to the lot-size or a multiple of it.

5. Projected Available Balance (PAB)

It is an inventory balance projected into the future for the item at the *end* of the period in which it is shown, PAB *includes* planned orders.

PAB at the end of current period
= PAB at the end of previous period
+ Scheduled Receipts for current period
+ Planned Order Receipts for current period
– Requirements (that is, Gross + Allocations) for current period

Projected available calculation is required because the projected on-hand, while simple to calculate, is very confusing in several respects. You can't really have a negative amount of goods on hand—either you have some or you have none. Furthermore, when projected on-hand is negative in some period, you are expected to do something about it, such as place an order to cover the projected shortage. And yet, that planned action is not reflected in the projected on-hand row of the MRP I record. Projected available is an alternative to projected on hand that reflects planned receipts as well as scheduled receipts. The projected available is not allowed to become negative by planning planned order receipts and releases to cover the projected shortages.

The timing convention for PAB is the *end* of the period; that is the PAB shown is the quantity *after* replenishment orders have been received and gross requirements have been satisfied. For this reason, the PAB row has an extra time bucket shown at the beginning. This bucket shows the balance at the present time. The PAB shown in a time bucket is actually available to meet the gross requirements in the next and succeeding time buckets. The term projected *available* balance is used, instead of projected *on-hand* balance, for a very specific reason. Units of the item might be on hand physically but not available to meet gross requirements, because they are already promised or allocated for some other purpose (Vollmann *et al.* 1997).

6. Net Requirements

It is the net number of items that must be provided and when they are needed. It appears in the same time period as gross requirements. It is calculated as follows:

Net Requirements
= Gross Requirements + Time-Phased Allocation in the Period
– [PAB (Excluding Safety Stock) from Previous Period
+ Scheduled Receipts]

The above formula assumes that the safety stock is not included in the PAB figures.

If PAB is to include safety stock, then the following formula should be used:

Net Requirements = Gross Requirements + Time-Phased Allocation in the Period + Safety Stock – [PAB (Including Safety Stock) from Previous Period + Scheduled Receipts]

Net requirements of an item must be met either by a scheduled receipt (open order due) or by a planned order receipt for that item. It should be noted that *a negative net requirement is considered as zero.*

7. Planned Order Receipts

It is the total quantity of items on replenishment orders that MRP I has *planned to receive* in a given period to cover net requirements. It is the same as net requirements adjusted for lot-sizing. If lot-sizing is seldom used or lead time is negligible, this row can be deleted from the matrix.

8. Planned Order Releases

It is always the same quantity as the planned order receipt but it is offset backward in time by the length of the lead time. This is called the lead-time offset. It shows when an order should be placed (that is released) so that items are available when needed. *Planned order releases* for an item at one level in the BOM generate *gross requirements* for the components going in that item at the next lower level in the BOM. That is, planned order release quantity of the parent multiplied by the usage factor of the child in the parent becomes the exploded requirements for the child in the same time bucket, of the MRP I tableau of the child. The recommended planned order releases must be released at the *beginning* of the period in which they are shown. The planned order releases are *not* shown in the scheduled receipt row because they haven't yet been released for production or purchasing. Planned orders are much easier to change than scheduled receipts. Not converting planned orders into scheduled receipts any earlier than necessary has many advantages.

To summarize, MRP I plans a part, period by period, starting with period 1. The MRP I procedure is to first balance each item record having a low-level code of zero. Then explode the parent planned order releases to all their components, multiplying by the usage factor shown in the BOM and adding the result into the gross requirements row of each component record in corresponding time periods. Then all items with a low-level code of one are balanced, followed by explosion to the next level of component gross requirements. This process continues, alternating between balancing and exploding until all levels have been processed. Once all gross requirements data are available for a particular record, the individual record processing logic is applied and the planned order releases for the part are passed down (with appropriate multiplier) as gross requirements to its components. The result is a schedule of planned order releases to cover predicted shortages for all items, both purchased and manufactured (John, 1993).

The MRP I records take proper account of gross to netting. They also incorporate back scheduling and allow for explicit timings, desired lot-sizing procedures, safety stocks, and part commonality. Even more important, however, is the independence of item planning. With MRP I approach, the person planning a parent item need not explicitly coordinate his planning with planning of the component parts. The MRP I system accomplishes the coordination and ensures that whatever is done to the MRP I records for the parent item, will result in a set of planned order releases, that the system will correctly pass down as gross requirements to its components.

2.16.17 Pegging and Where-Used File

Engineering frequently designs families of end products that have a common component. MRP I must pool the demands for these items. The contents of gross requirements buckets represent total requirements, the sources of which are obscure. Pegging is simply a way to trace *where the requirements come from* after the MRP I plan has been developed. It is a selective 'BOM in reverse' or a 'where-used' list. It would say, for example, that the locating pins go into the crankcases, which go into the engine, for which there is a particular order. Pegged requirements table shows for each component the quantity, the higher-level item and time period from which the requirements are coming. Thus, pegging relates all the gross requirements for a part to all the planned order releases or other sources of demand that created the requirements.

Pegging data represents a file of information that must be created during the requirements explosion to keep a record of the source of gross requirements figures in each period. That is, each gross requirement figure in each period for each record is 'pegged' to its origins. This information is very useful when using the output of MRP I to take action on exceptional conditions.

The pegging records contain the specific part number or numbers of the sources of all gross requirements. At level 0, for example, the pegging records might contain the specific customer orders that are to be satisfied by the gross requirements in the end item time-phased records. For lower-level part numbers, the gross requirements are most often pegged to planned orders of higher level items, but might also be pegged to customer orders if the part is sold as a service part.

Single-Level Pegging

When the requirements are pegged to their immediate source, the planned order release of the next higher parent, it is referred to as single-level pegging. An example is provided in Fig. 2.14 (left-hand side), in which the demand for item X comes from parent items A, C, and D, and from an interplant or service part order 38447. MRP I programs have fields labeled 'parent record' that contain file addresses of parents rather than merely item identification. This facilitates fast retrieval of these records, which is the key to effective use of pegged requirements by busy planners.

Full Pegging

The single-level pegging described above requires a succession of inquiries to trace lower-level item demand to the related specific end item in the MPS. In BOMs having many levels, this can be very tedious. Full pegging allows planners to link item demand to the MPS with a single inquiry. Thus although from single-level pegging reports, the pegs could be traced all the way up the product structures to the MPS end-item requirements that started the whole process, at times, a shortcut is desired to identify the end item directly. In this case, the computer can do the tracing, and a separate display or report is produced that directly pegs any item's gross requirements to the end item. This is called full pegging. Figure 2.14 (right-hand side) shows sample full pegging report.

Peg Record—Item X (Single-Level Pegging) (Adapted from Plossl, 1994)				Item No./Name:......................(Full Pegging) (Adapted from Oden, 1993)			
Period	Qty.	Parent Record (Source)	External Order	Period/ Date	Qty.	Source of Requirement	Order Number
1	20	Assembly A					
2	15	Assembly A					
3	20	Assembly C					
4	10		No. 38447				
6	15	Assembly D					

Fig. 2.14 *Pegging Reports*

Pegging versus Where-Used File

Pegging information can be used to go up through the MRP I records from a raw material gross requirement to some future customer order. In this sense, it is the reverse of the explosion process. Pegging is sometimes compared to where-used data, which indicate for *each* part number, the part numbers of *all* the items on which the part is used. These items may or may not be scheduled for production/purchase in particular time periods. Pegging, on the other hand, is a *selective* where-used file. Pegging shows only the specific part numbers that produce the specific gross requirements in each time period, that is, it pegs only to those items which are *on order* in each time period. Thus, the pegging information can be used to trace the impact of a material problem all the way up to the order it would affect.

Creating Pegging and Where-Used Reports

1. Pegging Method This method records the pegs as the MRP I explosion process is being carried out. As gross requirements for an item are being calculated, based on needs of the parents, and orders planned to cover those needs, the needs of the parents entering into calculations are retained and form the pegging report.

2. Where-Used Method This method uses the where-used logic associated with the BOM files whenever a query is received regarding pegging. The system determines what parents the item has, searches the MRP I to determine what orders are planned for the parents, and then shows how these orders created the requirements.

These two approaches provide similar, but not identical, results. The second approach indicates the part numbers of *all* items on which that part is used, that is, it creates the where-used list. The first approach is more selective; it shows only the specific part numbers that, being *on-order* in the MPS, produce the specific gross requirements in each time period, that is, it creates the pegging report.

2.16.18 Bottom-up Replanning

Bottom-up replanning is the process that closes the planning loop. Feedback, which is the opposite of 'silence is approval,' communicates problems that the users have with the current plans. The flow of information is going in the opposite direction, that is from the user, back to the planning system. Bottom-up replanning is also the process of using the pegging data to solve material shortage or other problems arising out of:

(i) unexpected sales,
(ii) unexpected scrap and rework in the factory,
(iii) poor quality,
(iv) equipment breaking down or tooling not available,
(v) inventory adjustments, or
(vi) late delivery from suppliers who cannot meet their commitments.

Using the pegging report, the planner determines the requirements that caused the order. With this information, he can explore alternatives at this level and at higher levels in the BOM. The pegging data also traces the component in short supply to all the orders that can possibly be affected by the shortage of that component. Based on this, the planner can decide which end items in particular orders can be allowed to be affected by the shortage, or what kind of corrective actions are possible. The planner can also apply certain priorities to certain orders. Without pegging, it would appear that 'everything is late,' and, because of lack of visibility, an order to replenish routine safety stock might share equal priority with a critical order to an important customer. The visibility provided by pegging can be used to establish the correct priorities.

When a material availability problem arises, analysis of all inter-dependencies—vertical and horizontal—is required. Some of the solutions to material shortage problem can be:

(a) Borrowing from Additional Stock of the Parent Use pegging list, to check the parent of the component in short supply to see if there is any additional stock, that can be 'borrowed' temporarily to meet production requirements. Availability of parents of those items can subsequently be pegged in the same manner, working up through the BOMs to the top.

(b) Borrowing from Service Parts. Ordinarily, there would not be any safety stock unless some of the items were stocked service items. By careful examination of pegging records, expected shortage conditions for manufacturing part requirements can sometimes be satisfied from available service parts. Conversely, critical service part requirements can perhaps be met with orders destined for higher-level items.

(c) Expediting Another solution to the problem is to contact the supplier or shop people to determine whether the needed components can be brought in on an expedited basis. Expediting is essentially compressing lead times. Lead times usually have some flexibility throughout the purchasing and production process, and they can be compressed throughout the product structure. In the short-term, this is usually accomplished by rescheduling component items and arranging for overtime. The planner should try to find whether some of the needs can be satisfied by rescheduling earlier completion of another order for the item that is already in process.

(d) There are many other things that can be done, for example:

- (i) Utilizing safety stocks.
- (ii) Cutting lot-sizes, many times shortage might be due only to the lot-sizing applied to the parent or component item; the necessary components for meeting the customer requirements might be available.
- (iii) Authorizing overtime.
- (iv) Employing subcontractor.
- (v) Approving air-freight shipments.
- (vi) Using alternative components.
- (vii) Buying a 'manufactured' part from a supplier (or even a competitor).
- (viii) Substituting other products and raw materials.
- (ix) Checking with other customers; although rare, there are times when the sales department can find a customer whose order can be delayed, solving the conflict.

(e) Changing the MPS This is the last resort, before reaching this stage all other solutions should have been applied. Here the scheduler has to decide which of the higher-level items can be delayed? Which cannot?

There is nothing new about borrowing from service stocks or releasing orders to expedite with the factory or with suppliers. With the MRP I database, however, these functions can be done quickly in an orderly and well-managed fashion. Proper communications are accomplished on a timely basis, with everyone in the company working on-line with the same data. Whatever action is taken will be 'cleaned up' automatically by the next MRP I planning run. Any borrowed items or shortages will have new requirements generated, and any overages will be carried forward for future use (Rice and Hatch, 1997).

Action to correct for changed conditions should occur as low in the product structure as possible—for example by expediting. Ideally, it is best to solve the due date or quantity inconsistencies at the lowest level possible in the structure hierarchy because that is where replanning actions begin. Where the problems cannot be resolved simply at the component level, it is necessary to move up a level to see which parent orders are affected and if the problem can be resolved just one level up. Pegging data are required for this process. Only in the worst case should the MPS itself be modified. Thus, replanning, to account for exceptions, should begin with the bottom levels and work its way up the structure until the problem can be resolved (John, 1993).

This process is accomplished by the planner (not the computer system), who evaluates the effects of possible solutions. When the MPS has to be changed, the customer needs have to be delayed or partial shipments made. In such situations, pegging and bottom-up replanning will provide advance warning of these problems so that customers likely to be affected can be warned in advance so that they can take appropriate actions. Bad news is never welcome, but giving advance warning accompanied by a realistic new promise date will in the long term earn more respect, than defaulting without warning.

2.16.19 Performance Reporting

Performance control outputs are comparisons of MRP I plans with actual performance, enabling management to monitor the performance of inventory planners, buyers, the shop, suppliers, and cost accounting.

Following are some specific MRP I/CRP performance measurements:

1. Inventory Performance Measurements

Aggregate or average inventory, inventory turnover (= Cost of Sales Average Inventory), Inventory record accuracy.

2. Customer Service Performance Measurements

Measuring how well customers are being served in terms of quantity and timing is a major indicator of success. There are a number of dates involved: date that the customer's order was entered, requested ship date, original promise date, and revised promise date, if one was necessary. Comparing actual ship date against each of these is important. Some companies also measure whether the shipment arrived at the customer's location when it was wanted. Some specific customer service performance measures are:

(a) Frequency of Stockouts.
(b) On-Time Delivery.
(c) Delivery Lead Times.

3. Productivity Performance Measurements.

Frequency of required expediting, Timeliness of supplier/factory order releases, Frequency of exception messages.

4. Capacity Performance Measurements

Amount of overtime, Capacity utilization (Actual Output ÷ Design Capacity), Queue management.

5. *Reports on Item Inactivity over a Selected Time or Obsolescence*

Inactivity (less than some preset low usage over a predetermined time) and obsolescence (no usage over a predetermined time) can easily be detected in time-phased MRP I programs and reported for corrective actions, including disposal (Plossl, 1994).

6. *Supplier Performance Measurements*

Delivery and Quality performance.

2.16.20 MRP I Program Health Monitors

Outputs of MRP I can be used to detect and help diagnose internal problems in operating programs. Finding problems of data integrity early and fixing them fast will keep planning valid. Following are the four classes of problems (Plossl, 1994):

1. Data Omissions and Errors MRP I programs commonly contain routines to report omissions of essential data in item master records. These data include lead times, lot-size rules, ABC class, costs, manufactured or purchase codes, order policy codes, and planner codes. Reports also can be generated showing 'manufactured items with no BOM' and 'purchased items with BOM.'

2. Number of Action Notices Rising or excessive numbers of order-release or expedite notices may indicate overloaded MPS. Increasing number of delay notices may show excessive capacity.

3. Transaction Errors These are routinely indicated by MRP I programs and include unplanned issues that report material moved from stores although MRP I had not planned it, returns to stores caused by BOM errors or excess issues by stores personnel, and errors reported previously but not yet corrected.

4. Record Errors These are often revealed by MRP I programs processing transactions. Negative balances in on-hand and open-order data are clear indicators of bad file data or poor handling of transactions. Quantities received on purchase or work orders in excess of ordered amounts show possible mistakes, sometimes in reporting the wrong unit of measure. Longstanding small balances in open purchase orders, may be, errors resulting from supplier deliveries on the low side of commercial tolerances.

2.16.21 The Basic Responsibilities of a Planner

The persons most directly involved with the MRP I system outputs are planners. They are typically in the production planning, inventory control, and purchasing departments. Planners have the responsibility for making detailed decisions that keep the material moving through the plant.

Material planner must manage the parts for which they are responsible. This means, not only releasing orders to purchasing and factory, rescheduling due dates of open orders, and reconciling differences and inconsistencies, but also finding ways to improve the system and removing the causes of potential errors. If the right components are to be in the right place at the right time, the planner must manage the process. The planner receives feedback from many sources such as:

(a) Supplier's actions through purchasing.
(b) Changes to open orders in the factory such as early or late completions or differing quantities.
(c) Management actions such as changing the master production schedule.

The planner must evaluate this feedback and take corrective action if necessary. The planner must try to maintain the correct due dates, to solve the problems due to changed conditions as low in the product structure as possible (that is, bottom-up replanning), and to reduce the system nervousness. Their range of discretion, however, is carefully limited, for example, in many companies, planners are not allowed to change plans for end items destined for customers without higher authorization.

Basic Responsibilities of a Planner

1. Launch (release) orders to purchasing or manufacturing.
2. Change order or requisition timings, or cancel them.
3. Reschedule due dates of open (released) purchase and production orders as required.
4. Reconcile errors or inconsistencies and try to eliminate root causes of these errors.
5. Solve critical material shortages by expediting or replanning.
6. Find key problem areas requiring action to prevent future crises.
7. Coordinate with other planners, master schedulers, Production Activity Control (PAC), and purchasing to resolve problems.
8. Approve requests for unplanned stock disbursements.
9. *Review action messages* Effective use of action messages allows focusing attention on the 'trivial many.' If a message is deemed significant then only the planner will try to respond to it. The first issue then would be whether the recommended action can be implemented or not. The planner might have to turn to other information and/or other people prior to arriving at the right response. For example, he may have to look up the availability of components or capacity, or check with people in the factory or purchasing to see how quickly they can change plans.
10. *Use the system to solve problems.* All decisions made to solve problems must be done within the system, and transactions must be processed to reflect the resultant changes. This would ensure that the actions taken to solve the problem are captured in the

records for the next processing. This means that the planner always works within formal MRP I rules, *not* by informal methods.

11. Indicate where further system enhancements (outputs, diagnostics, etc.) would make the planner's job easier. Identify procedural inadequacies in processing MRP I transactions and correct them to ensure that the material plans are accurate.

In addition to the execution tasks listed above, material planners also perform certain tasks related to the inventory files, such as:

1. *Maintain data* Many pieces of data, such as lead-times, order quantities, safety stock, and scrap and shrinkage factors, need to be analyzed and adjusted periodically by the planner.
2. Handle engineering changes in BOMs of items under their control.
3. Monitor inventory for inactivity and obsolescence, and recommend disposition.
4. Participate in analyses of cycle inventory counts.
5. Investigate and correct errors in inventory records (system records must be accurate and reflect the factory's physical reality if they are to be useful).
6. Analyze discrepancies in item requirements and coverage, and take appropriate corrective actions.
7. Request changes in MPS.

Each item is assigned to an individual responsible for planning the item. People performing these tasks are also known as *inventory planners, inventory analysts,* and *inventory controllers.* A very small company may have only one or two planners. A large company may have several large departments of planners. In this case, each planner is assigned a logical group of items (such as, mechanical parts, sub-assembly, raw steel materials, purchased electronic parts, electrical parts, or parts required for a particular distribution center). A number or name on the item master file, usually referred to as *planner number,* keys the item to a particular planner. This allows the MRP I software to sort and display the items for the individual to whom they are assigned. Of course, any item can be accessed at any time by anyone if a need to do so arises. After each planning cycle, the planner logs on to the system and brings up the requirements display screen. This is typically done each morning or at some designated time during the day (Rice and Hatch, 1997).

2.16.22 Sample MRP I Exercise

Figure 2.15 shows partial product structures for two end products, X and Y. (The complete structures would include additional components but the exercise is limited by showing just the common assembly B that is used in both X and Y.) Furthermore, B's components, 1 and 2, are also shown. All usages are assumed to be one. The Requirement data for both the products (X and Y) and for Assembly B has been shown in Table 2.7.

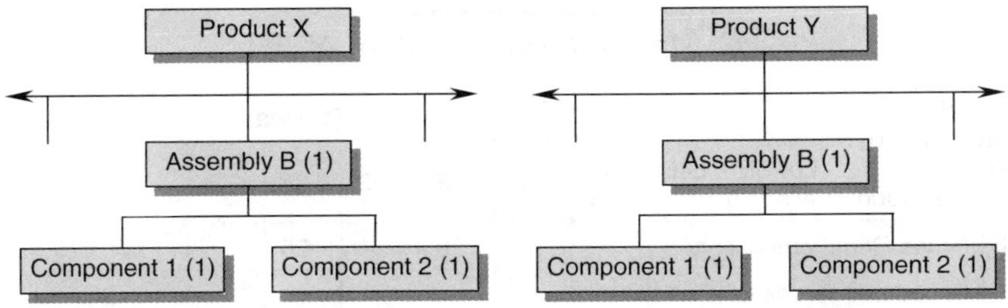

Fig. 2.15 *Partial Product Structure for X and Y* (Adapted from John, 1993)

Table 2.7

Requirements Data for Products X and Y, and for Assembly B

Period	*Requirements Product X*	*Requirements Product Y*	*Independent Demand Assembly B*
1	5	15	0
2	5	20	0
3	5	0	0
4	5	25	0
5	5	0	0
6	5	0	30
7	5	15	0
8	5	0	25

MRP I Exercise Solution (Tables 2.8.1 through 2.8.5)

Table 2.8.1

Material Plan for Item X

On-Hand: 25 Safety Stock: 10 Lead-Time: 3 Lot-Size: 10 Low-Level Code: 0 Alloc.: 0		Periods								
		Past Due	1	2	3	4	5	6	7	8
Item	Gross Requirements		5	5	5	5	5	5	5	5
	Scheduled Receipts									
X	Projected On-Hand	25	20	15	10	5	0	–5	–10	–15
	Projected Available		20	15	10	15	10	15	10	15
	Net Requirements					5		5		5
	Planned Order Receipts					10		10		10
	Planned Order Releases		10		10		10			

Table 2.8.2

Material Plan for Item Y

On-Hand: 20
Safety Stock: 0
Lead-Time: 1 Lot-Size: L4L
Low Level Code: 0 Alloc.: 0

			Periods							
Item		Past Due	1	2	3	4	5	6	7	8
Y	Gross Requirements		15	20		25			15	
	Scheduled Receipts									
	Projected On-Hand	20	5	–15	–15	–40	–40	–40	–55	–55
	Projected Available		5	0	0	0	0	0	0	0
	Net Requirements			15		25			15	
	Planned Order Receipts			15		25			15	
	Planned Order Releases		15		25			15		

Table 2.8.3

Material Plan for Item B

On-Hand: 14
Safety Stock: 0
Lead-Time: 4 Lot-Size: 50
Low Level Code: 1 Alloc.: 0

			Periods							
Item		Past Due	1	2	3	4	5	6	7	8
B	Gross Requirements		25		35		10	45		25
	Scheduled Receipts		48*				41*			
	Projected On-Hand	14	37	37	2	2	33	–12	–12	–37
	Projected Available		37	37	2	2	33	38	38	13
	Net Requirements							12		
	Planned Order Receipts							50		
	Planned Order Releases			50						

*Since the specified lot-size quantity of B is 50 units, the order release for B would always be 50 or a multiple of 50. Still, the scheduled receipts for item B are shown as 48 and 41 in the table above. One possibility is that the planner, for some reason, had intentionally released orders for quantities less than the lot-size using his authority to override the system through the use of FPOs or while releasing the order. Another possibility is that for some reasons the shop or the vendor would not be able to supply the exact ordered quantity and instead only 48 and 41 units would be delivered. This information must have been communicated to the planner *in advance* before the actual delivery and hence the planner must have manually modified the scheduled receipt figures. This ensures that the system can plan the required corrective action beforehand and avoid unpleasant surprises. Another possibility is that the planner has intentionally released order for quantities less than the recommended lot size using his authority to override the system logic through EPOs or while releasing the order.

Table 2.8.4

Material Plan for Item 1

On-Hand: 0 Safety Stock: 0 Lead-Time: 6 Lot-Size: L4L Low Level Code: 2 Alloc.: 0**		*Periods*								
		Past Due	1	2	3	4	5	6	7	8
Item	Gross Requirements			50						
	Scheduled Receipts			50						
1	Projected On-Hand	0	0	0	0	0	0	0	0	0
	Projected Available		0	0	0	0	0	0	0	0
	Net Requirements									
	Planned Order Receipts									
	Planned Order Releases									

Table 2.8.5

Material Plan for Item 2

On-Hand: 25 Safety Stock: 0 Lead-Time: 8 Lot-Size: L4L Low Level Code: 2 Alloc.: 0**		*Periods*								
		Past Due	1	2	3	4	5	6	7	8
Item	Gross Requirements			50						
	Scheduled Receipts			25						
2	Projected On-Hand	25	25	0	0	0	0	0	0	0
	Projected Available		25	0	0	0	0	0	0	0
	Net Requirements									
	Planned Order Receipts									
	Planned Order Releases									

**Component requirements for released orders (which are shown as scheduled receipts) are not shown as gross requirements, instead they are shown as allocations. Hence, the requirements for the released orders of B (shown as scheduled receipts of 48 and 41) must have been shown as allocations in Tables 2.8.4 and 2.8.5, but these Tables show allocations as zero. This implies that the allocated quantities have already been withdrawn from the stores.

2.16.23 Sample Daily Transactions and Planner's or System's Response

Table 2.9.1 summarizes the responses of the planner to various types of transactions. Tables 2.9.2 through 2.9.4 list the transactions that cross the planner's desk over three days in succession. There are four transactions each day. Tables 2.9.2 to 2.9.4 are worksheets, one for each of the three days, on which the planner records what happens and what actions are necessary by him in reaction to each transaction taken in turn.

Table 2.9.1

Summary of Actions to be Taken in Response to Various Types of Transactions

Sr. No.	*Type of Transactions*	*Action Required*
1.	Transactions as per expectations which do not affect planned orders.	▪ Simple updating required. ▪ No rebalancing. ▪ No BOM explosion.
2.	Transactions as per expectation or causing changes in Projected On-Hand, PAB, etc.	▪ Rebalancing required.
3.	If rebalancing or any other transactions cause changes in timing or quantity of planned orders.	▪ BOM explosion required.

2.16.24 Changes that Affect the MRP I Plan

Manufacturing organizations work in an environment where conditions keep changing. Using MRP I as a planning tool will not make this fact disappear, but it can provide a means for taking action in the light of changed conditions before they become crises. Reacting to feedback from actual events in execution system, so that plans may be changed as needed, is what is referred to as 'closing the loop' with MRP I (John, 1993). Following are some of the occurrences that would have an effect on MRP I plans.

1. MPS Revisions The MPS is revised within the cumulative lead-time, sometimes affecting near-term requirements when time fences cannot be observed.

2. Engineering Changes Engineering changes to bill of material (BOM) that must take effect immediately due to product safety, quality, or other critical needs.

Table 2.9.2

Transactions Happening on Day 1 and Actions Taken by the Planner (John, 1993)

Tr. No.	*Transaction*	*Part No.*	*Action Taken by the Planner or the System*	*Rebalance Needed?*	*Explosion Needed?*
1.	The scheduled receipt for 48 of item B has been received and placed into inventory.	Assy B	1. Delete scheduled receipt in period 1. 2. Add 48 to balance on hand making it (14 + 48 =) 62.	No	No
2.	A factory assembly order has been issued to the shop floor for 10 units of product X.	Prod X	1. Delete planned order release of 10 in period 1. 2. Create scheduled receipt in period 4. 3. The planned order receipt of 10 and net requirement of 5 in period 4 should also be deleted.	Yes (Because projected on-hand figures will change)	Yes (Because planned order release has changed)
		Assy B	4. Subtract 10 from gross requirements in period 1 5. Add 10 to allocated quantity making it (0 + 10 =) 10.	No	No

(contd.)

Tr. No.	*Transaction*	*Part No.*	*Action Taken by the Planner or the System*	*Rebalance Needed?*	*Explosion Needed?*
3.	A factory assembly order has been issued to the shop floor for 15 units of product Y.	Prod Y	1. Delete planned order release of 15 in period 1 and also delete planned order receipt of 15 in period 2. 2. Create scheduled receipt of 15 in period 2.	Yes	Yes
		Assy B	3. Subtract 15 from gross requirements in period 1. 4. Add 15 to allocated quantity making it (10 + 15 =) 25. (It was modified to 10 from 0 because of action 5 of Tr. 2.)	No	No
4.	A cycle count report shows that the quantity on-hand of item B is 60 units. This has been confirmed.	Assy B	1. Subtract 2 from quantity on hand making it (62 – 2 =) 60. [On-hand had been modified to (14 + 48 =) 62 because of action 2 of Tr. 1. This also results in the reduction in all on-hand and PAB figures by two. Hence, on-hand in period 1 is now 35.]	Yes (Because projected on-hand and PAB figures will change)	No

Table 2.9.3

Transactions Happening on Day 2 and Actions Taken by the Planner
(John, 1993)

Tr. No.	*Transaction*	*Part No.*	*Action Taken by the Planner or the System*	*Rebalance Needed?*	*Explosion Needed?*
1.	A report from the shop floor control system reports that operation number 30 for a quantity of 25 of component 2 has just been completed and sent along to operation number 40.	Comp 2	No MRP I action required. (This transaction might be of interest to some other system such as order tracking etc. but not to MRP I.)	No	No
2.	A quantity of 10 of product Y has been shipped to fulfill a customer order.	Prod Y	1. Subtract 10 from gross requirements (15) in period 1. 2. Subtract 10 from balance on-hand.	No	No
3.	15 nos. of assembly B were withdrawn from the inventory; so work could begin on the open order for product Y.	Assy B	1. Subtract 15 from allocated quantity, which was modified to 25 because of action 4 of transaction 3 of the first day, making it (25 – 15 =) 10. 2. Subtract 15 from balance on hand making it (60 – 15 =) 45. (On-hand had become 60 as a result of transaction 4 of the first day.)	No	No

(contd.)

Tr. No.	*Transaction*	*Part No.*	*Action Taken by the Planner or the System*	*Rebalance Needed?*	*Explosion Needed?*
4.	Production control has sent a notification that the lead time for assembly B has changed from four periods to five periods effective immediately.	Assy B	1. Change lead time to five. 2. Move planned order release of 50 from period 2 to period 1.	Yes	Yes
		Comp 1	3. Gross requirement for 50 moved from period 2 to period 1*. 4. Need to expedite*.	Yes	No
		Comp 2	5. Gross requirement for 50 moved from period 2 to period 1*. 6. Need to expedite*.	Yes	No

*It should be noted, that the planned order release quantity of 50 units of B in period 6 is due only to lot-sizing. Otherwise, the *real* requirement of B is only 14 units in that period (which was 12 in the original plan shown in Table 1.3, but was modified to 14 as a result of cycle count adjustment of 2 units in on-hand balance due to transaction 4 of day 1). This means, that the real requirement of both parts 1 and 2 is only 14 units (and not 50) in period 1. Out of this, since 25 units of item 2 are already on-hand, there is a need for expediting only 14 units of item 1 from period 2 to period 1 to achieve all higher level plans. This is all that bottom-up replanning is all about.

Table 2.9.4

Transactions Happening on Day 3 and Actions Taken by the Planner
(John, 1993)

Tr. No.	*Transaction*	*Part No.*	*Action Taken by the Planner or the System*	*Rebalance Needed?*	*Explosion Needed?*
1.	The plant has notified that the open order for a quantity of 25 of component 2 will be completed one week late. This is firm.	Comp 2	1. Shift scheduled receipt from period 2 to period 3. 2. Peg to assembly B.	Yes	No
		Assy B	3. Option 1: Change planned order for 50 in period 1 to firm* planned order of 25 and enter second firm* planned order for 25 in period 3. Or,	Yes	Yes
			Option 2: Change planned order for 50 in period 1, which was actually shifted from period 2 to period 1 as a result of Tr. 4 of day 2, to firm planned order of 50 in period 3. This will require expediting from a lead time of 5 periods to a lead time of 3 periods.	Yes	Yes
		Comp 1	4. Move scheduled receipt to period 3.	Yes	No
2.	An unforecasted customer order for product X has been received for a quantity of 10 with immediate shipment requested.	Prod X	1. Can ship immediately because of safety stock, reduce on hand by 10. 2. Increase planned order in period 5 and planned receipt in period 8** to 20 to replenish the safety stock.	Yes	Yes
		Assy B	3. Increase gross requirement in period 5 from 10 to 20.	Yes	No

(contd.)

Tr. No.	*Transaction*	*Part No.*	*Action Taken by the Planner or the System*	*Rebalance Needed?*	*Explosion Needed?*
3.	Nine nos. of assembly B were withdrawn from inventory so work could begin on the open order for product X.	Assy B	1. Subtract 9 from allocated quantity. 2. Subtract 9 from balance on hand. 3. Investigate*** residual of 1.	No	No
4.	A cycle count report shows that the quantity on-hand of item B is 36 units. This has been confirmed.	Assy B	No MRP I action required****.	No	No

**Transaction 1* In option 1 of action 3, both the orders of 25 units have to be firm since they violate the lot-sizing rule of 50 units. If they are not made firm they would be modified back to a single order of 50 units in the next MRP I run.

***Transaction 2* One option is to expedite the scheduled receipt of 10 units of X from period 4 (which has just been released in response to transaction 2 of day 1) to period 2; and immediately release a new order of 10 units of X to be received in period 4. However, it might be the view of the planner that it is not desirable to expend expediting effort just to replenish safety stock. A second option would be to increase the scheduled receipt quantity of 10 in period 4 to 20. The problem with either of these approach is that there will be an immediate demand for another 10 units of item B, which is already in short supply (as seen from transaction 1 of day 3). A third option that avoids this problem, and which is the selected option in Table 2.9.3, is to intentionally postpone replenishing the safety stock for X until period 8. This would, however, imply going without safety stock up to period 8.

****Transaction 3* Planner investigation is required to ascertain as to why only nine units of assembly B were withdrawn instead of ten. Perhaps, there is an intent to start the tenth unit of product X later and then there would be another inventory withdrawal of one. Or one unit of B was found somewhere so that the whole quantity of 10 units of X can be made even though only nine of B have been with drawn. If that be the case, then there should be a special transaction processed to reduce the allocation by one so that the record can be properly rebalanced to reflect the found B. Otherwise the allocation might stay in the record forever.

*****Transaction 4* No MRP I action is required in this case as the cycle count agrees with what the record shows. Remember that the on-hand quantity of B was changed from 62 to 60 (Day 1, Tr. No. 4) from that 15 units were again reduced (Day 2, Tr. No. 3) and finally 9 were deducted due to transaction 3 of day 4. Therefore, the on-hand balance should be (60 – 15 – 9 =) 36.

3. *Inventory Corrections* Errors are discovered in inventory data, perhaps as a result of cycle counting, such as loss due to spoilage or pilferage.

4. *Excessive Scrap* Scrap exceeds estimated quantities, resulting in less material completed than needed to cover requirements.

5. *Changes in Lot-Sizes* Lot-sizes change, perhaps because of revised assumptions regarding inventory carrying cost or order cost.

6. *Customer Order Revisions* Customers change their minds regarding quantities, timing, or configuration.

7. *Supplier or Production Problems* Supplier or production orders are not completed on time, are delivered in incorrect quantities, or do not have an acceptable quality level.

2.16.25 Keeping MRP I Up-to-date or Replanning: Regenerative and Net Change (Transaction-Driven) MRP I

Since conditions change and new information is received, the MRP I records must be brought up to date so plans can be adjusted. This means processing the MRP I records anew, incorporating current information. Two issues are involved in the processing decision: How frequently the records should be processed and whether all the records should be processed at the same time (Vollmann *et al.* 1997). There are two alternative ways of keeping MRP I up-to-date i) Schedule Regeneration (or Regenerative MRP I) and ii) Net Change or Transaction-Driven MRP I.

Regenerative MRP I (or Schedule Regeneration)

Processing all of the records in one computer run is called regeneration. This signifies that *all* item number records are completely reconstructed each time the records are processed. All the early MRP I systems are operated on the basis of 'Regenerative MRP I.' In this system,

(i) the orders that are worked on and released during a period are those that were calculated at the start of the period,
(ii) inventory, BOM, and MPS files are updated during a production period, and
(iii) at the end of the period, the MRP I program will be run to determine all the requirements based on the latest state of the MPS, BOM, and inventory (Childe, 1997). That is the entire MRP I process is performed periodically.

Regeneration process used in early computing systems reiterated all calculations once per week or less often and created huge stacks of paper output, that is why people used to say that MRP I stands for 'More Reams of Paper.' This also caused companies to think in terms of monthly or weekly time buckets.

Characteristics of the Regeneration Process

- All the requirements and projected rows are erased in the computer records, with only the scheduled receipts row showing time-phased open order due dates left intact. Regeneration or 'Regen' MRP I run then starts with a clean slate.
- The MPS data and component independent demand forecasts are loaded, all inventory and BOM data are updated as needed, and then the complete MRP I plan is computed.
- Every end-item requirement stated in the MPS is exploded through all levels in MRP I utilizing every active bill of material.
- The *inventory status data,* which includes *inventory data* (on-hand and on-order), and *requirements data* (gross and net requirements, and planned-order releases) of every active inventory item, are recomputed.
- In regenerative replanning, inventory data (on-hand and on-order inventory) are maintained by file-update processing of transactions at a relatively high frequency, usually daily (this process is called *intra-cycle file updating*). But their effect on requirements planning is not considered until replanning cycle.
- Requirements data (gross and net requirements, and planned-order releases) are reconstructed or reestablished at a lower frequency set by the regeneration cycle (this process is referred to as *requirements planning run*).
- In file updating, a transaction adjusts only one item record directly. A transaction may change parent item status, but file updating will not cause changes in component status. Thus, in regenerative MRP I, inventory transactions never trigger explosions into lower product levels, resulting in a gradual deterioration in the validity of components requirements data between successive regenerations.
- Each released order status is reviewed, and all planned order schedules are recalculated.
- Regeneration involves periodic, batch, sequential data processing and is inherently a massive data-handling task, which requires substantial computer processing time and generates voluminous printed output.
- Orders that have not yet been released or confirmed are overwritten by new ones if the data has caused them to be incorrect, such as, if a new customer order has appeared on the MPS, which will require more components to be bought.
- Orders may be released during the period between MRP I runs, but they will only be as up-to-date as the state of the information when the MRP I was last run.
- Computer programming is relatively straightforward, besides, batch

processing is inherently an efficient data processing technique. Yet, regenerative MRP I is fairly inefficient in terms of data processing, since it often repeats calculations which have already been made.

- Regenerative MRP I programs employ batch-processing techniques and replan periodically, typically in one-week intervals, passage of time triggers the process. The system is therefore said to be time driven.
- Four types of changes—(i) changes in requirements, (ii) coverage of these, (iii) product structure (engineering changes), and (iv) planning factors—affect inventory status and must therefore be reflected in replanning. Regenerative MRP I programs take a 'snapshot' of these factors at the time of each periodic replanning; the implicit assumption is that all relevant changes have been incorporated during the preceding interval; these programs deal periodically with static situations.
- Regeneration tends to purge errors caused by faulty data handling by MRP I users. Every run, the whole MPS and all active BOMs and inventory data are accessed and processed; the old plan is literally thrown away, including all old errors. These errors may reappear, however, until their cause has been removed. Most companies using net change regenerate the complete program to take advantage of its purging actions periodically, the frequency decreases as performance and discipline improve.

Regeneration Frequency Regeneration's massive data handling during the requirements planning run dictates that it be done only periodically, at some reasonable intervals. This delay causes MRP I to be increasingly out of date between replanning runs. How serious this may be depends on the environment in which MRP I must operate, and also on the uses of its output data. The appropriate frequency for processing the MRP I records depends on the firm, its products, and its operation. In dynamic, volatile environments, where there is a strong need for more timely response to change, the replanning frequency of the regenerative MRP I is not adequate. In a more stable and predictable environment, regeneration at weekly intervals may function satisfactorily.

The problem with processing less frequently is that the portrayal of component status and needs expressed in the records becomes increasingly out of date and inaccurate. This decrease in accuracy has both anticipated and unanticipated causes. As the anticipated scheduled receipts are received and requirements satisfied, the inventory balances change. As unanticipated scrap, requirements changes, stock corrections, or other such transactions occur, they cause inaccuracies if not reflected in all the time-phased records influenced by the transactions. Changes in one record are linked to other time-phased records as planned order releases become gross requirements for lower-level components. Thus, some change transactions may cascade throughout the product structure

(Vollmann *et al.* 1997). If these transactions are not reflected in the time-phased records early enough by frequent replanning, the capability of MRP I to keep released order due dates up-to-date and valid may be jeopardized.

Requirements Alteration (A Special Feature of Some Regenerative Programs) A special feature of some regenerative programs, called requirements alteration, is sometimes confused with net change. Requirements alteration is designed to process changes in the MPS, between the periodic MRP I, regeneration runs to get more frequent partial replanning without the effort and expense of a full explosion. Inputs to the program are new period MPS values (not net differences) for end items having changes. The explosion is limited to the changed end items and their components.

Requirements alteration handles only changes in the MPS, it does not react to changes caused by transactions on lower-level items as net change does. It updates the status of only those lower-level components related to MPS end items being processed. The status of all other items continues to deteriorate between regeneration runs. Requirements alteration is not the same as net change, which, in addition to processing changes to the MPS, also processes changes to inventory balances, BOMs, etc. through MRP I.

Regeneration in Conversational Mode Some regenerative systems can perform the MRP I process in *conversational* mode. As each low-level code of processing is completed, the human planner can review schedules for items on that level on a CRT. The planner may modify the plan, as he or she sees fit, by entering or modifying firm planned orders. MRP I includes these planned orders when it processes the items at the next lower level. Although this sounds interesting on paper, we know of no companies that actually use this technique (Oden *et al.* 1993).

Net Change, or Transaction-Driven MRP I

A more modern development of MRP I is the 'net change' approach, in which only those records affected by the new or changed information are reconstructed. This net change approach only creates a new item number record when a transaction makes the present component plan inaccurate. Advances in hardware and software now enable *net change MRP I* to constantly update and display the present system status on a computer screen with each new data entry, allowing companies to think in terms of buckets of one day or less and in terms of bucketless priority sequencing of operations. It is an approach in which the material requirement plan is continually retained in the computer. Whenever a change is needed in requirements, open order inventory status, or BOM, a partial explosion and netting is made for only those items that are affected by the change. This is a more sophisticated approach, which leaves existing planned orders in place, but updates them if required by a new order or a change in requirements. It is interesting to note that Joseph Orlicky, when he developed MRP I in early 1960s, originally envisioned MRP I as a real-time, transaction-driven replanning process. Only in recent years with increased hardware processing speeds, more sophisticated software, and highly trained users has this become possible.

- With net change, the requirements and projected data are never erased; they become a permanent part of each item database record.
- Every transaction that affects inventory, the BOM, or requirements information is processed, balancing records and reexploding whenever necessary, to maintain a constantly up-to-date material plan.

Transactions that will cause net change to replan a part include:

1. A change or extension in the MPS.
2. A change in gross requirements for an item, resulting from a change in the schedule for a higher-level item, service demand, or inter-plant shipments.
3. Unplanned inventory transactions such as
   (i) a correction in inventory balance (because of cycle count or physical inventory transactions).
   (ii) a lot being rejected in manufacturing.
   (iii) any unplanned receipt or issue (in some systems).
4. Engineering changes, such as changes in product structure, effectivity data or usage.
5. Change in inventory planning parameters, such as lead-time, safety stock, scrap factor, lot-sizing.

Some net change systems can run in real time, calculating the effects of any change as soon as it is keyed in. Others achieve good results by performing a daily update, calculating the effects of all the changes made that day. The calculation requires much less computer power since only the items which have changed need to be dealt with (Childe, 1997). Net change programs use inventory transactions to initiate replanning, more or less continuously. Net change programs are also *regenerated* occasionally to purge an accumulation of small errors.

Net change replanning has a much smaller scope, shorter duration, and far less output, but uses more *total* computer time because of inefficient computer processing. Net change explosions are partial in two senses:

(i) only part of the MPS is exploded at any one time, and
(ii) the effects of transaction-related explosions are limited to lower-level components.

Replanning requirements in MRP I involves level-by-level explosions. These cannot be eliminated, but net change makes them in high-frequency, consecutive, partial explosions. This minimizes the scope of replanning at any one time and thus, permits frequent processing.

Making only partial explosions automatically limits the volume of the resulting output. Some other characteristics of net change systems are as follows:

1. *Same Items May be Affected Many Times* The same items may be involved in rebalancing and explosion many times within a brief time span.

2. *Shorter but more Frequent Output* Output reports tend to be shorter but more frequent, and on-line access to all MRP I data may be provided through computer terminals.

3. *More Computer Resources Needed* Although the time to compute net change transactions is shorter than performing a complete regeneration, it is done more frequently, so that overall, more computer resources are needed.

4. *Computer Programs are more Complex* Computer programs are much more complex requiring a higher level of expertise.

5. *Transaction Driven* The net change system is transaction driven rather than time driven.

From the user's point of view, the greatest advantage of a net change MRP I program is more timely response to a wide variety of changes. On the other hand, it has four disadvantages:

1. *Lack of Self-Purging Capability* Net change lacks regeneration's self-purging capability of removing errors from files.

2. *Strict Discipline Required* It requires stricter disciplines in operating procedures.

3. *Inefficient Data Processing* Data processing is relatively inefficient in net change programs primarily due to multiple accessing of inventory record data when handling transactions and exploding requirements.

4. *Causes System Nervousness* Reacting to every possible transaction tends to make the system 'nervous' indicating actions to be performed one day with possibly counteracting actions called for the next day. The experience of almost every net change user has been frustration with the inability of people to react as quickly as computers (Plossl, 1994).

Most MRP I users start with regeneration because it is easier to understand and implement than net change, is more forgiving of lapses in discipline in data handling, and emits action notices only once a week, not continuously. Net change is adopted only after, considerable MRP I experience leads to a conviction that more up-to-date planning is needed. Few can handle the flood of signals emitted by a net-change MRP I, and most have to make compromises (Plossl, 1994). Table 2.10 compares the three methods of keeping MRP I up-to-date.

Table 2.10

The Differences Between Schedule Regeneration, Requirements Alteration, and Net Change

Aspect	*Schedule Regeneration*	*Requirements Alteration*	*Net Change*
Frequency of Replanning or MRP I Runs	Periodically (Longer Intervals)	Periodically (Short Intervals)	Periodically (Short Intervals)
Initiation of MRP I Run	When all MPS, revised or not, are fed to MRP I	When new period MPS values are fed	Whenever any transactions or changes to MPS are processed
Computer Run Time	Long computer runs	Short	Short
Processing Efficiency	High	Low	Low
Inventory, Open Order Files Updated	Periodically	Periodically	Continuously (Short Intervals)
All Master Production Schedules Exploded	Yes	No	No
Effects of Every Trans-action Analyzed	No	No	Yes

2.16.26 Reducing System Nervousness

System nervousness is the tendency for small changes in data requirements to occur frequently. The degree to which upper level change causes lower level action message is the system nervousness. Nervousness in an MRP I system is usually measured by the number of messages on the action reports, which indicate that the previous MRP I schedule has changed. Nervousness usually generates the following types of action messages:

1. Expedite or Deexpedite.
2. Cancel scheduled receipts.
3. Release planned orders in expedite mode.

This is especially exacerbated by net change systems that replan often and also by lot-for-lot order quantities.

Causes of Nervousness

1. Imperfect sales forecasts and/or changes in quantity and/or timing of customer orders.

2. Lot-sizing side effects, particularly when using dynamic lot-sizing techniques.
3. Long lead-times which result in more inaccurate forecasts for the longer periods involved.
4. Changes in scheduled receipts when the supplier and/or manufacturing are not able to honor their commitments.

Methods of Reducing System Nervousness

1. Maintain Small Safety Stock of Frequently Used Material One way to alleviate the effect of nervousness is to maintain small safety stock of frequently used material. Another way is to postpone taking immediate action on small changes and handle them periodically—every Friday afternoon, for example. Although it is a good thing for planners to be made aware of even small changes in the planning data, knee-jerk reactions should be avoided by waiting a reasonable amount of time to allow things to stabilize somewhat (John, 1993). Many companies dampen the effects of changes with rules like, 'Don't reschedule open orders earlier within one week,' or 'Don't reschedule open orders later within two weeks.'

(a) Not every change in inventory status calls for a reaction as minor changes in requirements often are absorbed by inventory surpluses These surpluses are created by safety stock, scrap allowances, lot-sizing, engineering changes, reduced requirements, supplier overshipments, early deliveries, and shop overruns. MRP I uses up such excess inventories quickly through the netting process. While inventory excesses are thus prevented from accumulating, they exist on some item or other at almost all times (Plossl, 1994).

(b) Planning safety stock does not dampen nervousness, yet it makes the shop or purchasing more stable as planners can afford to ignore the action messages Since, MRP I reacts to prevent the use of safety stock exactly the same way as it plans to prevent shortages. That is, even if there is safety stock available, the system will generate action messages to reschedule in or to create new orders to take care of shortages so that safety stock is not depleted. But the planner can ignore these messages because safety stock can take care of the shortages. Thus, although MRP I planning would show nervousness the actual shop floor or purchase would be stable due to safety stock.

2. Stabilize the MPS by Improving Demand Forecasting and Strictly Enforcing Time Fences Nervousness can be caused by changes in demand data such as forecast errors, changes in customer order delivery dates or quantities, and revised schedules of warehouse orders. Better communications with the customers reduces such changes by very significant amounts. One of the primary functions of sales/marketing is to avoid surprises from customers.

3. Constrain or Dampen Action Messages For example, do not permit any reschedule messages for one week or less, or, do not generate messages about increasing requirements unless the previous requirement is exceeded by a specified quantity. Although these dampers reduce the action messages, they do not reduce the underlying causes, and can cause operational 'surprises.' This is a bit like disconnecting the temperature gauge in a car so that we won't see that the engine is running hot (Oden *et al.* 1993).

4. Use Firm Planned Orders This will not decrease the number of action messages, because MRP I will still suggest rescheduling the FPOs to the date of actual need. This practice may, however, reduce the schedule nervousness at levels below the FPO. Its impact on other performance variables is unclear. MRP I will not plan inside the FPO, and thus will reduce the number of MRP I suggested changes there.

5. Use Lot-sizing Techniques Quantity nervousness may be reduced by employing a fixed-lot-sizing technique (either FOQ as EOQ), particularly at the upper levels. However, this will have other undesirable effects, such as creating inventory remnants, and inhibiting the reaction to customer demands.

6. Stabilize the Planning Data—Make Only those Changes that are Truly Needed.

7. Reduce Lead-Time.

The other major source of nervousness is poor performance in supply (purchase and manufacturing) activities. In the past, these upsets were believed to be natural in manufacturing, and unavoidable. Companies with world-class competitive status, however, have shown clearly that such problems can be eliminated or at least reduced to negligible effects. The only effective way to avoid excessive nervousness from causing myriad corrective action signals is to eliminate the causes.

2.16.27 Formalization of the Informal Process

MRP I is really just a formalization of the informal system. In MRP I, the computer does almost the same thing as a human planner/expediter does in an informal system. The computer takes an MPS, looks up the bill of material or formula to find out what material is required to manufacture the product, then checks the inventory to see if the material is on hand. If the material is not on hand, it tells the planner to order it and tells him when it will be needed. If the material is already on order, the computer will evaluate the due dates on that material and tell the planner if the due dates need to be changed. MRP I is simply a formal system for predicting the shortages and simulating what the expediter was really trying to do all along.

The combination of computers and software has given the planners an opportunity to be skilled as 'problem preventers' and more effective when unavoidable problems do occur. The computer's job consists of the following activities:

1. *Store Data* All the answers to the universal manufacturing equation are stored by the computer. Rather than having to recreate the answers, this feature permits the planners to focus on maintaining the answers.

2. *Display Data* The computer can display information in easily digestible formats and can also help in retrieving the information in a timely, convenient manner.

3. *Critique the Data* Through action messages, the computer can call attention to situations and recommend corrective action.

What computer cannot do is to separate significant from insignificant action messages nor make decisions. Differentiating between trivial and important needs to be done by skilled planners, decisions must be made by people if someone is going to be held accountable (Goddard, 1997).

Key Differences between MRP I and an Expediter Operating in an Informal System

1. *Fast and Easy Replanning* The MRP I system replans regularly, in contrast, the expediter would find it extremely impractical to put the items back on the shelf and start over again each week.

2. *Early Prediction of Shortages* The MRP I system also predicts shortages early enough so that something can be done to prevent them. It warns of a material shortage long before it happens so a planner can begin working on the problem today and resolve it in advance. Following are some of the possible options available for resolving the material shortage problem:

   (a) Moving scheduled receipts in to an earlier period.
   (b) Reducing lot-sizes.
   (c) Compressing lead-times.
   (d) Making use of material substitutions.
   (e) Changing the MPS.

An expediter simply cannot keep priorities properly updated and usually finds out about shortages too late. As a result, the informal system costs more and is less effective.

3. *Printing and Transmission of Orders, Reports, etc.* It is usual for the MRP I system also to print the actual paperwork, such as reports, purchase and production orders or schedules which are sent to suppliers or to the factory, or to transmit the information by electronic means, such as EDI, electronic mail, or the Internet.

2.16.28 Order Point versus MRP I

Order point and MRP I are two associated sets of techniques available for managing production inventories. *Order point (statistical order point* or *stock replenishment)* is defined *as a set of records, procedures, and decision rules intended to ensure continuous availability, even with uncertain demand, of all items comprising an inventory*. Order point is defined by APICS Dictionary as a set inventory level where, if the total stock *on hand plus on order* falls to or below that point, action is taken to replenish the stock. The order point is normally calculated as forecasted usage during the replenishment lead-time plus safety stock.

The order point assumes that the item's usage or consumption rate is fairly uniform. This is a fallacious assumption as the actual requirements of components in manufacturing occur in discrete 'lumps' caused by parent-order lot-sizes. Whenever a manufacturing order is to be executed, the total component requirements [= order quantity of the parent item × usage factor of the component ÷ (1 – scrap factor)] is issued to the shop. Thus, the demand for components is not continuous and not in small increments. Figure 2.16 compares inventories under order point and MRP

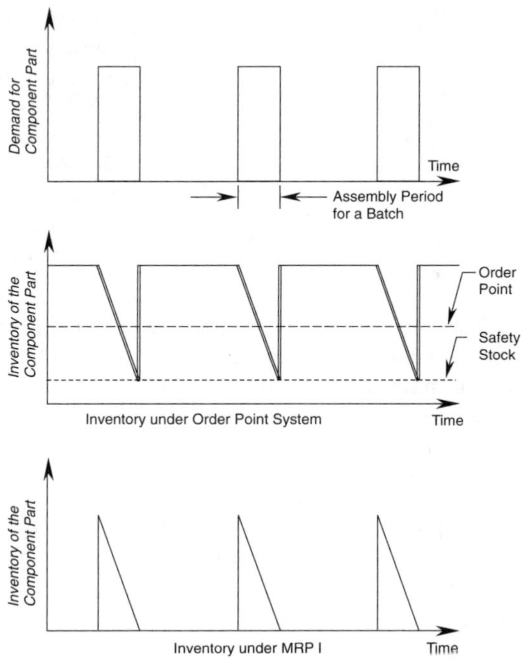

Fig. 2.16 *Comparison of Inventories under Order Point and MRP I*

I systems. The requirement for the component is lumped into batches whenever an order is released for the manufacture or assembly of a batch of the parent item. In case of order point system, an order is released as soon as the inventory level reaches the order point even though the actual requirement for the component would arise only when the order for the next batch would be released.

Thus, the maximum level of inventory is maintained virtually throughout the period. As against this, MRP I system releases order for the exact quantity of the required component only when there is an immediate requirement. Thus, inventory starts depleting because of usage in parent assembly as soon as it is delivered. Finally, OP system maintains a safety stock continuously. Table 2.11 compares OP and MRP I inventory systems.

Table 2.11

Comparison of Order Point and MRP I Inventory Systems

Aspect	*Order Point (OP)*	*MRP I*
Planning oriented towards.	Parts (components, sub-assemblies, and raw materials).	Products (end items). MRP I is the preferred technique to use when one item is a component of another, thus subject to dependent demand. MRP I is expressly designed for dealing with dependent, discontinuous, non-uniform demand, but it can also easily accommodate mixed dependent and independent demand.
Planning based on.	OP is history driven and plans on the basis of past consumption of the part (in isolation from all other items); it is reactive. Demand forecasts for components also are derived from each item's past usage and not calculated from the end-item quantities. OP systems do not consider horizontal or vertical dependencies. It knows nothing about future needs. Demand forecasting determines only the anticipated average amount not the timing of specific demands. To summarize, there is no understanding of future dependent requirements nor the time phasing of those demands.	MRP I is forward looking and plans on the basis of future requirements as defined by the MPS; it is proactive. MRP I does not use averages rather it calculates the exact requirements for each production batch. MRP I does not rely on forecast and *does* recognize *sets of items,* thus avoiding two major deficiencies of OP. MRP I also considers horizontal dependencies.

(Contd.)

Aspect	*Order Point (OP)*	*MRP I*
When inventory replenishment orders are released.	Orders are released whenever inventory is depleted below the order point quantity. Whether, when, and in what quantities the item is really required is not the consideration. The OP system reorders prematurely, far in advance of actual need, and excess inventory will be carried for significant periods of time. Since the schedule dates on the replenishment orders are wrong, this method results in low credibility of the system. It also results in the scarce capacity and materials being applied to the wrong items.	Orders are released only when there is a known or anticipated immediate requirement for the particular component. Replenishment orders are planned for the exact requirements and are released to the shop or vendor such that they would be delivered exactly when they are required—no later no sooner. Assuming both accurate inventories and BOM, an MRP I system will give valid due dates of all manufactured and purchased items required for planned end items.
Bills of material.	BOMs are not used to calculate order quantities.	BOMs are used to find the gross requirements.
Inventory levels.	Inventory is planned such that it is always maintained above the reorder point.	Inventory is planned such that it would be depleted immediately.
Order release.	Orders are released as soon as the inventory level goes below reorder point.	Orders are released to get replenishment just when they are needed.
Plans for future order releases	No future orders can be planned or shown.	Future orders can be planned and shown for the entire planning horizon.
Is replanning possible?	No	Yes
Accuracy requirements and ease of use	OP systems will call for orders when inventory reaches the order point even if the records are inaccurate. The reorder point is relatively easy to understand, calculate, and manage. If all usage was independent and used in a consistent uniform rate, this technique would work well in inventory management.	MRP I system can collapse if bills of material and/or inventory records are incomplete or inaccurate. A realistic MPS is also a prerequisite for effective MRP I. It is relatively difficult to understand, calculate, and manage.

Notwithstanding the difficulties experienced by many companies attempting to implement MRP I-based planning, MRP I has been the most successful innovation in manufacturing inventory management. It has demonstrated conclusively its operational superiority over alternatives in many companies, in many industries, and in many countries. In addition, it has provided students of inventory management with new insights into manufacturing planning and control. MRP I shows true interrelationships among activities affecting inventories and reveals the fallacies of previous concepts, pinpointing the causes of inadequacies of older methods (Plossl, 1994).

2.16.29 From MRP I to MRP II

MRP I represents the system that does the disaggregation of the MPS into schedules of purchase and production orders for components. MRP I was a major breakthrough in Manufacturing Planning and Control (MPC) made feasible by random access computers and database management systems. The first systems were often called 'launch and forget' systems, since there was initially no mechanism for updating MRP I information on orders already released.

Closed-Loop MRP

Because the environment in which MRP I works is very dynamic, that is, because order quantity and due date changes are not infrequent, there exists a need to update order due date and order status after the order has been launched (Fogarty *et. al.* 1991). Since MRP I could be used to update due dates for production and purchase through replanning (regeneration or net change), it was natural that it was also used as a dynamic priority-setting technique for scheduling and executing shop floor and vendor operations. Adding this feedback capability, which is nothing but the addition of replanning capability, to MRP I resulted in 'closed-loop MRP.' Similarly, when execution on the shop floor was improved, attention naturally turned to the front end to improve the process of establishing and maintaining a valid and realistic master production schedule. Incorporation of better master production scheduling in MRP-based MPC systems was another addition that facilitated conversion from MRP I to 'closed-loop MRP.' Today all systems are closed-loop systems, and the words 'closed-loop' need not be specifically mentioned.

Secondly, MRP I was initially developed without any capacity checks or input from other departments. Closed-loop MRP includes capacity checks which are used iteratively with MPS and MRP I, to generate *feasible* schedules.

Following *additional enhancements* were incorporated in closed-loop MRP:

1. Better capacity planning procedures at the front-end, engine, and back-end levels, which made plans more believable.

2. Inclusion of additional enhancements that convert a number of outputs of the detailed manufacturing planning and control into financial plans, such as inventory investment, labor budget, shipping budget, and vendor commitments. Decision-making is always better when people know the financial consequences.
3. Simulation capabilities to examine 'what-if' scenarios.

This overall vision of an MRP I system, which was at that time termed MRP, was so fundamentally different from the original concept that Oliver Wight coined a new term 'manufacturing resource planning' or 'MRP II.' Thus, MRP II is an extension of closed-loop MRP and includes financial planning and simulation capabilities. The fundamental concept behind MRP II is that MRP II should be able to provide virtually all the information that the financial management needs. MRP I, as we know, generates purchase commitment report which tells us when and how much we need to pay our vendors. MRP II provides the information on required staffing levels and overtime through its detailed Capacity Requirements Planning (CRP). Future payroll obligations can be determined based on this information. Finally, the receivable amounts can be predicated with great accuracy based on the information generated by MRP II on quantity and timing of materials delivered to the customers.

Today, the acronyms MRP, mrp, little MRP and MRP I typically imply the basic materials requirements budgeting calculations as described in Orlicky's book, and MRP II is the broader concept suggested by Oliver Wight for the integration of data, computations, and policies that extend materials and capacity planning into comprehensive manufacturing resources planning and control. Managers who understand the logic and discipline of MRP II recognize that it is an integrative approach that is not restricted to manufacturing, and hence the terms *managerial resources planning* (MRP III), *business requirements planning* (BRP) have been used, especially among consultants, to imply that it is a general approach to time-phased planning of the use of resources.

2.17 CAPACITY REQUIREMENTS PLANNING (CRP)

The output from the MRP I process that involves manufactured components constitutes a tentative schedule of factory order releases. Capacity Requirements Planning (CRP) considers these planned orders, together with work already in process, to evaluate the availability of resources—equipment and/or manpower—in the correct time periods to execute the plan in a more detailed fashion than is possible with rough-cut capacity planning. CRP produces a report that shows the capacity required by work center and by time period, to actually execute the MRP I plan. CRP looks at capacity in detail but because of computing, time limitations has traditionally only been used typically to look up to six months ahead.

CRP produces exception reports that show which work centers are overloaded and which are underloaded for each time period. Production planners use this information to see, in advance, the projected capacity problems so they can develop methods to solve the problems. For example, if a work center is underloaded in the third week but overloaded in the following (fourth) week, a production planner could pull some of the fourth week work to the third week in order to level the load between the two weeks. The production planner notifies the rest of the company of the change by revising the dates on the affected orders. When MRP I runs, it will plan to have the required material available to meet the new schedule (Oden *et al.* 1993).

The CRP calculations are nearly identical to the RCCP calculations. However, instead of using the production plan or MPS, as is done in RCCP, CRP uses the material plan, the scheduled receipts, and planned orders in MRP I. And instead of using a resource profile, CRP uses the routings (bill of processes) and work center files as inputs in the evaluation. The routings are kept updated by the engineering change module, and identify the resources required to make an item. Detailed capacity requirements are then calculated and accumulated by resource (work centers and labor skills) much as in RCCP. CRP involves the establishment of the individual capacity plans for each machine group, line, or labor skill for the full planning horizon based upon the detailed material plan. Included are the capacity needs for the work already in process as well as the capacity requirements of anticipated orders for all parts.

Based on the spreadsheets of material requirements, spreadsheets can be created for the capacity requirements plan that include labor, machine, capital, and requirements. A common procedure is to start with the MPS to calculate the material plan (MRP I), use the material plan to calculate the CRP, and observe whether sufficient capacity resources exist to achieve the material and capacity plan. CRP computes the projected manufacturing load by using the existing shop orders at their current state of completion, plus all the future shop orders planned by MRP I. CRP then places the load in each appropriate work center by reading the routing file for that order, and projecting when the order will be at each work center during its trip through the shop floor, based on its due date (Oden *et al.* 1993). If the tentative plan is not feasible, then the planner can simulate to iteratively revise the MPS and recompute MRP I and CRP until a feasible plan is discovered. This trial-and-error approach is called *infinite planning* because each trial starts by assuming infinite capacity exists, without regard to the actual capacity of work centers. Future computing advances will enable operations research techniques such as linear programming to be built into the on-line computations to enable *finite capacity planning* (Rice and Hatch, 1997).

The capacity requirements are displayed in a format similar to the one used in RCCP, and their use is nearly the same. Overloads and underloads are identified and planners work to resolve these. The primary difference is that rather than looking only at the key resources, as is done in RCCP, CRP looks at every work center on a detailed level (Landvater, 1997).

Capacity available can be established through historical information *(demonstrated capacity)* or by calculating *rated capacity,* that is, how many standard hours worth of work can be produced based on manning and equipment, efficiency rates, and time available for production (total hours reduced by time when the machine or person is not expected to be producing due to breaks, daily maintenance, planned slack time, etc.) (Crepeau, 1991).

Load is calculated by multiplying planned and released quantities by labor and equipment resources needed to do the work. Standard hours are a common denominator, although other measures may be used (pieces, tons, gallons, dollars, etc.). The important point is that the load must be calculated in a unit of measure that can be compared to the capability to produce in the same unit of measure (Crepeau, 1991).

Load and capacity must be brought into balance by adjusting capacity or by adjusting the load. If there are shortfalls or imbalances in the available resources, then, either the resources must be adjusted, the MRP I-planned orders must be revised, or the MPS may have to be changed (John, 1993). CRP procedure might involve the techniques of *Machine Loading* (now outdated) and *Production Smoothing.*

Machine Loading

In general, the capacity of production facilities is limited; its limit is the effective work time multiplied by the number of machines, expressed in terms of machine-hours, or, in the case of operative workers, man hours. Hence, it is necessary to adjust and optimally determine the workload for manufacturing products in every time period so as not to exceed the production capacity, in connection with due dates for those products. This decision is called machine loading.

Machine loading is the accumulation by work centers, machine, or machine group of the hours generated from the scheduling of operations for *released* orders by time period. Machine loading differs from CRP in that it does not use the planned orders from MRP I but operates solely from released orders. It may be of limited value because of its limited visibility of resources (Cox III and Blackstone Jr, 1998).

We have seen that machine loading provides capacity-planning information based only on released orders. However, for most companies, only the work to be done in the next few weeks exists in the form of released orders. But as you go further out in time, increasingly the workload is in the form of planned orders, hence the machine loading report gives the picture of diminishing workload over time. That, of course, is incorrect and hence machine loading is not an effective way to do capacity planning, and few companies use it now, making it an outdated technique.

Production Smoothing

It quite often happens, that in some periods workload exceeds the production capacity to meet due dates established by market demands. The function to level down the workload, so that it is within the production capacity, and to respond to fluctuations of the external environment of the manufacturing system is production smoothing..

Policies for Production Smoothing

1. *Redistribute Load*
   - Use alternate work centers.
   - Use alternate routings.
   - Adjust operation start date backward or forward.
   - Hold some work in production control for later.
   - Revise the MPS.
2. *Increase Capacity*
   - Schedule overtime during week, or work weekends.
   - Add extra shift.
   - Add equipment and/or personnel.
   - Subcontract one or more shop orders.
3. *Reduce Capacity*
   - Temporarily reassign personnel (JIT would suggest using this time to invest in worker education, or to maintain equipment and facilities).
   - Reduce length of shifts or eliminate shifts.
4. *Increase Load*
   - Release orders early.
   - Increase lot-sizes.
   - Make items that are normally bought items.
   - Increase the MPS.
5. *Decrease Load*
   - Buy some items that are normally make items.
   - Decrease the MPS.
   - Reduce lot-sizes.
   - Obviously, planners have far more options available when they can look farther into the future. For example, they can deal with a problem far out on the horizon by subcontracting, using new tooling, acquiring new equipment, or implementing a methods change. If they are trying to solve a problem this afternoon, their options list is reduced to overtime, extra shifts, or changing the MPS, none of which may be desirable (Landvater, 1997).

- CRP is important because it allows people to deal with problems quantitatively. Too often, companies stop at the fact that they have an overload or underload. What's the size of the underload? Five hours? Five hundred hours? It's important to quantify problems in order to develop practical, cost-effective solutions that can be implemented in the necessary time frame (Landvater, 1997).
- In the context of MRP II, CRP is medium range. Long-term adjustments to capacity (new facilities, procurement and installation of long-lead equipment, etc.) must be evaluated and planned at higher levels in the process (business planning, S&OP, and MPS). The broader term 'capacity management' thus, encompasses long-range capacity planning and control, as well as intermediate-range and short term capacity planning (Crepeau, 1991).

2.18 PLANT AND SUPPLIER SCHEDULING

After the plans have passed the validity test, they must be communicated to the people who will be accountable for making them happen. This is plant and supplier scheduling. The material plan is communicated to suppliers, and the material and capacity plans are communicated to the factory or plant. At this point, the top-level plans of the company have been broken down into specific tasks that people can be held accountable for achieving.

2.18.1 Plant Scheduling: Executing the Factory Plan

The execution of the manufacturing portion of the planned order release schedule whereby factory orders are released, monitored, and reported is called Shop Floor Control (SFC). SFC consists of Factory Coordination (FC) and Production Activity Control (PAC). In PAC, more detailed planning is performed. At this point, the top-level plans of the company have been broken down into specific tasks that people can be held accountable for achieving. In firms using MRP II systems, execution of the detailed material and capacity plans involves the scheduling of machine and other work centers. In these factories, scheduling must reflect such routine events as starting and completing orders for parts and any problem conditions, such as breakdowns or absenteeism. These schedules must usually be updated at least once per day in factories with complex manufacturing processes (Vollmann *et al.* 1997).

Activities Performed in SFC

1. The daily schedule of jobs to be run at each work center is prepared.
2. Material and capacity availability is checked prior to releasing manufacturing orders.
3. Orders are released to authorize production.

4. Manufacturing information is prepared for the shop.
5. Performance is measured and feedback provided so as to:
   (a) identify problems (such as breakdowns and absenteeism),
   (b) provide status information,
   (c) track costs, and
   (d) update remaining load, inventory balances, etc.
6. Corrective actions are taken to compensate for unanticipated problems, and adjustments are made to meet schedules and operate effectively (Crepeau, 1991).

- A PAC system provides the necessary functions to control the flow of products *within a manufacturing cell.* PAC calculates a target completion date for every operation, and provides facilities for progress to be recorded. A manufacturing order can go through several operations. Any delays which threaten the achievement of the final order completion date must be fed back to the planner (Kenworthy, 1997).
- Factory coordination, on the other hand, organizes the flow of products *throughout a factory* and ensures that the production in each cell is synchronized with overall production goals of the factory. Problem resolution should include efforts to identify and eliminate the causes so that they can be avoided in future (Crepeau, 1991).

Finite Loading

The finite loading technique is essentially a detailed CRP technique for the MRP I items, but it can be better viewed as a shop floor scheduling technique. It, more than any of the other capacity planning technique, makes clear the relationship between scheduling and capacity availability. Finite loading starts with a specified capacity level for each work center or resource grouping; this capacity is then allocated to work orders. Hence, finite loading is a method for scheduling work orders. It assigns no more work to a work center than the work center can be expected to execute in a given time period. The specific term usually refers to a computer technique that involves calculating shop priority revisions in order to level load operation by operation. The finite loading process requires linkages to the same files as the CRP technique, namely work-in-process, routing, scheduled receipts, and planned orders, as well as to file, which specify work center capacities.

Loading

It means deciding which jobs to assign to which work centers. Although the capacity planning system determines that sufficient *gross* capacity exists to meet the MPS, no *actual* assignment of jobs to work centers is made. Some equipment will generally be superior to for certain jobs, and some equipment will be less heavily loaded than other equipment. Thus, there is often 'best' (fastest or least costly) assignment of jobs to work centers (Shafer and Meredith, 1998).

Sequencing

Even after jobs have been assigned to work centers, the *order* in which to process the jobs—their *sequencing*—must still be determined. Unfortunately, even this seemingly small final step can have major repercussions on the organization's workload capacity and on whether or not jobs are completed on time. A number of priority rules have been researched (Shafer and Meredith, 1998).

Detailed (Short-Term) Scheduling

Once all this has been specified, detailed schedules itemizing specific jobs, times, materials, and workers can be drawn up. This is usually done only a few days in advance, however, since changes are always occurring and detailed schedules become outdated quickly. It is the responsibility of production planning and control to ensure that when a job is ready to be worked on, all the items, equipment, facilities, and information (drawings, routings, etc.) are available as scheduled (Shafer and Meredith, 1998).

Dispatching

It is the physical release of a work order from production planning and control department to operations.

Expediting and Deexpediting

Expediting is nothing but the production manager's task to get the job done on time once the shop order has been released. When jobs fall behind schedule, managers have historically tended to use expediters to help push these 'hot' jobs through operations. Of course, expediting can be done more proactively, by monitoring the progress of jobs to ensure that they are on schedule. One problem with the informal scheduling system, of course, was a lack of deexpediting (that is, delaying jobs that had dropped in priority) to reflect changes in required due dates and thus in priorities and schedules. Deexpediting has now been built into the computerized MRP II and scheduling systems (Shafer and Meredith, 1998).

Input/Output Control

It is a technique for capacity control where planned and actual inputs and planned and actual outputs of a work center are monitored. Planned inputs and outputs for each work center are developed by CRP and approved by manufacturing management. Actual input is compared to planned input to identify when work center output might vary from the plan because work is not available at the work center. Actual output is also compared to planned output to identify problems within the work center (Cox III and Blackstone Jr, 1998). It provides a method for monitoring the actual consumption of capacity during the execution of material plans. It is necessarily linked to the execution systems and database for shop floor control. Input/output analysis can indicate the need to update capacity plans as actual shop performance deviates from current plans, as well as the need to modify the planning factors used in other capacity-planning techniques (Vollmann *et al.* 1997).

2.18.2 Supplier Scheduling: Executing the Purchasing Plan

The result of the MRP I process that involves purchased components and raw materials is used by purchasing as a schedule of planned acquisitions. Supplier selection, contract negotiations, and material availability are then determined. This process has sometimes been referred to as *Vendor Requirements Planning (VRP)*. If acquisition problems exist, they must be resolved by purchasing, the MRP I-planned orders must be adjusted, or the master production schedule (MPS) may have to be revised.

The execution of the purchasing portion of the MRP is the actual disbursement and follow-up of purchase authorizations. Feedback from purchasing of exceptional conditions may require rescheduling at one or more of the higher planning levels (John, 1993). In essence, purchasing is the procurement of outside-work-center capacity. It must be planned and scheduled well to minimize the final customer's overall cost.

The purchasing lead-time must include the following:

1. Order preparation time.
2. Supplier lead-time.
3. Transit time.
4. Receipt-to-stock time.

It must be noted, that the due date for MRP I planning is the date to stock, that is, on this due date the items would be available for issue to the shop, whereas the due date to the supplier will be the date for delivery to the plant and hence, would be earlier than the due date for MRP I by the receipt-to-stock time. Further, reducing the purchasing lead-time will not decrease work-in-process, but will reduce the critical path, that is the cumulative lead-time, making the system more flexible and able to operate to a shorter forecast horizon (Toomey, 1996).

Supplier (Vendor) Scheduling

Good purchasing systems typically separate the procurement activity (buying) from routine order release and order follow-up (supplier scheduling). Procurement (or buying) is a highly professional job involving contracting for vendor capacity and establishing ground rules for order release and order follow-up (supplier scheduling). Supplier scheduling is a purchasing approach that provides suppliers with *schedules* rather than with individual hard-copy purchase orders. A purchase order is a contract and a schedule rolled into one document. With supplier scheduling, the contract and the schedule are separated. The contract is arranged ahead of time. In a typical case, the company and supplier negotiate the contract once a year.

Blanket Orders

Use of MRP I for forward planning of requirements has considerably changed the conception of the purchasing function in an organization. By performing 'rough-cut' planning, it is possible to estimate the approximate quantities of all purchased items that will be required over a planning horizon of up to one year ahead. This information can be used as the basis of *blanket ordering* from an individual supplier. Thus, purchased items that have repeat usage are negotiated for future requirements and controlled through a blanket purchase order. The blanket order is a long-term commitment with short-term releases to follow (Toomey, 1996). The schedule, on the other hand, is constantly changing. As a result, supplier schedules are communicated frequently—at least weekly and sometimes more often.

Blanket ordering is effectively a means of giving a supplier planning information about one's approximate future requirements to assist him in his own planning. The idea here is that if one places a blanket order for a year's supply, it is generally possible to negotiate an agreement whereby one does not take delivery of a year's supply in one batch, one simply orders individual deliveries to be made at very short notice, once the precise quantities and requirement dates have been fixed. The extra long-term planning information and the commitment given to the supplier enables him to firm up his own demand forecasts and cut his lead-times drastically for individual deliveries, particularly if he is operating his own MRP I system. There can be considerable advantages to both parties by exchanging the information and giving the commitment required to operate in this manner. These advantages outweigh the disadvantage of tying oneself to a single supplier for a particular item. It does, however, require the identification and cultivation of individual suppliers whose prices and standards of reliability and quality assurance are acceptable. MRP I used in conjunction with blanket ordering has been termed *scheduling the outside factory* since it almost regards the supplier as being subject to extension of one's own production planning and control system (Gibson *et al.* 1995).

Normally, a supplier scheduling system will include a business agreement (contract) for each supplier, a weekly (or more frequent) schedule for each supplier extending for some time into the future, and individuals called *supplier schedulers.* Also required is a formal priority (material requirements) planning system that works well, because it is essential in this arrangement to provide the supplier with valid due dates (Cox III and Blackstone Jr, 1998).

Salient Features of Blanket Ordering

1. Beginning and ending dates are specified in the agreement to establish its effectivity.
2. Buyer has to commit to buying a minimum quantity, or total value, during the agreement period. Each time he releases an order, he increments a running total to ensure he is on schedule for meeting the minimum quantity requirements for each ending date—suppliers charge penalties if minimums are not achieved.
3. Sometimes suppliers require that the order does not exceed a certain maximum amount during the agreement period. This is to discourage the buyer from asking for more than the vendor has committed to supply based on his capacity. This also provides the supplier an opportunity to adjust pricing for further supplies. If the supplier's costs have increased by the end of the agreement period he would certainly *not* want to sell excess amount at the old rates.
4. If there is a maximum amount specified, the running total, once it crosses the minimum, is compared to it. However, the decision whether to enforce the max quantity restriction or not is left to the supplier.

IT Tasks in Blanket Order

1. Maintain running totals of all released quantities of a purchased item and make them available to the buyers on demand.
2. Assign release number to every order issued against blanket orders.
3. Provide ability to generate several releases in a single order while scheduling receipts over time, if desired.
4. Generate error messages for releases issued outside the effective dates.
5. Generate error messages for releases that causes blanket maximums to be exceeded, either in terms of volume or value, whether for an individual product or for the entire blanket order.
6. Automate all clerical aspects of the process.

The *buyer, supplier scheduler* and *supplier* are most directly involved with supplier scheduling. The responsibilities of these people are described next.

Buyer (Purchasing Agent)

Buyer is responsible for negotiating with the suppliers. He is an individual whose functions include supplier selection, definition of specifications and quality levels, price negotiation, measurement and control of supplier performance, value analysis, and evaluation of new materials and processes. In some companies, the functions of order placement and follow-up are handled by the *supplier scheduler* (Cox III and Blackstone Jr, 1998).

However, in those companies where there is no separate position of *supplier scheduler* order placement and supplier follow-up are also the responsibility of the buyer.

Value Analysis (VA) One of the functions performed by buyer listed above is value analysis. APICS Dictionary defines Value Analysis *as the systematic use of techniques that identify a required function, establish a value for that function, and finally provide that function at the lowest overall cost*. This approach focuses on the functions of an item rather than the methods of producing the present product design. Another related term is Value Engineering, which is defined by APICS Dictionary *as a disciplined approach to the elimination of waste from products or processes through an investigative process that focuses on the functions to be performed and whether such functions add value to the good or service.*

When a description of the need is being developed, tremendous benefits can result from collaboration and teamwork among the requestor, the buyer, and frequently the supplier. It is the buyer's responsibility to verify that a need actually exists and to help determine if the most cost-effective means of satisfying that need have been identified. The systematic technique of Value Analysis (VA) is one method that can be used by the team to maximize the value of a particular material, part, or service. The VA method asks, 'What is the function to be performed by the material, part, or service?' Once the function is identified, the next question is, 'What is the value of this function?' After the functions and its value have been identified, it then becomes the team's responsibility to determine how to match the value of the function with the cost of providing it. Such systematic exploration of the association between value and cost often leads to creative and more cost-effective alternatives to the original part, material, or service that was requested (McKeller, 1997).

Supplier (Vendor) Scheduler

He is a person whose main job is working with suppliers regarding *what* is needed and *when*, and is responsible for managing the supplier schedule. Supplier schedulers are in direct contact with both MRP I and the suppliers. They do the material planning for the items under their control, communicate the resultant schedules to their assigned suppliers, do the follow-up, resolve problems, and advise other planners and the master scheduler when purchased items will not arrive on time to support the schedule. Just as the manufacturing work center is controlled with a dispatch list, supplier control can be maintained by a *supplier due date list* of all open items. The supplier schedulers are normally organized by commodity, as are the buyers. By using the supplier scheduler approach, the *buyers* are freed from day-to-day order placement and expediting, and therefore, have the time to do cost reduction, negotiation, supplier selection, alternate sourcing, etc.

Supplier

The supplier's role in using the supplier schedule is to meet the due dates or communicate any problems to the supplier scheduler. In the event of a significant change to the supplier schedule, the supplier scheduler will typically call the supplier to discuss the situation prior to sending a revised schedule.

- In effect, the buyer sets up the buy, and the supplier scheduler works out the individual deliveries.
- The buyer typically doesn't come back into the process unless the supplier scheduler is having a serious problem with the suppliers.

Providing scheduling information to suppliers is one aspect of working *with* them rather than *against* them. The supplier schedule is a document that gives your suppliers your best information about what you need and when you need it. If you can help your suppliers run their business better, everyone in the loop will operate more efficiently and enjoy the accrued cost savings. Surveys have shown that companies using supplier scheduling consistently get much higher delivery performance (95% as against 76% for hard-copy purchase orders) and much lower purchase costs (9% less) than companies that don't.

Customer-Supplier Linking

Customer linking is going right to the source of demand—the customer. What does the customer's own internal system say about what's needed and when? Customer's know their on-hand balance, they have a prediction of demand for item, they know if they are planning to make any changes in design that would affect the demand for the item, and so on. They have more information than anyone else about the anticipated demand pattern for the item. The customer not only has the information to make good predictions about future demands, but he is already making his own predictions about how many of your items he will need and when he will need them. Customer linking is taking that prediction and making it available to you.

Today most companies have the planning system to create required supplier schedules (that is valid material plans) and also they are well equipped to exchange this amount of information electronically through Electronic Data Interchange (EDI), widespread e-mail, and inexpensive access to the Internet.

Multiple-Source Arrangement

For an item for which there are multiple suppliers, a source decision is required every time a purchase order for that item is to be released. This is similar to each manufactured part having alternate routings and one of the routings have to be manually chosen at the time of work order release. The MRP I logic can accept only one supplier at a time. This supplier has to be identified in the item master file. Thus, MRP I will recommend that the order be released to the supplier identified in the item master file at the time of replanning. However, at the time of converting the planned order into an FPO or at the time of order release the planner can manually make whatever modifications he wants and the order can be released to another supplier for full quantity or to a number of suppliers for partial quantities.

Supplier Kanban

Where JIT is being implemented or used, kanbans are an important tool for controlling the delivery of material from suppliers. A kanban can be a faxed page—sometimes called 'faxban,'—or a reusable container, or a card. Each serves as an authorization for the supplier to deliver material. When used with supplier schedule, kanban controls the timing of deliveries and the supplier schedule provides planning information out in the future. The supplier looks at the supplier schedule to see the overall demand rate and plans material and capacity to support this rate. The timing of individual quantities to be manufactured, however, is determined by the kanban triggers. It is important to realize that the extremely short lead-times used in supplier kanban and faxban situations are made possible by the supplier schedule. If the supplier had no idea what its customer needed in the future, short delivery times of a few hours or a day would not be possible. If a supplier has been using a supplier schedule to produce the items needed, then the kanban trigger to ship is just the last step in the process.

2.18.3 Supplier Scheduling and Control (SSC) System

1. Requisition Tracking The SSC system tracks a material requisition from the time it is initiated to the time it is covered by a released purchase order. The SSC system should also provide the requisitioner (MRP I planner) with a list of all open requisitions in release date sequence. This list can be used by the MRP I planner/buyer to follow up in order to assure that all open requisitions receive the appropriate attention.

2. Priority Dispatching of Open Requisitions The SSC system provides each buyer, on a daily basis, with a backlog report, in priority sequence, of all open requisitions not yet placed. The priority should reflect the relative difference between the time left to achieve on-time delivery and the supplier's quoted lead-time (the critical ratio).

3. Release-to-Dock-to-Stock Tracking To facilitate tracking, the SSC system must have the ability to assign a *purchase routing*, which can vary by item. The operational steps to be tracked could include any of the following:

(a) Supplier acknowledgement of delivery date and quantity.
(b) Acknowledgement of receipt at dock (no count).
(c) Receipt count.
(d) Acceptance by quality and/or inspection.
(e) Storage in a specific location.

(JIT attempts to eliminate the need for inspection, counting, and material movement activities.) The purchase routing can be maintained in the computer files and can be automatically assigned to each purchase order. Individuals update the SSC system activity by entering the operation number and associated data at a terminal. The SSC system allows the status or location of any purchase order to be determined by on-line enquiry. It also accumulates labor hours by operation, and will thus allow direct costing of all dock-to-stock activities.

4. Matching Supplier Invoices The final step in the SSC system is to match accounts payable entries to purchase order and receipts history to make sure that the quantity, price, terms, FOB point and so on, match to the receipt and any subsequent dock-to-stock activity. (FOB, that is *free on board,* is included in the terms of sale and identifies where the title or ownership passes to the buyer.) This comparison provides automatic feedback to accounting indicating that it is 'OK to pay,' or listing discrepancies that must be resolved before payment can be made (Oden *et al.* 1993).

2.19 CLOSED-LOOP APPROACH

Closing the loop in an MRP I system involves providing the system with feedback about progress. Feedback is provided by the execution functions so that the planning can be kept valid at all times. Feedback from production operations may require rescheduling at any of the higher planning levels. Notice that on the total manufacturing system model there are looping processes at several levels, especially from the execution of both the purchasing plan and the factory plan, whereby plans may have to be revised at higher levels based on the practical realities of the acquisition and production processes. These feedbacks are often referred to as 'closing the loop' in the system model (John, 1993).

Once an acceptable plan has been established, the system monitors progress against the plan. This is done in terms of the expected times of the work done. Thus, a job which is expected to occupy a work center for one hour is shown as a load of one hour in the capacity plan, and a value of one hour's worth of work is considered to be achieved when the job is finished, *irrespective of how long the operation really took.* In this way it is possible to measure progress against the plan (Childe, 1997). When the manufacturing and purchase schedules are not realized, the MRP II would be informed about it through the changed values of scheduled receipts which planners enter through transactions. The corrective action will come only after the replanning run. Thus, closing the loop in MRP II simply means that the planners provide the feedback to the system by entering transactions informing the system about what actually happened and then trigger a replanning run so that the system can provide a revised plan based on existing conditions.

It must be noted that manufacturing systems are basically *people systems.* Computer hardware and software provide excellent tools to update information, perform calculations, summarize data, and identify exception conditions—but human beings provide the plans and input, evaluate alternatives, select courses of action, and make things happen (Crepeau, 1991).

With a closed-loop system in place, people are willing to be held accountable because they have been part of the planning process and part of the decisions to make changes where necessary. In other words, they have ownership in the plan. Further, with a closed-loop system in place, accountability, which is fundamental to managing any business including manufacturing, can be maintained. People can be measured. Overall management effectiveness goes up dramatically when the informal system no longer runs the business.

Another advantage of closed-loop approach is that it helps you manage change better. Even the best-laid plans may have to be changed when unexpected events happen:

(a) *within the company* (a strike, equipment breakdown, etc.),
(b) *within the market place* (a competitor goes out of business, a new technology makes your product obsolete, etc.), or
(c) *within the political / economic arena* (currency rates change, new regulations affect products, etc.).

As thousands of companies have learned, with an effective closed-loop system, the business can quickly see the impact, adjust for unexpected changes, and position itself to meet increased or decreased customer demand. It can also count on information that will lead to informed decisions likely to benefit the company.

2.20 PROBLEMS ASSOCIATED WITH MRP II

A successful MRP II implementation means that the top management runs the company through MPS, inventory is under control and delivery performance is excellent. Although companies that have carefully implemented MRP II have seen dramatic improvements in customer service levels (33% improvements), inventory levels (40% reductions), and production costs (20% reductions). Still, MRP II has never fully achieved the successes that were promised in the early 1970s. First, the expectations for what MRP II could accomplish were inflated by software vendors and consultants, who seemed to assure managers that all the problems of manufacturing could be solved if they purchased certain hardware and software products. Overselling of MRP I as a 'system,' rather than just the priority planning element, diverted attention from the implementation of Master Production Scheduling (MPS) to drive it and CRP to plan the resources needed to support it. MRP II depends heavily on efficient data feedback and highly disciplined staff.

The major reasons behind the MRP II problems have been associated with:

(a) the implementation process, and
(b) the mismatches between certain MRP II concepts and the realities of manufacturing.

2.20.1 MRP II Implementation Problems

1. Underdeveloped Management Systems

MRP II software can relate the many information needs of all the functions and departments, including purchasing, accounting, payroll, distribution, marketing, engineering, production activity control, and general managerial planning, measurement, evaluation, and control. The involvement of so many users of the MRP II system requires a comprehensive policy development by management and the sophisticated understanding and execution at all levels of a company, which many users find difficult to implement. Some users find that 'MRP II doesn't work for us,' which usually implies that their company does not have their management systems developed to the point that they are ready for a computerized management information system (Rice and Hatch, 1997).

2. Lack of Top Management Commitment

The introduction of MRP II needs considerable changes to an organization and these require commitment from all areas.

3. Lack of MRP II Education for the Users of the System

4. Inaccurate Data

Automated systems, especially those dealing with large amounts of information, are only as good as the input provided to them. Inaccurate data can ruin an MRP II system. Thus, one of the major drawbacks of MRP II system is that it requires a lot of information, and the information should be accurate and timely. One of the main causes of MRP II failure that have been identified is inaccurate data, particularly BOM data and inventory data. It was because of these data errors that the term MRP I meant 'More Ridiculous Priorities' to many users (Plossl, 1994). Further, data entry and maintenance take up much time.

5. *Poorly Managed MPS*

MRP II is a 'top-down' system, which depends heavily on a valid MPS. Master scheduling has long been the weakest element in core systems. Management's failure to understand the role of MPS, to resolve conflicts among organization groups, and to insist on realism hamstrings MRP I, in making valid plans.

6. *Over-Sophistication/Addition of Fancy Options*

MRP I program designers, obsessed with the potential power of computers and software, attempted to build into MRP I capabilities to cope with every eventuality in manufacturing and to include every known technique, however little use this would be (Plossl, 1994). Adding fancy options, such as lot-sizing and safety stock, and unrealistic assessment of capacity are common sources of problems (Russell and Taylor III, 1998). However, which options are fancy, and which are not, is very much a debatable issue, and would vary from organization to organization.

7. *Lack of User Control*

The MRP I planner/buyer must respond correctly and expeditiously to the outputs provided by both MRP I, CRP, and all other modules of the MRP II system, as well as information external to the system. However, to respond appropriately, the planner/buyer must first understand the reason why a response is required in the first place, the various alternatives available, and also the costs and benefits of each alternative (Oden *et al.* 1993). Inexperienced planners, who do not know how to read output reports or foresee the consequences of overriding MRP II, naively accept computer recommendations. Lack of credibility of MRP II outputs, causes many capable planners to refuse to use them, depending on-side records and subsystems. Failure to take important actions, such as rescheduling open orders, negates the benefits MRP II makes possible (Plossl, 1994).

8. *Time-Consuming Implementation Process Delayed Payback*

The creation of databases, and the installation and tailoring of the MRP II systems to the particular manufacturing company is time consuming. Consequently, planning and implementing MRP II can take years. Hence, MRP II systems generally do not bring any significant results in the first two years of their implementation.

9. *Behavioral Problems*

As a common database for the entire firm, MRP II tends to be perceived initially by those in different functional areas as an intrusion into their turf. Although eventually, the common access to information improves managerial decision making, but there is an adjustment period while managers learn to work together.

Secondly, MRP I, being the first computerized system used on the factory floor, was quite disruptive. Shop supervisors, used to making their own daily schedules for their workers, found it difficult to adjust to the computer-printed schedules that laid out worker assignments for the whole week in advance. Similarly, workers accustomed to long queues of work waiting to be processed, instinctively slowed down their production rate as queues dwindled because of tight MRP I scheduling (Russell and Taylor III, 1998). Senior and middle management may find that the discipline of an MRP II system conflicts with their habitual freedom of action.

To summarize, poor implementation strategies, lack of support by top management, and lack of commitment by other company personnel have been the major problems with the implementation of MRP II. As with many new technologies or concepts, the technical issues of MRP II were solved far in advance of the behavioral issues. Fortunately, the passage of time and the educational efforts have helped to smooth the way for effective implementation of MRP II. To make an MRP II system operate effectively, certain disciplines are required, such as inventory record accuracy, database integrity, employee understanding, realistic master plans, and so forth.

2.20.2 Mismatch between MRP II Concepts and the Realities of Manufacturing

1. MRP II Plans for Material Requirements First, Capacity is an Afterthought

The iterative procedure used for leveling the load on a machine center and making an MRP schedule workable is not very efficient. Too many manual adjustments are required. Furthermore, in some industries, the approach is wrong. If there is a particular process that constrains the system or other capacity constraints that are difficult to relax, then they should drive the schedule rather than the availability of materials.

2. Lead-Time in an MRP II System is Fixed

The lead-time figures used in MRP II are fixed and are typically average of past practice, whether that practice produced good or bad manufacturing performance. Fixed lead-time assumes that lot-sizes will continue unchanged or that they have no bearing on lead-time. Further, fixed lead-times ignore current shop loads. Consider two jobs whose processing time is five hours each. If these jobs require different resources and the resources are available, then the lead-time will be five hours for each job. If, however, both jobs require the same resource (which is currently available), then one job will have a lead-time of five hours, and the other will have a lead-time of ten hours.

Thus, lead-time is a function of capacity, priority and the level of load on a particular facility at any instant. Most experts agree that the shop floor control problems experienced by MRP II users are directly related to inaccurate lead-times. Incorrect lead-time information results in orders being issued at the wrong time which, in turn, result in parts being completed late or in excess waiting time which is wasteful. The basic MRP II system has no means of taking the variations in the lead-times into account, beyond placing the best estimates of lead-time in the item master file.

Some closed loop systems update the system lead-times. The latest recorded lead-time may be used, or the program may calculate a moving average which is adjusted by each new figure. If load fluctuates then lead-times will change with load (lead-times grow at times of heavy load and shrink at times of light load). In plants that are always very busy, this causes a cycle of ever increasing lead-times because almost everything is late, causing the system to extend its lead-time. The level of work-in-process inventory therefore increases, which in turn leads to increased queues. By increasing planned lead-times in response to increasing lead-times the situation is worsened, not improved. Thus, the main drawback of MRP II, apart from the necessity of large amounts of accurate data, is that it can lead to extended lead-times and high inventory levels. It is easy for MRP II to get out of control.

3. The Reporting Requirements of MRP II are Excessive

MRP II tries to keep track of the status of all jobs in the system and reschedules jobs as problems occur. In a manufacturing environment of high speed processing and small lot-sizes, this is cumbersome. It might take as long to *record* the processing of an item at a workstation as it does to process the item. Bar code technology has helped alleviate this problem somewhat, but the issues still remain: How much processing detail is really needed in the common database? How much control is enough?

4. Execution Capabilities of MRP II are Limited

Perhaps the biggest problem with MRP II has been its limited success in the execution of shop floor schedules at a time when just-in-time (JIT) has excelled in this endeavor. Manufacturing experts are now encouraging managers to take advantage of the planning capabilities of MRP II and the execution capabilities of finite scheduling or JIT.

2.21 BENEFITS OF AND PROSPECTS FOR MRP II

Excellent Planning Capabilities

Typically, companies that lack MRP II do not properly master schedule or plan; they simply react to customer orders. Such companies are often dominated by a sales philosophy that assumes that the production department should be 'pounded on' to deliver whatever the customers ask for, as scheduled by the customers. Such philosophy is flawed in several respects. Unplanned, reactive production is generally more costly than planned production. And customers typically ask for unrealistic delivery dates when they distrust the abilities of producers to schedule. Effective MRP II allows effective managerial planning and control through cost-benefit analysis in managing the master schedule (Rice and Hatch, 1997). Some industries, primarily companies that produce goods in standard batches, can successfully use MRP II as is. However, for most companies, the real benefits of MRP II are obtained from its planning capabilities and its ability to coordinate a company's strategy among different functional areas. Not only can an MRP II system detail what should be ordered and when, but also it can indicate how and when late items will affect other aspects of production. It can signal how tardiness will alter the existing production schedule. MRP II improves materials flow performance and frees management time and so forth.

Centralization and Coordination

The philosophy behind MRP II is one of centralization and coordination, this allows considerable effort to be saved by everyone following the same system. Traditionally, accounting figures were not based on the same information as the manufacturing figures. This was because the manufacturing information was not believed to be accurate for accounting purposes—it was prepared by engineers, not accountants. Production departments would often independently calculate the value of work in process measured in terms of standard hours. Advanced users of closed-loop MRP systems realized that if the accuracy of the inventory records was good enough to support an MRP I system then it would be good enough to be used for financial purposes. This allowed the financial and production departments to work with the same, accurate, information, thus allowing them to keep the business plan up-to-date (Childe, 1997). Prior to MRP II data was maintained and sometimes modified by each organizational function to suit their own limited needs. With the computerized MRP II system one common set of data that serves all users has evolved. This has resulted in benefits like less maintenance, more accurate data with only one set of numbers, and less stored information.

Simulation Capability

The common database and simulation capabilities are very useful in responding quickly to 'what-if?' questions at various levels of detail. MRP II systems can be used to simulate different strategies for the use of manufacturing resources. Many of them are sufficiently powerful to allow the whole of the MPS to be calculated and the orders put through finite capacity planning, with calculations performed on the data to determine the expected financial performance of the factory over a given period. This can be performed repeatedly to allow the best scenario to be planned, taking into account the needs of customers, the creation of a smooth cash flow, the control of inventory etc. This can be used as a 'game plan' so that all departments can pull together effectively to achieve clear aims (Childe, 1997).

Standard requirements for Manufacturing Information Systems Provided

In its simplest form MRP II can be thought of as a companywide communication mechanism that translates high-level plans into specific tasks. A wealth of information useful to managers, employees, customers, and suppliers is automatically generated from the MRP II system. MRP II system provides complete information on current inventory balances, demand requirements, planned orders, open-order status, and recommended order action for all manufactured and purchased items. Further, this information is readily accessible, on-line, throughout the company and is the result of relatively few data inputs. BOM processors, purchasing modules, and customer order entry are standard requirements for manufacturing information systems. They are especially helpful in monitoring design quality, vendor quality, and customer service (Russell and Taylor III, 1998). In addition, many people throughout the organization will benefit from information generated by the MRP II program. Employees in the order-entry department, materials management, scheduling, purchasing, engineering, and factory floor will have the responsibility for inputting, updating, and maintaining data to support MRP II.

Greater Discipline

MRP II can provide a discipline so that key areas as MPS, BOM, lead-times with suppliers, etc. are reliable, accurate, relevant, and known to all parties, which is essential to any well-run management information systems.

Greater Transparency (Removal of Functional Boundaries)

MRP II starts with a business plan and ultimately will involve every function within the organization. The transparency of MRP related decisions to different areas of the firm is invaluable in building trust, teamwork, and better decisions. This ultimately results not only in a reduction of waste but also in a reduction in finger pointing.

Better Cash Flow Planning

The financial tie-ins can produce superior fine-tuning of cash flow planning and profit/cost projections (Russell and Taylor III, 1998).

Increased Responsiveness to Customers Needs

The major benefit from MRP II system is that when it is combined with

JIT execution techniques it makes the entire manufacturing planning, execution, and control system very agile so that it can respond to changing customer needs faster.

Improved Communications with Customers

When a customer calls to inquire about placing a new order, or to check on an existing order for a product, service part, or accessory, the information is readily available from an on-line computer. The customer service representative can call up the item immediately on an inquiry screen to answer questions on the phone about availability of product or parts or about current schedule dates. This is done with consistent reliability. The ability to provide customers with quick and reliable information builds trust and confidence among the customers.

Reduction in Cost of Money Investment and Space

MRP II helps in reducing raw material, in-process, and finished goods inventory levels resulting in lesser amount of money and space requirements.

Indirect Benefit

An effective MRP II ultimately results in number of indirect benefits, such as:

1. Sales revenues increase due to improved customer service resulting from better-managed production and delivery.
2. Costs are reduced by better management of capacity.
3. Reliable delivery times can be quoted at short notice.
4. Customers become more realistic in lead-time negotiation when producers demonstrate reliability.
5. Lead-times usually shrink.
6. Customer service level increases while inventory levels decrease (Rice and Hatch, 1997). Higher stock turnover can be achieved.
7. The capability to deexpedite assists in inventory reduction and workload control.
8. Utilization of facilities improves as materials are always available when needed. Better planning also results in less expediting and emergency orders. Thus, productivity increases due to less part shortages and reduced expediting.
9. Obsolescence is reduced due to forward visibility.
10. Quality improves due to schedule stability (Toomey, 1996).

All the above benefits may be real but quite difficult to calculate due to other variables. For example, the determination and auditing of a sales increase due to improved service can be most difficult. A 10% sales increase may be partially due to improved service, but market conditions, advertising, and quality improvements may have also contributed. In a market downturn, sales may actually drop by 10% after MRP II implementation, but it might be that the improved customer service due to MRP II had actually limited the drop to 10% rather than 15%. Estimated benefits that cannot be measured are of questionable value (Toomey, 1996).

The earliest efforts towards computerizing manufacturing management sought to computerize individual company procedures. It was soon discovered that the approach should be reversed and that company management procedures should be altered to fit into MRP II. Early resistance to MRP II included contentions that 'our company is different, and generalized MRP II will not work for us.' The evolution of MRP II has shown that few companies are really 'different' in the principles of management and that only small adjustments arc necessary from company to company. Today the nearly universal approach is to select from commercial software packages that which best fits a particular company and, with minimal revision of software, to adapt company management procedures to be consistent with the software (Rice and Hatch, 1997).

This has led to the benefit that MRP II computer software has been developed to a high degree by vendors and is now sold 'off the shelf' for implementation by knowledgeable users. However, it has likewise led to an error of thinking that MRP II is a computer system; some managers fail to understand that MRP II is instead a management system that is computerized (Rice and Hatch, 1997).

2.22 MRP II AND JIT

MRP II and JIT are Compatible and can be Used Together in Most Situations

MRP II is a push system that pushes material into production to meet anticipated future needs. A master schedule is constructed based on future forecasts/orders. This MPS then drives the MRP I process and determines what components and parts are to be ordered and pushed through production. JIT, on the other hand, is a pull system, in which materials are pulled through production by the subsequent work centers. Because of the lack of forward visibility, a repetitive master schedule is required for a pull system to work.

In spite of this major difference, there is nothing to stop MRP II being used as the planning system, then for the tools and techniques of JIT to be used to actually pull the materials only when needed. At any rate, there must be some sort of master plan for a given time period in order for the firm to know what is to be made in a particular time. MRP II can therefore be used as an exhaustive management tool by which numbers of products and, consequently, sub-components can be determined and tracked throughout the process. MRP II should not be used to push components or materials onto a workstation before they are required.

The ideals of JIT, namely *Zero Breakdowns, Zero Defects, Zero Delays, Zero Lead-Time, Zero Inventory, Zero Paper, Zero Accidents,* are also compatible with MRP II. Thus, the only conflict between MRP II and JIT appears to be in the implementation of kanban in production and procurement. But this conflict is not real and can be very easily resolved as follows:

To implement Kanban in production, the SFC module of MRP II is eliminated and the entire execution process except for the final assembly is done using the pull system of JIT. The MPS determines the number of final assemblies to be produced. Then the final assembly pulls the components from all the preceding work centers. MRP II provides the much needed forward visibility for planning of capacity and cash flow requirements.

To implement supplier kanban in procurement, following steps are required:

1. Identify the Items to be Procured Through Supplier Kanban Here the cost of the item might be a possible consideration—for example, a company might decide that it will procure the most expensive items through traditional method only.

2. Identify the Best Supplier(s) for the Component so that He/They can be Considered to be the Single Supplier or for Supplier Reduction. Here the vendor rating capability of the MRP II is of great help.

3. Negotiate a Blanket Order with the Supplier A fairly good estimate of future requirements in terms of aggregate quantities to be procured over contract period and approximate delivery rate is necessary for these negotiations. MRP II with its forward planning capability can provide these figures

4. Design Kanban Containers and/or Cards and Implement the System

5. Use Backflushing for Accounting Purposes This requires that the MRP II system should have the backflushing capability.

In summary, it can be said that MRP II and JIT are not only compatible but they also help each other in being more effective. Table 2.12 summarizes the application of MRP II and JIT in different manufacturing environments. Further, with the implementation of JIT in the execution process, the performance of production planning (MRP II) system also improves because:

(a) lead-times are reduced making the forecast—the chief driver of MRP II—to be more accurate;
(b) with the implementation of kanban system in both production and procurement a number of processes, such as ordering, incoming inspection, storing, issues and accounting are either eliminated or reduced in number and/or scope; and
(c) less number of transaction are to be entered as the entire SFC module can be eliminated in case of repetitive manufacturing.

Table 2.12

Application of JIT and MRP II in Different Manufacturing Environments
(Adapted from Schroeder, 2000)

	Type of Manufacturing		
	Pure Repetitive (Mass) Manufacturing	*Batch Manufacturing (or Repetitive Manufacture that Shifts from one Product to Another)*	*Non-Repetitive Job or Batch Manufacturing*
Characteristics	■ Master Schedule is the same from day-to-day and level-loaded.	■ Master schedule will not be identical for all periods but will have repetitive elements within fixed periods.	■ Master schedule is different for every period.
The Suitable Systems	■ ***JIT.*** The pull system of JIT works well. Breakdowns—which form the only element of uncertainty—are handled by Total Productive Maintenance (TPM).	■ A hybrid of JIT and MRP II (with SFC eliminated). If the system keeps shifting from repetitive manufacture of one product to another from time-to-time, but is still repetitive within each fixed planning horizon, then MRP II is needed for *major event planning*—that is, for planning changes in the future master production schedule. Here, the JIT system, which is used for execution of the plans on the shop floor, is adjusted for major events by the MRP II system.	■ MRP II to plan and control production. ■ The pull system of JIT (Kanban) will not work in this case. ■ However, certain other elements of JIT, such as setup time reduction, flexible workers and equipment, problem solving and continuous improvement by every employee, and supplier partnership, can be effectively used.

(*contd.*)

	Type of Manufacturing		
	Pure Repetitive (Mass) Manufacturing	*Batch Manufacturing (or Repetitive Manufacture that Shifts from one Product to Another)*	*Non-Repetitive Job or Batch Manufacturing*
Reasons	■ The end item and hence the component parts needed each day are the same and there is no need to predict future production requirements hence MRP is not required.	■ MRP II is required to push the material into factory and to plan capacity.	■ Material must be pushed into the factory to meet future demands that are different for each order.
		■ JIT (pull system) is used for execution on the shop floor. This makes it possible to eliminate the SFC element of MRP II with its substantial work-in-process tracking.	■ A capacity-planning and SFC system is needed to regulate the flow of materials through production.

2.23 ROLE OF MRP II IN PROCESS MANUFACTURING

For many years during early development of the production and inventory control body of knowledge, so-called process industries were convinced that they were different from metal, wood, and plastics working and that APICS teachings could not be applied to them. While it is true that very few process industry facilities fall into job shop and intermittent job lot categories, the distinction between repetitive and continuous production for specific plants is often Arbitrary—based more on tradition than reality. Even in the process industries, few plants run continuously as the petroleum refineries do, making the same products; most are repetitive, modified flow facilities. The first *APICS Special Interest Group (SIG),* formed in 1980, undertook as their first task the definition of Production and Inventory Control (PIC) system framework for process industries. The diagram they produced was obviously applicable to all types of production, showing the same core system elements as previous charts for other industries (Plossl, 1997).

In reality, the similarities between process and other industries far outnumber their differences. Planning and control of production operations are based on data involving the use of facilities and movement of materials. The system structure and use of data within one of the above production categories is influenced very little by differences in materials, processing, and markets served. The core planning and control system needed is common to all categories regardless of what is made, how it is made, and for whom it is made (Plossl, 1997). However, there are a few minor differences in MRP II functionalities as applied to discrete and process industries, some of which are highlighted in Table 2.13 and discussed below.

Continuous Process Manufacturing Requires MRP I Mostly for Purchasing.

This with its fixed routings, shallow bills of material (formulas), and short manufacturing lead-times do not require the time-phased dependent demand logic of MRP I for scheduling and WIP control. If there are finished goods calling for a high degree of raw material commonality or raw material with long procurement lead-times, MRP I systems can be useful in purchasing management.

Batch Process Production Requires MRP I

When process manufacturing is in a batch mode, with four or more BOM levels, longer manufacturing lead-times, routing variations, or work-center-dependent lot-sizes, planning and scheduling complications can be similar to those of jobshop environments and MRP I logic is needed.

Further, shop floor data collection in continuous processing environments will be for quality and cost control purposes rather than WIP planning and control (Toomey, 1996).

Table 2.13

Differences Between MRP II Functionalities as Applied to Discrete and Process Manufacturing

Sr. No.	*Function*	*Discrete Manufacturing*	*Process Manufacturing*
1.	MPS	Unit driven production.	Process driven family.
2.	Constraint	Material availability.	Capacity utilization.
3.	Planning Issues	Schedule attainment.	Process continuity and yields.
4.	MRP I	Primary tool for detailed planning.	Secondary tool, or not used for detailed planning.
5.	CRP	Secondary tool for detailed planning.	Primary tool for detailed planning.
6.	BOMs/ Formulas	1. Deep and complex BOMs. 2. BOM shape is generally either 'A' or 'X'. 3. BOM models the product breakdown. 4. Quality specifications are tight (for example, acceptable dimensions 10 ± 0.002 mm). 5. Process outputs are accurately predictable.	1. Shallow and simple formulas 2. Bill shape could be 'A,' 'V,' 'X,' or 'I.' 3. BOM models the manufacturing process. 4. Quality specifications are much wider (for example, acceptable water content between 2 and 7%). 5. Process outputs may vary due to variation in the input specifications, such as potency (a measurement of active material in a specific lot), concentration, or purity.
7.	Material Specifica-tions	Item number, quantity, and location are enough in most cases.	Additional specifications, such as unit of measure, lot/batch number, sublot number, expiry and/or best-before date, container type and ID, potency, etc. are required.
8.	Routings	Many with alternatives.	Fixed by process.
9.	Work centers	Discrete and flexible.	Fixed and dedicated.
10.	Planners	Commodity and part focused.	Process focused

References

Childe, Stephen J, *An Introduction to Computer-Aided Production Management*. London: Chapman & Hall, 1997.

Correll, James G, 'Reengineering the MRP II Environment', in *APICS: 37th International Conference Proceedings,* American Production and Inventory Control Society, Falls Church, V A, 1994, pp 1–4.

Cox III, James F, and John H Blackstone Jr, *APICS Dictionary* (ninth edition), American Production and Inventory Control Society, Falls Church, V A, 1998, pp 1–4.

Crepeau, Eugene A, *Production Activity Control: (APICS) Certification Review Course, Instructor Guide*, American Production and Inventory Control Society, Falls Church, VA, 1991.

Crepeau, Eugene A, 'CRP: Manage Your Capacity . . . or It Will Manage You!', in *APICS: 38th International Conference Proceedings*; American Production and Inventory Control Society, Falls Church, V A, 1995, pp 134–136.

Dougherty, John R, 'Master Planning: A Key to Satisfying Customers in the 90s', in *APICS: 38th International Conference Proceedings*, American Production and Inventory Control Society, Falls Church, V A, 1995, pp 80–84.

Fogarty, Donald W, John H Blackstone Jr, and Thomas R Hoffmann, *Production and Inventory Management* (second edition), South-Western Publishing Co, Cincinnati, Ohio, 1991.

Gibson, P, G Greenhalgh, and R Kerr, *Manufacturing Management: Principles and Concepts,* Chapman and Hall, London, 1995.

Goddard, Walter E (Editor), *Production and Inventory Control Information Requirements.* In James H, Greene, *Production and Inventory Control Handbook* (third edition), McGraw-Hill, Inc, New York, 1997.

Hitomi, Katsundo, *Manufacturing Systems Engineering*, Taylor and Francis Ltd, London, 1996.

John, Ralph E St, *Material and Capacity Requirements Planning: (APICS) CPIM Certification Review Course, Student Guide*; Revision 3, American Production and Inventory Control Society, Falls Church, V A, 1993.

Kenworthy, John, *Planning and Control of Manufacturing Operations,* John Wiley & Sons, Inc, New York, 1997.

Landvater, Darryl V, *World Class Production and Inventory Management* (second edition), John Wiley and Sons Inc, New York, 1997.

McMahon, Chris and Jimmie Browne, *CADCAM: From Principles to Practice,* Addison-Wesley Publishing Company, Wokingham, England, 1993.

McKeller, John M (Editor), 'Purchasing', in James H Greene, *Production and Inventory Control Handbook* (third Edition), McGraw-Hill, Inc, New York, 1997.

Noujaim, Selim (Editor), 'Professionalism', in James H Greene, *Production and Inventory Control Handbook* (third Edition), McGraw-Hill, Inc, New York, 1997.

Oden, Howard W, Gray A Langenewalter, and Raymond A Lucier, *Handbook of Material and Capacity Requirements Planning*, McGraw-Hill, Inc, New York, 1993.

Plossl, George, *Orlicky's Material Requirements Planning* (second Edition), McGraw-Hill, Inc, New York, 1994.

Plossl, George (Editor), 'The Production and Inventory Control Environment', in James H Greene, *Production and Inventory Control Handbook* (third Edition), McGraw-Hill, Inc, New York, 1997.

Rice, James W and Michael F Hatch, 'Manufacturing Resource Planning', in James H Greene, *Production and Inventory Control Handbook* (third Edition), McGraw-Hill, Inc, New York, 1997.

Russell, Roberta S and Bernard W Taylor III, *Operations Management: Focusing on Quality and Competitiveness* (second Edition), Prentice-Hall International, Inc, New Jersey, 1998.

Sari, F John, 'Master Production Schedule', in Greene, James H, *Production and Inventory Control Handbook* (third Edition), McGraw-Hill, Inc, New York, 1997.

Sartori, Luca G, *Manufacturing Information Systems*, Addison-Wesley Publishing Co, Wokingham, England, 1988.

Schroeder, Roger G, *Operations Management: Contemporary Concepts and Cases*, Irwin McGraw-Hill, Boston, Burr Ridge, I L, 2000.

Shafer, Scott M and Jack R Meredith, *Operations Management: A Process Approach with Spreadsheets*, John Wiley & Sons, Inc, New York, 1998.

Stonebraker, Peter B, *Master Planning: (APICS) CPIM Certification Review Course, Student Guide*, Revision 3, American Production and Inventory Control Society, Falls Church, V A, 1994.

Taylor, Sam G and Steven F Bolander (Editors), 'Production Planning and Scheduling for Process Flow Manufacturers', in James H Greene, *Production and Inventory Control Handbook* (third Edition), McGraw-Hill, Inc, New York, 1997.

Toomey, John W, *MRP II: Planning for Manufacturing Excellence,* Chapman and Hall, New York, 1996.

Vollmann, Thomas E, William L Berry, and D Clay Whybark, *Manufacturing Planning and Control Systems* (fourth Edition), McGraw-Hill, Inc, New York, 1997.

Wemmerlov, U, 'Assemble to Order Manufacturing: Implications for Materials Management', *Journal of Operations Management,* 4(4), pp 347–368, 1984.

Chapter 3

Strategic and Business Planning

3.1 STRATEGIC PLANNING

What is a strategy? Unfortunately, strategy is both an underdefined and overused word. In the most basic sense, a definition of strategy is the *science* and *art* of the military *command* exercised to meet the enemy in combat under *advantageous conditions.* Current management literature tailors this definition to the business environment. For example, Hayes and Wheelwright (1984) suggest that:

> *Business strategy specifies the scope of the business in a way that links the strategy of the business to that of the corporation as a whole, and describes the basis on which that business unit will achieve and maintain a competitive advantage.*

Manufacturing, marketing, R&D, finance, and human resources must all create strategies that collectively and synergistically comprise the overall strategy for the business. These strategies are referred to as *functional strategies.* Each may have a slightly different emphasis on achievement of one or more of the objectives of the business strategy (Beckman *et al.* 1997).

Characteristics of Strategy

1. Strategy is concerned with decisions which tend to have a long-term impact and not just short-term, quick-fix solutions.
2. Strategy is formed by senior managers, although other levels may be involved. It was believed, particularly before 1980, that strategy was an elitist activity, devised at the highest levels of the firm

by the senior managers only, without any bottom-up involvement or consideration, and then it was simply passed down for implementation. This view has changed to some degree, and now the strategy process in the firms involves various functional areas and people from levels other than just the highest levels.

3. Strategy serves to focus the organization's aims—these can then be part of its mission statement.
4. Strategy is about generating possible options (planning) which will then become narrowed down to specific choices (actions for implementation).

Strategic plans cover long time horizons (typically 3 to 5 years) and include the following:

1. A statement of the charter of the business.
2. A brief summary of the type of business it is and will become.
3. Strategies for the market places entered.
4. Types of products to be manufactured.
5. Manufacturing practices to be followed.

3.2 INTENDED AND ADAPTIVE STRATEGIES

Strategy can be both proactive (intended or planned) and reactive (adaptive or unplanned).

Intended or Planned Strategy

It comprises deliberate and purposeful actions which managers have carefully plotted out in advance.

Adaptive or Unplanned Strategy

It comprises of as-needed reactions to unanticipated developments and fresh competitive pressures. Adaptive strategies may be reactions to changing circumstances, such as important technological developments, rival's successful new product introductions, newly enacted government regulations and policies, widening consumer interest in different kinds of performance features, or whatever.

Future business conditions are sufficiently uncertain so that managers cannot plan every strategic action in advance and pursue a preplanned or intended strategy without any need for alteration. A company's actual strategies end up, therefore, being a combination of intended strategy (planned actions and business approaches) and adaptive strategy responses (on-the-spot adaptive reactions to freshly developing industry and competitive events). A company's actual strategy is something managers must shape and reshape as events transpire outside and inside the company. It is quite normal, therefore, for a company's actual strategy to differ from management's planned strategy as new strategy features are added and others subtracted to react and adapt to changing conditions (Thompson Jr and Strickland III, 1998).

3.3 NEED FOR DEVISING COMPANY STRATEGIES

Managers devise company strategies because of two very compelling needs:

1. *To Proactively Shape How a Company's Business will be Conducted*

Passively allowing strategy to drift along as the by-product of ongoing business approaches, occasional proposal for improvement, and periodic adjustments to unfolding events is a surefire ticket for inconsistent strategic actions, organizational drift, competitive mediocrity, and lackluster business results. Rather, it is management's responsibility to exert entrepreneurial leadership and commit the enterprise to conducting business in a fashion shrewdly calculated to produce good performance. Drawing a carefully reasoned conclusion about what the company's long-term direction should be pushes managers to take a hard look at the company's present business and form a clearer sense of whether and how it needs to change over the next five to ten years. A strategy provides a road map to operate by, a prescription for doing business, a game plan for building customer loyalty, and winning a sustainable competitive advantage over rivals.

2. *To Develop a Companywide Coordinated Game Plan*

Strategic planning can help in molding the independent decisions and actions initiated by departments, managers, and employees across the company into a *coordinated*, companywide game plan. Absent a strategy, managers have no framework for weaving many different action initiatives into a cohesive whole, no plan for uniting cross-department operations into a team effort. A company's activities necessarily involve the efforts and decisions of many divisions, departments, managers, and employees. All the actions and initiatives being taken in such areas as production, marketing, customer-service, human resources, information systems, R&D, and finance need to be mutually supportive if a companywide game plan, that makes good business sense is to emerge (Thompson Jr and Strickland III, 1998).

Good strategy and good strategy execution are the most trustworthy signs of good management. Competent execution of a mediocre strategy scarcely merits enthusiastic applause for management's efforts.

3.4 MAKING DECISIONS IN AN ORGANIZATION

Some decisions are very important to an organization with consequences felt over many years. Other decisions are less important, with consequences felt over days or even hours. We can use these levels of importance to classify decisions as *strategic, tactical,* or *operational* (Waters, 1996).

I. *Strategic decisions* are made by senior management, they are long-term, use many resources, and involve high risk.
II. *Tactical decisions* are made by middle management, they are medium term, use fewer resources, and involve less risk. Tactics are the methods or techniques of supporting and executing the objectives designated by the chosen strategy.
III. *Operational decisions* are made by junior management, they are short term, use few resources, and involve low risk.

Every organization makes decisions at all levels. For a manufacturer, a decision to build a new factory five or 10 years in the future is strategic; a decision to introduce a new product one or two years in the future is tactical; a decision about the number of units of a product to make next week is operational. Table 3.1 differentiates between strategy and tactics.

Table 3.1

Differences between Strategy and Tactics

(Adapted from Hitomi, 1996)

Sr. No.	*Strategy*	*Tactics*
1.	Priority on external issues of the system where matters are uncertain.	Priority on internal issues involving only matters of certainty..
2.	Deals with long-term issues in the wide territory of the system environment.	Treats problems in the short-term in the narrow area inside the system.
3.	Involved in decision making for issues in the future (futuristically oriented).	Involved in decision making for present problems.
4.	No routine method or means for obtaining solutions only intuition and experience is helpful.	Decision making is rather programmed and optimum solutions are obtainable.

The time scales of, and the expenditures, associated with, decisions vary widely between organizations. A strategic decision for a power corporation might involve the building of new power stations. This decision looks 20 years or more into the future and involves huge expenditure. A strategic decision of a retail shop might look one or two years into the future and involve a relatively insignificant expenditure. The important point is that decisions are always needed at different levels.

Senior managers make the strategic decisions that set an organization on its course. These decisions are not an end in themselves, but are the beginning of a planning process that filters down through the entire organization. Strategic decisions set the environment in which lower level decisions are made; they are passed down the organization to middle management and give the objectives and constraints for more detailed tactical decisions. These, in turn, are passed down the organization to give the objectives and constraints for the detailed operational decisions made by junior managers. The result is a hierarchy of decisions as shown in Fig. 3.1.

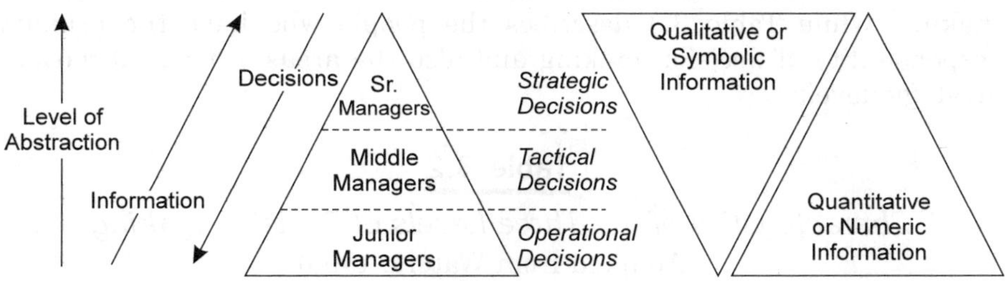

Fig. 3.1 *Flow of Information and Decisions, and Symbolic and Numeric Information and their Amounts at Different Levels*

While decisions are passed downwards through the management hierarchy, information about actual performance and other feedback is passed upwards. This information must be filtered as it passes upwards, or top managers would be swamped by too much detailed, irrelevant information. They need enough details to be useful, but not so much that overall patterns are hidden (Waters, 1996). Although this information flow is complex, two generic classes can be distinguished (McMahon and Browne, 1993):

1. Qualitative (or Symbolic) Information

The first class concerns qualitative information, which supports highly abstract statements and therefore is predominant in strategic decision making (McMahon and Browne, 1993). Many companies perceive strategy to be broadly based and are content with arriving at all-embracing but essentially unclear statements (Hill, 1995).

2. Quantitative (or Numeric) Information

The second class consists of numeric information. Although its use is not limited to operational layers, this class represents more or less a quantitative translation of goals and strategies, which are elaborated in symbolic terms.

Figure 3.1 shows an overall representation of the amount of information type at different levels.

- The importance of symbolic information decreases as we move from the strategic management level down to the operational level.
- Alternatively, the importance of numeric information grows as we proceed from the higher to the lower level.
- Each layer of the hierarchy translates 'some' symbolic information to 'some' numerical information. The translated 'chunk' of information corresponds to the result of a decision making activity.

Table 3.2 summarizes characteristics of these three levels of decision making; while Table 3.3 describes the people who have the primary responsibility of decision making and also the areas of focus at various strategic levels.

Table 3.2

Characteristics of the Three Levels of Decision Making
(Adapted from Waters, 1996)

Aspect	*Type of Decision*		
	Strategic	*Tactical*	*Operational*
Level of Manager	Senior	Middle	Junior
Importance	High	Medium	Low
Resources Used	Many	Some	Few
Time Scale	Long	Medium	Short
Risk	High	Medium	Low
Uncertainty	High	Medium	Low
Amount of Detail	Very General	Moderate	Very Detailed
Data Available	Uncertain	Some	Certain
Structure	Unstructured	Some	Structured
Focus	Whole Organization	Divisions in the Organization	Individual Units

Table 3.3

Primary Responsibility and Areas of Focus at Different Strategy Levels

Strategy Level	*Primary Responsibility*	*Areas of Focus*
Corporate	CEO, Other Key Executives	Determining a portfolio of business units; long-term investment decisions; reviewing, revising, unifying major strategic approaches.
Strategic Business Units	General Managers	Determining areas of competitive advantage; forming responses to external changes, seeking opportunities.
Functional Strategies	Functional Managers	Determining strategies for key areas such as production, marketing, finance, human resources, R&D.
Operating Strategies	Unit Leaders	More time-specific strategies aimed to support the broader strategies of senior managers.

3.5 THE FIVE TASKS OF STRATEGIC MANAGEMENT

The strategy-making and strategy-implementing process consists of five interrelated managerial tasks (Thompson Jr and Strickland III, 1998), which has been shown in Fig. 3.2.

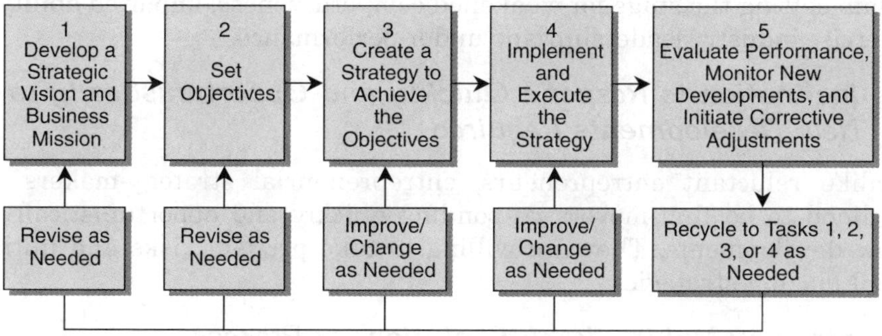

Fig. 3.2 *Five Tasks of Strategic Management* (Adapted from Thompson Jr and Strickland III, 1998)

1. Forming a strategic vision of what the company's future business makeup will be and where the organization is headed. This provides long-term direction, delineates what kind of enterprise the company is trying to become, and infuses the organization with a sense of purposeful action.

2. Setting objectives. Converting the strategic vision into specific performance outcomes for the company to achieve.
3. Crafting a strategy to achieve the desired outcomes.
4. Implementing and executing the chosen strategy efficiently and effectively.
5. Evaluating performance and initiating corrective adjustments in vision, long-term direction, objectives, strategy, or implementation in light of actual experience, changing conditions, new ideas, and new opportunities.

3.6 CHARACTERISTICS OF STRATEGY-MAKING PROCESS

1. Strategy Making is Fundamentally a Market-Driven and Customer-Driven Entrepreneurial Activity

Venturesomeness, business creativity, an eye for spotting emerging market opportunities, keen observation of customer needs, and an appetite for risk-taking are inherent to the task of crafting company strategies.

2. Good Strategy Making is More Outside-in than Inside-out

Inside-out strategies are dictated by inside considerations, such as what is philosophically comfortable, what is acceptable to various internal political coalitions, and what is safe, both organizationally and careerwise. These strategies, while not disconnected from external developments, nearly always fall short of being truly market-driven and customer-driven, once again setting the stage for weakened competitiveness, impaired ability to exercise industry leadership, and under-performance.

3. Inclination to Respond Quickly and Opportunistically to New Developments Required

Unlike reluctant entrepreneurs, entrepreneurial strategy-makers are inclined to be first-movers responding quickly and opportunistically to new developments. They are willing to take prudent risks and initiate trail blazing strategies.

4. Strategy-Making is an Evolutionary Process

A company's strategy is dynamic, forming in bits and pieces and then reforming as managers see avenues for improvement or a need to adapt business approaches to changing conditions. Frequent fine-tuning and tweaking of a company's strategy, in various functional areas are quite normal. Occasionally, however, quantum changes in strategy are called for—when a competitor makes a dramatic move, when technological breakthroughs occur, or when crisis strikes and managers are forced to make radical strategy alternatives very quickly.

3.7 STRATEGY-MAKING AS A GROUP TASK

Many companies are today involving teams of managers and key employees in strategy-making exercises. Frequently, these teams include line and staff managers from different disciplines and departmental units, a few handpicked junior staffers known for their ability to think creatively, and near-retirement veterans noted for being keen observers, telling it like it is, and giving sage advice. Customers and suppliers are also, sometimes, involved in these teams (Thompson Jr and Strickland III, 1998). Following are the reasons for making strategy-making a group task:

1. Since many strategic issues cut across traditional functional and departmental lines, involvement of all the key employees is important.
2. This gives an opportunity to tap into the ideas and problem-solving skills of people with different backgrounds, expertise, and perspectives.
3. This also gives a greater number of people an ownership stake in the strategy that emerges and wins their wholehearted commitment to implementation.

3.8 STRATEGIC VISION (Organization's Values and Aspirations) AND MISSION (Organization's Purpose or Reason for Existence)

Strategic Vision/Vision Statement

A strategic vision is a roadmap of a company's future—the direction it is headed, the business position it intends to stake out, and the capabilities it plans to develop.

Mission/Mission Statement

Mission is defined by APICS Dictionary as the overall goal(s) for an organization set within the parameters of business (Cox III and Blackstone Jr, 1998). Many organizations have an overall statement of their purpose, which defines their mission. This gives the purpose of the organization—its reason for existence. Often, what a company is currently seeking to do for its customers forms its mission. A mission statement is useful for putting the spotlight on what business a company is presently in and the customer needs it is presently endeavoring to serve. For example, the mission of a manufacturer might be to supply high quality products to a wide market, while making a reasonable profit.

The mission statement is sometimes very brief. But more often the mission statements are longer and describe the fundamental beliefs and aims of the organization. They answer questions like, 'What is our purpose?' 'What business are we in?' and 'What are our overall objectives?' Sometimes they describe the organization's responsibilities to its shareholders, employees, customers and community. Some statements also include views on productivity, markets, technology used, public image, company philosophy, and a range of other topics. A mission statement that is too specific can be confining for the firm, on the other hand, a mission statement that is too broad tends to defocus the business.

Many organizations do not have a formal statement of their mission. But there are benefits in having a precise statement of overall purpose. The mission statement is important because a well-thought-out mission statement controls the scope of the business of the firm (Pannesi and O'Brien, 1992). This gives a focus to make sure employees are working towards the same aims, that there are no conflicts, and that resources are sensibly shared out. A mission statement is the starting point for all other decisions in the organization—it sets the context that the organization works in.

- Vision statements are used to express the organization's values and aspirations. Mission statements express the organization's purpose or reason for existence (Shafer and Meredith, 1998).
- In practice, actual company mission statements tend to exhibit more concern with 'what our business is now' than with 'what our business will be later' (Thompson Jr and Strickland III, 1998).
- In the event a company's mission statement not only sets forth a clear definition of the present business but also indicates where the company is headed and what its business will become in the years ahead, then the concepts of company mission (or mission statement) and strategic vision merge into one and the same—in other words, a strategic vision and a future-oriented business mission amount to essentially the same thing.
- Some organizations may choose to combine the vision and mission statements into a single statement. Regardless of whether separate statements or combined statements are developed, the intent is to communicate the organization's values, aspirations, and purpose so that employees can make decisions that are consistent with and support these objectives (Shafer and Meredith, 1998).
- Effective vision and mission statements tend to be written using language that inspires employees to high levels of performance. Further, to foster employees' commitment, it is advisable to include a wide variety of employees in the development of the vision or mission statement, rather than enforcing top management's view by edict.
- Once the vision and mission statements are developed for the organization as a whole, divisions, departments, process teams,

project teams, work groups, and so on can develop individual vision-mission statements that support the organization's overall statement (Shafer and Meredith, 1998).

Examples of Company Mission and Vision Statements

McDonald's Corporation McDonald's vision is to dominate the global foodservice industry. Global dominance means setting the performance standard for customer satisfaction while increasing market share and profitability through our Convenience, Value, and Execution Strategies.

Otis Elevator Our mission is to provide any customer a means of moving people and things up, down, and sideways over short distances with higher reliability than any similar enterprise in the world.

Microsoft Corporation One vision drives everything we do: A computer on every desk and in every home using great software as an empowering tool.

American Red Cross The mission of the American Red Cross is to improve the quality of human life, to enhance self-reliance and concern for others, and to help people avoid, prepare for, and cope with emergencies.

Eastman Kodak To be the world's best in chemical and electronic imaging.

Intel Corporation Intel supplies the computing industry with chips, boards, systems, and software. Intel's products are used as 'building blocks' to create advanced computing systems for PC users. Intel's mission is to be the preeminent building block supplier to the new computing industry worldwide.

Compaq Computers To be the leading supplier of PCs and PC servers in all customer segments.

APICS—The Educational Society for Resource Management

Vision of APICS To inspire individuals and organizations toward life-long learning and to enhance individual and organizational success.

Mission of APICS To be the premier provider and global leader in individual and organizational education, standards of excellence, and information in integrated resource management.

3.9 CORPORATE OBJECTIVES

Corporate objectives are what the firm seeks to achieve through its operations. They are yardsticks for tracking an organization's performance and progress. Because the mission statement is vague, it is difficult to know whether or not it has been achieved. The purpose of setting objectives is to convert managerial statements of strategic vision and business mission into *specific performance targets,* something the organization's progress can be measured by. Operations objectives provide carefully defined, measurable goals that help the firm achieve its mission. Objectives are the 'ends,' and strategy is the 'means' of achieving them.

The challenge of trying to achieve bold, aggressive performance targets pushes an organization to be more inventive, to exhibit some urgency in improving both its financial performance and its business position, and to be more intentional and focused in its actions. Setting objectives that require real organizational stretch and disciplined effort helps build a firewall against the possibility of the organization being content with making slow, low-grade improvements or merely dragging on.

- The objectives must be defined carefully because these terms are often used loosely.
- Secondly, the objectives should be clearly measurable so that management knows whether they are meeting their goals or not. It is quite possible, and perhaps desirable, to use more than one measure for each objective (for instance, warranty cost and parts-per-million defective for the quality objective).
- Thirdly, the objectives should be ranked as well, so that managers can give priority to the correct objective when it is necessary to make trade-offs.
- Finally, the objectives should be treated as dynamic—to be changed with the changing conditions. For instance, as a new product begins full-scale production, the firm may emphasize flexibility (to design changes) and on time delivery so that market share is not lost to competitors. Over time, as the product design stabilizes, the emphasis may change toward quality of conformance and cost (Silver *et al.* 1998).

Although a full range of objectives is often expressed (such as those related to customer satisfaction, quality consciousness, corporate citizenship, social responsibility, and the like), some objectives seem to be more pervasive than others and in fact occupy more time and interest in the firm. From a companywide perspective, following are the more important and very distinct objectives in the firm:

1. Financial performance objectives/yardsticks
2. Strategic performance objectives/yardstick

3.9.1 Financial Performance Objectives/Yardsticks

Achieving acceptable financial results is crucial. Without adequate profitability, a company's pursuit of its vision, as well as its long-term health and ultimate survival, is jeopardized. Neither shareowners nor lenders will continue to sink additional monies into an enterprise that can't deliver satisfactory financial results.

Financial Objectives Signal Commitment to Such Outcomes as:

1. *Earnings*
2. *Sales Growth*
3. *Return on Investment (ROI) or Economic Value Added (EVA)* EVA is profit over and above the company's cost of debt and equity capital. More specifically,

   Economic Value Added (EVA)
   = Operating Profit – [Income Taxes + Cost of Debt
   + Cost of Equity Capital (Usually 15%)]

   The EVA can be interpreted to mean that the company's management has generated profits well in excess of the benchmark (15 percent) equity cost needed to justify or support the shareholder investment in equity capital—all of which represent wealth created for the owners above what they could expect from making investment of comparable risk elsewhere. Such companies as *Coca-Cola* and *AT&T* use EVA as a measure of their profit performance (Thompson Jr and Strickland III, 1998).
4. *Return on Assets.*
5. *Profit and Dividend Growth.*
6. *Stock Price Appreciation (or Market Value Added—MVA)* MVA is defined as the amount by which the total value of the company has appreciated above the amount of money actually invested in the company by shareholders.

   MVA = Company's Current Stock Price × The Number of Shares Outstanding – Shareholder's Equity Investment

   - It represents the value that management has added to shareholder's wealth in running the business.
   - If shareholder value is to be maximized, management must select a strategy and long-term direction that maximize the market value of the company's common stock.
   - In recent years, MVA and EVA have gained widespread acceptance as valid measures of a company's financial performance.
7. *Good Cash Flow*
8. *Credit Worthiness*

3.9.2 Strategic Performance Objectives/Yardstick

In addition to financial performance, attention also has to be paid to a company's strategic well-being—its competitiveness and overall long-term business position. Strategic objectives serve notice that management not only intends to deliver good financial performance but also intends to improve the organization's competitive strength and long-range business prospects. Unless a company's performance reflects improving competitive strength and a long-term market position, its progress is less than inspiring and its ability to continue delivering good financial performance is suspect. Strategic objectives direct efforts toward such outcomes as:

1. Winning additional market share.
2. Growing niche markets.
3. Overtaking key competitors on product quality, customer service, or product innovation.
4. Achieving lower overall costs than rivals.
5. Boosting the company's reputation with customers.
6. Winning a stronger foothold in international markets.
7. Exercising technological leadership.
8. Gaining a sustainable competitive advantage.
9. Capturing attractive growth opportunities.
10. Introducing new products.

The need for both financial performance and good strategic performance calls for management to set short- and long-term financial objectives and strategic objectives.

Examples of Strategic and Financial Objectives

Domino's Pizza To safely deliver a hot, quality pizza in 30 minutes or less at a fair price and a reasonable profit.

Ford Motor Company To satisfy our customers by providing quality cars and trucks, developing new products, reducing the time it takes to bring new vehicles to market, improving the efficiency of all our plants and processes, and building on our teamwork with employees, unions, dealers, and suppliers.

General Electric To become the most competitive enterprise in the world by being number one or number two in market share in every business the company is in. To achieve an average of ten inventory turns and a corporate operating profit margin of 16% by 1998.

3M To achieve annual growth in earnings per share of 10% or better, on average; a return on stockholders' equity of 20–25%; a return on capital employed of 27% or better; and have at least 30% of sales come from products introduced in the past four years

3.10 CORPORATE STRATEGY (Business or Organizational Strategy)

Once the vision-mission statement has been drafted, senior managers can make a series of other strategic decisions specifying how the vision and mission will be accomplished. These might identify the organization's core competencies, describe the methods of competition, define the contributions of each product and function to the goals of the business, and allocate resources among products and functions. It might also design the structure of the organization, where it operates, what its relations are with other organizations, how it treats customers, and so on. These strategic decisions which refer to the whole organization form the corporate strategy. Corporate strategy is the foundation upon which all other functional strategies of the firm are built. Some other terms are also used for this, particularly *business strategy, organizational strategy,* and *long-term plans.*

Development of a corporate strategy, a necessary prerequisite to generation of a manufacturing strategy, is a difficult task and iterative process. It depends on inputs from many different sources. External sources (customers, competitors, and the economic environment) should drive some objectives, while an assessment of internal strengths and weaknesses should drive some others (Beckman *et al.* 1997).

Core Competencies

One important result of developing a business strategy is identifying the organization's core competencies. According to APICS Dictionary, core competencies are the bundles of skills or knowledge sets that enable a firm to provide the greatest level of value to its customers in a way that is difficult for competitors to emulate and that provides for future growth. Core competencies are embodied in the skills of the workers and in the organization. They are developed through collective learning, communication, and commitment to work across levels and functions in the organization and with the customers and suppliers. For example, a core competency could be the capability of a firm to coordinate and harmonize diverse production skills and multiple technologies (Cox II and Blackstone, 1998). Typically, core competencies center on an organization's ability to integrate a variety of specific technologies and skills in the development of new products and services. Although core competencies are not directly related to the product or market, they, in effect, provide the basis for developing new products and services, and they are a primary factor in determining an organization's long-term competitiveness.

- Often, it is more useful to think of an organization in terms of its portfolio of core competencies rather than its portfolio of businesses or products. For instance, Sony is known for its expertise in miniaturization; 3M for its knowledge of substrates, coatings, and adhesives; Black and Decker for small electrical motors and industrial design; Boeing for its ability to integrate large-scale complex systems; and Honda for engines and power trains. Had Sony initially viewed itself as primarily a manufacturer of Walkmans, rather than a company with expertise in miniaturization, it might have overlooked several profitable opportunities such as entering the camcorder business. As another example, Boeing has successfully leveraged its core competency related to integrating large-scale systems in its production of commercial jetliners, space stations, fighter-bombers, and missiles (Shafer and Meredith, 1998).

- Core competencies are often used to gain access to a wide variety of markets. For example, Honda's core competencies in engines and power trains are the basis for its entry into other businesses: automobiles, motorcycles, lawn mowers, and generators.
- A core competence should be strongly related to the key benefits provided by the products or services provided by the organization. In Sony's case, its expertise in miniaturization translates directly into important product features such as portability and aesthetic designs.
- Core competency should be difficult to imitate. Clearly, no sustainable competitive advantage is provided by a core competence that is easily imitated.

Outsourcing The topic of core competence is strongly related to outsourcing. Outsourcing is defined by APICS Dictionary as *the process of having suppliers provide goods and services that were previously provided internally. Outsourcing involves substitution—the replacement of internal capacity and production by that of the supplier.* It is an approach that is increasingly used and involves subcontracting out certain activities such as the production of parts to an external supplier. For example, a manufacturer might outsource the production of certain components, the management and maintenance of its computer resources, or the processing of its payroll. Activities that are strategically unimportant are candidates for outsourcing. Strategically unimportant activities that might be ideal candidates for outsourcing include the parts and activities that:

(i) are not strongly linked to key product characteristics,
(ii) do not require highly specialized knowledge and skill,
(iii) do not need special physical assets, and
(iv) belong to an area in which the organization does not have (and is not likely to have in future) the technological lead.

Thus, if it is beneficial to outsource these parts or activities—perhaps because of lower cost or higher quality—no loss in competitiveness should result. On the other hand, when most of the company's strategically important parts and production activities are outsourced, particularly to a foreign supplier, the firm becomes *hollow*. If the firm outsources strategically important parts and activities as well, it soon finds that the engineering design talent follows the production of the part outside the firm, too. Then, the firm has been *hollowed out,* becoming merely a distributor of its supplier's products. In such cases, after a while, when the supplier has improved on the process and the original manufacturer has forgotten how to make the complex parts, the supplier is likely to start producing the end products himself and drop the original manufacturer as a customer. This happened extensively in the television industry, where the Japanese learned first how to produce, and then how to engineer black-and-white and later colour television sets. They then started tentatively introducing their own brands, to see if American customers would buy them. Their products were inexpensive, were of high quality, and caught on quickly in free-enterprise American markets. The Japanese now virtually control this industry (Shafer and Meredith, 1998).

Corporate Strategy Concerns

1. How to Grow the Business

- (a) Moves to diversify the company's revenue base and enter altogether new industries or businesses.
- (b) Efforts to alter geographic coverage, integrate forward or backward, or stake out a different industry position.
- (c) Actions to merge with or acquire a rival company, form strategic alliances, or collaborate closely with certain industry members.

2. How to Satisfy the Customers and Outcompete Rivals

- (a) Fresh offensive moves to strengthen the company's long-term competitive position and secure a competitive advantage.
- (b) Efforts to broaden/narrow the product line, improve product design, alter product quality, alter performance features, or modify customer service.
- (c) Defensive moves to counter the actions of competitors and defend against external threats.

3. How to Respond to Changing Market Conditions/Environment

- (a) Actions to respond to changing environment, such as shifting customer preferences, new government regulations, the globalization of competition, exchange rate instability, and entry or exit of competitors.
- (b) Actions to capitalize on new opportunities, such as new technologies, product innovation, new trade agreements that open up foreign markets.

4. How to Manage Each Functional Piece of the Business and Develop Needed Organizational Capabilities

- (a) Moves and approaches that define how the company manages R&D, manufacturing, marketing, finance, and other key activities.
- (b) Actions to strengthen the company's resources base and competitive capabilities.

5. How to Achieve Strategic and Financial Objectives A corporate strategy is the set of plan that makes sure the mission is achieved. These plans, according to Waters (1996), include decisions about:

- *Organizational Structure* This is how the organization is divided, what each division does, who works there.
- *Geographical Locations* This relates to where operations are done, which markets are important, where administration is done.
- *Dominant Orientation* A company may be oriented towards a market, to materials or products, or to a particular technology. Market-oriented companies may set out to understand fully and provide for a particular market. Material- or product-oriented companies may be tied to a particular product and will develop new uses for their material in a variety of markets (Childe, 1997).
- *Competitive Position / Priorities* This describes how the organization stands in relation to its competitors. Does the company prefer to maintain a high profit margin by selling a premium product in a narrow market, or does it prefer the low-price high volume business. Does it compete on the basis of tailoring a product to a customer's individual requirements, or by being able to bring new products to the market quickly (Childe, 1997).
- *Amount and Pattern of Diversification* This describes how wide a range of products the organization makes. Will the company widen the scope of its activities into new markets or new technologies, or will it concentrate on a more narrow set of business activities?
- *Acquisition Policy (Vertical and Horizontal Integration)* This is how the organization might expand: by growing or by buying other organizations. In case of growing by buying other organizations, there are two possibilities: a) *vertical integration* by acquiring suppliers (backward integration) or by acquiring customer organizations (forward integration); and b) *horizontal integration* by acquiring competitors both in terms of competitive products and complementary products.
- *Formation of Strategic Alliances* Rather than trying to acquire other organizations, an organization can form strategic alliances with others based on licensing agreements, buyer-seller relationships, R&D agreements, marketing agreements, through franchising, joint ventures, consortia between companies, or joint access to technology and markets. Such alliances might provide massive benefits as they allow the firms to concentrate on their core competencies, and also as they provide a means of organizational learning.
- *Profitability* This describes the relationship between income generated and costs.
- *Productivity* The amount of products made in relation to resources used.
- *Resources Used* The relative amount of various resources used by the organization.
- *Innovation* This describes how the organization changes the way it works.

Design of a Corporate Strategy

The design of a corporate strategy starts by considering three factors. The first is the *mission* itself, which sets the overall goals or corporate objectives. The second is the *environment* in which the organization is working. The scanning of the environment helps in identifying the opportunities and threats for the organization in the environment. The third is the organization's *distinctive competence,* which allows it to succeed in this environment. It is decided based on the analysis of the strengths and weaknesses of the organization as shown in Fig. 3.3.

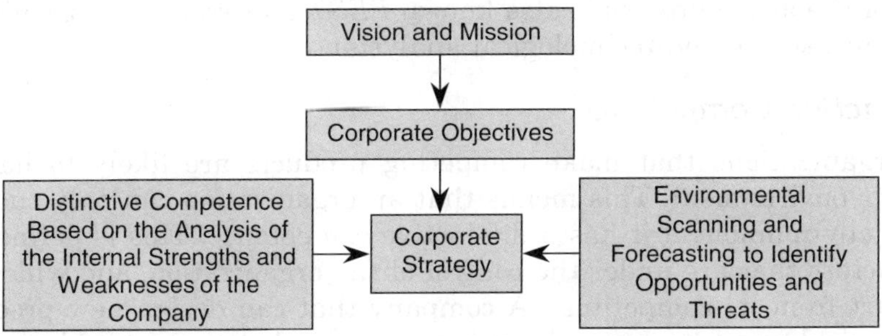

Fig. 3.3 *The Factors on Which a Corporate Strategy is Based*

Environment The environment includes all the factors that affect an organization, but over which it has no, or very little, control. These include (Waters, 1996):

- *Customers* their expectations, attitudes, and demographics.
- *Market conditions* size, location, and stability of markets, if the firm is involved in exporting then conditions in foreign markets and exchange rates.
- *Technology* current availability, likely developments, and rate of innovation.
- *Economic conditions* gross domestic product, rate of growth and inflation; unemployment levels and disposable incomes.
- *Legal restraints* trade restrictions, liability, and employment laws.
- *Competitors* number, ease of entry to the market, and strengths.
- *Shareholders* their target return on investment, objectives, and profit needed.
- *Interest groups/other players in the market* their objectives, their strength, and amount of support they have [For each market niche, both buyers and suppliers have various strengths. In some cases, especially when there is a worldwide limit on the capacity (and thus availability) of some resource or supplier part, those firms that have more powerful positions may be able to preempt supplies (Pannesi and O'Brien, 1992).

Thus, an analysis of relative balance of power between buyers and suppliers is vital for strategic planning.]

- *Social conditions* customers' life styles, changing demands, significant trends and feelings of the communities toward the firm.
- *Political conditions* stability, amount of governmental control, external relations.

Nearly all of the above issues require not only some assessment of the current situation but some analysis of future possibilities. Thus, in many cases, some forecasting of external environmental conditions is required. Part of the above process is also known *PEST analysis* (that is, political, economic, social, and technological analysis).

Distinctive Competence

All organizations that make competing products are likely to have a similar environment. This means that an organization can only succeed in its environment if it has a distinctive competence. This is defined by the factors that are under the control of the organization, and which set it apart from its competitors. A company that can design new products very quickly, and has enough resources to market these, could say this ability to innovate is part of its distinctive competence. A distinctive competence develops from an organization's assets, which, according to Waters (1996), include:

- *Customers* their demands, loyalty, and relationships.
- *Employees* their skills, expertise, and loyalty.
- *Finances* capital, debt, and cash flow.
- *Products* quality, reputation, and innovations.
- *Facilities* capacity, age, and value.
- *Technology* currently used, planned, special types.
- *Suppliers* reliability, service, and flexibility.
- *Marketing* experience, reputation.
- *Resources* patents, ownership.

SWOT Analysis

When managers want to think about their corporate strategy, a useful way of summarizing ideas is a 'SWOT Analysis.' Most firms usually have certain areas in which they excel and other areas in which they are weak. SWOT analysis should be done for all functional areas of the firm. It lists different aspects of the organization according to:

- *Strengths* *What the organization does well*—features it should build on to seek competitive advantage.
- *Weaknesses* *What problems the organization has*—areas it should improve to ensure that current weaknesses will not hurt strategic initiatives.

- *Opportunities* — *Which can help the organization*—openings it should seize.
- *Threats* — *Which can damage the organization*—hazards it should avoid.

IBM could, for example, start a SWOT analysis by listing the following:

- *Strengths* — Market dominance, customer service, and large resources.
- *Weaknesses* — Lack of innovation, high costs.
- *Opportunities* — Growing demand for computers and new technologies.
- *Threats* — Manufacturers copying IBM products, low cost competitors.

A full SWOT analysis would add many other factors to these lists, as shown in Table 3.4.

Table 3.4

Typical Listing of a Firm's Strengths, Weaknesses, Opportunities and Threats

Internal Considerations Related to Past and Present Capabilities		*External Considerations Related to Present and Future Possibilities*	
Strengths	*Weaknesses*	*Opportunities*	*Threats*
Distinctive competencies. Financial resources. Acknowledged as market leaders. Insulation from competition. Cost leaders. Product innovation skills. Superior technological know-how.	No clear direction. Obsolete plant. Poor track record of strategic implementation. A too narrow product line. Weak market image. Weak distribution network.	Additional customer groups. New markets or market segments. Opportunity for expanding product line and diversification. Falling trade barriers in foreign markets. Faster market growth.	Entry of low-cost competitors. Introduction of substitute products by competitors. Slow market growth. Adverse exchange rate condition. Changing buyer needs and tastes. Adverse demographic changes.

Wheelam and Hunger's Model for Formulation of Corporate Strategy

The internal strengths and weaknesses of a firm are considered simultaneously with external environmental factors to develop a corporate strategy. Wheelam and Hunger (1993) have presented the following model for formulation of corporate strategy:

1. Evaluation of a corporation's current performance.
2. Examination and evaluation of the current mission, objectives, and policies.
3. A scan of the external environment (opportunities and threats).
4. A scan of the internal corporate environment (strengths and weaknesses).
5. Analysis of the strategic factors and revision of mission and objectives.
6. Generation and evaluation of alternative strategies, and selection of the best alternative strategy.
7. Implementation.
8. Evaluation and control.

Alternate strategies are often formulated (Step 6) and each is tested against certain criteria or questions, some of which are:

1. Is the strategy consistent with the mission and corporate objectives of the firm?
2. Does it exploit the opportunities available and guard against possible threats?
3. Does it exploit the current strengths of the company and take measures to safeguard against the vulnerability of the company to the current weaknesses?
4. Whether the assumptions made during the strategy formulation are realistic?
5. What changes are required in the firm to implement the strategy? Is the firm capable of bringing about the required changes? Will the current organizational structure facilitate or hamper the change? Are the current resources sufficient to make the changes?

Contingency Plans

In all cases, contingency plans must be developed for the chosen strategies based on a number of possible and probable scenarios for the internal and external environment. When contingency plans are put in place before the corporate strategy is finalized, the firm is able to react much more quickly to change and not be caught unawares, unable to respond when changes disrupt the strategies of the firm (Pannesi and O'Brien, 1992).

Corporate strategy provides a basis for the coordination of decisions and activities throughout the organization. The company must focus its activities on one particular set of goals, which must be communicated throughout the organization so that everyone pulls in the same direction. If there is no coordination, different parts of the company might work against each other. Where strategy is poorly defined, it is quite possible that factory manager might plan to reduce the cost of manufacturing through the increased use of automatic equipment, which is cheap to run, but expensive to reset for each product variant. At the same time, the sales manager might accept orders for all kinds of special products with different features for each customer to increase the sales. Thus the factory manager sees the sales force chasing the wrong types of orders, which do not suit the factory, and the sales manager sees the factory buying the wrong type of equipment for the orders coming in. The two managers are both trying to do the best for the business (as each sees it), but the lack of coordination results in excess costs due to unsuitable machinery or orders, depending upon which way you look at it (Childe, 1997).

3.11 COMPETITIVE STRATEGY

An important part of an organization's corporate strategy concerns its ability to compete. All organizations have competitors—both the organizations that already supply similar products, and those that might start making similar products in the future. The part of the corporate strategy that deals with competition is sometimes separated in a *competitive strategy* or *competitive positioning* (Waters, 1996). When an organization is developing a competitive strategy it must look at its own strengths and weaknesses in relation to its competitors and try to gain competitive advantage over its competitors. The competitive advantage of the firm is that characteristic of its goods, services, or operations which causes customers to choose the firm over its competitors. If a firm can establish an advantage over its competitors, then it will influence customers to purchase from the firm and not its competitors allowing it to increase its market share and profitability.

- If a company is good at making a high quality product, while most of its competitors are aiming for lower quality, its strategy is clear — it should produce the best product available.
- If there is a trend for very large supermarkets on trading estates, one competitive strategy is to build small corner shops.

This illustrates the principle behind a competitive strategy, which is *'search for a **favorable position** in the industry.'*

Cost Leadership and Product Differentiation

There are basically two ways of finding such favorable positions—*cost leadership* and *product differentiation*.

1. Cost Leaders The strategy is to make the same product more cheaply. This means the organization lowers its unit costs by having efficient operations, often making large numbers of a standard product.

- A computer manufacturer, for example, must decide whether to make a standard computer at a low price, or a computer with features that distinguish it from the competition. Ford, for example, makes a large number of similar, inexpensive cars, while Rolls Royce makes a few different, expensive cars.

2. Product Differentiation The strategy is to make different products which have noticeable advantages to buyers, such as improved features or higher reliability. This means the organization makes products that are significantly different to competitors' and hence, they can charge higher prices.

- It might appear that some form of compromise would give better results—using some aspects of cost leadership with other aspects of product differentiation. But there is a widespread agreement that this approach does not work. Companies with 'average' products and 'average' costs do markedly worse than those that concentrate on either low costs or unique products.

Companies in 1990s and beyond have to be Both Low Cost and Differentiated Some argue that whilst being focused on competing either on cost or differentiation serves as a point of clarity and does indeed help to focus the firm. In reality companies in 1990s and beyond have to be both low cost and differentiated in particular market segments. Thus, although there is clearly a price war in certain segments within the car industry, each company is offering perceived differentiation—however small—for their products as well. One of the problems with this is that perceived differentiation is often very easily copied—for example, features such as airbags in cars—and, therefore, the perceived differentiation is lost after relatively short periods of time. Thus, in the 1990s and beyond, it becomes increasingly difficult to simply choose one or other strategy and to believe that this is the formula for success. Other competitive factors, such as flexibility, delivery speed, and reliability, and product innovation are more prominent than ever before and provide the biggest indication of production/operations' contribution.

- An overemphasis on low-cost efficiency can cause a firm to be out of touch with the changing requirements of customers, low cost production is relatively easy to copy and as industries mature, the concept of the experience curve becomes invalid.
- Similarly, if the perceived distinction is easily imitated, then rivalry will quickly switch to cost factors, costs incurred in providing differentiation might not be paid back by prices which are unattractive to customers.

An organization can compete with cost, quality of design and conformance, service, reliability, availability, flexibility, delivery speed, location, image, and a number of other features. The essential point about a competitive strategy is that it compares an organization's operations with those of its competitors. Then it might ask who are the competitors? How many competitors are there? What are the competitors' strengths? What are our strengths? What flexibility do we have? What are our future prospects? Who are the customers?

The answers to these questions suggest a range of more detailed questions like, what products should we concentrate on? What volumes should we produce? What quality should we give? Do we supply low or high cost products? Are our products reliable? Do we give fast deliveries? Who are our biggest customers? Do we have adequate financing?

Categorizing Competitive Strategies on the Basis of New Product Introduction Timings

One approach to categorizing an organization's competitive strategy is based on its timing of introductions of new products. Maidique and Patch (1988) have suggested the following four categories:

1. First-to-Market Organizations that use this strategy, attempt to have their products available before the competition expecting to get higher profit margins, larger market share, or both as a result of initially having the market to themselves. To achieve this, strong applied research and, consequently, large investments in product development are needed to stay ahead of the competition. If a company is first-to-market, it has to decide if it wants to price its products high and thus skim the market to achieve large short-term profits or set a lower initial price to obtain a higher market share and perhaps larger long-term profits. The strategy is somewhat risky because a competitor may end up beating them to the market. Also, even if a company succeeds in getting to the market first, it may end up simply creating an opportunity for the competition to learn from its mistakes and overtake it in the market. To illustrate, although Sony introduced its Betamax format for VCRs in 1975, JVC's VHS format —introduced the following year—is the standard that ultimately gained widespread market acceptance.

2. Second-to-Market Organizations that use this strategy try to quickly imitate successful outputs offered by first-to-market organizations. This strategy requires less emphasis on applied research and more emphasis on fast development. Often, firms that use the second-to-market strategy attempt to learn from the mistakes of the first-to-market firm and offer improved or enhanced versions of the original products.

3. Cost Minimization or Late-to-Market Organizations that use this strategy wait until a product becomes fairly standardized and is demanded in large volumes. They then attempt to compete on the basis of costs as opposed to features of the product. These organizations focus most of their research and development on improving the production system, as opposed to focusing on product development.

4. Market Segmentation This strategy focuses on serving niche markets with specific needs. Applied engineering skills and flexible manufacturing system are often needed for the market-segmentation strategy.

Key Points for Successful Competitive Strategy

1. The firm must know where it will specifically attack in a market (by products, by region and so on) and must then focus its resources and capabilities to succeed.
2. The firm must anticipate and be prepared for retaliation from other competitors once it has made its attack on a particular market segment.
3. Equally, the firm must know where it will *not* attack and to stay out of dangerous areas in which it cannot hope to compete.
4. Everybody in the organization must know and understand the strategy, feel that they own any changes that will be made and be committed to them.
5. A sense of timing is vital—a long-term strategy does not mean neglecting short-term urgency. In many markets, especially high-tech manufacturers, speed is vital both for delivery speed of existing products and rapid product innovations.
6. Alliances are an invaluable means of providing competitive advantage for the manufacturing firm. They allow the firm to concentrate on what it does best rather than being involved in areas in which it does not have particular expertise. Alliances with other manufacturing firms provide excellent means of organizational learning. Alliances also act as a means of combining a cluster of firms with the capability of competing against other clusters of firms.

Focused Organization

McKinsey & Company, a top management consulting firm, studied 27 outstanding successful firms to find their common attributes. Two of the major attributes reported in *Business Week* are directly related to the formulation of the competitive strategy:

1. Stressing One Key Business Value At Hewlett-Packard, it is developing new products; at Dana Corporation, it is improving productivity.

2. Sticking to What They Know Best All the outstanding firms define their core competencies (or strengths) and then build on them. They resist the temptation to move into new areas or diversify.

When an organization chooses to stress one or two key areas of strength, it is referred to as a *focused organization*. For example, IBM is known for its customer service, General Electric for its technology, and Procter and

Gamble for its consumer marketing. In general, most but not all areas of focus relate to product supply processes. Some firms, such as those in insurance industry, focus on financial strength and others on marketing strengths. Kenner Toys, for example, considers its strength to be a legal one: the ability to win contracts for exclusive production of popular children's toys.

Areas of Focus

Following are the areas of focus that organizations commonly choose when forming competitive strategy:

1. *Innovation* Bringing a range of new products and services to market quickly.
2. *Customization* Being able to quickly redesign and produce a product or service to meet customer's unique needs.
3. *Flexibility of Products and Services* Switching between different models or variants quickly to satisfy a customer or market.
4. *Flexibility of Volume* Changing quickly and economically from low-volume production to high volumes and vice versa.
5. *Performance* Offering products and services withunique, valuable features.
6. *Quality* Having better craftsmanship or consistency.
7. *Reliability of the Product or Service* Always working acceptably, enabling customers to count on the performance.
8. *Reliability of Delivery* Always fulfilling promises with a product or service that is never late.
9. *Response* Offering very short lead times to obtain products and services.
10. *After-Sale Service* Making available extensive, continuing help.
11. *Price* Having the lowest price

Mass Customization

Customization refers to offering a product or service exactly suited to a customer's desires or needs. Until recently it was widely believed that producing low-cost standard products required one type of transformation system and producing higher-cost customized products required another type of system. However, the continuous improvement programs of 1980s resulted not only in improved operating efficiency but also in an unexpected improvement in flexibility. Indeed, prior to this, efficiency and flexibility were thought to be trade-offs. Increasing flexibility meant that efficiency had to be sacrificed, and vice versa. Many manufactures initiated efforts to reduce the amount of time required to set up (or change over) equipment from the production of one product to another. Obviously, all time spent setting up equipment is wasteful, since the equipment is not being used during this time. Reductions in equipment setup times also resulted in improved flexibility. Specifically, with shorter equipment setup times, manufacturers could produce economically in smaller-size batches, making it easier to switch from the production of one product to another. In response to the discovery that efficiency and flexibility can be improved simultaneously and may not have to be traded off, the strategy of mass customization emerged. Organizations pursuing mass customization seek to produce low-cost, high-quality outputs in high variety.

Categories of Mass Customization

James Gilmore and Joseph Pine II (1997) identified the following four categories of mass customization:

1. *Collaborative Customizers* These organizations establish a dialogue to help customers articulate their needs and then develop customized outputs to meet these needs.
2. *Adaptive Customizers* These organizations offer a standard product that customers can modify themselves to suit their needs.
3. *Cosmetic Customizers* These organizations produce a standard product but present it differently to different customers. For example, it might simply be packaging the product in different containers for different customers.
4. *Transparent Customizers* These organizations provide custom products without the customers' knowing that a product has been customized for them. For example, an on-line computer service may track how each customer uses its service and then suggest additional features that the customer may find useful.

3.12 FUNCTIONAL STRATEGIES

It deals with how separate functions (such as manufacturing, marketing, R&D, finance and human resources) support the business in meeting its objectives. Each may have a slightly different emphasis on achievement of one or more of the objectives of the corporate strategy but all of them must collectively and synergistically comprise the overall strategy for the business. At the functional level, the principal focus of strategy is on the maximization of resource productivity and the development of distinctive competencies.

Business Process Strategy In organizations structured on the basis of business processes, separate functional strategies specifying how each functional area will support the overall business strategy are not needed. Rather, a strategy for how each business process supports the overall business strategy is developed. Along these lines, a business process strategy is a set of objectives, plans, and policies describing how a particular business process will support the business strategy of the organization (Shafer and Meredith, 1998).

3.13 MANUFACTURING STRATEGY

Manufacturing strategy is a plan that describes the way to produce and distribute the product. It is defined by APICS Dictionary as

> *A collective pattern of decisions that acts upon the formulation and deployment of manufacturing resources. To be most effective, the manufacturing strategy should act in support of the overall strategic directions of the business and provide for competitive advantages (edges).*
>
> *Cox III and Blackstone Jr, 1998*

The manufacturing strategy must describe the contribution that manufacturing makes to the cost, quality, availability, and features objectives of the business. It must also provide direction for achieving that contribution. At this macro, or strategic, level, following structural issues are considered:

(i) Number of Facilities and their Locations, Focuses, and Capacities Plant focus aims to reduce the complexity of managing a plant by focusing the activities around a limited set of tasks. APICS Dictionary defines a focused factory as a plant established to focus the entire manufacturing system on a limited, concise, manageable set of products, technologies, volumes, and markets precisely defined by the company's competitive strategy, technology, and economics (Cox III and Blackstone Jr, 1998). Product focus means the plant will confine its activities to a single product or product family. Focus on a process means the plant will do only those activities that are connected with a certain process. A foundry, for example, produces multiple products from a single process. Focus on order winners means that a plant will concentrate on providing its output in such a way that a certain order-winning criterion is achieved. A plant that is organized to provide lowest cost and correspondingly lowest price in the market niche is an example of an order-winner focused plant (Pannesi and O'Brien, 1992).

(ii) Degree of Vertical Integration Vertical integration focuses on the decisions to manufacture more or less of the bill of material and to adopt activities that move the firm either closer to the supplier side (backward

integration) or closer to the customer side (forward integration) of the business. That is, the firm decides to make parts or ingredients that it previously purchased. It might try to acquire raw material or component suppliers, machine manufacturers, designers, and possibly transportation. Or, it decides to take over the tasks of its customers by trying to acquire distribution outlets, repairs and servicing activities, marketing companies, and transportation outlets. A manufacturer can achieve increased control over supply and demand, and also lower costs for the total product. However, it might complicate the running of the operations of the company making it ineffective. It would also result in loss of 'focus' of the firm (Pannesi and O'Brien, 1992).

(iii) Choice of Process Technology The alignment of process to what the firm is trying to accomplish in the marketplace (competitive advantage sought) and to the current and expected volumes/varieties of product offerings is absolutely critical. Process choices are made infrequently and put machinery and systems in place. These must be compatible with the market needs otherwise the firm might not be able to achieve its strategic objectives (Pannesi and O'Brien, 1992).

(iv) Overall Organizational Structure Organizational structures fall into three general categories: centralized, decentralized, and matrix. In centralized organization, all support functions for the entire firm are done by one central authority. Examples are corporate purchasing and corporate cost accounting. In decentralized organization, support functions for the firm are performed at the plant level. Examples are the plant production and inventory management function and plant purchasing (Pannesi and O'Brien, 1992). The matrix organizational structure is defined by APICS Dictionary as an organizational structure in which two (or more) channels of command, budget responsibility, and performance measurement exist simultaneously. For example, both product and functional forms of organization could be implemented simultaneously, that is, the product and functional managers have equal authority, and employees report to both managers (Cox III and Blackstone Jr, 1998). This type of structure has been shown over the years to be of limited value in most production organizations with the exception of the project-oriented company.

Common Characteristics of Manufacturing Strategy

1. Manufacturing strategy must support the corporate strategy.
2. It should describe allocation of resources within the manufacturing function in a way that allows achievement of the manufacturing objectives.
3. It will be reflected in the patterns of actual decisions (such as capacity expansion, automation) made by the manufacturing function.

3.14 MANUFACTURING TACTICS

At the tactical, or more micro, level within manufacturing organization itself the tasks can be categorized into three interrelated activities:

1. *Control of the Manufacturing Resources* The right resources—people, materials, and tools—should be in the right place at the right time.

2. *Control of the Manufacturing Process* The right processes should be executed properly at the right time.

3. *Control of Manufacturing Information* The information to enable and improve execution should be available just as needed to enable appropriate, localized decision making.

The emphasis in executing any of these activities, however, must be on the areas of concentration dictated by the manufacturing strategy—cost, quality, flexibility/responsiveness, or innovation/technology. Manufacturing tactics, then, specify the methods by which process, resource, and information control are executed to optimize their contribution to cost, quality, availability, and features of the product (Beckman *et al.* 1997).

3.15 DEVELOPING A MANUFACTURING STRATEGY

In broad terms, there are two important ways in which manufacturing can strengthen a company.

(a) By Providing a Marketing Edge Through Unique Technological Developments The first is to provide manufacturing processes that give the business a distinct advantage in the marketplace. In this way, manufacturing will provide a marketing edge through unique technological developments in its process and manufacturing operations that competitors are unable to match. This is quite rare and examples are hard to find. One such example is Pilkington's float-glass process. The development of the float-glass process in the 1950s by Pilkington, a UK glass manufacturer, was a remarkable step forward in the technology of float-glass making. The costly grinding and polishing operations in the conventional manufacture of glass were eliminated and the result was plate glass production at a fraction of the cost.

(b) By Providing Manufacturing Support to Help Company Products Win Orders The second is to provide coordinated manufacturing support for the essential ways in which products win orders in the market place, that is, better than such support provided by the manufacturing functions of its competitors. Manufacturing must choose its process, and design its infrastructure (for example, controls, procedures, systems, and structures) in ways that help products win orders, and that choice and design must

be adaptable to changing business needs. Most companies share access to the same processes, and thus technology is not inherently different. Similarly, the systems, structures, and other elements of infrastructure are universal. What is different is the degree to which manufacturing matches process and infrastructure to those criteria that win orders. In this way, manufacturing constitutes a coordinated response to the business needs that embraces all those aspects of a company for which manufacturing is responsible (Hill, 1994).

Order Winners, Order Qualifiers, and Non-issues

An essential part of manufacturing's task is to recognize and apply the concept of order winners, order qualifiers, and non-issues.

1. Qualifiers These are the criteria that a company must meet for a customer to even consider it as a possible supplier. These criteria qualify the product to be there in the marketplace.

For example, customers increasingly require suppliers to be registered under the ISO 9000 series. Therefore, suppliers who are so registered have only achieved the right to bid or be considered. Furthermore, they will need to retain the qualification in order to stay on the short list or be considered as a competitor in a given market. However, providing or attaining these criteria alone do not win orders.

2. Order-Winners These are the criteria that win orders in the marketplace. These are the characteristics that make your customers prefer you over your competitors.

To provide qualifiers, companies need only to be as good as competitors; to provide order-winners; they need to be better than competitors. However, qualifiers are not less important than order-winners—they are different. Both are essential if companies are to maintain existing share and grow. Manufacturing, therefore, must provide the qualifying criteria to get into or stay in the marketplace. But these alone will not win orders. They merely make a company eligible to compete in the marketplace. Once the qualifying criteria have been achieved, manufacturing then has to turn its attention to the ways in which orders are won and ideally to provide these better than anyone else. It has been argued that delivery and flexibility are the most important objectives in some industries because cost-cutting programs and quality improvement programs have 'leveled the playing field' on cost and quality. These firms then compete on 'time'—rapid introduction of new products and rapid delivery of existing products. The phrase '*time-based competition*' has been used to describe this phenomenon (Silver *et al.* 1998). In other words, in some industries good quality and low cost are now merely order qualifiers, flexibility (product variety) and rapid delivery are the order winners. With the globalization of business in recent years, firms are sourcing and distributing components and finished goods across the globe. Customers want their products quickly and reliably, even if they are coming from another continent, and they are often willing to look for other suppliers (Silver *et al.* 1998). This makes rapid and reliable delivery performance an extremely important order winner.

3. Non-issues Finally, for any market niche, there are some characteristics that do not play a significant role in the competitive picture. These are called non-issues. For example, it may be that image is a non-issue or that flexibility is not a significant factor in a particular niche (Pannesi and O'Brien, 1992).

Knowing which competitive characteristics are qualifiers, which are order winners, and which are non-issues is vital to the formulation of manufacturing strategy.

3.16 NECESSITY OF INVOLVING MANUFACTURING IN THE CORPORATE STRATEGY DEBATE

To provide order qualifiers and winners effectively, manufacturing needs to be involved throughout the whole of the corporate strategy debate to explain, in business terms, the implications of corporate marketing proposals and, as a result, be able to influence strategy decisions for the good of the business as a whole. Too often in the past, manufacturing has been too late in this procedure (Hill, 1994).

Many companies have not integrated manufacturing into the strategy process; instead, corporate strategy is developed around marketing issues, and then the strategy is given to manufacturing to implement. Unfortunately, the strengths of manufacturing may not match the marketing strategy, and a mismatch may result in failure (Narasimhan *et al.* 1995). The inability of the manufacturing to influence decisions has often led to its seemingly endless complaints about the unrealistic demands placed upon production. In many western firms, one of the common areas of corporate mismanagement has been the omission of contribution of production/operations from the decision-making at corporate level. If the ultimate decision-makers know little or nothing about manufacturing then not surprisingly the nature of decisions could be detrimental to the manufacturing firm. The criterion might shift to being purely financial in nature, where short-term approaches might override long-term investment in new processes, plant and technology. For manufacturing firms, there would seem to be two alternative methods of incorporating manufacturing inputs into corporate strategy (Brown, 1996):

1. The ultimate decision-makers must be senior managers who are themselves knowledgeable about manufacturing as well as other business considerations.
2. The ultimate decision-makers themselves might not necessarily know a great deal about manufacturing but will involve those

personnel within the company who do—thus allowing the manufacturing managers to be a key input into shaping and deciding corporate strategy for the manufacturing firm.

A key aspect of linking the manufacturing strategy to the corporate strategy is to have manufacturing team members determine the contribution that manufacturing can make in each of the areas of customer need: cost, quality, availability, and features. Companies require a strategy not based solely on marketing, manufacturing, nor any other function, but one that embraces the interface between markets and functions. Thus, the markets, which the business serves or intends to serve, should and must form the link between functional strategies. For example, both marketing and manufacturing must have a common understanding and agreement about company markets. Only then can the degree of fit between the proposed marketing strategy and manufacturing's ability to support it be known at the business level and objectively resolved within corporate perspectives and constraints (Hill, 1994).

For this to take place, relevant internal information explaining the company's manufacturing capabilities needs to be available within a business, as well as the traditional marketing information, which is primarily concerned with the customer and the market opportunities associated with a company's products. In addition, the ownership of its use must be vested in top management. As with other functions, manufacturing strategy is not owned by manufacturing. It requires corporate ownership. Senior executives need to understand all the strategic inputs in the corporate debate, for without this understanding the resolution between conflicting or non-matching functional perspectives cannot be fully investigated, hence, individual functions must handle the trade-offs involved. Developing and agreeing strategy is top management's key task. In fact, it is the one task that cannot be delegated.

A business must develop a manufacturing strategy not only because of the critical nature of manufacturing within corporate strategy but also because many of the decisions are structural in nature, that is they are hard to change. If the business does not fully appreciate the issues and consequences, it can be locked into a number of manufacturing decisions that will take years to change. These can range from process investment on the one hand to human resource management practices and controls on the other. Decisions not in line with the needs of the business can contribute significantly to a lack of corporate success. To change them is costly and time-consuming. But even more significant, they will come too late.

The development of a corporate policy consisting of a coordinated set of main function inputs will mean that a business would be able to go in one consistent, coherent direction based on a well-argued, well-understood, and well-formed strategy. This is achieved, in part, by moving away from argument, disagreement, misunderstanding, and short-term, parochial moves based on interfunctional perspectives, to the resolution of these differences at the corporate level. Currently, marketing-led strategies leave the aftermath to be resolved by manufacturing, who without adequate, appropriate guidance or discussion and agreement at the corporate level, resolve the issues as best they can largely from their unilateral view of what is best for the business as a whole. In the majority of cases, manufacturing is simply not geared to a business's corporate objectives. The result is a manufacturing system, good in itself, but not designed to meet company needs.

One clear consequence of this is the need for the production managers to change from a reactive to a proactive stance. The long-term inflexible nature of manufacturing means that the key issues involved in process choice and infrastructure development need to be reflected in business decisions, with the business being made aware of the implications for manufacturing of proposed corporate changes. When this is achieved, the strategy decisions that are then taken reflect what is best for the business as a whole. So manufacturing management's attention must increasingly be toward strategy. This does not mean that operations are unimportant. But management must appreciate the impact on business performance of strategy and operations while recognizing that both need to be done well.

Top management have, by and large, perceived corporate improvements as coming mainly through broad decisions concerning new markets, takeovers, and the like. However, the building blocks of corporate success are to be found in creating effective, successful businesses where manufacturing supports the market requirements within a well-chosen, well-argued, and well-understood corporate strategy.

3.17 STRATEGIC INTEGRATION—LINKING MANUFACTURING TO CORPORATE PLANNING AND MARKETING

There are five basic steps that provide an analytical and objective structure in which the corporate debate and consequent actions can be taken.

1. Define corporate objectives.
2. Determine marketing strategies to meet these objectives.
3. Assess how different products qualify in their respective markets and win orders against competitors.
4. Establish the most appropriate process to manufacture these products (process choice).
5. Provide the manufacturing infrastructure to support production.

While each step has substance in its own right, each has an impact on the others—hence the involved nature of strategy formulation. This is further exacerbated by the inherent complexity of manufacturing and the general failure to take account of the essential interaction between marketing and manufacturing strategies.

The suggested approach to linking manufacturing with corporate marketing decisions is schematically outlined in Table 3.5. It is presented in the form of a framework to help to outline the stages involved. The approach provides the key to stimulating corporate debate about the business in such a way as to enable manufacturing to assess the degree to which it can support products in the marketplace. It has been researched and tested successfully in many industries and businesses of different sizes (Hill, 1994).

Table 3.5

Framework for Reflecting Manufacturing Strategy Issues in Corporate Decisions

(Hill, 1994)

Corporate Objectives	*Marketing Strategy*	*Order Winning Criteria*	*Manufacturing Strategy*	
			Process Choice	*Infrastructure*
Survival Growth Profit Return on investment Other financial measures	Product markets and segments Range Mix Volumes Degree of standardization and customization Level of Innovation Leader versus follower alternatives	Price Quality Delivery: speed and reliability Demand increases Product range Design leadership Technical support being supplied	Choice of alternative processes Trade-offs embodied in the process choice Process positioning Capacity: size, timing, and location Role of inventory in the process configuration	Function support Manufacturing planning and control system Quality assurance and control Manufacturing systems engineering Clerical procedures Payment systems Work structuring Organizational structure

Although the steps to be followed are given as finite points in a stated procedure, in reality the process will involve statement and restatement, for several of these aspects will impinge on each other.

The objective of using this framework is to produce a manufacturing strategy for a business (Steps 4 and 5). In all instances this will include a review of existing products plus a review of proposed product introductions. Furthermore, the review will be based on current and future market expectations. This is because manufacturing needs to support a product (that is, with after-sales service and spare parts supply) over the whole and not just a part of its life cycle, and hence the business needs to address this total decision. Figure 3.4 shows how Corporate, Marketing and Manufacturing Strategies have been linked through the two models.

TRADITIONAL MODEL

Corporate Strategy

- Created by an elite group whose criterion is essentially financial.
- Often little or no manufacturing presence, expertise, or knowledge.

Marketing Strategy

- To determine new products, new market segments, pricing and distribution policies in order to satisfy corporate strategy.

Manufacturing Strategy

- Meant only to react to plans already in place.
- Manufacturing viewed as tactical function that responds to corporate and marketing strategies.
- Manufacturing planning centers around areas such as scheduling, materials handling, and day-to-day activities.
- Unfocused strategy, often based on short-term, non-holistic view.

ENLIGHTENED MODEL

- Long time horizons for corporate, marketing, and manufacturing strategies.
- Plans are made in a holistic approach by jointly agreeing on internal capabilities and matching these to market opportunities.
- Market requirements of cost, flexibility, delivery speed and reliability, together with quality requirements are clearly understood by all and resources are allocated to meet these requirements.
- Decisions concerning manufacturing investment, in terms of new process or technology, are made on long-term manufacturing advantage and not narrow financial criteria such as Return On Investment (ROI) or immediate cost savings.
- Focused strategy based on long-term, holistic view

Fig. 3.4 *Models for Linking Corporate, Marketing and Manufacturing Strategies*

Step 1 Corporate Objectives

Corporate objectives provide the basis for establishing clear, strategic direction for the business. They also define the boundaries and mark the parameters against which the various inputs can be measured and consistency established, thus providing the hallmark of a coherent corporate plan.

- For each company, the objectives will be different in nature and emphasis. They will reflect the nature of the economy, markets, opportunity, and preferences of those involved. The important issues here, however, are that they need to be well thought through, hold logically together, and provide the necessary direction for the business.

- Typical measures concern profit in relation to sales and investment, together with targets for both absolute growth and market-share growth. Businesses may also wish to include employee policies and environmental issues as part of their overall set of objectives.

Step 2 Marketing Strategy

It is linked closely to the corporate objectives and will often include the following stages:

1. Establishment of Market Planning and Control Units Their task is to bring together a number of products that have closely related market targets and that often share a common marketing program. This will help to identify a number of manageable units with similar marketing characteristics.

2. Situational Analysis of Product Markets This includes:

(a) Determining current and future volumes.
(b) Defining end-user characteristics.
(c) Assessing patterns of buying behaviour.
(d) Examining industry practices and trends.
(e) Identifying key competitors and reviewing the business's relative position.

3. Identification of the Target Markets and their Objectives This will include both a broad review of how to achieve these and the short-term action plans necessary to achieve the more global objectives involved.

4. Service Support Planning In addition, the company should agree on the level of service support necessary in each market and assess the investments and resources needed to provide these throughout the business.

The outcome of this will be:

(i) A declaration to the business of the product markets and segments that the strategy proposes.
(ii) Identification of the range, mix, and volumes involved.
(iii) Determination of the degree of standardization or customization involved within each product range.
(iv) Determination of the level of innovation and product development proposed.
(v) Decision as to whether the business should be a leader or follower in each of its markets.
(vi) Determination of the extent and timing of these strategic initiatives.

Step 3 Order Winning Criteria

Manufacturing's strategic task is to provide, better than the manufacturing functions of competitors, those criteria that enable the products to win orders in the marketplace. The company as a whole needs to agree on the markets and segments within these markets in which it decides to compete. In no way can these critical decisions be the responsibility or prerogative of a single function. Typically, however, most companies develop strategy through a marketing perspective. As a function, marketing will have an important and essential view. But it is not the only view, and in no way should it be allowed to dominate corporate strategy resolution. *Functional dominance, no matter of what origin, is detrimental to today's business needs and must be avoided.*

An essential perspective of a firm's market has to come from manufacturing. This perspective is established by determining those order-winners and qualifiers that manufacturing needs to provide. The procedure is, in reality, to ask the marketing function questions that require manufacturing answers. This step, therefore, is the essential link between corporate marketing proposals and commitments and the manufacturing processes and infrastructure necessary to support them. The relevant order-winners and qualifiers have different levels of importance in different markets. Similarly, the degree of importance of order-winners and qualifiers will also change over time.

Step 4 Process Choice

Manufacturing can choose from a number of alternative processes to make the particular products. The key to this choice is volume and the order-winning criteria involved. Each choice, therefore, needs to reflect the current and future trade-offs involved for the various products.

Step 5 Infrastructure

Manufacturing infrastructure consists of the non-process features within production. It encompasses the procedures, systems, controls, compensation systems, work-structuring alternatives, organizational issues, and so on that are involved in the non-process aspects of manufacturing.

The test of manufacturing strategy is consistency with the overall corporate strategy and the contribution that it will make to the enterprise. Although production/operations' contribution will vary according to specific markets, there are general improvements that any manufacturing company can make in order to be competitive. The steps that a manufacturing change should follow are:

1. First, produce high quality.
2. Then, achieve delivery reliability.
3. Then, lower production cost.
4. Finally, increase production flexibility.

The Characteristics of Japanese Approach to Manufacturing Strategy

Following are the characteristics of the enlightened Japanese manufacturing strategy:

1. Short-term profitability is sacrificed for longer-term commitment to winning market share and customer loyalty.
2. Investment is made in manufacturing in all aspects, including process technology.
3. Long-term relationships are built with the suppliers.
4. Human resources are continuously developed through training etc.

3.18 MANUFACTURING STRATEGY DEVELOPMENT PROCESS DEVELOPED AT HEWLETT-PACKARD

Hewlett-Packard (HP) has developed a five-step strategic manufacturing planning process that can help a company reestablish manufacturing's position as a critical contributor to the success of the company's business strategy. According to the developers (Beckman *et al.* 1997), it combines many of the techniques proposed by the academic community, tools from real-world manufacturing, and the collective US manufacturing experience to create an understandable process that works. It has been proved successful by HP and some of its major customers from aerospace, food processing, telecommunications, and other industries.

The five-step approach to manufacturing strategy development at HP is as follows:

1. Start with the Business Strategy More specifically, understand why customers will prefer your product or service to the competitors'.

2. Create a Linked Manufacturing Strategy Create a manufacturing strategy that is linked to the business strategy. In other words, specify *manufacturing's contribution* to making customers choose your product (or service) instead of your competitors'.

3. Identify Manufacturing Tactics to Execute the Strategy This requires understanding how to manage and control the people, processes, materials, and information needed to deliver products or services in a way that meets the objectives of the strategy.

4. Organize for Manufacturing Success Organization design, including structure and performance measurement, must match strategic needs, or success will be limited.

5. Measure the Results and Initiate Further Changes Strategies must be altered continually to meet the needs of a constantly changing environment. Feedback loops are critical to the continuous improvement process.

This process must begin at the top. The highest levels of management must commit the corporation to a strategic direction that is clearly defined and well communicated throughout, and conduct periodic reviews to ensure that its direction is being followed.

3.19 BUSINESS PLANNING—MANAGEMENT LEVERS

Business Planning Business plans have a shorter time horizon (typically 1 year) and focus on product families to meet strategic goals. Business planning provides the plans that are necessary to drive the sales, manufacturing, and financial activities of an organization. These plans define:

1. Business mission of the company (the products to be manufactured).
2. The markets to be addressed (what the company expects to sell in monetary units, for example, in dollars or rupees).
3. Required volumes, resources, and profit objectives.
4. Financial resources (such as, amount of money to be invested in R&D).

Here, we will take a brief overview of business planning and focus exclusively on the manufacturing planning and control aspects of this important topic. In business planning a manufacturing strategy is formed by decisions taken in, and in connection with, manufacturing for the following purposes:

(a) To direct a company's manufacturing resources in a manner that supports the competitive ability of that company.
(b) To support certain Order Winning Criteria (OWC) which in today's business environment are based on delivery performance and product customization. Price and quality are now considered to be necessary prerequisites (qualifiers) for continued existence in the market place.

Management Levers Although corporate objectives provide measurable goals, they do not indicate how a firm should pursue those goals. Management levers provide the tactical steps necessary to achieve the goals. Following are the seven decision categories or management levers involved in the manufacturing aspect of business planning:

3.20 MAKE OR BUY DECISIONS—VERTICAL INTEGRATION

Make or buy (or vertical integration) decisions are concerned with the choice of which products are to be manufactured in-house and which are to be subcontracted to external suppliers. It is a very important decision within a manufacturing strategy since it dictates such choices as:

(i) capacity,
(ii) facilities,
(iii) process, and
(iv) (to some extent), infrastructure.

The aim is to allow the manufacturing firm to concentrate on

(i) those products which add most value,
(ii) those technologies which enhance its OWC (Order Winning Criteria) competitive position while removing unnecessary management complexities.

3.21 CAPACITY DECISIONS

This involves decision regarding type, size, and timing of potential capacity and can be expressed in terms of:

(i) production equipment,
(ii) factory floor space, and
(iii) human resources.

These decisions involve various risk levels attached to them. The ability to strike a balance between these capacity decisions is critical to the success of the business planning function. The main factors influencing the capacity decisions are as follows:

(a) Anticipated future customer demand,
(b) The cost of expanding capacity,
(c) The technology currently available and the rate of technological change, and
(d) The manner in which the competitors are likely to react.

In addition, the capacity decisions are also influenced by the intended utilization of the facilities in terms of number of shifts to be operated, amount of overtime to be allowed, and amount of the work to be subcontracted. Capacity decisions interact with the facility location decisions.

Following three basic questions need to be answered:

(a) When should capacity be added?
(b) How much capacity should be added?
(c) Where should capacity be added?

3.22 FACILITIES DECISIONS

Facilities decisions concern *the location and focus of factories and other facilities.* More specifically, they are concerned with the assignment of specific products, customers, and markets to individual facilities. These decisions are related to the size, location, specialization, and organization of facilities. These decisions are made to answer the following questions:

1. Where does a company locate its factories and distribution centers to meet the needs of its customers?
2. Is there a need for multiple facilities, or, is a single location sufficient?
3. Does each facility perform all functions, or is one facility focused on a particular market, a particular process, or a particular product?

Many manufacturing companies have one plant that acts as a 'parent plant' that is responsible for odd-ball parts and new product introduction. When products reach high-volume manufacture, they are moved to a satellite plant, where efforts are focused on excellence in that particular product and/or process. It is critical to evaluate each plant manager based on the specific charter for his or her facility.

- Facilities decisions are closely related with the capacity decisions. Make or buy, capacity, and facilities decisions are primarily made for the long-term because they involve bricks and mortar, and significant investment.
- Facility decisions are concerned with the problems of:
  - (i) Availability of the shop floor space in a plant in anticipation of future increase in demands.
  - (ii) Layout of the plant.
  - (iii) Location of additional plants.
- Factors influencing the facilities decisions in general include:
  - (a) Forecast increase/decrease in demand.
  - (b) Technology change.
  - (c) Change in the external environment (including changes in Government policies, shifts in public opinion).
  - (d) Opportunities to make significant impact on the company's competitive position.
- Factors influencing the plant location decision include:
  - (a) *Cost factors,* such as facility costs, taxes, local labor rates, utilities costs, and transportation costs.
  - (b) *Qualitative factors*, such as proximity to customers and suppliers, availability of labor, transportation, power, and utilities, quality of life, legal and environmental issues, special incentives by local governments, and community attitudes.
- Analysis of facility costs is critical—especially to the firm that wants to compete on price.
- With the advent of JIT manufacturing, proximity to customers has become even more important as it can help in a number of strategic

initiatives such as flexibility, prices, and delivery speed. Proximity to suppliers results in lower costs, especially when raw materials are bulky, or easily damaged in transit. Other advantages include close interaction during design and quality control stages, quicker deliveries, and flexibility.

3.23 PROCESS AND TECHNOLOGY DECISIONS

These are related to choosing a particular process or technology and level of automation for a manufacturing plant. The process and technology lever encompasses the choice of a production process and the level of automation. The process or technology adopted should be in alignment with what the firm is trying to accomplish in the marketplace and to the current and expected volumes/varieties of product offerings. All too often, process decisions and the acquisition of new equipment are made because someone thinks 'this is the latest, greatest stuff' or 'we need a robot because everyone is using them' or 'our competitors are using these new high-tech machines and if we don't, we'll fall behind.' These are not sound reasons for acquiring new processes and associated equipment. Process-choice decisions should be made for strategic reasons that represent far more than simply evaluating financial payback. Present value analysis, the heart of current payback calculations that are being done to justify new equipment, is used to decide between alternative process choices but the decision to acquire is made for strategic reason (Pannesi and O'Brien, 1992). When determining a process choice for a product line the issues that need to be considered include:

1. Order qualifiers and winners.
2. Current and expected volumes/varieties (level of customization).
3. Required and available capabilities.

   These considerations set the stage for the process-choice decisions and lead to an understanding of the needed capabilities of the process. If and when it is decided to go for the new process, following issues need to be considered for the specific process selected:

4. Make sure that the selected process is really the overall best process, that is, make sure that there is no other process that is better, cheaper, faster and yet more effective.
5. Make sure that the new process helps in enhancing the strategic advantages.
6. Check the feasibility of implementing the new processes in terms of human and capital resources required and the completion time. Further, consideration of the impact of those processes and technologies on the information system in general, and the planning and control system in particular, is mandatory (Pannesi and O'Brien, 1992).

(a) The Level of Customization The level of customization influences process choice. Figure 3.5 shows the relationship between the level of customization and the process decisions.

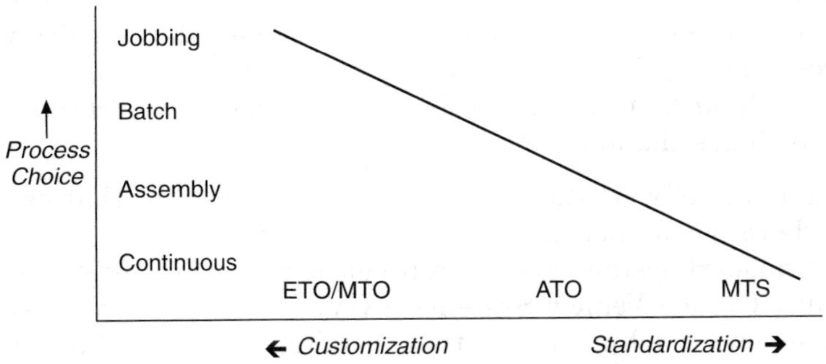

Fig. 3.5 *Process Decisions Based on Level of Customization*
(Adapted from McMahon and Browne, 1993)

- If a firm's marketing strategy is to provide quality and a high degree of customized product, then the highly flexible nature of jobshop processing is more attractive than the other alternatives.

(b) The Expected Volume of Product The volume of market demand also influences a process design. Figure 3.6 shows the relationship between the volume of products produced and the process employed.

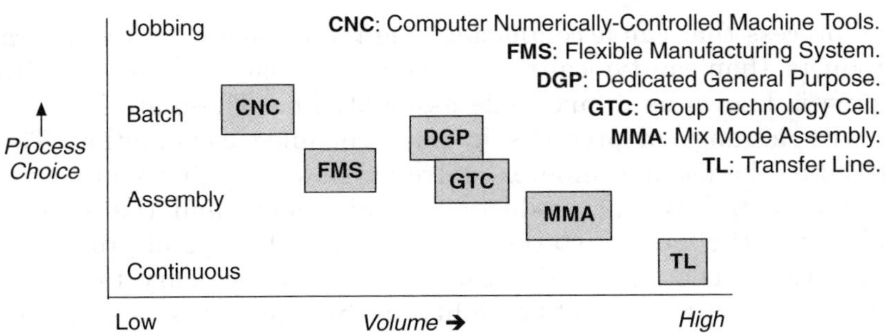

Fig. 3.6 *Process Decisions Based on Volume*
(Adapted from McMahon and Browne, 1993)

Flexible Manufacturing Systems (FMS)

Flexible manufacturing systems are a combination of standard and special Computer Numerical Control (CNC) machine tools, automated material handling, and computer control in the form of Distributed Numerical Control (DNC) for the purpose of extending the benefits of

CNC to mid-volume manufacturing situations. Whereas, CNC machine tools are best suited to low volumes, an FMS is appropriate to mid-volume requirements. FMS too is designed to complete a given number of operations on an item before it leaves the system. However, rather than the item being contained in a single center, in an FMS the workpiece is transferred from one process to the next automatically. Besides volume differences, the physical dimensions of the items to be machined are typically much larger than those completed by machining centers. Standard FMS includes the following:

1. A number of workstations, such as CNC machines that perform a wide range of operations.
2. A transport system, such as automatic conveyors and/or Automatically Guided Vehicle Systems (AGVS), which will move material from one machine to another, loading and unloading stations, where completed or partially completed components will be housed and worked upon.
3. A comprehensive computer control system, which will coordinate all the activities. The activities will include:

   (a) The control of each workstation.
   (b) Distribution of control instructions to workstation.
   (c) Production control.
   (d) Traffic control.
   (e) Tool control.
   (f) System performance monitoring.

Mix Mode Assembly Lines

A line process that can accommodate the requirements of a wider range of products, than can typically be made on a classic line process without stopping is known as a mix mode assembly line. These are designed to cope with a range of products in any scheduled combination. This is achieved by the use of computer-controlled flow lines that schedule work according to the overall production requirements and the short-term workloads at the various stations. The increased range of products that can be accommodated justifies the investment necessary to achieve a higher level of this aspect of flexibility. In this way, a mix mode assembly line is an alternative process, to batch, that could be used to meet the required level of volume of the products in question.

(c) The Requirements of the Process These refer largely to the level of skill and experience that the personnel of the firm are required to posses.

(d) Other Factors

(i) Ability of the firm to integrate the process with existing processes and equipment.
(ii) Required investment.
(iii) Firm's competitive ability.

(iv) Technical and commercial risks involved in employing a particular process.

Most of the decisions described above are structural or 'hard' decisions. These decisions deal with bricks and mortar or with equipment, they are made infrequently (although they may be reviewed frequently or regularly), and they are usually made by the upper management. In general, these decisions will put tangible resources on the factory floor. The next two categories of decisions are soft or infrastructural decisions. These decisions tend to be evolutionary and are usually made by middle or lower management. Top management of the firm should, however, be just as aware of the infrastructural choices being made as it usually is of the structural choices

Table 3.6

Comparison of Various Types of Processes

	Process Choice				
	Project	*Job*	*Batch*	*Line**	*Continuous Process**
Dominant Utilization	Essentially labour	Labour	Labour/plant	Plant	Plant
Process Technology	General Purpose	Universal	Universal/dedicated	Dedicated	Very dedicated
Process Flexibility	High		Medium	Low	Inflexible
Marketing Strategy	Customization for each customer. Design and innovation capabilities are important as new products are required on an ongoing basis.		Initially products are mapped to job or line characteristics, then the plant is focused to support these cells of manufacture.	High volume standard products—introducing new product is difficult.	
Order-Winning Criteria	Delivery speed, quality-design capability.		Delivery speed/Availability, Customization.	Price	
Order-Qualifying Criteria	Price		Quality and price.	Delivery speed, quality-design capability.	
Process Layout	Fixed	Process	Process, for low-volume/high variety production. Product, for high-volume/low variety production.	Product	Product

* One of the main differences between a line process and continuous process is that a line process can be stopped at a particular stage and the product will be at *that* stage of production. Stopping of continuous processes is extremely difficult and very costly.

The capacity, facilities, and process decisions are part of the decision about investment in technology, and hence, are interlinked. Process choice should include layout as part of the planning process, and technology links into, but is not limited to, process choice or layout as shown in Fig. 3.7.

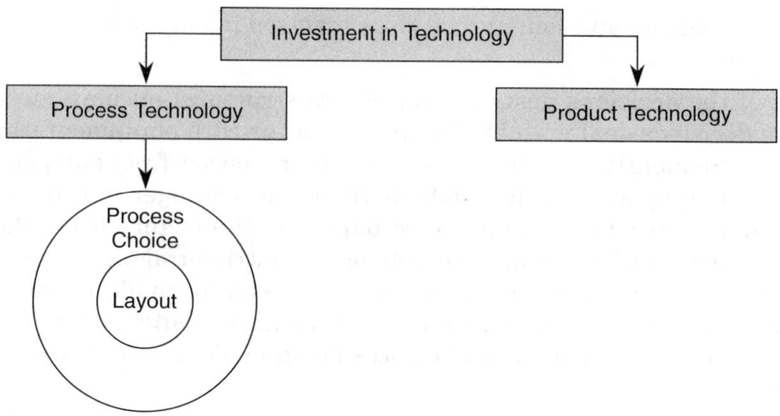

Fig. 3.7 *The Interrelationship between Technology, Process Choice, and Plant Layout* (Adapted from Brown, 1996)

3.24 MANUFACTURING INFRASTRUCTURE DECISIONS

These involve policies, procedures, and organization by which manufacturing accomplishes its work. Manufacturing infrastructure consists of the systems which support the manufacturing activity. It is the network of nonphysical support system that enables the technical structure to operate. It is defined as policies, procedures, and organization by which manufacturing accomplishes its work, it includes:

(i) Production Planning and Scheduling Systems Forecasting and demand management, MRP II, JIT, order point systems, etc.

(ii) Inventory Management Lever This encompasses decisions regarding purchasing, distribution, and logistics, and specifically addresses when and how much to order.

(iii) Quality Management Lever This encompasses the tools, programs, and techniques used to achieve the quality goals. It is distinct from the quality objectives, in that the quality objective specifies the definition and measurable targets, and the quality management lever specifies the means to achieve the targets. Quality management includes such things as Statistical Process Control (SPC), Taguchi methods, quality circles, type, degree, and frequency of inspection, and so on.

(iv) Information Systems This includes types of information, degrees of access, degree of technology, types of reports, etc.

(v) Workforce Management Policies

(vi) Organizational structure his includes types and number of people needed, degrees of skills required, and degrees of empowerment.

3.25 HUMAN RESOURCES DECISIONS

Human resources lever involves the selection, promotion, placement, reward, and motivation of the people that make the business work. These decisions are to formulate the following:

(i) The design of the organization: vertical or horizontal, centralized or decentralized, matrix, number of levels.
(ii) The design of the reward system.

3.26 NEW PRODUCT INTRODUCTION

This management lever involves the procedures and organizational structure that facilitates new product introduction. Most firms now employ multifunctional teams, composed of design engineers, marketing personnel, manufacturing managers, and production line workers. The new-product lever specifies the reporting relationships as well as any procedures for setting milestones in the development process.

3.27 SUPPLY CHAIN RELATIONSHIPS

This lever focuses on relationships with vendors and customers in the firm's supply chain. This, for example, might involve developing a 'virtual market' where parts specifications are put on the Internet forming a 'trading process network,' so that a large number of suppliers can bid on the job. Or, strategic alliances can be maintained with a small number of vendors or customers. Vendor-managed inventory (VMI) is a coordination mechanism in which a supplier manages the inventory on its customer's shelf, deciding when and how much to reorder.

References

Beckman, Sara L, William A Boller, Stephan A Hamilton and John W Monroe, 'Using Manufacturing as a Competitive Weapon: The Development of a Manufacturing Strategy', in: Patricia E Moody (Editor), *Leading Manufacturing Excellence: A Guide to State-of-the-Art-Manufacturing',* John Wiley & Sons Inc, New York, 1997, pp 47–67.

Brown, Steve, *Strategic Manufacturing for Competitive Advantage: Transforming Operations from Shop Floor to Strategy*, Prentice Hall, London, 1996.

Cox III, James F and John H Blackstone Jr, *APICS Dictionary* (ninth edition), American Production and Inventory Control Society, Falls Church, V A, 1998.

Childe, Stephen J, *An Introduction to Computer-Aided Production Management*, Chapman & Hall, London, 1997.

Gilmore, James H and B Joseph Pine II, 'The Four Faces of Mass Customization', *Harvard Business Review,* January–February 1997, pp 91–101.

Hayes, Robert H and Steven C Wheelwright, *Restoring Our Competitive Edge: Competing Through Manufacturing,* John Wiley & Sons, New York, 1984.

Hitomi, Katsundo, *Manufacturing Systems Engineering,* Taylor and Francis Ltd, London, 1996.

Hill, Terry, *Manufacturing Strategy: Text and Cases* (second edition), Richard D Irwin Inc, Burr Ridge, Illinois, 1994.

Maidique, M A and P Patch, 'Corporate Strategy and Technological Policy', in M L Tushman and W L Moore (Editors), *Readings in the Management of Innovation* (second edition), Harper Business, 1988, pp 236–248.

McMahon, Chris and Jimmie Browne, *CADCAM: From Principles to Practice,* Addison-Wesley Publishing Company, Wokingham, England, 1993.

Narasimhan, S L, D W, McLeavey and P J Billington, *Production Planning and Inventory Control* (second edition), Prentice-Hall Inc, Englewood Cliffs, N J, 1995.

Pannesi, Ronald T and Helene J O'Brien, *Systems and Technologies, APICS Certification Review Course, Student Guide*, American Production and Inventory Control Society, Falls Church, V A, 1992.

Shafer, Scott M and Jack R Meredith, *Operations Management: A Process Approach with Spreadsheets,* John Wiley & Sons, Inc, New York, 1998.

Silver, Edward A, David F Pyke, and Rein Peterson, *Inventory Management and Production Planning and Scheduling* (third Edition), John Wiley & Sons, New York, 1998.

Thompson Jr, Arthur A and A J Strickland III, *Crafting and Implementing Strategy: Text and Readings* (tenth Edition), Irwin/McGraw-Hill, Boston, Massachusetts, 1998.

Waters, Donald, *Operations Management: Producing Goods and Services.* Addison-Wesley Publishing Company, Harlow, England, 1996.

Wheelam, T and J Hunger, *Strategic and Business Policy* (fourth Edition), Addison Wesley, Reading M A, 1993.

Chapter 4

Sales and Operations Planning (S&OP)

4.1 INTRODUCTION

In business, as in war, failures often stem from lack of coordination between essential functions. The company whose sales force is out booking orders and promising delivery dates without the concurrence of finance, engineering, and manufacturing is likely to fail. While coordination among business functions is obviously essential, it does not just happen, but needs a formal mechanism to ensure that it occurs. For most companies, that mechanism is 'Sales and Operations Planning (S&OP).' S&OP is a formal process for managing change related to product demand. It is a new process that has been around for only a dozen years or so (Proud, 1994). S&OP is also known as *production planning, aggregate planning,* or *game planning.*

The best way to determine what to make and when to make it is by using a process in which all departments meet regularly to assess future market demand and determine whether the factory or plant can meet this demand and if so, how. This process is known as S&OP; and it brings people together. With S&OP people are able to relate the changes throughout the company without one manager dictating a change that nobody else agrees on. They come out of the S&OP process with consensus and confidence. Further, S&OP is the proper place to make changes because it ensures that all departments are in agreement. Moreover, because S&OP is the top-level driver and links the high- and low-level plans, changes made at the S&OP level are automatically passed down and translated into detailed plans.

Sales and operations planning (S&OP) is the process that provides management, the ability to strategically direct its business to achieve competitive advantage on a continuous basis by integrating customer focused marketing plans for new and existing products with the management of the supply chain. *S&OP provides the framework within which the MPS is developed.*

Once the strategic plan is in place, a sales and operations plan must be put in place for the next 18 months. Simply stated this plan balances the demand (both actual and forecasted) with the capacity of the company at a product family level. It must be noted that capacity is not just in manufacturing but in all aspects of the company (Correll, 1994).

The S&OP process typically begins with an updated sales forecast covering the next one year or more. Any desired increases or decreases in inventory or backlog levels are added or subtracted, and, the result is the aggregate production plans as shown in Fig. 4.1. The key functions

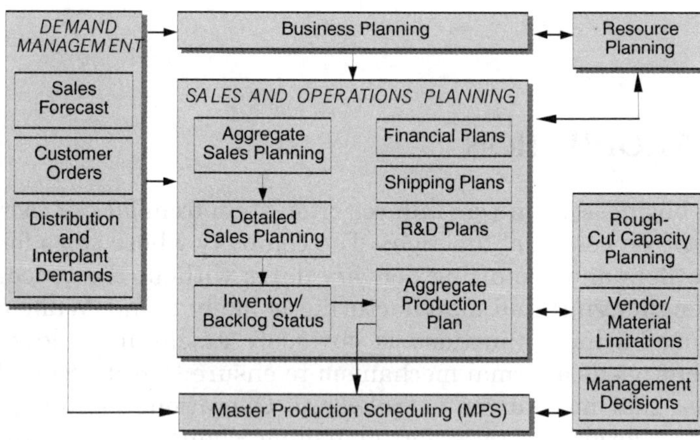

Fig. 4.1 *Sales and Operational Planning (S&OP)*

of the organization are all involved in the S&OP process. Each provides input and develops its respective plans on understandings reached in the process. The S&OP process ensures that individual plans are in sync with each other. Thus, the production plan is coordinated with the sales plan, the financial plan is reconciled with the production, sales, and business plans, and so forth. In addition to serving as a means of generating other plans, S&OP is a process for developing a 'budget' for the master schedule.

S&OP generally has a planning horizon of one year or more (the average is 18 months), and deals with sales and production volumes at the product family level. Unlike the overall business plan of the company, which is stated in financial terms, S&OP speaks in the language of sales and production: bookings, demand, supply, standard hours, units, tonnages, etc.

One of the principal benefits of the S&OP process is its focus on 'workable' plans. Natural optimism and business units planning in isolation are factors that lead to plans that cannot be fulfilled. Workable plans can be created only by systems that have built-in reality checks. The S&OP process offers those checks by starting at the top and providing verification at lower levels. The participation of all major functions is mandated by the process, and the feasibility of each function's plans is scrutinized by the others. Feedback loops ensure that plans are adjusted to the capabilities of related functions.

S&OP is the ability to state future sales and operations plan by product families. It is an aggregate planning process aimed at keeping aggregate balance between supply and demand, and making shipments to meet customers' needs with a minimum time, energy, and effort.

S&OP is defined by *APICS Dictionary* as follows:

- Sales and Operations Planning is a process that provides management the ability to strategically direct its businesses to achieve competitive advantage on a continuous basis by integrating customer-focused marketing plans for new and existing products with the management of the supply chain.
- The process brings together all the plans for the business (sales, marketing, development, manufacturing, sourcing, human resource, and financial) into one integrated set of plans.
- It is performed at least once a month and is reviewed by management at an aggregate (product family) level.
- The process must reconcile all supply, demand, and new-product plans at both the detail and aggregate level and tie to the business plan.
- It is the definitive statement of the company's plans for the near, to intermediate term covering a horizon, sufficient to plan for resources and to support the annual business planning process.
- Executed properly, the sales and operations planning process links the strategic plans for the business with its execution and reviews performance measures for continuous improvement (Cox III and Blackstone Jr, 1998).

To perform the necessary communication role, the production plan must be stated in commonly understood, aggregated terms, such as the monetary value of total monthly/quarterly output. Some companies prefer to break this total output down by individual factories or by product families etc. The key requirement is that the production plan be stated in some commonly understood homogeneous units that thereafter can be kept in concert with other plans (Vollmann *et al.* 1997).

S&OP is the process whereby the management of the company provides direction, resolves conflicts, and manages the operations of the business. It is the tool that links the business plans to the more specific objectives of the organization ensuring that all the divisions, departments, and other organizations within the company are pulling in the same direction at the same time toward the same goals (Palmatier and Shull, 1989).

Product Families

The production plan needs to be expressed in meaningful units, but it also needs to be expressed in a manageable number of units. Experience indicates that 5 to 15 family groups seem to be about right for a top management group to handle. Hence S&OP is done in aggregate by dividing a company's product line into product families. A typical company will have between 5 and 12 product families in which it does S&OP. The APICS Dictionary defines a product family in the following manner (Cox III and Blackstone Jr, 1998):

> *A Product Family is a group of products with similar characteristics, often used in production planning.*

Each product family has to be considered in terms of expectations on sales, manufacturing, and resultant inventories and backlogs.

4.2 OBJECTIVES OF SALES AND OPERATIONS PLANNING

One of the primary objectives of S&OP is to establish production rates that will achieve management's objective of maintaining, raising, or lowering inventories or backlogs, while usually attempting to keep the workforce relatively stable. It must extend through a planning horizon sufficient to plan the labor, equipment, facilities, material, and finances required to accomplish the production plan. As this plan affects many company functions, it is normally prepared with information from marketing, manufacturing, engineering, finance, materials, etc.

Following are the objectives of S&OP:

1. To develop and agree upon, for each family of product:
   (i) a sales plan, and
   (ii) a rate of production.
2. To ensure that the plans are realistic and valid, that is, ensure that proper amounts of resources (capacity, material, etc.) are available to satisfy customers in a cost effective manner and that the plans support the strategic and business plan.
3. To integrate manufacturing with other activities of the business plan.

4. To focus on corporate production goals.
5. To identify management objectives for production rates, inventories and/or backlogs, capacity, workforce management, and resource management.

4.3 FACTORS THAT AFFECT S&OP

1. Management's Perception of the Business S&OP reflects top management's intuition concerning direction of the business as well as its judgments regarding the likely impact of macroeconomic influences.

2. Market Conditions S&OP is also affected by market research, which indicates customer needs, competitor actions, and so forth.

3. Financial and Budgetary Planning Activities of the Company The S&OP process enables the business to translate manufacturing activities into monetary terms so that its financial impact can be understood (Vollmann *et al.* 1997).

All of the plans that are derived from the annual business plan must be updated and synchronized, if a formal planning process is to function effectively. From an input/output standpoint S&OP process would look like Fig. 4.2.

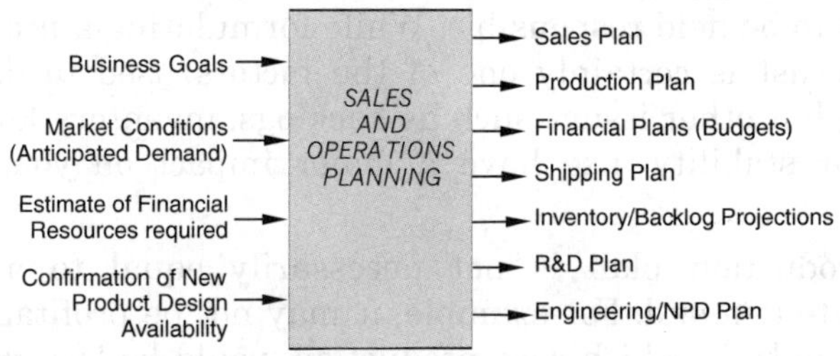

Fig. 4.2 *S&OP Viewed as an Input/Output Process* (Adapted from Stahl, 1995)

Some of the important points to be noted about the S&OP process are described below:

- The Production Plan is a rate of production (rate per month or week for a family of products) and does not include the timing or quantities of individual production lots. Thus the primary objective of S&OP is to develop rates of production for families, not schedules for individual items.

- In concept S&OP can be defined *as the monthly update to the annual business plan* (Stahl, 1995).
- The overall context within which trade-offs are made and the production plan is developed is increasingly called *Game Planning*. The game plan reflects the *strategy* (such as, increased market share) and *tactics* (such as, increased inventory for improved service) that are feasible for the firm. *The manufacturing part of the game plan is the (aggregate) production plan.*
- People are involved in developing the high level plans and reconciling them to create a company game plan. This company game plan balances the needs of sales and marketing against the capabilities of the factory to produce. (This plan might not be the best marketing plan, or the best manufacturing plan, or the best engineering plan).
- The process must reconcile all supply, demand, and new product plans at the aggregate (product family) level and tie to the business plan.
- It is the definitive statement of what the company plans to do for near to intermediate term covering a horizon sufficient to plan resources and support the annual business planning process (Ling, 1995).
- The (aggregate) production plan is *not* a forecast of demand. It is the planned aggregate production for which manufacturing management is to be held responsible. While formulating a production plan the forecast is certainly one of the factors used to determine an answer, but other issues such as backlogs, inventory levels, and the need for stability also have a major impact on your production needs.
- The production plan is not necessarily equal to a forecast of aggregate demand. For example, it may not be profitable to satisfy all demands, in which case production would be less than forecast. Conversely, a strategic objective of improved customer service could result in aggregate production in excess of aggregate demands. These are important management trade-offs (Vollmann *et al.* 1997).
- S&OP connects strategy and operations. Executed properly, it ties the company's high-level business and strategic plans to the operations of each department, guaranteeing that they are in harmony. The S&OP process also reviews performance measures for continuous improvement.
- With financial simulation capability it develops into an integrated business planning process. When S&OP is used in this manner it becomes a process that gives management the ability to strategically direct its business to achieve competitive advantage, by inte-

grating customer focused marketing plans for new and existing products with management of the supply chain (Ling, 1995).

- Management is responsible for developing and executing both the sales plans and the production plans for the company, and they must have direct control over these plans. The computer can critique and evaluate these plans by using some simple rules, and produce recommendations, but it should never add to, delete from, or change the S&OP developed by people (Landvater and Gray, 1989).
- S&OP/Production Planning process is one of the least understood processes in the modern manufacturing firm. It is the production planning that links the other company activities to manufacturing, and it is the production plan that states how production will support the company game plan. It has been referred to as *management's handle on the business*. Production plans are the communication media for integrating all functional areas of the company with manufacturing activities.

4.4 ISSUES TO BE ADDRESSED BY S&OP

The monthly S&OP process must address the following issues properly:

1. *Variances* Review the past month's actual to plan for sales, production, inventory, and/or backlog. If the variance is greater than expected, investigate and explain, revising the future plans, as appropriate.

2. *Market Conditions* Provide a forum for all disciplines to present and discuss their view of the market, so as to bring about cross-functional understanding of market dynamics.

3. *New Products* Present and discuss the status and impact of any pending new product introductions.

4. *Alternate Plans* Present, evaluate, discuss, and decide upon alternative plans and actions.

5. *Conflict* Raise and constructively resolve any policy conflict that exists within the organization. This often involves a change in traditional culture from being, 'the person who presents a problem is the problem.'

6. *Contingency Plans* Develop necessary contingency plans as appropriate to the situation.

7. *Accountability* Identify and agree upon necessary actions and changes to plans, creating accountability for results that support the company's goals.

This list implies that S&OP is part strategic and part tactical in nature, that is, it deals with both policy issues of strategy, as well tactical issues dealing with the execution of the company's strategy. This is shown graphically in Fig. 4.3.

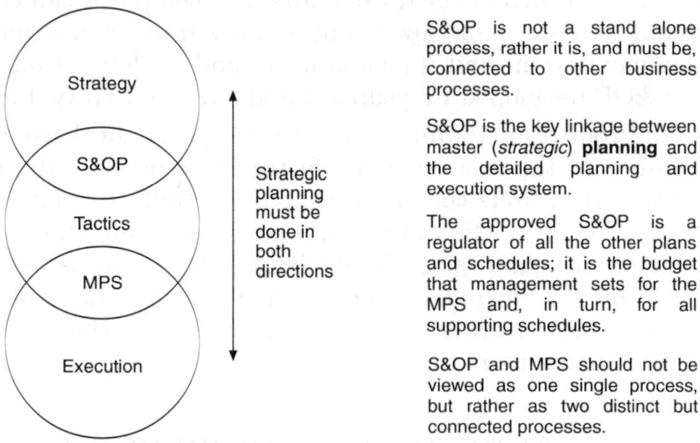

Fig. 4.3 *S&OP is Part Strategic and Part Tactical in Nature* (Adapted from Stahl, 1995)

4.5 S&OP AND MPS

Table 4.1 compares Master Production Scheduling (MPS) and the Sales and Operations Planning (S&OP).

Table 4.1

Factors That Distinguish Between S&OP and MPS
(Adapted from Stahl, 1995)

S&OP	*MPS*	*S&OP*	*MPS*
Volume	Mix	Months/Quarters	Days/Weeks
Aggregate	Detail	Rate of Production	Sequence of Product
Business Focus	Plant Focus	Market Perspective	Customer Perspective
Centralized	Decentralized	Executive Responsibility	Middle Management Responsibility

4.6 KEY MANAGEMENT CONSIDERATIONS TO MAKE S&OP A SUCCESS

Following are the key management considerations to ensure that S&OP properly reconciles and drives inter-functional interaction in a company:

- Everyone's participation (to maximize communication between sales and marketing, manufacturing, product development, finance, human relations, etc.).
- Premeeting preparation (maintain meaningful information).
- Decision making (involving what if?) and change management.
- Conflict resolution and consensus development (team building).
- Process (business) under control.
- Ongoing realistic improvement.

Managers from sales and marketing, manufacturing, product development, finance, human relations, etc. must participate in S&OP process as each of these functions has an impact or can be impacted by shifts in plans on what will be sold and what needs to be made. Involving the highest level of management of each of these functions on at least a monthly basis is critical. They must review trends, identify opportunities and approve changes in plans to keep the overall strategic objectives and business plans of the business fully connected to the tactical schedules and initiatives being pursued at lower levels.

Involving middle and upper levels of management means that there should be a defined process where a lot of the detailed data review and development is done prior to meetings, to make the best use of people's time.

4.7 THE S&OP PROCESS

The essence of S&OP process as shown in Fig. 4.4 is decision making. For each product family reviewed, a decision is made based on:

(a) Recent history.
(b) Recommendations from middle management.
(c) The executive team's knowledge of business conditions.

The decisions can be any of the following:

(a) Change the sales plan.
(b) Change the production plan.
(c) Change the inventory/backlog plan.
(d) None of the above, the current plans are okay.

- These decisions represent the agreed-upon plans by the president and all involved vice presidents and are documented and disseminated throughout the organization.

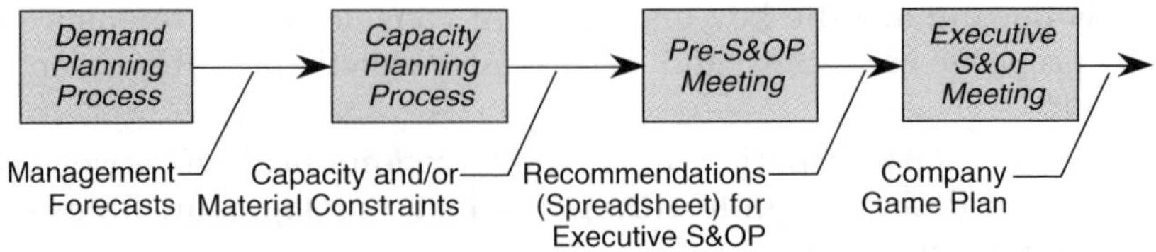

Fig. 4.4 *The S&OP Process: The Stages and the Output of Each Stage* (Adapted from Stahl, 1995)

- These decisions form the overall orders for the marketing, sales, manufacturing, materials, finance, and product development departments. These groups break down the aggregate plans from S&OP into the necessary level of detail: individual products, customers, regions, plants, and materials.
- New product plans are reviewed within S&OP in terms of their impact on the demand/supply picture.

4.7.1 Pre-S&OP Process

S&OP is not a point event that occurs in a two-hour executive meeting each month. Rather preliminary work begins shortly after month end and continues for several days. This phase, called pre-S&OP involves middle management, functional senior managers, and others throughout the company. Following are the objectives of pre-S&OP meeting process:

1. Review performance and impact on future.
2. Build preliminary consensus.
3. Identify best alternatives.
4. Reduce top management time in S&OP meeting.
5. Eliminate surprises in S&OP meeting.

Pre-S&OP Process Includes the Following Activities

(a) Preparing agenda for the S&OP meeting.
(b) Updating the sales forecast for product families.
(c) Reviewing the impact of these changes on the production plan and determining whether adequate capacity and material will be available to support them.
(d) Identifying alternatives where problems exist.
(e) Formulating agreed-upon recommendation for top management regarding overall changes to the plans and identifying areas of disagreement where consensus is not possible.

(f) Communicating this information to top management with sufficient time for its review before the executive S&OP meeting.

Pre-S&OP process involves two separate pre-S&OP meetings as described below:

1. *Sales and Marketing Pre-S&OP Meeting* This involves following activities:

- Review sales and forecast performance for previous period.
- Highlight changes in demand (volume, mix, or timing) and consider impact.
- Analyze assumptions and vulnerabilities and make requests for changes to manufacturing or inventory plans.

2. *Manufacturing Pre-S&OP Meeting* This involves following activities:

- Review last month's manufacturing performance.
- Consider any proposed manufacturing or inventory changes.
- Evaluate capacity (detailed or rough-cut) and material requirements to support the new plan.
- Cost all realistic options and prepare recommendations.

Thanks to pre-S&OP, the executive S&OP meeting should not take long; two hours or less is the norm with companies that understand the process.

Each department must do its pre-S&OP meeting 'homework.' S&OP can only work if each department has assembled its respective data and analyzed past performance to plan.

Assumptions and Vulnerability Analysis

In addition to projecting future customer demand, sales and marketing must also document their assumptions about:

1. Why will things change or stay the same?
2. What's happening in the economy?
3. What are our competitors doing?
4. What promotions, advertising, and other sales activities are we pursuing?
5. Why are we selling less or more than planned?
6. Which of our assumptions is incorrect?

Similarly, manufacturing has the pre-S&OP meeting task of assessing material and capacity needs for the current plan or any proposed changes in the plan. This is especially important if the plan calls for an increase in production.

4.7.2 S&OP MEETING

The focus of S&OP process is a meeting that is typically held monthly. Some volatile business may need to hold S&OP meetings more frequently, or, a special situation that can affect production, such as product recall or competitor on strike, will necessitate an unscheduled meeting. In any case, the experience of a broad range of companies shows that it is not effective to hold S&OP meetings less frequently than once a month.

The Major Activities Performed in S&OP Meeting

1. Review last month's sales, manufacturing, inventory, and financial performance.
2. Consider any new recommendations to support the new plans and decide actions based on realistic expectations.
3. Authorize changes to medium long-term future plans and fix the manufacturing plans for the next month.

S&OP Meeting Agenda

1. Special issues.
2. Performance Review.
3. Review assumptions and vulnerabilities.
4. Family by family review.
5. New products.
6. Special projects.
7. Review of minutes.

S&OP Decisions

The S&OP meeting is attended by the general manager (or president) and staff (vice-presidents). These people make decisions about issues such as:

1. Adding a shift.
2. Subcontracting on a large scale.
3. Responding to customer service concerns.
4. Increasing production on a particular product to ensure high customer service levels in a new market.
5. Reducing production on one or more product families because sales are lower than expected or because manufacturing cannot support the higher rate.
6. Setting new targets for finished goods inventories or backlogs by product family.

4.7.3 The S&OP Spreadsheet

One of the most important elements in this business process is the S&OP spreadsheet. Using the tool, key people review each major product family. The information—displayed in aggregate rather than in detail—shows sales performance to forecast, production performance to plan, and the resultant inventory levels and/or customer order backlogs. See Fig. 4.5 for an example of how the information is displayed.

SALES AND OPERATIONS PLAN FOR JUNE 1998

Family: Medium Widgets (Make-to-Stock) | Unit of Measure: 1,000 Units

Target Customer Service Level: 99% | Target INV Level: 10 Days on Hand

Sales	*Last 3 Months*			*First 3 Months*			*Second 3 Months*			*3rd three months*	*4th three months*	*Next twelve months*	*Totals*	
	Mar	*Apr*	*May*	*Jun*	*Jul*	*Aug*	*Sept*	*Oct*	*Nov*				*Fiscal year*	*Business plan*
Forecast	100	100	100	105	105	105	110	110	110	345	345	1.335	$ 12,770	$ 12,700
Actual	110	98	114											
Diff. +/–	+10	–2	+14											
Cum. Diff.		+8	+22											
Production														
Plan	100	100	100	105	110	115	115	115	100	345	350	1,365	1,272	1,300
Actual	100	102	100											
Diff. +/–	0	+2	0											
Cum. Diff.		+2	+2											
Inventory														
Plan	50	50	50	30	35	45	50	55	45	45	50			
Actual	40	44	30											
Days on Hand	8	9	6	6	7	9	10	11	9	9	10			
Customer service %	97	98	89											

Demand Issues and Assumptions

1. Forecast assumes no competitor price changes.
2. Accelerated export demand will continue through middle of next year.

Supply Issues

1. June and July production plan represents maximum effort due to vacations.

Fig. 4.5 *S&OP Spreadsheet*
(Adapted from Wallace, 1997)

- In this example, for a make-to-stock product family called 'medium widgets,' the demand supply strategy specifies a target customer-service level (shipment on time and complete) of 99 per cent. The target finished inventory necessary to support this shipping performance has been set at a 10 day supply.
- The company feels it needs 10 days of finished goods inventory because its experience over the recent past has shown that 10 days is the minimum level necessary to provide 99 per cent customer service. However, this 10 day supply target should not be considered as constant or fixed far into the future. The principle of continuous improvement should drive this company to improve its sales, production, and logistics processes so that 99 per cent customer service is attainable with only, say, 8 day supply. And then a 7 day supply. And then 5. But for now the realities of the life are that it takes 10 day's worth in inventory to provide the 99 per cent service level.
- In the upper part of the table, actual sales are compared to those estimated. For the past three months, sales have been running ahead of the forecast by 22,000 units.
- Actual production performance to the plan is evaluated in the middle portion of the table. It's close to perfect.
- The lower portion of the table shows inventory performance to plan and the actual customer service performance. There is no *right* level of finished goods inventory. Inventory management is only a process of managing inventory in response to the constant fluctuations of the marketplace. For example, if sales are higher or lower than anticipated, inventory will fluctuate above or below the desired levels. Therefore, managing finished goods inventories is not so much a matter of deciding on the right number as it is of responding to the marketplace. The same thing is true for backlogs in a make-to-order product. S&OP is the right place for making decisions about finished goods inventories or backlogs and the factors that affect them, namely *sales and production.* This is because the right people are available to make the best decisions, the planning process is structured, and management's expectations are known. In the spreadsheet, we can see a serious problem developing here, because the forecast was oversold, the actual inventory dropped below plan. The result is that customer service has dropped to 89 per cent for May, far below its 99 per cent target.
- Showing the customer-service-level numbers next to the inventory number is strongly recommended, because it helps people avoid suboptimal decisions. Many companies lurch from 'the inventories are too high; cut the inventories' to 'build inventory; we have too many backorders.' Understanding the interplay between finished inventories and customer service is aided by viewing this situation organically, which can help greatly in maintaining the demand/supply balance.

- The new sales forecast is shown in the top-right part of the table. Toward the right of the table are shown a total for the next 12 months and totals in units and dollars for the fiscal year, ending in December in this example. As a result, these totals are made up of both sales history (January through May) and sales forecast (June through December).
- The forecasted dollars in the business plan are also shown for an easy comparison between the business plan and the S&OP forecast for the fiscal year's volume. Based on these data, the top-management team will probably elect to change the business plan accordingly.
- The future production plan, based on the new forecast and other considerations is shown.
- The future inventory projections for finished good are also shown. It shows that the target inventory of 10 days on hand will not be reached until September.
- The assumptions underlying the demand forecast, and the relevant supply (production/procurement) issues are listed at the pages on 270.
- The intent is to have all of the relevant information for a given product family on one piece of paper. This enables each family's situation, both in recent past and future outlook, to be viewed completely.
- The information contained in the spreadsheet represents the recommendations coming from the pre-S&OP team to top management regarding 'medium widgets.' It's then top management's job to make a decision: accept the plan as submitted or change it. For example, in Fig. 4.5, the below-target inventory levels between June and August may be viewed by top management as unacceptable, because the company will not be providing good customer service. The company may decide to incur additional, and perhaps substantial, costs of paying for extreme overtime, subcontracting, or bringing in temporary help, so that inventory—hence customer service—can get back on target sooner than September.

4.7.4 Need for Top Management's Direct Involvement in S&OP

Senior management's involvement is critical to ensure that the process results in decisions being made when necessary and appropriate. Rather than reacting after the fact to changes, the focus should be on anticipating possibilities, opportunities, and vulnerabilities. Available data should be used to examine different alternatives and scenarios based on 'what if' simulations and then choosing the best course of action. At times, different functions or people will disagree on the best course of action. It is in these cases that top management needs to resolve the conflicts and drive a consensus, cooperative team approach to picking one path and ensuring that every function is working towards the same end (Dougherty, 1995).

Many of the decisions made in S&OP affect the financial plan for the current year, and top management owns that business plan. It has a stewardship responsibility for it, and only top management can make decisions to change the plan.

When the business plan is not changed to reflect the new sales and operations plan there is a discrepancy between the financial numbers expected by the top management and the forecasts and production plans being used to operate the business. Best-in-class performance in this area means that the business is managed—at all levels—using only one set of numbers. Through their direct involvement, the leaders of the business make a strong statement that S&OP is the process that they use to manage this highly important part of the business, bringing demand and supply into balance and keeping them there.

Responsibility for the S&OP Process cannot be Delegated

Although the detail work and research can be conducted by middle managers and their subordinates, the department heads must be accountable for the plans, and they must attend the formal meetings. The general manager must also be willing to set aside the time to participate in the monthly meetings — the presence of the general manager indicates the seriousness of the process and serves as an incentive for others to make it a high priority.

4.7.5 Attendees for S&OP Meeting

Table 4.2 lists *mandatory attendees* who make up the *core team* at monthly S&OP meetings as well as attendees who might be called in to provide information during special situations.

Table 4.2

Attendees List for S&OP Meetings

Department	*Mandatory Participation (Core Team)*	*Potential Participants Who Might be Called in to Provide Information during Special Situations*
—	General Manager/President	—
Sales	Department Manager/ Vice-President	1. Customer Service Manager 2. Distribution Manager 3. Service Parts Manager 4. Demand Manager
Marketing	Department Manager/ Vice-President	1. Chief Forecaster 2. Product Manager(s)

Department	*Mandatory Participation (Core Team)*	*Potential Participants Who Might be Called in to Provide Information during Special Situations*
Manufacturing	1. Department Manager/ Vice-President 2. Materials Manager	1. Manufacturing Manager 2. Master Scheduler 3. Purchasing Manager 4. Quality Assurance Manager
Engineering	Department Manager/ Vice-President	1. Drafting Manager 2. Engineering Scheduler 3. Manager—Design Engineering
Finance	Department Manager/ Vice-President	1. Budget Manager 2. Cost Accounting Manager
Human Resources	Department Manager/ Vice-President	—
Program/ Special Projects	Appropriate Managers	—

- The schedule of the meetings for the entire year should be distributed well in advance so that people can arrange their travel and other plans *around* the S&OP sessions.
- Meetings should start on time and should not be interrupted by telephone calls.
- Prior to the formal meeting, an S&OP kit should be distributed. The S&OP kit contains the following:
  1. All of the premeeting plans created by the various departments.
  2. Analysis of the plan's assumptions and vulnerabilities.
  3. An agenda, with the following topics:
     (i) Special issues
     (ii) Company Performance Review.
       (a) Review of Customer Service.
       (b) Review of financial and business performance.
       (c) Department-by-department (sales and marketing, engineering or R&D, and manufacturing) review in aggregate.
     (iii) Review of assumptions and vulnerabilities.
     (iv) Family-by-family review.
     (v) New product discussion.
     (vi) Review of special projects.
     (vii) Review of meeting decisions.
     (viii) Critique of processes.

4.7.6 S&OP Process as a Monthly Cycle

Figure 4.6 below shows a typical monthly business cycle and the activities that are addressed at different points in time to maintain an effective S&OP process.

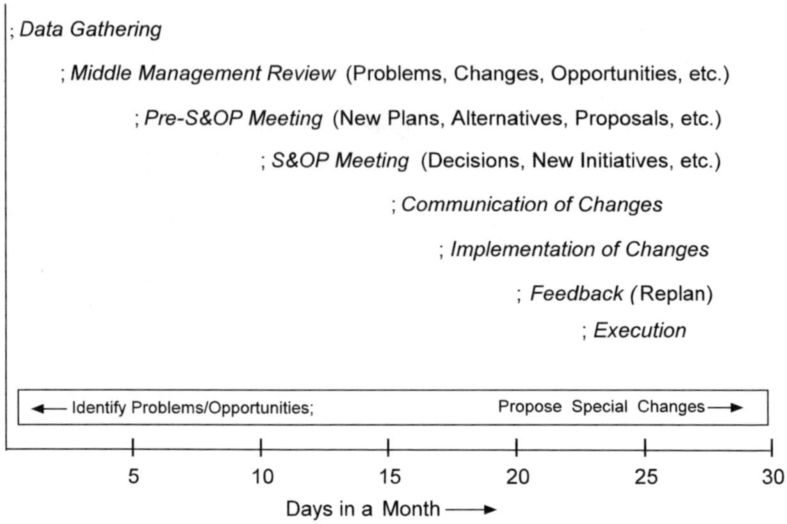

Fig. 4.6 *The S&OP Process: A Monthly Business Cycle* (After Dougherty, 1995)

Time Required for Executive S&OP Meeting One 2-hour meeting per month can often replace several meetings and thus result in a net reduction in meeting time. For the president (or CEO) preparation time is zero.

Benefits of S&OP to Top- and Middle-Management The net result of S&OP for the top- and middle-management group should be: fewer meetings, less time spent in meeting, more productivity in its decision making process, and a higher quality of worklife.

Software Requirements Virtually all companies generate their S&OP spreadsheets via simple spreadsheet software.

Time Required to Attain Proficiency It has taken most companies between six months and a year to become proficient with S&OP.

Cost Implications S&OP is relatively inexpensive to implement. It does not cost a lot and relatively few people are involved.

The factors that are used to develop the production plan include the following:

1. The sales plan.
2. Vendors and capacity limitations.
3. The current and desired inventory (for a make-to-stock product).
4. The current and desired backlog (for a make-to-order product).
5. Both inventory and backlog for a family that contains some MTS products and other MTO products.

The production plan is often set at a rate different from the sales rate (for example, to increase or reduce inventory or backlog. Following are some of the formulas used in S&OP process:

Production Plan = Sales Plan + The Adjustment in Backlog or Inventory*

*The adjustment in Backlog/Inventory

= + Proposed Reduction in Backlog, or
– Proposed Increase in Backlog, or
+ Proposed Increase in Inventory, or
– Proposed Reduction in Inventory.

Inventory = Starting Inventory + Production Plan – The Sales Plan

Backlog = Starting Backlog + Sales Plan – The Production Plan

Resource planning is directly related to production planning. In the long-term, the marketplace should be the driver and the factory resources should meet the needs of the marketplace. Hence, in the long-term planning, the extended production plans should be translated into capacity requirements. In the short run, however, the limitations of the available (factory) resources provide a set of constraints to production planning and thus determine the rates of production.

4.8 INTEGRATION ISSUES

(a) Engineering or R&D Plan

The major issue here is the need to integrate the new product development plans into the S&OP process as well as the material planning system to address such questions as, when should material be ordered for the new products? S&OP tells everyone when the new product lines will be introduced. The New Product Development (NPD) system informs the development people as to when their specific tasks and activities must be completed. And the material planning system lets the purchasing people and the plant know what to buy or make, and when. Following are the two methodologies for integrating NPD and manufacturing activities:

1. Some companies use their MRP II system to plan and schedule both the manufacturing activities and the non-manufacturing activities (design, drawings, methods, toolings, etc.). Material and capacity planning is done for all these new product introduction activities as well as for the activities related to manufacturing the new products.

2. Other companies use scheduling systems developed for NPD and integrate these with their MRP II system. The systems communicate on dependent events.

Either method is workable, and successful companies use both.

(b) Employee Empowerment

Ultimately S&OP is a group process. Although the general manager/VP is accountable, if the group cannot work effectively as a team, the results are going to be disappointing. Therefore, another integration issue is how well the group works together. If your company has been subjected to 20 years of autocratic management in which everyone waits to see what the highest-ranking person in the room says before speaking, then you will probably need to do some work on employee empowerment and group process skills. In such a situation, the S&OP process is an excellent forum for developing these skills.

4.9 S&OP MEASUREMENTS

1. Performance Relative to Production Plan

The single most important S&OP measurement is performance relative to the production plan, which should be 98% or better. In other words, a company should be able to perform within 2% of its goal by product family. Some of the individual items may be off by more than 2%. One might be off by more than +5% and others by –5%. In total, though, a deviation from the goal by 2% or less by family should be achieved on a monthly basis.

2. Performance Relative to Sales and Finance Plans

It is not possible to provide hard-and-fast performance criteria for these because the nature of different industries varies so much. For example, in the high fashion business, 30% accuracy relative to forecast may be acceptable. But in a stable industrial market, such as replacement parts for grinding machines, you would expect greater accuracy.

3. Performance Relative to Engineering and R&D

The integration and integrity of these plans is critical in today's increasingly competitive markets. Most companies measure these plans in the same way that they measure manufacturing and purchasing on-time performance.

4. Group Consensus

Finally, a non-quantitative measurement of the S&OP process is the ability to achieve consensus as a group—consensus on the company plan and on specific issues regarding the marketplace, the economy, the competitive arena, and how fast the company can adapt to a changing world. The ability to come to consensus and work as a team is a vital measurement of how well the S&OP process is working.

4.10 BENEFITS OF S&OP

S&OP Provides a Common Plan for the Entire Company

When the company has a single game plan, each department functions in harmony with the other departments. If the company doesn't have a common plan, each group of people works very hard at its plan while other groups work equally hard at their own plans. Because the plans are different, however, each group is effectively working on the 'wrong' plan.

S&OP Yields a Workable Plan

In a world of competitive pressures, unreasonable customers, and increasing expectations from top management, there's a tendency to overstate what can be realistically done. This, although done with the best of intentions, invariably results in plans that are unworkable. To make a plan workable, a series of reality checks is needed, starting with the S&OP process. At the top level, a rough-cut capacity analysis must be done to verify that the plan is attainable. If the plan passes the rough-cut test the detail material and capacity plans need to be checked to verify that they are attainable as well. Without such reality checks, unrealistic plans are the norm, not the exception, leading to confusion, additional costs, and reduced customer service. Any time more work is due today than can actually be done, someone has to decide what gets done and what doesn't. That decision should be made by someone who can see the 'big picture' in terms of customers, products, due dates, and so on.

4.11 SUMMARY

1. S&OP gives executives and managers better control of the business. It provides a window into the future and thus helps top management make better decisions.
2. S&OP enhances the company's flexibility to respond to customer needs and to ship complete orders on time.
3. Top management's role is crucial. For the process to work, the president or general manager must be engaged directly via his or her leadership at the executive S&OP meeting.
4. Changes in production rates can be made less radically and hence with lower costs and less negative impact on the workforce.
5. Inventories and order backlogs can be managed *proactively* to support the company's delivery and lead-time objectives.
6. The sales and operations plan can be converted to monetary units (such as rupees or dollars) to test the validity of the business plan, which may have been set many months earlier.
7. S&OP provides a more holistic view of the business, because sales demand, production, inventories, and customer lead times are viewed simultaneously for a given product family and set of

resources. Thus, S&OP enhances teamwork at both the top- and middle-management levels.

8. S&OP has been shown to help companies improve customer service, reduce inventories, shorten customer lead-times, and control costs.

References

Correll, James G, 'Reengineering the MRP II Environment', in *APICS: 37th International Conference Proceedings,* American Production and Inventory Control Society, Falls Church, V A, 1994, pp 1–4.

Dougherty, John R, 'Master Planning: A Key to Satisfying Customers in the 90s', in *APICS: 38th International Conference Proceedings,* American Production and Inventory Control Society, Falls Church, V A, 1995, pp 80–84.

Cox III, James F, and John H Blackstone Jr, *APICS Dictionary* (ninth Edition), American Production and Inventory Control Society, Falls Church, V A, 1998.

Landvater, Darryl V, and Christopher D Gray, *MRP II Standard System: A Handbook for Manufacturing Software Survival*, Oliver Wight Limited Publishing Company, Essex Junction V T, 1989.

Ling, Richard C, 'Beyond Sales and Operations Planning: An Integrated Global Business Planning Process', in *APICS: 38th International Conference Proceedings,* American Production and Inventory Control Society, Falls Church, V A, 1995.

Palmatier, George E and Joseph S Shull, *The Marketing Edge,* Oliver Wight Publications, Inc., Essex Junction, V T, 1989.

Proud, John F, *Master Scheduling: A Practical Guide to Competitive Manufacturing,* John Wiley & Sons, Inc, New York, 1994.

Stahl, Robert A, 'Sales and Operations Planning: A Fundamental that Still Works', in *APICS: 38th International Conference Proceedings,* American Production and Inventory Control Society, Falls Church, V A, 1995, pp 215–218.

Wallace, Thomas F, 'Sales and Operations Planning: Top Management's Handle on the Business'. in Patricia E Moody (Editor), *Leading Manufacturing Excellence: A Guide to State-of-the Art Manufacturing',* John Wiley & Sons Inc, New York, 1997, pp 112–125.

Vollmann, Thomas E, William L Berry, and D Clay Whybark, *Manufacturing Planning and Control Systems* (fourth edition), McGraw-Hill Inc, New York, 1997,.

Chapter 5

Demand Management

5.1 INTRODUCTION

Demand management is the function of recognizing all demands for goods and services to support the marketplace. It involves doing what is required to help make the demand happen and prioritizing demand when supply is lacking.

Proper demand management facilitates the planning and use of resources for profitable business results. It encompasses the activities of forecasting, order entry, order promising and determining branch warehouse requirements, interplant orders, and service parts requirements.

APICS Dictionary

Demand management is the process most closely related to the marketplace. It serves as the link between production and the market for both the input of resources and the output of product. Demand management is thus a gateway module in Manufacturing Planning and Control (MPC/MRP II) system providing the link to the marketplace. Activities performed here provide coordination between manufacturing and the marketplace, sister plants, and warehouses. It is through demand management that a channel of communication is maintained between the MPC systems and their 'customers.' Specific demands initiate actions throughout MPC, which ultimately result in the delivery of products and the consumption of materials and capacities. The external aspect of the demand management module and its interfaces with other modules are depicted in Fig. 5.1.

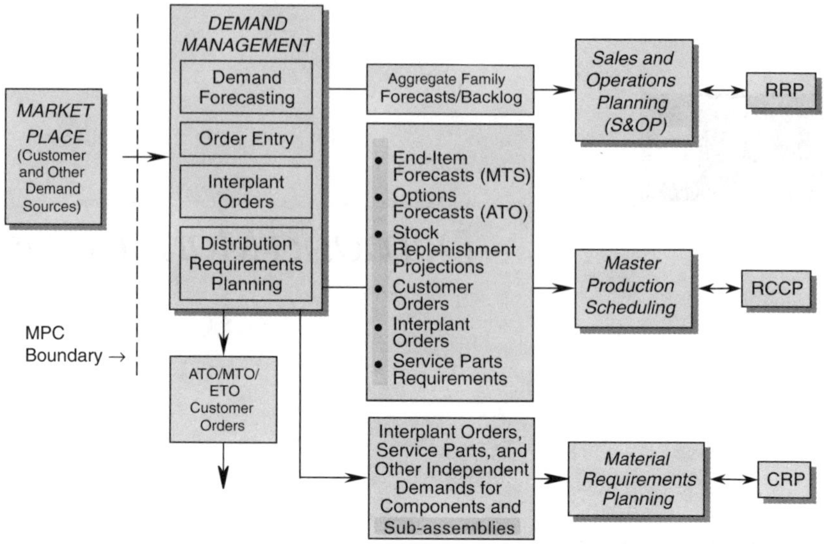

Fig. 5.1 *Demand Management Interfaces*

5.1.1 Demand Management–An Information-Centered Activity

Demand management is very much an information-centered activity, involving gathering information from customers, maintaining information on product availability, and monitoring the match between supply and demand. This is the process that provides information to match the production rate with the output required.

Service Levels

It is in demand management that service levels and resultant safety stocks are explicitly defined. Service levels are adversely affected by forecast error, purchasing or manufacturing lead-time error, and quality problems. If these errors cannot be corrected, the service level must be protected through safety stock or safety lead time. If the problem is forecast error, safety stock should be used, but when the problem is lead-time error, safety lead time is more practical. Both increase inventory levels.

The requisite degree of flexibility for responding to mix or engineering design change is set here as well. This is done through the determination of hedge quantities and timings. The master scheduler is then responsible for maintaining the required level of buffer stocks and timings. Further, as changes occur in the marketplace, demand management can and should routinely pick them up, indicating when managerial attention is required.

Demand generally comes from outside the company and is, thus, beyond the full control of management. Based on this there are some that argue that demand cannot be managed. It would be convenient to proceed with the notion that demand should be left to rise and fall of its own accord, with all of management's attention directed toward supply. This notion, according to Proud (1994), fails on several counts:

- Few production facilities are so flexible with respect to volume that they can operate efficiently with low output in one period and high output in the next. This violates the basic principle of load leveling.
- Not all demand is external to the company—at least in the larger sense. Much of modern production simply creates components for use in final products manufactured by the same company or its subsidiaries. Thus, even though the final customer decision is external, demand is not entirely created from outside of the company.
- Demand can be created or its timing shifted through marketing.

Thus, the idea of managing demand is reasonable and necessary if the sales and the company's capabilities are to be kept in balance. Not only can demand be planned, but it can also be shifted when it is of value to the customer, or if major economies can be achieved. The key point is that demand cannot simply be allowed to happen and then have the company react. It must be anticipated and planned for explicitly (Vollmann *et al.* 1997).

Demand management is an important activity because it relates the factory to the customer, and synchronizes manufacturing accordingly. Through demand management, all potential demands on manufacturing capacity are collected and coordinated. This activity manages the day-to-day interactions between customers and the company. A well-developed demand management module within Production and Inventory Management (PIM) brings significant benefits. Proper planning of all externally and internally generated demands means that:

- Capacity can be better planned and controlled.
- Timely and honest customer order promises are possible.
- Physical distribution activities can be improved significantly.

The approach to demand management depends on two important factors:

(i) the degree of product customization, and
(ii) the level of service to be provided to the customer.

- When products are sold on a Make-to-Stock (MTS) basis, forecasting how many of each product will be demanded over some time period is the essential task.
- In Assemble to Order (ATO) firms, customers communicate the product variants and options they would like and a delivery date

must be provided to them. Thus, the customer orders identify future requirements for the final assembly schedule in case of ATO type of production. However, to reduce the lead-time between order entry and delivery the variants and options have to be manufactured and carried in inventory before receipt of customer orders and hence the demand for the modules have to be forecast.

- In Make-to-Order (MTO) companies it might be necessary to work out the exact specifications of the product, even while it is being manufactured. Thus all the activities except the design of components would be done after the receipt of customer orders.
- In Engineer-to-Order (ETO) companies even the actual design would be undertaken after the receipt of customer orders.
- In both MTO and ETO companies forecasting is still required to plan the manufacturing and engineering capacity.

5.2 FUNDAMENTAL REQUIREMENTS OF DEMAND MANAGEMENT

Demand management has four fundamental requirements (Proud, 1994):

1. Prediction

Maintaining a balance of supply and demand requires some ability to know the level of incoming orders *in advance,* especially in ATO and MTS environments.

2. Communications

Successful companies know that they will have a chance of preparing for incoming demand if they maintain their own listening posts near the customers. Typically this is done through the field sales force, which visits customer facilities, talks with purchasing managers, and otherwise tries to gauge the level and timing of future orders.

3. Influence

Communications lead to knowledge and knowledge leads to influence. Production works to level the load on the factory. The master scheduler uses his influence with sales and marketing to negotiate, where necessary, the shifting of customer demand to produce a better situation on the manufacturing floor—one that makes better use of fixed assets and human resources. This might take the form of a phone call to marketing or to the sales representative asking: 'Do you really need this big order in October?' 'Would it be helpful to you if we shipped a third in September and two-thirds in October?' Or 'Would it be a problem for you if we shipped half in October and half in November?' Marketing can also influence demand, both its quantity and timing, using advertising, pricing, and incentives to dealers, sales representatives, and customers (Proud, 1994). Pricing is the traditional mechanism for influencing demand. Order management is not passive but recognizes that demands not only result from product attractiveness in the eyes of customer, as addressed by marketing and design, but are also influenced by expectations created as a result of actions and policies taken by operations. For example, customers may be discouraged by long queues or by unacceptable quoted delivery dates (Becheno and Elliott, 1997).

Marketing has an important role to play in managing orders in ways which result in greater stability. It may take actions which result in greater stability, for example, by encouraging off-peak demand, but it may also create instability, for example, by promotion which is timed just when capacity is stretched or materials are in short supply. Marketing should not work independently of operations in sales promotions or making promises to customers. Equally, operations should not take actions which reduce costs at the expense of customer service. Both should have an interest in improved service and lower costs to customers. Cooperation through order management is one of the keys.

4. Prioritization and Allocation

The idea behind demand management and master scheduling is to satisfy all customer demand. However, if a situation presents itself that not enough product exists as requested, or that the materials and resources needed to produce the required product are not available, then a decision must be made as to which customers get their orders filled as requested and which need to wait. This decision is the responsibility of sales and marketing. Allocation is the process used when the company cannot produce enough product to cover the demand, and hence some business may have to be turned away. In this case, the available product needs to be allocated so that the company does not oversell and overcommit its ability to produce.

5.3 FUNCTIONS OF DEMAND MANAGEMENT

Demand Management encompasses the areas of:

1. Advertising, promotions, pricing, adding and deleting products.
2. Forecasting anticipated business.
3. Customer Order Servicing which involves the following activities:
   (a) receiving and responding to the customer inquiries, and
   (b) interacting with the master scheduler on availability of products.
4. Order delivery date promising.
5. Customer order entry (entering new orders and processing changes to existing ones).

6. Distribution Requirements Planning (DRP) and interplant orders. (DRP may be thought of as an upward expansion of the MRP I logic from the MPS through the distribution network to the ultimate customers.)
7. Other customer-contact-related activities.

Demand management also concerns other sources of demand for manufacturing capacity, such as:

1. Intracompany Requirements, which include:
   (a) Inventory changes,
   (b) Promotion requirements or special exhibition requirements,
   (c) Quality assurance needs,
   (d) R&D,
   (e) Engineering testing,
   (f) Pilot project requirements, etc.
2. Pipeline Stock (Pipeline Inventory/Transportation Inventory) [It is the inventory in the transportation network and the distribution system, including the flow through intermediate stocking points. The flow time through the pipeline has a major effect on the amount of inventory required in pipeline. Time factors involve order transmission, order processing, scheduling, shipping, transportation, receiving, stocking, review time, etc. (Cox III and Blackstone Jr, 1998).]

Demand management is an integral part of the high level planning and scheduling process and acts as a gateway module that provides the link to the marketplace. A communication channel between the customer and MPC system is maintained through demand management.

- Activities performed in demand management provide coordination between manufacturing and the marketplace, sister plants, and warehouses.
- In a manufacturing company, forecasts and the backlog of customer orders are the starting point for company business plans, S&OP process, and the MPS.
- In businesses where physical distribution is critical and the factory must support a warehouse replenishment program, which can differ significantly from the pattern of final customer demand, the demands from physical distribution play a major role in the development of valid production plans and MPS.
- Demands, as input to the S&OP and MPS processes, are combined in a process of human judgment and evaluation to produce a valid S&OP and individual master production schedules.

5.4 LINKAGE OF DEMAND MANAGEMENT AND PRODUCTION PLANNING

The exact linkage of demand management (DM) and production planning depends to some extent on how the firm handles production planning. Following are the functions that demand management should perform for the production planning function.

1. *DM should synchronize demand plan stated in sales units for shorter time periods with the quarterly production plan stated in monetary units* If the production plan is a quarterly statement of output in monetary units or in some other financial measure, then the key requirement for demand planning is for synchronization with this target.
2. *DM should communicate market/distribution activities to the production planning function* For example delivery timings for customer orders must be communicated to production planning when they are significant enough to affect the production plan. Similarly a major change in distribution inventory policy might influence the production plan, and hence, should be communicated to the production planning function.
3. *DM should assure the completeness of demand information to be incorporated in production and resource requirements planning* All sources of demand for manufacturing resources, such as spare parts demand, intercompany transfers, promotion requirements, pipeline buildups, quality assurance needs, exhibition or pilot project requirements, and even charitable donations, etc., must be identified and incorporated in the production and resource requirements planning processes. Sometimes this is more difficult than it appears. The difficulty seems to be greatest for companies with a significant number of interplant transfers. We have often heard plant managers complain that their worst customer is a sister plant or division (Vollmann *et al.* 1997).

5.5 INTERACTIONS BETWEEN DEMAND MANAGEMENT AND MPS

Interactions between Demand Management and MPS are frequent and detailed. Details vary significantly between MTS, ATO, and MTO environments. In all instances, however, the underlying concept is that *forecasts are consumed over time by actual customer orders.*

1. For MTS environments, there are very few actual customer orders, since demand will be generally satisfied from inventory. Thus, the MPS task is one of providing inventory to meet forecasted future customer orders.

2. In ATO environments, the MPS task involves the use of *Available-to-Promise (ATP)* concept for each module (variant/option) to manage *the conversion from forecast to booked orders.* Here the MPS task is to provide viable customer promise dates. Usually, there are customer orders already booked for several periods into the future, the master scheduler uses the ATP concept for each module or customer option to manage the conversion from forecast to booked orders.
3. In MTO environments, the MPS task is concerned with controlling *custom orders*, some of which are not completely specified and engineered, as they progress through all steps in the process. This involves engineering activities, which affect resource requirements as well as manufacturing. In addition, all this has to be coordinated with customers as the orders become completely specified.

5.5.1 Providing Stability through Time Fences

In each company environment, the demand management module's objective is to bridge the firm and the customer. This is facilitated by the *time fences*: the *Demand Time Fence (DTF)* and the *Planning Time Fence (PTF)*.

Time Fences

Time fences are means of recognizing the degree of stability required in the MPS. Frequently, they are merely statements of policy. A time fence is defined by APICS Dictionary as a policy or guideline established to note where various restrictions or changes in operating procedures take place. For example, changes to the MPS can be accomplished easily beyond the cumulative lead time, while changes inside the cumulative lead-time become increasingly more difficult to a point where changes should be resisted. Time fences can be used to define these points (Cox III and Blackstone Jr, 1998).

Demand Time Fence (DTF)

(1) It is the point in time inside of which the forecast is no longer included in total demand and projected available inventory calculations; inside this point, only customer orders are considered. Beyond this point, total demand is a combination of actual orders and forecasts, depending on the forecast consumption technique chosen. (2) In some contexts, the demand time fence may correspond to that point in the future inside which changes to the MPS, must be approved by an authority higher than the master scheduler. Note, however, customer orders may still be promised inside the DTF without the higher authority approval if there are quantities Available-To-Promise (ATP). Beyond the DTF, the master scheduler may change the MPS within the limits of established rescheduling rules, without the approval of higher authority (Cox III and Blackstone Jr, 1998).

Planing Time Fence (PTF)

It is a point in time, denoted in the planing horizon of the master scheduling process that marks a boundary inside of which changes to the schedule, may adversely affect component schedules, capacity plans, customer deliveries, and cost. Planned orders *outside* the PTF can be changed by system planning logic. Changes *inside* the PTF must be manually changed by the master scheduler.

The two time fences—DTF and PTF—divide the MPS planning horizon into three areas termed as *firm* (or *frozen*), *flexible* (or *slushy*), and *free* (*unrestricted* or *liquid*) as shown in Fig. 5.2.

Firm/Frozen From the present until the demand fence, the schedule is *firm* (sometimes called *frozen*). Resources have been committed to production, no changes can occur unless approved at the highest level, and the costs of changing the schedule must be borne by whoever requires the change.

Flexible/Slushy Between the demand fence and the planning fence, the schedule is *flexible* or *slushy*. Changes are permitted but are evaluated

MPS Time Fences

DTF *PTF*

Emergency Changes	Trading Area	Adding and Changing
FIRM (Frozen)	*FLEXIBLE* (Slushy)	*FREE* (Unrestricted/Liquid)
Stabilize	Material Ordered Capacity in Place	Future Planning
■ Forecast is not included in total demand and projected available inventory calculations; only customer orders are considered. ■ Changes to the MPS must be approved by an authority higher than the master scheduler.	■ Total demand is a combination of actual orders and forecasts depending on the forecast consumption technique chosen. ■ Changes to the schedule may adversely affect component schedules, capacity plans, customer deliveries, and cost.	■ Planned orders can be changed by the system planning logic itself. That is, computer is allowed to schedule, reschedule, or cancel orders as the requirement figures change.
■ Customer orders may be promised without the higher authority approval if there are quantities available-to-promise (ATP).	■ Master scheduler may change the MPS *manually* within the limits of established rescheduling rules.	

← *Planninghorizon* →

Fig. 5.2 *MPS Time Fences*

to prevent costly disruptions. That is, it can be changed but may require the approval of master scheduler or top management to do so. Though tangible resources have not been committed to a particular requirement, scheduling and planning efforts have already been spent.

Free/Unrestricted/Liquid Beyond the planning fence, the schedule is liquid. That is, changes can be made with little or no cost because neither tangible nor intangible resources have been committed to the requirement.

The authority for making changes, as well as the way the MPS is stated, differs among these areas. Fences provide guidelines to help the master scheduler to govern schedule modifications as forecasts become actual orders. Changes to the MPS can take place with relative ease if they occur beyond the cumulative lead time. They are much more difficult and costly, however, if they occur within the cumulative lead time. MPS Time fences for different manufacturing environment has been shown in Table 5.1.

Table 5.1

MPS Time Fences for Different Manufacturing Environment

	Firm	*Flexible*	*Free*
MTS	Customer order processing and shipping time.	Component lead-time + subassembly and final assembly lead-time.	Beyond Flexible Period.
ATO	Customer order and shipping lead time and final assembly time.	Raw material and component procurement and subassembly lead-time.	
MTO	Component lead-time, final assembly lead-time, and order processing and shipping lead-time.	Common raw material procurement lead-time.	

The actual timefence decisions are dependent on the manufacturing environment.

The underlying idea is that manufacturing should be responsive to marketplace needs, but it cannot do so without any constraints. In order to have some degree of stability in manufacturing, which allows for efficient operations, the company needs to place some restrictions on responsiveness and define the ways in which trade-offs are to be evaluated (Vollmann *et al.* 1997).

Facilitating the Buffering Against Demand Uncertainties

One aspect of relationship between demand management and MPS is facilitating the buffering against different types of demand uncertainties.

1. In MTS case, uncertainty is largely in the demand variations

around the forecast at each of the inventory locations. Here, levels of safety (and/or safety lead-time) must be set in order to provide the service levels required.

2. In the ATO case, the uncertainty involves not only the quantity and timing of customer orders but product mix as well. Safety stock options can be used, and product-mix hedging with components and/or materials is a valuable technique.
3. In MTO case, the uncertainty is often not the timing or quantity of the customer order, but rather how much of the company's resources (capacity) will be required as the engineering is finally completed and exact requirements are determined.

5.6 THE SPARE PARTS ISSUE

The demand management task of specifying all sources of demand on the facility impacts master production scheduling in several ways. Some of the sources will be handled directly in the MPS, such as pipeline inventory buildup, special exhibition requirements, and interplant transfers. Others such as spare parts may not be. The spare parts issue can be quite complex in many firms. The demand for spare parts is typically forecast on an item-by-item basis and added to the gross requirement data in the corresponding MRP I records. The actual spare part demand will, of course, vary around the forecast. The variability can be buffered with safety stocks. The point is there may not be an explicit treatment of each spare part by an MPS.

A key problem might be raised if spare part demand is not incorporated at the MPS level; the resulting capacity requirements will not be reflected in rough-cut capacity planning models. The requirements would, however, be seen by the detailed capacity requirements planning models that are driven from MRP I records. To adequately perform rough-cut capacity planning, the service part demand will have to be included. This could be done on a rough basis by estimating the service part demand in monetary units and using some base to convert it to the capacity measure. Alternatively, one could prepare a bill of capacity for each service part and multiply by the forecast service part demand levels. The principle is clear: The capacity must be provided for *all* sources of demand (Vollmann *et al.* 1997).

5.7 OBJECTIVES OF DEMAND MANAGEMENT

1. To develop the most reasonable projection of future requirements.
2. To update this projection of future requirements when change is warranted.

3. To avoid trivial changes to S&OP and MPS by properly managing the different demand streams.
4. To identify any meaningful differences in the marketplace (such as trends) at the earliest possible moment so that effective action can be taken.

Demand management is a people process and the essence of effective demand management is:

(i) good communication,
(ii) quick feedback, and
(iii) properly defined accountabilities.

The computer logic that is part of an effective demand management system should support these important activities.

5.8 FORECASTING SYSTEM

A forecast is a prediction of what will occur in the future. Companies attempt to predict how much of their product will be sold in the future. A forecast of product demand is the basis for most important planning decisions. Planning decisions regarding scheduling, inventory, production, facility layout and design, work force, distribution, purchasing, and so on, are functions of customer demand. Similarly, long-range strategic plans by top management are based on forecasts of the type of products consumers will demand in the future and the size and location of product markets (Russell and Taylor III, 1998). A forecast is, in effect, a formal request from sales and marketing, to manufacturing, to have product or capacity available to meet perceived demand. Forecasts of end items, spare parts, and other independent demands should be a part of the front end modules of the Manufacturing Planning and Control (MPC) system. A key objective is to provide one, and only one, source for forecast data; this source is to be unbiased and usable by all areas in the firm.

Among the sources of demand that place requirements on manufacturing are sales forecasts prepared by the marketing and sales functions. Because of lengthy lead times encountered in manufacturing and purchasing, sales forecasts are an important source of demand information in making production and inventory decisions. Forecasts are needed for almost every management decision. All decisions become effective at some point in the future, so they should be based on circumstances not as they are at present, but as they will be when decisions become effective. Thus, forecasts are a primary input to the Business Plan, Sales and Operations Plan, and the Master Production Schedule. It is not only demand for the output that can be forecast. The tools of forecasting can also be used to predict the development of new technology, national and international economic conditions, and even many factors internal to the organization such as changes in lead time, scrap rates, cost trends, personnel growth, and departmental productivity.

5.8.1 Uses of Forecast

Forecasts are used in organizations for four primary purposes:

1. To Decide Whether the Demand is Sufficient to Justify Entering the Market If demand exists but at too low a price to cover the costs that an organization will incur in producing an output, then the organization should reject the opportunity.

2. To Determine Long-term (2- to 5-year) Capacity Needed to Design Facilitie An overall projection of demand for a number of years in the future serves as the basis for decisions related to expanding, or contracting, capacity to meet the demand.

3. To Determine Midterm (3- to 18-month) Fluctuations in Demand, in Order to Avoid Shortsighted Decisions that will Hurt the Company in the Long Run.

4. To Ascertain Short-term (1-week to 3-month) Fluctuations in Demand for the Purposes of Production Planning, Workforce Scheduling, Materials Planing, and Other Such Needs (Shafer and Meredith, 1998).

Forecasts are used throughout an organization—so they should certainly not be prepared by any isolated group of specialists. Neither should you think that forecasting is a job that is done once and is then finished. Forecasting is continuous and there is no stage at which it is ever finished. As time moves on, actual circumstances are compared with forecasts, original forecasts are updated, plans are modified, decisions are revised, and so on (Waters, 1996).

Forecasting is guaranteed to be the most contentious activity. Attempting to predict future business warrants skilled people, good analytical tools, reliable history of past demands, and close communications among customers, sales, marketing, and PIC. Even then the results will be inaccurate. In industries, such as fashion, which are characterized by highly volatile demand, the approach to forecasting is to determine what can and cannot be predicted well. Products in the 'predictable' category are made farthest in advance, saving manufacturing capacity for the 'unpredictable' products so that they can be produced closer to their actual selling season (Shafer and Meredith, 1998).

5.9 THE STRATEGIC ROLE OF FORECASTING IN SUPPLY CHAIN MANAGEMENT AND TQM

A company's supply chain encompasses all of the facilities, functions, and activities involved in producing a product or service from suppliers (and their suppliers) to customers (and their customers). Supply chain functions include purchasing, inventory, production, scheduling, facility location, transportation, and distribution. All these functions are affected in the short-run by product demand and in the long-run by new products and processes, technology advances, and changing markets.

Forecasts of product demand determine how much inventory is needed, how much product to make, and how much material to purchase from suppliers to meet forecasted customer needs. This in turn determines the kind of transportation that will be needed and where plants, warehouses, and distribution centers will be located so that products and services can be delivered on time. Without accurate forecasts large stocks of costly inventory must be kept at each stage of the supply chain to compensate for the uncertainties of customer demand. If there are insufficient inventories, customer service suffers because of late deliveries and stockouts. This is especially hurtful in today's competitive global business environment where customer service and on-time delivery are critical factors.

Long-run forecasts of technology advances, new products, and changing markets are especially critical for the strategic design of a company's supply chain in the future. In today's global market if companies cannot effectively forecast what products will be demanded in the future and the products their competitors are likely to introduce, they will be unable to develop the production and service systems in time to compete. If companies do not forecast where newly emerging markets will be located and do not have the production and distribution system available to enter these markets, they will lose to competitors who have been able to forecast accurately.

Continuous Replenishment

A recent trend in supply chain design is *continuous replenishment* wherein continuous updating of data is shared between suppliers and customers. In this system, customers are continuously being replenished, daily or even less, by their suppliers based on actual sales. Continuous replenishment, typically managed by the supplier, reduces inventory for the company and speeds customer delivery. Variations of continuous replenishment include quick response, JIT, VMI (Vendor-Managed Inventory), and stockless inventory. Such systems rely heavily on extremely accurate short-term forecasts, usually on a weekly basis, of end-use sales to the ultimate customer. The supplier at one end of a company's supply chain must forecast the company's customer demand at other end of the supply chain in order to maintain continuous replenishment. The forecast also has to be able to respond to sudden, quick changes in demand. Longer forecasts based on historical sales data for six to twelve months into the future are also generally required to help make weekly forecasts and suggest trend changes.

Total Quality Management

Forecasting is crucial in a Total Quality Management (TQM) environment. More and more, customers relate goodquality service with having a product when they demand it. This holds true for manufacturing and service companies. An accurate forecast of customer traffic flow and product demand enables a service company to schedule enough servers, to stock enough goods, and to schedule goods production to provide high quality service. An inaccurate forecast causes service to breakdown, resulting in poor quality. Similarly, for manufacturing operations, especially for suppliers, customers expect parts to be provided when demanded. Accurately forecasting customer demand is, thus, a crucial part of providing the highquality service. Further, TQM requires a fincly tuned, efficient production process, with no defects, minimal inventory, and no waste to reduce the costs. Accurate forecasting is essential for maintaining this type of process.

Strategic Planning

There can be no strategic planning without forecasting. The ultimate objective of strategic planning is to determine what the company should be in the future—what markets to compete in, with what products, to be successful and grow. To answer these questions the company needs to know what new products its customers will want, how much of these products customers will want, and the level of quality and other features that will be expected in these products. Forecasting answers these questions and is a key to a company's long-term competitiveness and success. The determination of future new products and their design subsequently determines process design, the kinds of new equipment and technologies that will be needed, and the design of the supply chain, including the facilities, transportation, and distribution system that will be required. These elements are ultimately based on the company's forecast of the long-run future.

5.10 CLASSIFICATION OF FORECASTING METHODS

Forecasting methods can be classified in several ways. The first concerns the time in the future covered by forecasts.

Long-Term or 'Economic' Forecasts

These look ahead several years—the time typically needed to build a new plant—and are generally concerned with strategic decisions. Long-term forecasts (that is, 2 to 10 years) are used more for facility and capacity planning rather than for any scheduling function. They are normally used to establish long-term goals, plan new products for changing markets, enter new markets, devclop new facilities, plan extensions of existing facilities, develop technology, design the supply chain, implement strategic programs such as TQM, and company's general economic future.

Medium-Term or Intermediate Forecasts

These look ahead between three months and two years—the time typically needed to replace an old product by a new one—and are generally concerned with tactical decisions, such as forecasting total demand for the company's products over the next 12 to 18 months. It is used in aggregate planning for setting overall inventory and production programs and financial budgets.

Short-Term Forecasts

These cover the next few weeks—describing the continuing demand for individual product items—and are concerned with operational decisions. They are primarily used to determine short-term production and delivery schedules and to establish inventory levels. The short-term forecasts are the basis for a firm production schedule for the coming month, a tentative production schedule for the month after, and an even more tentative schedule for the third month.

The time horizon affects the choice of forecasting method because of the availability of historical data, how relevant this will be for the future, the time available to make the forecast, the cost involved, the seriousness of any errors, the effort considered worthwhile, and so on (Waters, 1996).

Another classification of forecasting methods shows the difference between qualitative and quantitative approaches as shown in Fig. 5.3.

Informal or Intuitive Forecasting

Informal methods involve intuition, spur-of-the-moment guesses, and seat-of-the-pants (intuitive) predictions.

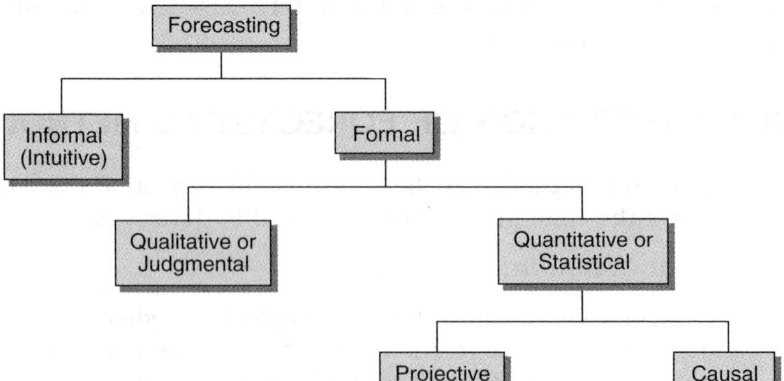

Fig. 5.3 *Qualitative and Quantitative Forecasting Methods* (Waters, 1996)

Qualitative, Subjective, Judgmental Forecasting or Prediction

These methods are subjective methods based on intuition, judgment, opinion, past experience, or best guesses of experts. In general, qualitative forecasting methods are often used for long-range forecasts, especially when external factors may play a significant role. They are also of use when historical data are very limited or nonexistent (Shafer and Meredith, 1998). They are used for forecasting new products, services, processes, or markets, and also for adjusting quantitative forecasts. Also, whenever there is reason to suspect that the future will not be like the past, a prediction (qualitative or judgmental forecasting) is preferable to quantitative forecasts. They are the most common type of forecasting method for the long-term strategic planning process. Qualitative forecasts are made using information such as telephone or mail *surveys* of consumers' attitudes and intentions, consumer panels, test marketing in limited areas, expert opinion and panels, and historical demand for similar products or services. There are five widely used methods of qualitative or judgmental forecasting:

1. Personal Insight,
2. Panel Consensus,
3. Market Survey,
4. Historical Analogy,
5. Delphi Method, and
6. Life-Cycle Analysis.

Quantitative or Statistical Methods

These methods involve the assumption that current trends will continue into the future, and consider forecasting to be an objective procedure, using data collected over time. The procedures are based on mathematical formulas. If an organization is already making a product it will have records of past demand, and will know the factors that affect this. Then it can use a quantitative method for forecasting future demand. There are two ways of doing this:

1. Projective Methods (Autoprojection, Intrinsic, or Time Series Methods) These methods involve 'reaching back and throwing forward,' and are based on a computational projection of historical patterns of the forecasted item's historical data. Projective methods extrapolate time series data. A time series is a historical record of the past activity. These look at the pattern of the past demand and extend this into the future—predicting the future using information from the past. A fundamental assumption for extrapolation is that the future is related to the past in some way. This assumption does not require tomorrow to be just like today, it only requires *stable relationships.* Even in today's rapidly changing environment, fundamental relationships usually hold, at least in the short term (Fogarty *et al.* 1991). If demand in the last four weeks has been 10, 20, 30, and 40, we can project this pattern into the future and suggest that demand in the next week will be around 50. Simple projection techniques are obviously limited to, and primarily used for, very short-term forecasting. Such approaches often work well in a stable environment but cannot react to changing industry factors or changes in the national economy.

2. Causal Methods (Extrinsic or Regression Methods, Associative Forecasting) These methods analyze the effects of outside influences and use these to forecast. They are based on the computational projection of numerical relationships of external patterns. The productivity of a factory might depend on the bonus rates paid to the employees. Then it would be more reliable to use the planned bonus rate to forecast productivity than to project figures achieved in the past few months. Reports from sales representative who keep an 'ear to the ground' can be a valuable source of information. Any special advertising, promotional or price-cutting activities will also have an effect. From outside the company, the general economic condition of the community expressed in terms of certain standard economic indicators can provide advance warning of possible trends in demand, as can activities of competitors or new legislation by governments. Causal methods are usually quite complex and may include histories of external factors and employ sophisticated statistical techniques.

Regression Analysis of Demand against a Market Indicator

This is one of the causal methods and involves investigating the statistical correlation between the demand and the market indicator concerned (such as, unemployment ratio, inflation rate, etc.) and attempting to use any correlations found in a predictive sense. The simplest form of this is to plot demand against the indicator concerned and, provided reasonable correlation exists, to use a least squares fit to draw the regression line. The expected change in demand resulting from a given change in the economic indicator is obtained from the slope of the regression line. *Econometrics* is the term given to the extensive use of this type of technique, and it is most commonly used for longer-term aggregate forecasts and predictions which are more relevant to government departments and very large companies than to the small or medium sized firm (Gibson *et al.* 1995).

Associative Forecasting

It is another term used for causal methods of forecasting. Associative forecasting links external information with internal information in an attempt to establish a relationship which can be used for forecasting. When a relationship exists it can be measured by a correlation coefficient. One example would be to link ice cream sales with temperature, and then to use long-term weather forecast to forecast the sales of ice cream. The number of students registered at the university may be correlated with meals sold in the student canteen. Governments and research organizations publish various indicators on, for example, unemployment, inflation, and housing which may be correlated with company sales of relevant products or services (Bicheno and Elliot, 1997).

Both of these methods rely on accurate numerical data. Suppose that an organization is introducing an entirely new product, there are obviously no past demand figures the organization can project into the future, and it does not know what outside influences affect demand. This means that a quantitative method cannot be used—so the only alternative is a qualitative method. Such methods are generally called judgmental, and they rely on subjective views and opinions.

5.10.1 Factors Influencing the Choice of Forecasting Method

If the data are available one of the quantitative forecasting methods can be used. Otherwise, non-quantitative techniques are required. The demand history need not be long or complete, but some historical data should be used if at all possible. Secondly, the greater the limitation on time or money available for forecasting, the more likely it is that an unsophisticated method will have to be used. In general management wants to use a forecasting method that minimizes not only the cost of making the forecast but also the cost of an inaccurate forecast. Costs of inaccurate forecasting include the cost of over or understocking an item, the intangible and opportunity costs associated with loss of goodwill because a demanded item is not available. With the advent of computers, the cost of statistical forecasts based on historical data and the time required to make such forecasts have been reduced significantly. It has therefore become more cost-effective for organizations to conduct sophisticated forecasts. Finally, if the forecasts must be very accurate, highly sophisticated methods are usually called for (Shafer and Meredith, 1998).

5.11 PROJECTIVE METHODS (TIME SERIES METHODS)

5.11.1 Simple Moving Average Method

The simple moving average method uses several demand values during the recent past to develop a forecast. This tends to *dampen*, or *smooth out*, the random increases and decreases of a forecast that uses only one period. If there are no radical changes anticipated for the future and the activity is relatively constant, a simple moving average may be used. The principle is to use the average of only the last n periods' demand observations as the best prediction of the next period's demand (n is often in the range of 4 to 7). When the current period's demand is known, the oldest demand observation is dropped from the average, and the current period's demand is substituted. The sample of demand periods over which the average is taken thus 'moves' as we move forward in time, and we are only considering the most recent data in the average. The value of n, known as the order of the moving average, is usually based on the expected seasonality in the data, such as four quarters or 12 months in a year. If n must be chosen arbitrarily, then it should be based on experimentation, that is the value selected for n should be the one that works best for the available historical data.

- *Naive or Intuitive Forecast* A time series forecast can be as simple as using demand in the current period to predict demand in the next period. This is sometimes called a *naive* or *intuitive* forecast.
- The simple moving average is a compromise between using the last period's demand as the forecast and using the simple average. It is useful for forecasting demand that is stable and does not display any pronounced demand behavior, such as, a trend or a seasonal pattern (Russell and Taylor III, 1998).
- The number of periods to be averaged in the moving average depends on the specific situation. If too few periods are included in the average, the forecast will be similar to that obtained when only the last period's demand had been used. Using too many periods in the moving average will result in a forecast similar to that obtained when the simple average was used.
- If we use the most recent data so as to be responsive, only a few periods will be included in the forecast, but if we want stability, large numbers of periods will be included. The only approach to deciding how many periods to include is to experiment with several different numbers and evaluate each on the basis of its ability to produce good forecasts and to smooth out random fluctuations.
- The disadvantage of the moving average method is that it does not react to variations that occur for a reason, such as cycles and seasonal effects. Factors that cause changes are generally ignored. It is basically a 'mechanical' method, which reflects historical data in a consistent way (Russell and Taylor III, 1998).
- Moving average has the advantage of being easy to use, quick, and relatively inexpensive. In general, this method can provide a good forecast for the short run, but it should not be pushed too far into the future (Russell and Taylor III, 1998).

5.11.2 Weighted Moving Average Method

A refinement of the moving average approach is to weigh the older or, more commonly, the newer data more heavily, rather than use equal weights. If the most recent history seems to give a more accurate view of the future, a weighted average with the recent periods receiving greater weight can be used. Thus, in case we are using the demands of only the last four periods, instead of using equal weights of ¼ or 0.25, we might use 0.10, 0.20, 0.30, and 0.40, which sum up to 1.0. The weights should be between 0 and 100% and sum of all the weights applied should be equal to 100% or 1. Determining the precise weights to use for each period data usually requires some trial-and-error experimentation, as does determining the number of periods to include in the moving average.

5.11.3 Exponential Smoothing

It is a method based on the weighted-average technique but requiring the maintenance of two numbers only—the last exponential forecast and the actual demand of the last period.

The formula for basic or first-order exponential smoothing is

Exponential Forecast for Period $(t + 1)$
= (α) Actual Demand for Period t
+ $(1 - \alpha)$ Exponential Forecast for Period t

which means that our forecast for the next period is based on a fraction α of the current *actual* demand and $(1 - \alpha)$ times of past demand. The past demand is in the form of exponential forecast F_t, since F_t itself is derived in a similar fashion from previous demands and forecasts.

That is,

$$F_{t+1} = \alpha A_t + (1 - \alpha)F_t$$

$$= \alpha A_t + F_t - \alpha F_t$$

$$= F_t + \alpha A_t + \alpha F_t$$

$$= F_t + \alpha(A_t + F_t)$$

= Exponential Forecast for Period t
+ α (Actual Demand for Period t
− Exponential Forecast for Period t)

= Exponential Forecast for Period t
+ α (Forecast Error Made in Period t)

Now since, $$F_{t+1} = \alpha A_t + (1 - \alpha)F_t$$

and $$F_t = \alpha A_{t-1} + (1 - \alpha)F_{t-1}$$

Substituting the value of F_t in the equation for F_{t+1}

$$F_{t+1} = \alpha A_t + (1 - \alpha)[\alpha A_{t\ \ 1} + (1 - \alpha)F_{t-1}]$$

Thus the data point A_{t-1} receives a weight of $(1 - \alpha)\alpha$, which, of course, is less than α. Since, this process is iterative, we see that exponential smoothing automatically applies a set of diminishing weights to each of the previous data points and is therefore a form of weighted averages. Exponential smoothing derives its name from the fact that the weights decline exponentially as the data points get older and older. In general, the weight of the nth most recent data point is $\alpha(1 - \alpha)^{n-1}$.

- The exponential smoothing forecast formula shows that the new forecast developed for the next period is equal to the forecast for the current period plus some percentage of the forecast error for the current period.
- The parameteris the weighing factor, referred to as the *smoothing constant* that must be between zero and one. For $\alpha = 1$, of course, the forecast for a period is the same as the actual demand for the previous period; and for $\alpha = 0$, the forecast from one period to another remains unchanged.
- A value of α close to zero (say 0.1) gives much less weight to last actual demand and can be used to smooth turbulent data. Thus it will smooth out the random fluctuations (or noise) but will only respond very slowly to changes in demand; that is, the forecasting model will be somewhat *insensitive* to fluctuations in the current data.
- A value of α close to 1 (say 0.9) gives much weight to last actual demand and forecasts the next period to almost duplicate the last observation. Thus, if a large value of α is selected, the forecast will be very sensitive to the current demand value. With a large value of α, exponential smoothing will produce forecasts that react quickly to fluctuations in demand. This, however, is irritating to those who have to constantly change plans and activities on the basis of forecasts.
- A high value of α, thus, results in a quick response to changes in demand level but it will not be very effective at smoothing out random fluctuations (or noise) in the demand. Thus we need to choose a value ofwhich makes the desired trade-off between smoothing of random fluctuations and response to changes in the absolute level of demand (Gibson *et al.* 1995).
- As with *n*, the number of periods to be included in forecasting, the appropriate value of α is usually determined by trial and error, values typically lie in the range of 0.01 to 0.30. One method of selecting the best value is to try several values of α with the existing historical data (or a portion of the data) and choose the value ofthat minimizes the average forecast errors. Spreadsheet programs can greatly speed the evaluation of potential smoothing constants and the determination of the best value of α (Shafer and Meredith, 1998).

The advantage of exponential smoothing over simple moving average is the simplicity of calculation as well as avoiding the necessity of carrying historical demand data because in exponential smoothing only the current demand is needed to forecast the following period. The formula itself 'remembers' data stretching back indefinitely but with exponentially decreasing weights. Most importantly, exponential smoothing has a good track record of success. It has been employed over the years by many companies that have found it to be an accurate method of forecasting (Russell and Taylor III, 1998).

5.12 SEASONAL ADJUSTMENTS

A seasonal pattern is a repetitive increase and decease in demand. Many demand items exhibit seasonal behavior. For example, clothing sales follow annual seasonal patterns. Seasonal patterns can occur on a yearly, monthly, weekly, or even a daily basis. One of the methods for reflecting seasonal patterns in a time series is by using a *seasonal factor*, which is a numerical value that is multiplied by the normal forecast to get a seasonally adjusted forecast. One method for developing a seasonal factor (S_i) is to divide the actual demand (A_i) for each (ith) seasonal period by total annual demand (ΣA), according to the following formula:

Seasonal Factor for the ith Seasonal Period (S_i)
= Seasonal Demand in the ith period (A_i) ÷ Annual Demand (ΣA)

The resulting seasonal factors between 0 and 1.0 are, in effect, the portion of total annual demand assigned to each seasonal period. These seasonal factors are multiplied by the annual forecasted demand to yield adjusted forecasts for each season (Russell and Taylor III, 1998).

5.13 ERROR MEASUREMENTS

5.13.1 Standard Deviation

The calculated standard deviation forecasts the variation (+ or –) to be expected from the forecast.

$$\text{Standard Deviation} = \sqrt{(\Sigma D^2/n)}$$

where, the deviation, D = (Actual – Forecast) for a period
and n = the total number of periods

5.13.2 Mean Absolute Deviation (MAD)

A calculation simpler than standard deviation, which can also give similar variation forecast, is the Mean Absolute Deviation (MAD), which is the average of the absolute deviations.

$$\text{Mean Absolute Deviation (MAD)} = (\Sigma\ |D|)/n$$

where | | = absolute value.

Since MAD sums only absolute values of errors, both positive and negative errors add to the sum and the average size of the error is determined. This gives the manager a sense of the accuracy of the forecasting model. A manager might use this knowledge by saying, 'Our forecasts are typically accurate only within 10 per cent; we had better maintain an extra stock of 10 per cent or so to hedge against this potential error' (Shafer and Meredith, 1998). The smaller the value of MAD, the more accurate the forecast, although viewed alone, MAD is difficult to assess. If the data values are small, the, MAD value is also small. Whereas, if the data values are large, the MAD value for the same accuracy would be very large. The point is one cannot compare a MAD value for one set of data with another set of data and say which one is more accurate since MAD values depend also on the relative magnitude of data.

5.13.3 Mean Absolute Percent Deviation (MAPD)

To take care of the problem of relative magnitude of the data the absolute error is expressed as percentage of the total actual demand rather than as a per period value. This measure (the forecast error as a percentage of total actual demand) is called the *mean absolute percent deviation (MAPD)*, and is calculated as:

$$\text{MAPD} = (\Sigma\ |D|)/\Sigma\ A$$

where, $\Sigma\ A$ is the sum of all actual demands in the periods considered.

5.13.4 Cumulative Error

Cumulative error is computed simply by summing the forecast errors, as shown in the following formula:

$$CE = \Sigma\ D = \Sigma\ (A_t - F_t)$$

A large positive CE value indicates that the forecast is biased low, probably consistently lower than the actual demand. A large negative value implies the forecast is biased high, probably consistently higher than actual demand.

5.13.5 Bias or Average Error

$$\text{Bias} = (\Sigma\ D)/n = \Sigma\ (A_t - F_t)/n$$

It indicates bias or the tendency of the actual observation to be either on one side or the other of the forecast; that is, it indicates whether the forecast is typically too low or too high, and by how much. A positive value indicates that the forecast is biased low. A negative value indicates that the forecast is biased high. A value close to zero implies a lack of bias. A manager knowing this might say, 'On average, we are forecasting 10 per cent too high; we can safely make 5 to 10 per cent less to hedge against this potential error.' That is, MAD shows only the average size of the error, whereas bias shows the average total error and its direction (Shafer and Meredith, 1998).

5.13.6 Tracking Signal and Forecast Control

The ratio of cumulative error (CE) to mean absolute deviation (MAD) is also some times used as a 'tracking signal' in certain *adaptive* forecasting models.

$$\text{Tracking Signal} = \text{Cumulative Error/MAD} = (\Sigma\, D)/\text{MAD}$$
$$= n \times (\Sigma\, D)/(\Sigma\, |D|)$$

The tracking signal is recomputed each period, with updated 'running' values of cumulative error and MAD. The movement of the tracking signal is compared to *control limits;* as long as the tracking signal is within these limits, the forecast is in control.

Forecast errors are typically normally distributed, which results in the following relationship between MAD and the standard deviation σ of the distribution of error.

$$\text{MAD} \bullet \; 0.8\sigma$$

This enables us to establish statistical control limits for the tracking signal that corresponds to the more familiar normal distribution. For example, statistical control limits of $\pm 3\sigma$, corresponding to 99.7 per cent of the errors, would translate to ($\pm 3\sigma/0.8 =$) ± 3.75 MADs. Control limits of ± 2 to ± 5 MADs are used most frequently (Russell and Taylor III, 1998).

Tracking Ratio

Another term, sometimes called *Tracking Ratio,* is defined as the ratio of bias to mean absolute deviation, that is,

$$\text{Tracking Ratio} = \text{Bias/MAD} = (\Sigma\, D)/(\Sigma\, |D|) = \text{Tracking Signal}/n$$

Tracking ratio will have its maximum value when all deviations are positive and hence $(\Sigma\, D) = (\Sigma\, |D|)$ resulting in a maximum value of +1; it will have its lowest value when all deviations are negative and hence $(\Sigma\, D) = -(\Sigma\, |D|)$ resulting in a minimum value of –1.

5.13.7 Adaptive Exponential Smoothing

Tracking ratios approaching –1 indicate that all or most of the forecast errors tend to be negative (that is, the forecasts are too high). Ratios approaching +1 indicate that all or most of the forecast errors tend to be positive. When tracking ratio exceeds some threshold, called the *trip value,* an action message is printed indicating that forecast bias is suspect. When a tracking ratio trips, some action must be taken to reevaluate the data. It is likely that a shift in the trend line has occurred. The usual corrective action is to discard older data that represents past trend in order to properly define the current trend. This evaluation process, if done manually, consumes valuable managerial or analyst time.

A procedure that takes this corrective action automatically would be useful, especially for low dollar volume items that merit little managerial attention.

Adaptive exponential smoothing is a technique that performs automatic correction. Adaptive forecasting models self-adjust by increasing or decreasing the smoothing constantwhen the tracking ratio becomes too large. Since the tracking ratio can range only from –1 to +1, its *absolute* value is used as the smoothing constant, α. Thus, if the forecast becomes biased, the tracking ratio's absolute value, and hence the value of smoothing constant α, increases, thereby placing more weight on recent data and less weight on older data. This is precisely the desired correction. On the other hand, if the forecast is unbiased, the tracking ratio is close to zero. If a random fluctuation occurs, the lowα value causes little weight to be given to the random fluctuation. However, should the random fluctuation become a new trend, then the value of α will increase, causing an automatic correction (Fogarty *et al.* 1991).

5.13.8 Demand Filter

Data collected for forecasting review are monitored through a demand filter. Demand filter is a calculation that will highlight any readings that are beyond reasonable expectations based on expected variances. It will highlight for review any reading above the established value. For example, any reading greater than 3.2 standard deviations or 4 MAD will only happen 0.07% of the time, hence (forecast ± 3 SD) or (forecast ± 4 MAD) could be made as demand filters. Any reading above (forecast + 4 MAD) or below (forecast – 4 MAD) should be reviewed first for data entry error; and if the reading is then determined to be valid and reasonable, the forecast should be reviewed.

To avoid too much manual intervention, we must establish appropriate limits as a compromise between chasing down nonexistent errors and allowing erroneous data to enter the system. In choosing a value for a demand filter, one should use a fairly wide band for unimportant items, and a narrow band for expensive and high volume items. The reason for a variable band width is that for low dollar volume items it is cheaper to carry safety stock to cover the effects of the error than it is to spend valuable management time correcting the error. For high dollar volume items, the reverse is true (Fogarty *et al.* 1991).

5.14 QUALITATIVE, SUBJECTIVE OR JUDGMENTAL FORECASTING

These methods are subjective methods based on intuition, judgment, opinion, past experience, or best guesses of experts. Following are the six widely used methods of qualitative or judgmental forecasting:

1. Personal Insight

This uses a single person who is familiar with situation to produce a forecast based on his own judgment. This most widely used method relies entirely on one person's judgment, as well as his opinions, prejudices, and ignorance. The major weakness of this method is its unreliability. Comparisons of forecasting methods clearly show that someone who is familiar with a situation, using experience and subjective opinions to forecast, will consistently produce *worse* forecasts than someone who knows nothing about the situation but uses a more formal method.

2. Panel Consensus

One person can easily make a mistake, but collecting together a group of people should give a consensus that is more reliable. Senior managers accumulate a vast store of knowledge which this technique tries to tap into. Ideally, the group should be multidisciplinary, and should not feel constrained in expressing opinions. Usually the group talks through differences in opinion until consensus is reached. If there is no secrecy and the panel members talk freely and openly, a genuine consensus may be found. On the other hand, there may be difficulties in combining the views of different people when a consensus cannot be found.

Although it is more reliable than one person's insight, panel consensus still has the major weakness that everybody even experts can make mistakes. There are also problems of group working, where 'he who shouts loudest gets his way,' everyone tries to please the boss, some people do not speak well in groups, some people have a 'hidden agenda' or are involved in personal power politics, and so on. Overall panel consensus is an improvement on personal insight, but you should be cautious about the results from either method. Panel consensus is least expensive, often fast, but results may be highly variable depending on the experience of the managers and the degree of internal politics.

3. Market Survey

Sometimes, even groups of experts do not have enough knowledge to give a reasonable forecast. This happens, for example, with the launch of a new product. Experts can give their views, but more useful information can be collected from potential customers. It is a type of forecasting or statistical sampling and may involve quantitative data. Market surveys collect data from a sample of customers, analyze their views and make inferences about the population at large. It can take various forms: *focus groups, consumer surveys,* and *complaints analysis.* Often, for forecasting purposes, the collected information is subjected to adjustments to correlate past experiences with anticipated developments (Bicheno and Elliot, 1997). Market surveys can give useful information but they tend to be expensive and time consuming. The survey instrument has to be carefully designed. They are good for short-term, less good for long-term forecasts.

They are most effective where participants understand the change, but can be useless for new developments (Bicheno and Elliot, 1997). Market surveys are also prone to errors as they rely on:

- A sample of customers that accurately represents the population.
- Useful, unbiased questions.
- Fair and honest answers.
- Reliable analysis of the answers.
- Valid conclusions drawn from the analysis.

4. Historical Analogy

This uses past experience of special events as a guide to what may happen when a similar event takes place. If an organization is introducing a new product, it might have a similar product that was launched recently, and assume that demand for the new product will follow the same pattern. Historical analogy can only be used if a similar product has been introduced fairly recently. In practice, it is often difficult to find such similar products, or to fit the characteristic life-cycle curve. Historical analogy is low-to-moderate in terms of cost and time, depending on the degree of involvement by the company in the previous situation. Typically, it has low forecast accuracy, but may result in some valuable insights into the wider problem (Bicheno and Elliot, 1997). One example of historical analogy was the use of demand data for black-and-white television sets to predict the demand curve for color TV sets.

5. Delphi Method

A special type of expert panel is called Delphi. The RAND Corporation developed the Delphi method as a group technique for forecasting the demand for new or contemplated products or services. The name for this method comes from the ancient Greek *oracles* (that is, the priests who answered questions about future) of *Delphi* who forecast future events. This is the most formal of the judgmental methods and has a well-defined procedure. It is actually a more structured form of group consensus involving outside experts who should have a special knowledge of the forecast area. Delphi can be conducted by several experts who are located at different sites and may not even know the rest of the panel. Further, Delphi is conducted over several iterations. The intent here is to eliminate the undesirable effects of interaction between members of the group (such as loud and dominating individuals) while retaining the benefits of their broad experience and knowledge.

One possible way of using Delphi method is that secret forecasts are collected from each panelist and the average determined. Panelists responsible for results, which are significantly different from the average, are asked to either change their forecast to bring it in line with the group or to provide information on why they believe their original estimate was correct. Perhaps a non-conforming panelist has information not known to others. Then all panelists are provided with the new information and the process is repeated until a sufficient level of consensus is reached (Bicheno and Elliott, 1997).

Another way of using Delphi method is that a number of experts are contacted by post/e-mail and each is given a questionnaire to complete. The individual written forecasts along with any supporting arguments and assumptions and replies from these questionnaires are submitted to a Delphi researcher, who edits, clarifies, and analyzes the data. Summaries of the analyses are passed back to the experts along with a second round of questions. Each expert is then asked to reconsider their original reply in the light of the summarized replies from the others. Each reply is anonymous so that undue influences of status and the pressures of face-to-face discussions are avoided. This process of modifying responses in the light of replies made by the rest of the group is repeated several times—usually between three and six. By this time, the range of opinions should be narrow enough to help with decisions.

Delphi can be moderately expensive, and a typical time to undertake is several weeks, especially if world experts are involved. Results are variable but this may be all that is possible (Bicheno and Elliot, 1997). The Delphi method may be useful in technological forecasting, that is, in predicting the general state of the market, economy, or technological advances five or more years from now.

6. Life-Cycle Analysis

Experienced managers who have introduced several products are often able to estimate how long a product will remain in each stage of its life cycle. This forecast, coupled with other market information, can produce reasonably accurate estimates of demand in the medium to long range (Shafer and Meredith, 1998).

5.15 USE OF MANAGER'S VIEWS TO UPDATE FORECASTS

The classification of forecasting methods described above does not mean that each must be used in isolation. Managers should look at all available information and then make the decision they feel best. This means that any forecasts should have a subjective review before they are used as shown in Fig. 5.4.

Preparation of forecasts is an integrative effort involving a number of company areas. Following factors are incorporated into subjective forecasting by the sales and marketing staff:

1. Known customer requirements,
2. Sales force expectations,

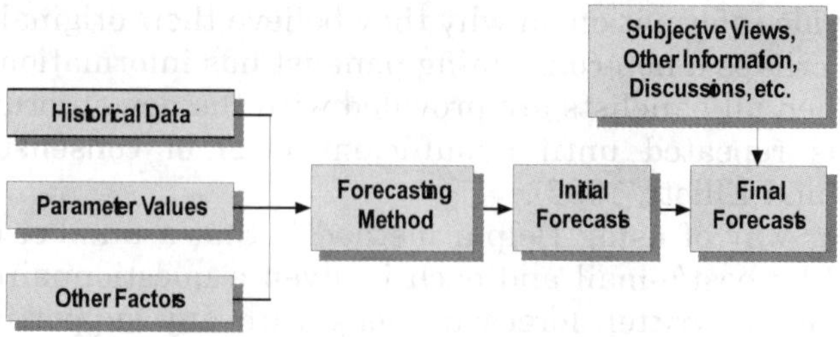

Fig. 5.4 *Use of Manager's View to Update Forecasts* (Waters, 1996)

3. Potential market demand,
4. Recent trends in customer order entry (or sales quoting) activity, and
5. Planned sales promotions.

In addition to the above factors, forecasts are also influenced by certain other considerations as described below:

- The impact of macroeconomic factors is frequently introduced by the corporate economist.
- New product developments and introductions are indicated by the research and development staff.
- Information regarding financial targets and budgetary objectives often influences development of sales forecasts.
- Finally, where companies face critical shortages in purchased material, sales forecast information is often converted to component requirements and checked against supplier capabilities.

Forecasting is a management process with specific accountabilities, although the computer can and does provide significant assistance in developing and updating the forecasts, in the end, though, people evaluate and approve forecast and develop sales plans to make the forecast happen.

5.15.1 Four Basic Functions of a Comprehensive Forecasting System

1. *Several simple forecasting techniques and a method for evaluating them.*
2. *A way to take seasonal factors into account.*
3. *A way to break down and allocate (disaggregate) forecasts for families of items by warehouse, configuration (modules), package size, etc.*

   To be useful to the master scheduler, forecasts must be expressed as items (or product families), quantities, and dates. A forecast of $ 10 million in sales revenue is of little value to the multiproduct company when it needs to schedule its production. A forecast of 1,000 A, 2,000 B, and 1,500 C is more useful.

For the company with multiple product lines, the forecast may be developed in the aggregate but must then be broken down into manufacturable segments. It should, for example, be able to break-down nationwide forecasts of product families into individual item and distribution center forecasts. This breakdown would typically use historical percentages. Similarly, the system must provide a method for reconciling the aggregate forecast for a product family with the sum of the individual forecast by item.

4. *A way to review and approve the forecast before updating the system.*
   The key to effective forecasting system is to combine simple forecasting techniques with good human judgment. A human being will be responsible for the forecast, and a human being needs to review it before it is updated. A person must have the opportunity to change the forecast before it goes into the system so that he or she can be held accountable for the forecast being used. Forecasts should be routinely updated with the help of exception reporting to produce updated projection.

5.16 AGGREGATING AND DISAGGREGATING DEMAND FORECAST

Often forecasts are prepared for a variety of business decisions by many different people in a company, some of them are as follows:

1. *Forecasts for individual products* (in units) are used in scheduling operations,
2. *Forecasts of overall sales activity* (in monetary units) are utilized by top managers to plan and prepare budgets, and
3. *Forecasts for groups of products,* [such as, by horsepower (in units), package size (in kg or pounds), product brand (in monetary units), and so on], are used by product managers in developing marketing plans for sales promotions, advertising, and distribution.

Good planning and communication at the master planing level require the two aggregation and disaggregation processes shown. The forecasting process should provide the ability to aggregate the item forecasts to ensure consistency with the family forecasts. Conversely, it is necessary to be able to disaggregate a family forecast into item forecasts. Companies derive forecasts both ways. Some begin at the item level and aggregate to the family. Others forecast the family and disaggregate to the items. Either way, it is important to have consistency. A similar aggregation-disaggregation process exists between the production plan

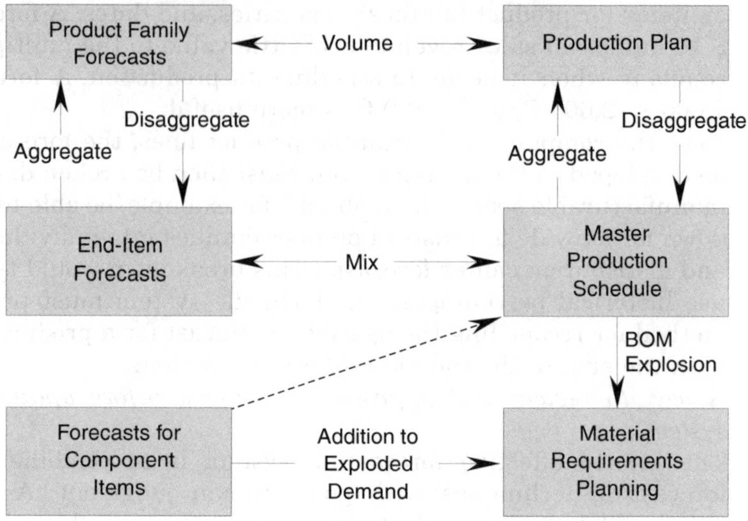

Fig. 5.5 *Demand Interfaces* (Adapted from Sari, 1997)

and the MPS. The agreed-to family rate of manufacture must be disaggregated to the item detail of the MPS. Super family bills of material often serve this purpose. Bottom-up MPS summaries are used to aggregate the MPS to ensure that it agrees with the production plan within reasonable tolerances. Forecasts of lower-level demands often interface directly to MRP I, especially in companies with significant service-parts activity or interplant supply activity.

The fact that sales forecasts are used for so many different decision-making purposes at different levels in the planning hierarchy presents a significant coordination problem in many firms. Often these *independent* forecasts are never brought together within the organization to ensure that they are consistent and the whole equals the sum of the parts. An important aspect of demand management process is providing forecasts at the appropriate level of detail. It might also imply constraining forecasts at a particular level to meet certain overall requirements of the company. Techniques for aggregating and disaggregating demand, such as *pyramid forecasting,* can facilitate this process. Following are two of the methods used to provide a consistency of forecasts:

5.16.1 Disaggregating through the Space Family Bill of Material

One means of providing a consistency of forecasts for various purposes is through the bill of material (BOM). Using the BOM to pull together forecasts for different purposes may require special structuring of the BOM. The 'super family bill' is one such example. The bill of materials forces a consistency to the forecast by explicitly calling out the relationships. The constraint on forecasts for differing 'levels' in the bill is a concept that lies behind other techniques for providing forecast consistency.

5.16.2 Pyramid Forecasting Approach

Another method of 'constraining' the forecasts is pyramid forecasting. It provides a means of coordinating, integrating, and forcing consistency between the forecasts prepared in different parts of the organization. These consistent forecasts can then be used to meet the needs in marketing, manufacturing, distribution, and so on.

The procedure used in implementing the pyramid forecasting approach begins with individual item forecasts at the lowest level (say at level 3), which are rolled up (aggregated) into forecasts for groupings of individual products at the next higher level (level 2). The forecasts for product groupings are then aggregated into a total forecast (in monetary units) at the highest level (level 1). This total forecast is taken into consideration by the top management to arrive at the final aggregate forecast (plan), which then becomes the constraint for the lower level (disaggregated) forecasts and is used to constrain the product grouping and individual item forecasts, so that they are consistent with the final plan.

5.17 THE PRINCIPLES RELATED TO FORECAST ACCURACY

1. *Forecast will be more Accurate for Groups* Total units or money sold are easier to forecast than to forecast by specific products. The more detailed the items to be forecasted, the less accurate the forecast.

2. *The Forecast will be more Accurate for the Short-term* The farther you go in time the more inaccurate you are.

3. *It is Possible to have an Estimate of the Forecast Error* The forecasts will be wrong. Although there will be forecast error, it is most important to have an estimate of that error. Through mathematical techniques, it is possible to estimate the probability of error (Toomey, 1996).

4. *Test the Forecasting Method before using (Focus Forecasting)* There are many models to use for forecasting and it is recommended to test the various techniques based on the same past history. The technique or model, which worked best in the past, will most likely work best for the future. With computer technology, it is now possible to test a number of forecasting technique to determine which technique would have been the most accurate in forecasting the last period. The most successful technique for the past is then used for the future forecast. This is termed as focus forecasting (Toomey, 1996). Thus, focus forecasting is not really a forecasting technique but rather a *simulation* or *trial-and-error* procedure. It uses several different forecasting methods, and applies them to

past data using many different values for the required parameters, finds out which of the methods and set of parameter values would have worked best for the latest period, and then uses this same rule for the next period. The remarkable thing about focus forecasting is that it has often outperformed more sophisticated models (Bicheno and Elliott, 1997).

5. The Forecast is no Substitute for Actual Demand As there will be a degree of error in any forecast, reducing lead time as much as possible, so as to allow actual demand to have a greater impact on the MPS, is most desirable.

There is one final point about forecast errors, many efforts to reduce them are wasted—since they will not go away. A better alternative to improved forecasting found by many firms is improved systems for *responding* to forecast errors. The gross-to-net logic of MRP I prevents errors from compounding, since the net requirements are simply changed from period to period as the actual inventory deviates from the projected inventory.

5.18 BOOK DATE AND QUANTITY VERSUS DEMAND/SHIPMENT DATE AND QUANTITY

- Since most forecasts are made by marketing personnel, the forecasted date is typically expressed by booking (or 'order') date—that is, when the order is to be received. This is the red-letter day for marketing, as getting orders is the reason for its existence. But master scheduling is concerned with satisfying demand with a complete product, so it needs to know the demand (or customer's desired 'shipping') date.
- Many forecasting errors have been made through failure to recognize the difference between orders and shipments. For example, orders and shipments differ in timing. In make-to-order environments, orders precede shipments by manufacturing lead-time. Further, quantities shipped might be less than quantities ordered for a variety of reasons. Partial shipments may be made over a period of time to fill one order. Shipments may exceed orders because spare parts or allowances for defects may be included. Whatever the reason, the distinction between orders and shipments must be taken into account when using historical data to forecast (Fogarty *et al.* 1991).
- The degree to which booking and demand dates differ depends upon the nature of the business and its manufacturing strategy.
- In a make-to-stock (MTS) business, for example, the two dates may be separated by just a few hours or a day—just long enough to pack, invoice, and ship the items from finished goods.
- In a make-to-order (MTO) business, a greater gap typically exists

because the items are in some incomplete stage of production when the orders are booked. Here, a lead-time is added to the booking date to get the true demand date. Figure 5.6 illustrates the difference for make-to-stock and make-to-order environments. For general learning purposes, an MTO situation with a *single blanket order* and *multiple shipments* is also shown.

Bookings Versus Demand/Shipment Dates	*Period*					
	1	2	3	4	5	6
MTS						
Bookings	100	150	125	150	125	
Demand	100	150	125	150	125	
MTO (1-Period Lead Time)						
Bookings	100	150	125	152	125	
Demand		100	150	125	150	125
MTO—Scheduled Shipments, Blanket Order (2-Period Lead Time)						
Bookings	1,000					
Demand			250	250	250	250

Fig. 5.6 *Booding Dates and Quantities versus Demand/Shipment Dates and Quantities* (Adapted from Proud, 1994)

5.19 ORDER ENTRY SYSTEM

An order entry system is used to add, change, and delete customer orders. The demand management function of order entry is the process of accepting and translating a customer order to the internal language and definition used by the manufacturer. This can range from translating a verbal description to a part number to a protracted engineering process. The order-promising function assigns delivery dates to the order. Available-to-Promise (ATP) tools and procedures provide the necessary information to make good customer promises. Order entry can be as simple as creating shipping documents for finished goods in a product line, or it might be a more complicated series of activities, including engineering effort for MTO products.

The customer demands are presented to either of the following two systems:

1. To the Master Production Schedule (MPS) Netting and Exception Checking Logic If the item ordered by the customer is controlled through MPS. Usually, end items in case of MTS type of production, and major subassemblies (modules) in case of ATO type of production are controlled through MPS.

2. To the Material Requirements Planning (MRP I) Netting and Exception Checking Logic If the item ordered by the customer is controlled through MRP I. Usually, dependent demands for components and sub-assemblies are controlled through MRP I.
The order entry system must be able to create multiple line items on a single customer order, for example, different items on the same delivery date, or different items on different delivery dates, or the same item on different delivery dates.

Following five pieces of information are required for each customer order item:

1. Item Number.
2. Quantity (used in the netting and exception checking logic).
3. Customer Request Date [needed for (i) available-to-promise check, (ii) forecasting, (iii) forecast consumption calculations, and (iv) performance measurement.]
4. Promised Date (used in the netting and exception checking logic).
5. Customer Order Number (needed for pegging).

- The customer request date is the date the customer originally specified for the line item, and may be different from the promised date that is being used by MPS and MRP I. The customer request date is needed for an available-to-promise check when the order is entered, and can be helpful for performance measurement, forecasting, and forecast consumption.
- Once a customer order has been created, the information must be available for changing promised date and quantity for each line item.
- When a customer order is deleted, the system should close the order and remove each of the line items regardless of the order status.
- When a customer order line item is deleted, the system should remove the line item, and if all line items are completed or closed, mark the order complete.

5.20 AVAILABLE-TO-PROMISE CALCULATION

Available to promise is defined as the *uncommitted portion of a company's inventory and planned production maintained in the master schedule to support customer order promising.* Available to promise is calculated from the MPS and is used for customer order promising. The ATP figure thus shows the current picture of product available to customers based on current customer commitments and current build plans (MPS). It is important to protect existing customer commitments when new customer orders are promised. Thus, new customer orders can be slotted for shipment based on the existing MPS by promising customer orders using the MPS available-to-promise calculation. Many companies now rely on this information for quoting delivery dates instead of the older practice of quoting standard delivery lead-times. Hence, the order entry process should include an available-to-promise check for every line item on the customer order. There are four different types of ATP calculations, which are described in detail in the next chapter on master production scheduling. Here, only one type of ATP, namely *Cumulative ATP Without Lookahead (Cum. ATP-WOL)* is described.

Cumulative Available-to-Promise-WOL
= Prior Period ATP + Any MPS Production/Receipts Due
– All Customer Orders Due till the Next MPS Supply

- Capacity managed businesses such as foundries and engineered-to-order companies may base customer order promises on ATP capacity. Incoming orders are converted to capacity demands through an estimating process. These demands compete for ATP capacity with other demands.
- Many forms of make-to-order (MTO) businesses routinely employ customer promising based on ATP data because they are continuously dealing with changing backlogs of customer orders scheduled for future shipments (Sari, 1997).

There are a number of different ways to do an available-to-promise check:

1. In cases where the customer promise is made before the order is entered, the available to promise has to be checked prior to the actual entry. It is done either:
   (i) using the MPS report or display, or
   (ii) using special on-line promising displays to automate the checking and promising process.
2. In cases where the available to promise is checked as the order is being entered, the items not available on the requested date are highlighted and shown with both the requested date and the available date for review by person.
3. In cases where the available-to-promise check is run after the order is entered, the items not available on the requested date are listed on a report or in an on-line display for promising.

An automated available-to-promise check normally works by comparing the cumulative available to promise on the request date with the quantity required on the customer order. If the available to promise is greater than or equal to the requirement, then the line item can be promised. If the available to promise is less than that required, then a person has to make a decision about the importance of changing the MPS for this single customer order. A good MPS policy recognizes that there is an expense in direct and setup labour, inventory, overhead, and confusion for each change to the MPS. Yet changes can and should be made to the MPS. The policy should be to provide an evaluation of the effects and cost of each change prior to the change being made.

5.21 METHODS FOR ANALYZING CUSTOMER ORDER: ABNORMAL DEMAND OR EXTRA DEMAND

Unanticipated bookings are made as sales representatives locate new customers and obtain orders. Marketing's efforts at trade shows and with direct mail sometimes result in huge new accounts. Occasionally an order may be received from a new customer who was not considered as a possibility at the time the forecast was made. These unanticipated orders, called 'abnormal demand,' or 'extra demand' are every salesperson's dream, but they can be every master scheduler's nightmare. Whether or not such an order should be accepted and a delivery promise offered are matters of commercial judgment. If such an abnormal order is accepted, it should not consume forecast. Since it is *additional* to the requirement expected when the forecast was made. If these orders enter the system as actual demand and consume the forecast, big overbooking problems can result when the forecasted orders do appear. In fact, it can be stated that a master scheduling system *will not work* effectively if normal and abnormal demand cannot be differentiated. Without this ability, total demand cannot be determined with sufficient accuracy to produce a reliable Projected Available Balance (PAB) line of the MPS matrix. And it is from this line that the master scheduling system generates the all-important action messages. If the projected available balance line cannot be calculated correctly, the generated action, or exception, messages could be misleading and could cause the master scheduler to make bad decisions.

5.21.1 Identifying Abnormal Demands

Abnormal demands are orders that can be treated as a one-time deviation (increase) to the forecast. Some method is needed to assist in analyzing incoming orders to identify abnormal demands (and give them some type of special handling) and to pick up trends in the market place. This has two parts:

1. Computer logic to compare existing customer orders to the forecast for the purpose of identifying demands that may be abnormal, or indicative of major shifts in the marketplace.
2. A way to code demands that have been identified as abnormal.

The abnormal demands are identified by setting up some simple rules, such as:

(i) if customer orders exceed a certain percentage of the forecast for a period, or

(ii) if the customer orders are less than a certain percentage of the forecast for a period, then it is possible that the demand is abnormal or indicative of certain trend and hence should be subject to human evaluation.

Some other signs of abnormal demands are:

- A new customer account,
- The seasonal pattern is wrong,
- A one-time order, and
- An order that comes through a nontraditional distribution channel.

Customer orders in excess of the limits can be coded as abnormal demands automatically (this is not a recommended practice), or better listed on an exception report for a review by someone who can evaluate the situation. A person would be responsible for reviewing the situation and deciding if additional actions should be taken. This way early communication with the customer may avert bad commitments and future service problems. Similarly, in situations where the customer orders are less than the lower limits, the item should be listed on an exception report for human review.

Abnormal demands should be:

1. *Excluded from forecast consumption logic* (described later), that is receipt of a *normal* order must consume *both* the ATP and the sales forecast; abnormal order should only consume ATP leaving the sales forecast unchanged.
2. *Added to the remaining forecast*, other customer orders, etc. to compute total demand against MPS.
3. *Excluded from the forecasting process* because they represent events not likely to happen in the future.

5.22 FORECAST CONSUMPTION LOGIC

5.22.1 Introduction

In the distant future, the expectancy of demand is based entirely on the forecast. However, as the time period comes closer and closer, customer orders are received, and they increasingly *'consume'* the forecast. If originally the forecast for a period is ten items with no orders yet received, the total requirement is ten. Although we have not so far received any orders, we can still expect that we shall receive the orders for a total of ten units in due course of time. If we now receive an order for four units from a regular customer whose business was expected when the forecast was made, the total expected requirement does not increase to fourteen but remains at ten, that is, the order (or consumed forecast) for four plus remaining (or unconsumed) forecast of six. The order for four has consumed four of the forecast. A forecast consumption mechanism of this sort is essential if the routine arrival of an expected order is not going to change the requirements on the MPS (Kenworthy, 1997). As more firm orders come in, they continue to replace or consume the forecast, as shown in Fig. 5.7. The unconsumed forecast is the basis for computing

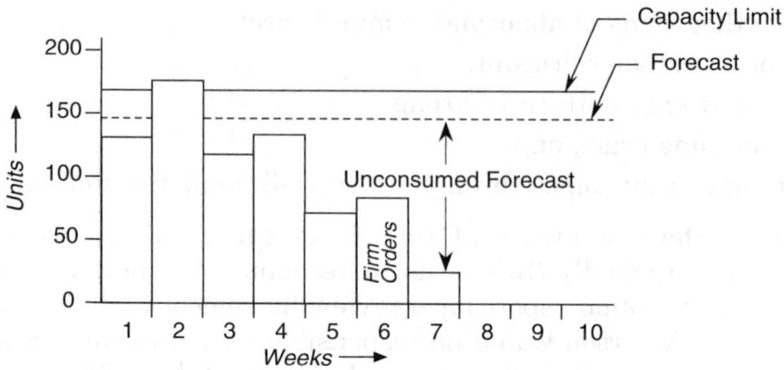

Fig. 5.7 *Normal Orders Consume the Forecast*

Available-To-Promise (ATP) to be used by the sales for promising new orders. Forecast consumption logic monitors the match between actual and forecast demand in the demand management module.

Consider Fig. 5.8, which contains a one-month forecast of 400 units. This demand came from the S&OP process; the breakdown of this

Timing versus Demand Problems

Aggregate One-Month Demand Forecast from S&OP	400 for the 4-week period			
Periods (Weeks)	1	2	3	4
Disaggregated Weekly Product Demand Forecast	100	100	100	100
Actual Demand	85	0	90	50
Unconsumed Forecast	15	100	10	50
Abnormal Demand				
Total Demand	100	100	100	100

Fig. 5.8 *Consumption of the Forecast*
(Adapted from Proud, 1994)

aggregate figure of 400 into weekly periods was accomplished through collaboration between production and marketing. Thus, weeks (periods) 1 through 4 are each forecasted at 100 units, which both parties deemed reasonable in terms of past order patterns and future expectations.

Actual Demand

The table contains a line for actual demand, which is the quantity of product for which the company has firm customer commitments, actual demand excludes abnormal demand. In the table, each week has less actual demand than what had been forecasted. The actual demand line is updated as confirmed orders are received. In most systems, updating of actual demand may be accomplished automatically through the company's order-entry system. These quantities remain in this line, however, until those items are produced and shipped.

Forecast Consumption

As actual demand appears and is entered into matrix, that demand 'consumes' part of the forecast. Thus, the 85 units of actual demand in period 1 consume that same amount of the forecast, leaving 15 units as unconsumed forecast. Notice, however, that *total demand* remains the same as the original forecast of 100 per week. Here an assumption has been made that the actual demand represents normal demand already anticipated during the development of the forecast.

Time passes, and as the end of week 1 is reached, the forecasting system automatically drops that column and shifts the remaining 3 weeks to the left. But if no new orders come in to consume the remaining 15 units of the original forecast, what would we do with those 15 units of the original forecast, what would we do with those 15 units? This question highlights a perennial problem for supply and demand managers: determining whether the forecast was inaccurate in quantity—and those orders will never appear—or whether the forecasted orders are simply delayed. Hence it is necessary to provide logic to handle forecasts and customer orders that fall past due. It is incorrect to drop off the unconsumed forecast at the end of each week. In most situations, this has the effect of changing the forecast, even though no change was intended or indicated. Here two options are available:

1. Assume that orders for the 15 units will *never* come in, and drop them entirely.
2. Assume that the orders for the 15 units are merely *delayed* and carry them over as part of the unconsumed forecast.

The first option of dropping the unconsumed forecast can cause the following problems:

1. Dropping the forecast involves change in the aggregate forecast—from 400 to 385—which may require consultation with other parties in the company (marketing and finance in particular). Many forecasting and master scheduling software systems will automatically drop the 15 unsold units as they update the records with the passage of time. This may not be prudent, however, and

the demand manager and/or master scheduler would be advised to check with sales about any variance between actual demand and the forecast before the period is closed out.

2. The MPS, which was created to satisfy this demand, would now appear to produce excess product. When the MPS and MRP I are rerun, the new opening stock will be higher than expected and MPS/MRP I logic would immediately generate a 'reschedule out' message to avoid production of excess quantity of the product. If the planner acts according to the message it would cause system nervousness.
3. If this forecast were wrong in timing only then the dropped unconsumed forecast would appear as a delayed order (or orders) in some future period(s). Obviously the system would not be in a position to satisfy this demand from the inventory. Consuming the forecast, on the other hand, will delay this and allow some time to see whether this unconsumed forecast appears as delayed order(s). If it does, the item shall be in inventory.

If the second option is chosen—that is, the missing 15 unit order is merely delayed—no change to the overall forecast of 400 units is made, and the unsold 15 units are 'rolled' forward. But how? There are many possible approaches to this problem, and Fig. 5.9 shows just a few:

Timing versus Demand Problems

Rolling the Forecast	*Periods*			
	Past Due	1	2	3
Original Forecast	100	100	100	100
Unconsumed Forecast	15	100	100	100
New Forecast (Option 1)		115	100	100
(Option 1)		110	105	100
(Option 1)		105	105	105
(Option 1)		100	100	115
Etc.		Etc.		

Fig. 5.9 *Rolling the Forecast*
(Adapted from Proud, 1994)

In the first option, the entire 15 units are 'front loaded' into the next period; subsequent options spread the 15 units in other ways. Master schedulers should know that there is no 'best way' of rolling the forecast that fits all companies and all situations. Ultimately, they must exercise judgment based upon demand patterns experienced by their own companies and the input of knowledgeable and affected parties.

Obviously the unconsumed forecast cannot be rolled forward indefinitely, after some time the unconsumed forecast should be assumed to be lost. The simplest way to handle this problem is to specify the number of weeks for which the unconsumed forecast should be accumulated before being dropped. Usually, this might be equal to the aggregate forecasting period. For example, four weeks could be specified. Unconsumed forecasts would accumulate in the past for a month, and then be dropped automatically. In most situations, an exception report should list the forecasts that were dropped automatically this way, a person can review the situation and add back a forecast quantity if it was dropped in error.

A general rule is:

Orders that fail to materialize in an aggregate forecasting period should not be automatically rolled into the next forecasting period.

For example, in Fig. 5.9, the 15 units that failed to materialize in the past due week might be rolled forward into *new* weeks 1, 2, and 3 by any means viewed as reasonable. However, a new aggregate forecasting period would take effect in the *new* weeks 4 through 7 (corresponding to the old weeks 5 through 8), and it should not be burdened by any inaccuracies that manifested themselves in the previous forecasting period.

Prudence dictates that—barring reliable information to the contrary—demand that fails to materialize in weeks 1 through 4 could be assumed to be lost, and therefore not rolled forward into weeks 5 through 8. Instead, it could be dropped. If that demand were still lurking in the market, one would expect that the new forecast would have picked it up. The best way to handle this situation is to have someone in marketing and demand management review the prior period's forecast and make a determination as to whether the unconsumed forecast should be rolled forward or dropped. By doing this, the accountability for forecast accuracy is not lost.

5.22.2 The Concept of Forecast Consumption

- Forecast consumption logic is a method to reduce the forecast by *normal* customer orders that are part of the forecast. Abnormal demand is not considered in forecast consumption.
- The objective of forecast consumption logic is to most accurately represent the needs of the marketplace based on current projections and actual orders. Forecast consumption is useful in updating the demand plan in the light of the actual customer orders received till the time of MPS regeneration.
- By consuming the forecast properly it is possible to compute total customer demand for each period, taking into account timing

differences between the actual customer orders and the forecast. These timing differences can occur when customer demands match the forecast quantities, but are for different dates than those anticipated.

- Forecast consumption, of course, cannot correct the effects of discrepancies between forecast and actual sales in the past periods. Those discrepancies are taken care of by the netting process. The additional inventory caused by the unconsumed forecast would be available for the netting against the demand in future periods.
- In the forecast consumption logic that is part of the system, the forecast should be reduced in the period in which the customer order is requested. If demand exceeds the forecast in the period, the system can reach ahead or back to reduce forecast in other time periods.

When Total Order Quantity in a Past Time Bucket is Less Than the Forecast

- This might be a case of delayed demand and hence this unconsumed forecast would be appearing as an order (over and above the forecast quantity) in a future period. When we apply forecast consumption logic to the period where the excess demand is occurring, the unconsumed forecast would be consumed in the *backward forecast consumption.* It is also possible that this demand has happened earlier, and hence it might have already appeared as an order above forecast quantity in a previous period. When we apply forecast consumption logic to the period where excess demand is occurring the unconsumed forecast would be consumed in the *forward forecast consumption.* Both the cases would be taken care of by the forecast consumption logic. Thus in case of a delayed demand, we would be consuming this forecast when we apply backward forecast consumption logic to an excess order. And in case of a demand happening earlier, the unconsumed forecast in a future period would be consumed when we apply the forward forecast consumption logic.

When the Total Quantity of Normal Orders Received in a Period is More Than the Forecast

- Whenever the forecast consumption logic finds that the total quantity of orders received in a period is more than the forecast in that period it takes the *excess* amount, that is *(total order quantity – forecast quantity),* and goes backward by a period and checks whether there is any unconsumed forecast in that period. If there is no unconsumed forecast in this period, the logic will go backwards to the next period. However, If there is unconsumed forecast in this period, the logic consumes it (reduces it) by as many units as possible from the excess order quantity. If the excess order quantity from the forward period is less than the unconsumed forecast of this period, the unconsumed forecast would be reduced by the excess amount and the process will terminate as no more units of the excess order quantity remain to be consumed. If however some excess units remain after consuming the whole of the unconsumed forecast in the period, the logic will go backwards to the next period and repeat the procedure. This continues until the excess order quantity gets exhausted or the last period as defined by the *backward forecast consumption* parameter is reached.

- If some excess order quantity still remains, the consumption logic starts working in the forward direction from the day or period where the excess quantity had occurred and repeats a similar procedure in the forward direction till the time the excess quantity gets exhausted or the last period as defined by the *forward forecast consumption* parameter is reached.
- If there still is a balance order quantity over and above the forecast quantity which has not been exhausted in either backward or forward consumption, then these units are considered as the error in the forecast quantity of the period and not as the order quantity that has merely occurred late or has happened earlier than anticipated. This then is the additional demand in that period. The *total excess order quantity* appears as a revision to the demand plan of the particular period. The part of this excess quantity, which is used to consume the forecast represents the timing errors, and this quantity does not get added to the aggregate forecast. The part which is not used to consume the forecast is the correction to the forecast quantity and gets added to the aggregate demand plan.
- As long as the additional order quantity is consumed in either backward or forward consumption, it implies that aggregate forecast quantity is correct, there are only timing problems.
- If the unconsumed order quantity is added as a revision to the forecast, it implies that aggregate forecast was less than the actual, hence forecast has also gone wrong in terms of quantity. This additional quantity, included in the customer order should be considered for forecasting. However, when the additional quantity is added as an abnormal demand it is not to be considered in forecasting as it is only a one-time demand.
- The unconsumed forecast in some period that remains unconsumed in either backward or forward, forecast consumption logic should be dropped after a certain period of time. The forecast consumption does this simply by ignoring the unconsumed forecast occurring in the period which does not fall inside the period specified by the backward forecast consumption parameter. The unconsumed forecast in a past period is available for consumption as long as it falls within the specified backward forecast consumption period *from the*

current period. Once it is outside this period it is simply ignored. In other words, *backward forecast consumption parameter* is the period for which an unconsumed forecast is available for being rolled forward; after this the unconsumed forecast is no longer available for being rolled forward and thus is as good as dropped.

The Revised Demand Plan and Demand History

The Revised Demand Plan for a Period
= Normal Customer Orders for that Period (that is orders already received) + Unconsumed Forecast in that Period (that is orders likely to be received) + Abnormal Demand (that is orders received but not included in the forecast)
= Unconsumed forecast + Total Actual Orders Received

The demand history to be used for forecasting is actual orders received *excluding* the abnormal/extra demand.

Demand to be Considered for Forecasting
= Total Quantity of Actual *Normal* Orders Received

5.22.3 Forecast Consumption Procedure

1. Forecast consumption logic starts from the period where the total quantity of normal orders received (excluding abnormal or extra demand) is higher than the forecast for the period.
2. The assumption of the consumption logic is that the excess quantity is more likely to be, first a delayed demand, or, secondly a demand that has happened earlier, that is the forecast has gone wrong in timing only not in the aggregate forecast quantity. When this assumption is proved wrong then only we accept that the forecast has gone wrong in aggregate quantity as well.
3. If this excess quantity is a delayed demand then there should be a period or periods *earlier* to this period where the demand (normal customer orders received) is less than the forecast giving an unconsumed forecast; hence, the forecast in that period is consumed in the backward forecast consumption procedure. How far the backward forecast consumption procedure should go backward searching for an unconsumed forecast in the earlier periods is obviously equal to the maximum period by which the demand can get delayed. This period, in number of days, becomes the *'backward forecast consumption'* parameter.
4. On the other hand, if this is a demand that has happened earlier than expected, then there must be some period or periods *after* this period where the total order quantity is less than the forecast giving an unconsumed forecast. Hence, this excess quantity should consume the unconsumed forecast in that period in forward consumption procedure. How far the forward forecast consumption

procedure should go forward searching for an unconsumed forecast in later periods is obviously equal to the maximum period by which the demand can happen earlier. This period, in number of days, becomes the *'forward forecast consumption'* parameter.

5. If we believe that in our environment the demand is more likely to be delayed than happening earlier, then we apply backward forecast consumption procedure first and forward consumption procedure later. However, if the demand is more likely to be happening earlier than getting delayed, then the forward consumption procedure should be applied first and backward forecast consumption later.
6. If these excess units are not consumed in either backward or forward consumption, then it is accepted as an error in the forecast quantity. These units will automatically get added to the aggregate forecast in the form of excess normal order quantity which could not be used to consume some unconsumed forecast. Note that excess normal order quantity is an *addition* to the aggregate forecast, whereas consumption of unconsumed forecast is, in effect, a reduction to the aggregate forecast. Thus, as long as the excess quantity is used to reduce unconsumed forecast in some period, there is no addition to aggregate forecast quantity because there is an addition and a corresponding reduction to the aggregate forecast quantity. But whatever excess quantity is not used in consuming some unconsumed forecast is an addition to the aggregate forecast because there is no corresponding reduction to the aggregate forecast.
7. The unconsumed forecast is added to the other demands to compute total demand for the item. The total *excess* order quantity appears as a revision to the *forecast in the period.*

Revised Demand Plan for a Period

= Normal Customer Orders for that Period (that is orders already received) + Unconsumed Forecast in that Period (that is orders likely to be received) + Abnormal Demand (that is orders received but not included in the forecast)

Forecast consumption logic can be done in either of two places:

1. As Customer Orders are Entered In this case, the unconsumed forecast is updated as each customer order is entered. The system finds the proper unconsumed forecast quantities and reduces them by the customer order quantity.

2. In a Separate Calculation that is Run Periodically In this case, the forecast consumption logic is part of a separate calculation and not part of the order entry process. The customer orders are combined with the original forecast to complete the unconsumed forecast in a separate computation, either as part of the MPS report or a separate batch run.

- Both the original forecast and unconsumed forecast are stored in the system. The original forecast is needed so that customer orders can be reanalyzed periodically to identify potentially abnormal demands and trends.
- Generally, because the forecast is designed to predict based on a past pattern, the forecast would include only 'normal and expected' orders. Hence, 'extraordinary and unexpected' orders would *not* consume the forecast but would be scheduled in addition to the forecast.

Setting Backward and Forward Forecast Consumption Parameters

Obviously, we cannot go backward or forward any number of periods to consume the unconsumed forecast. There has to be some limits as to how many periods we can go *backward* into the past to consume the forecast and how many periods we can go *forward* into the future to consume the forecast. These limits are defined by two parameters, namely backward forecast consumption, and forward forecast consumption. Both these parameters, usually defined in number of days, are determined based on following considerations.

'Backward Forecast Consumption' Parameter The people in marketing and/or production have the idea about *the maximum number of days by which an order can possibly get delayed* in their environment. This is obviously based on their experience and judgment. This figure—in number of days—becomes the value of the parameter 'backward forecast consumption.' All unconsumed forecasts which are earlier to the *current date* by more than the backward consumption parameter are no longer available for consumption and, hence, are automatically dropped.

'Forward Forecast Consumption' Parameter Similarly, the maximum number of days by which an order can be received earlier than forecasted becomes the value for the parameter 'forward forecast consumption.'

5.22.4 Forecast Consumption Example

This example is based on an MPS with an MPS plan period length of 1 day and working days Monday through Friday. The backward and forward consumption of the demand forecast is 9 and 7 days respectively. The demand forecast for each weekday is 10 units. On Wednesday of week 2 there is a customer order of 50, which is in excess of the forecast quantity of 10. The orders received on other days are as shown in Fig. 5.10. The MPS then looks as shown in Fig. 5.10.

- *Backward forecast consumption,* which is *entered in calendar days* (including the day on which the order is to be delivered), enables you to determine to what extent an actual customer order or sales

Weeks	Week 1						Week 2							Week 3			
Days	M	T	W	T	F	S	S	M	T	W	T	F	S	S	M	T	W
Demand Forecast	10	10	10	10	10			10	10	10	10	10			10	10	10
Customer Orders	05	08	10	04	10			06	10	50	07	10			05	04	0
Consumption	0	02	0	06	0			04	0	10	03	0			05	06	0
Unconsumed Forecast	05	0	0	0	0			0	0	0	0	0			0	0	10
Abnormal Demand				12													15
Revised Demand Plan	10	08	10	16	10			06	10	50	07	10			05	04	25

|← 9 days →|

|← 7 days →|

Fig. 5.10 *Backward and Forward Forecast Consumption*

order (or possibly dependent demand) may consume an unconsumed demand forecast in the past.

- *Forward forecast consumption,* which is *entered in calendar days* (including the day on which the order is to be delivered), enables you to determine to what extent an actual customer order or sales order (or possibly dependent demand) may consume a later unconsumed demand forecast.

1. Backward/forward consumption of the demand forecast always takes place from the relevant customer orders and backward/forward from the planned or actual delivery dates of these customer orders.
2. The demand forecast for the plan period of the MPS in which the relevant customer order must be (or must have been) delivered is always consumed first, independent of the value entered in this field.

Demand forecast is consumed as follows:

1. Independently of the aforementioned parameters for backward and forward consumption of the demand forecast, the demand forecast is *always* consumed first for the MPS plan period to which the relevant actual or planned demand or customer order for the planned item relates.For example, in case of weekly time buckets, the demand forecast which is nearest to the delivery or requirement *date* within the week of the day of actual demand is always consumed first.
2. Subsequently a backward consumption of the demand forecast is performed, based on the delivery or requirement date of the actual demand or customer order and the value of the field 'Backward Forecast Consumption.'
3. Next, a forward consumption of the demand forecast is performed,

based on the delivery or requirement date of the actual demand or customer order and the value of the field 'Forward Forecast Consumption.'

Revised Demand Plan for a Period
= Normal Customer Orders for that Period (that is, orders already received) + Unconsumed Forecast in that Period (that is, orders likely to be received) + Abnormal Demand in that Period

Out of the total order quantity, 10 units consume the forecast for the same day as the date of receipt of the order, 12 units are consumed during backward forecast consumption, and 14 units are consumed in forward consumption. Thus, total (10 + 12 + 14 =) 36 units from the order quantity of 50 are consumed. The remaining order quantity of (50 – 36 =) 14, which cannot consume any of the forecast in the periods within backward and forward consumption parameters, gets automatically added to the demand plan as a quantity over and above the forecast quantity. This can be easily verified from the above.

The revised demand plan is the sum of customer orders already received and unconsumed forecast, which is nothing but the forecast orders we can still expect to receive if our forecast is correct. Thus, the MPS should be planned based on this revised demand plan. The total of the *initial* forecast from Tuesday of the first week to the Tuesday of the third week, that is, for the period falling within the backward and forward consumption parameters, is 110 units. Whereas the sum of revised demand plan for the same period excluding abnormal orders is (8 + 10 + 4 + 10 + 6 + 10 + 50 + 7 + 10 + 5 + 4 =) 124 = (110 + 14). Thus, as long as an order quantity can consume some forecast quantity, the aggregate quantity is not changed. This is in keeping with the assumption behind forecast consumption logic that the forecast might have gone wrong only in timing not in the quantity. But when an order quantity itself remains unconsumed after backward and forward consumption, it gets added to the aggregate forecast implying that the forecast has gone wrong in terms of both timing and quantity.

The unconsumed forecast in periods falling before the 'consumable' periods, that is not falling within the backward consumption parameter from the current day are no more available for consumption, that is, these unconsumed forecasts are dropped assuming that they could not possibly get delayed beyond the backward forecast consumption parameter.

References

Bicheno, John and Brian B R Elliott, *Operations Management: An Active Learning Approach.* Blackwell Publishers Limited, Oxford, UK, 1997.

Cox III, James F, and Blackstone Jr, *APICS Dictionary* (ninth edition).

American Production and Inventory Control Society, Inc, Falls Church VA, 1998.

Fogarty, Donald W, John, H Blackstone Jr, and Thomas R Hoffmann, Production and Inventory Management (second edition) South-Western Publishing Co, Cincinnati, Ohio, 1991.

Gibson, P, G Greenhalgh, and R Kerr, *Manufacturing Management: Principles and Concepts*. Chapman & Hall, London, 1995.

Landvater, Darryl V and Christopher D Gray, *MRP II Standard System: A Handbook for Manufacturing Software Survival*. Oliver Wight Limited Publishing Company, Essex Junction V T, 1989.

McMahon, Chris and Jimmie Browne, *CADCAM: From Principles to Practice*. Addison-Wesley Publishing Company, Wokingham, England, 1993.

Plossl, George, *Orlicky's Material Requirements Planning* (second edition). McGraw-Hill, Inc, New York, 1994.

Proud, John F *Master Scheduling: A Practical Guide to Competitive Manufacturing*. John Wiley & Sons, Inc, New York, 1994.

Russell, Roberta S and Bernard W Taylor III, *Operations Management: Focusing on Quality and Competitiveness* (second edition). Prentice-Hall International, Inc, New Jersey, 1998.

Sari, F John 'Master Production Schedule.' In James H. Greene, (Editor) *Production and Inventory Control Handbook* (third edition). McGraw-Hill Inc, New York, 1997.

Shafer, Scott M and Jack R Meredith, *Operations Management: A Process Approach with Spreadsheets*. John Wiley & Sons, Inc, New York, 1998.

Stonebraker, Peter B, *Master Planning: (APICS) CPIM Certification Review Course, Student Guide;* Revision 3, American Production and Inventory Control Society, Falls Church, V A, 1994.

Vollmann, Thomas E, William L Berry, and D Clay Whybark, *Manufacturing Planning and Control Systems* (fourth edition). McGraw-Hill Inc, New York, 1997.

Waters, Donald, *Operations Management: Producing Goods and Services*. Addison-Wesley Publishing Co, Harlow, England, 1996.

Chapter 6

Master Production Scheduling (MPS)

6.1 INTRODUCTION

Master Production Schedule (MPS) is defined as an anticipated build schedule for manufacturing end items or product options by quantity per planning period (called the *time buckets*). It represents what the company plans to produce expressed in specific configurations, quantities, and dates. The authorization to produce the product must be translated into the MPS. The MPS is a listing of:

(i) the end items (products) or major sub-assemblies (standard modules) that are to be produced,
(ii) the quantity of each item to be produced, and
(iii) when they are to be ready for shipment.

The sales and operations planning team considers products by aggregate families and by monthly periods, and it is the job of master-scheduler to breakdown those aggregate build rates into detailed, weekly production schedules for each item. In this way, S&OP drives and guides the master-schedule.

MPS is not a control technique or a system. Rather, it is a logical representation of information for decision making. The MPS highlights conflicts that can be solved only by people. If the MPS is done properly, the rest of the system can be harnessed to reach the desired objectives of management (Narasimhan *et al.* 1995)

- MPS is what the firm or plant expects to produce in the short run. MPS is the plan for providing the supply to meet the demand. It is the best among several alternative production plans.
- MPS applies only to the items assigned to the master-scheduler. Units of months, weeks, or even days are used to specify the deliveries on the master-schedule.

- What is commonly called 'the master production schedule' is really *a set of individual schedules* for each item planned at this level. Each individual MPS has a supporting BOM defining the components needed to produce the planned quantities of the end item. The aggregate of all master production schedules determines the total future loads and inventory investment (Plossl, 1994).
- Master production schedules are the inputs that 'drive' MRP I. It is the first step in the implementation of the overall manufacturing program of a plant, represented by data in MPS. These data are the wellhead of the flow of manufacturing logistics planning information. MRP I depends on the validity and realism of the MPS for its effectiveness.
- The validity of MPS establishes how well management's policies and goals will be met. Their realism determines how well the plans will be executed, and what level of performance will be reached in customer service and profitability. Downstream systems are unable to compensate for deficiencies of their input. MRP I will carry out its function of priority (material) planning and provide data for capacity planning with great efficacy only if it is presented with realistic and valid MPS to be processed.

6.2 THE NEED FOR MASTER SCHEDULING

When companies first began using MRP I in 1960s, they drove it with forecast and/or customer orders (demand). In other words, to calculate material requirements, computer multiplied the latest demand numbers by the quantities required in the BOM. The problem with this approach was that it blindly assumed that the resources would be available to manufacture a product in sufficient quantities just as it was sold. Unfortunately, manufacturing rarely produced each product as it was sold. And as demand numbers inevitably changed over time, material requirements changed with them. With the rapid replanning capabilities of the computer-driven MRP I, it was frequently impossible for the plant or suppliers to react to the overwhelming changes in plans caused by forecast changes. This meant that the information in the system was often in chaos, and so was the production line.

Some of the MRP I pioneers quickly realized the following factors:

- They realized that their formal systems were of little value, if they failed to predict and control the resources needed to support the way production was actually scheduled. That is, shop and vendor capacity should be assessed *before* the plans were finalized.
- They also realized that they had left the computer too much decision-making power, nowhere in the process was there a human being who ensured a true balance between supply (factory and supplier resources) and demand (customers).

- Thirdly, they also discovered how disastrous the direct linkage between production and demand could be. When demand forecast drives production directly there is no intermediate gearing, no decoupler, to keep the forecast from causing gyrations in the production and hence, MRP I takes the full shock of every fluctuation in the demand forecast. It, in turn, causes production fluctuations, sometimes two, three, even ten times faster than the initial change in the demand forecast. It became obvious that a more formal tool had to be the 'driver' of these powerful MRP I systems, which would uncouple the MRP I and the demand management system.

These insights led to the development of a 'master-schedule' that controlled all other schedules: factory, purchase orders, and so forth. Equally important, a new position was created: that of the master-scheduler. This development allowed MRP I to work effectively. It was not until the advent of master-scheduling computer software in the mid-1970s and the practical implementation of master-scheduling that MRP I achieved its full potential. In fact, the addition of MPS was a key ingredient in the evolution of MRP I to MRP II.

The master-scheduling function has the intelligence and experience to interpret the signals it gets from sales and the forecast, to think of alternative means of satisfying customer needs, and to make the adjustments in capacity, inventories, and so forth that allow the company to serve the customer without causing supply and demand imbalance. In so doing, it helps the company avoid manufacturing chaos and fulfills the overarching strategy of profitably satisfying customers (Proud, 1994).

6.3 THE OBJECTIVES OF MASTER SCHEDULING

1. Develop data (MPS) to drive detailed planning (MRP II).
2. Provide devices to reconcile customer's desires with plant, material, and vendor capabilities.
3. Furnish means to make reliable delivery promises and to evaluate the effects of schedule changes.
4. Coordinate plans and actions of all organization functions and to measure their performance.
5. Provide management with means to authorize and control all resources needed to support integrated plans.
6. Schedule production orders and purchase orders for MPS items.

6.4 FUNCTIONS OF MPS

MPS have two principal functions:

1. *In the short horizon*, to serve as bases for planning material

requirements, production of components, order priorities, and short-term capacity requirements. This function relates to the 'firm' (front end) period of the MPS, which is usually equal to the sum of production lead-times from release of purchase requisition to completion of MPS level items. Following are the functions performed by the MPS:

(a) *RCCP check* The MPS process limits the work to what can be realistically done.
(b) *Production smoothing* It buffers the forecast by 'smoothing' demand over time, reducing capacity-constrained peaks, and raising low-demand idle periods.
(c) *Netting* It subtracts existing inventories of products from the forecast so that the shop doesn't make products that already exist in inventory.
(d) *Lot-sizing* MPS batches demands over time into convenient and economical groups for production so that scarce facilities and equipment are not wasted.

2. *Over the long horizon*, to serve as bases for estimating long-term demands on company resources such as people, plant, and equipment, warehousing, and capital.

6.5 MANAGEMENT AND THE MPS

Master Production Schedules properly can be called 'management's handles on the business.' They are planning devices, not execution tools. They are statements of what should and can be made, balancing customers' needs against manufacturing capabilities. They provide means for coordinating the activities of all functions and measuring their performance, as well as giving management control over manufacturing resources. By altering the MPS, management can alter inventory levels, lead-times, capacity demands, and so on. MPS are not sales forecasts, customer orders, or assembly, packaging, and shipping schedules.

Coupled with modern MRP I programs, MPS constitute means for the solution of many problems often neglected in manufacturing. To realize their full potential benefits, users must understand fully the necessity of maintaining close correspondence between MPS and realities on the manufacturing floor. The key to this is management's willingness to make MPS realistic. Resource planning driven by invalid data (instead of a realistic MPS) is the single most common reason why expensive MRP I programs have failed to help manufacturing companies realize their enormous potential benefits.

Management's role (with MPS driving modern MRP I) should be to:

1. Understand their function of linking company policies and goals with detailed plans.

2. Set policies and guidelines for all functional groups involved with MPS, and measure their performance.
3. Insist that MPS be realistic (and not just top management's 'wish list').
4. Resolve conflicts among user groups.
5. Review MPS regularly and rigorously.

6.6 MPS: THE COMMUNICATION LINK BETWEEN STRATEGIC AND TACTICAL PLANNING

- MPS resides at the interface between the strategic and tactical planning levels and forms the basic communication link between manufacturing and business planning. An effective MPS integrates the production plan (expressed in top management terms) into the actual production of products (expressed in manufacturing terms). Without this linkage in place, priorities, utilization of capacity, and customer satisfaction will all suffer.
- MPS is stated in product specifications, that is, in part numbers for which BOM exist. Because it is a build schedule, it must be stated in terms that can be used to determine component-part needs and other requirements.
- The availability of a flexible MPS framework is a fundamental requirement in a state of the art manufacturing planning and control system.
- The demands coming from business planning are translated at the MPS level into demands on the manufacturing system.
- MPS becomes an anticipated statement of production from which all other schedules at the lower levels are derived.
- On a day-to-day basis, marketing and production are coordinated through the MPS in terms of *order promising*. This is the activity by which customer order requests receive shipment dates. The MPS provides the basis for making effective customer delivery promises, utilizing the capacity of the plant effectively, attaining the strategic objectives of the business, and resolving trade-offs between marketing and manufacturing.
- The MPS is the basis for key interfunctional trade-offs. The most profound of these is between production and marketing in terms of exact product definition in the MPS. A request to increase production for any item usually results in the need to reduce production of some other item. If production for no item can be reduced, then by definition the production plan and resultant budget for production must be changed.
- The master production schedule is regulated by the Sales and Operations Plan (Production Plan). The production plan is the

budget that management sets for the MPS. Consequently, the MPS for the specific items within a family of products, when totaled, must equal the production plan for the family.

- Even though the production plan regulates how many units can be in the MPS, the detailed sales plans indicate which items to make first, second, third, etc. In effect, the MPS is prioritized by the detailed sales plans. The sales plan is the anticipated build demand for the item and these projected requirements can be netted against inventory and any existing MPS, to determine the items that are needed and the sequence.
- Because the MPS becomes the basis for the manufacturing budget, it follows that financial budgets should be integrated with production planning and MPS activities. When the MPS is extended over a time horizon sufficient to make capital equipment purchases, a better basis is provided for capital budgets. On a day-to-day basis, both cash flow and profits can be better forecast by basing these forecasts on the planned production output specified in the MPS.
- Material Requirements Planning (MRP I) uses MPS statement of production to calculate all supporting plans. The MPS is exploded down to all lower levels in the product structure, and so it has tremendous leverage on the items found there. Errors in the MPS have the potential to be magnified many times as they are passed down through the product structure. When this happens, the effectiveness of the entire system is destroyed.

6.7 MASTER SCHEDULING: AN ITERATIVE PROCESS

The actual scheduling is usually iterative, with a preliminary schedule being drawn up, checked for problems, and then revised. After a schedule has been determined, the following points are checked:

1. Does the schedule meet the production plan?
2. Does the schedule meet the end item demand forecasts?
3. Are there conflicts in the schedule involving priority (materials) or capacity?
4. Does the schedule violate any other constraints regarding equipment, lead-times, supplies, facilities, and so forth?
5. Does the schedule conform to organizational policy?
6. Does the schedule violate any legal regulations or organizational or union rules?
7. Does the schedule provide for flexibility and backups?

Problems in any one of these areas may force a revision of the schedule and a repeat of the previous step. The result is that the master-schedule then specifies *what end items* are to be *produced in what periods* to *minimize costs* and gives some measure of assurance that such a plan is *feasible* (Shafer and Meredith, 1998).

6.8 ITEMS THAT ARE TO BE INCLUDED IN THE MPS

'Items' in the MPS are generally the end items, defined as highest-level entities—not components of any parent. However, many times, items other than end items might also be included in the MPS. Following are the items that may be included in the MPS:

1. (Shippable) products.
2. Major assemblies.
3. Planning modules.
4. Group of components covered by pseudo bills.
5. Individual parts used at the highest level in the product structure or subject to demand from sources external to the plant.

The nature of the manufacturing environment (MTS/ATO/MTO) is a critical determinant of the MPS process. Figure 6.1 shows the various

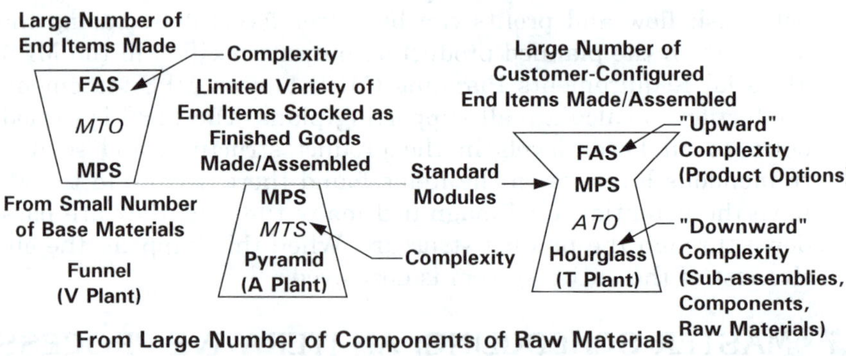

Fig. 6.1 *MPS is Done at the Narrowest Level of the Product Structure*

patterns of product structures, namely *funnel (V plant), pyramid (A plant),* and *hourglass (T plant)* associated with different manufacturing environments. It also shows the levels in the product structure at which master-scheduling is most effective. In each of the product structure diagrams, the top portion represents finished goods, and the bottom represents the raw materials/base items. Because master-scheduling is highly leveraged activity, the idea is to pick the point in the product structure that gives the maximum control with the fewest items to master-schedule.

In order to pick the most effective point for master-scheduling, it's helpful to visualize the entire product structure. This allows a company to select the point that:

(a) requires the minimum number of items, and
(b) provides the maximum amount of control.

Control is defined as setting a master-schedule for one item and then determining the schedule for many items.

6.9 VAT ANALYSIS

VAT analysis is the methodology used to describe the logical flows of raw materials to finished products. Bills of materials and part routings are combined to reflect the logical product flows through a plant's resources. Basic flows are classified by shapes including V (Funnel), A (Pyramid), and T (Hourglass). These logical product structures provide information for plant design, staffing, planning and control, and information system design. Each logical product structure has control points that simplify production planning, scheduling, and control.

1. Funnel (Inverted, or V Plant) In this pattern, a small number of raw materials are common to a wide variety of individual products. Here a relatively small number of raw materials and components at the bottom of the product structure can be used to make a large number of end items at the top of the product structure resulting in an 'upward' complexity. A V logical structure consists of a logistics flow where a few raw materials flow through fixed routings to produce one of many similarly finished products. A primary characteristic of a V structure is product divergence. At each divergent point, a decision has to be made as to what product is to be produced. General characteristics found with most V structures are the following:

(a) A few raw materials that diverge to many finished products.
(b) Possibility of misallocation of material at divergence points.
(c) Planning and control information usually too late to be effective.
(d) Poor performance measures of department efficiencies and utilization.

A good example of divergent process is found in a steel rolling mill, where a given piece of steel may be rolled into any of a large number of different thicknesses, which in turn would be heat-treated into a large number of combinations of desired strength and hardness characteristics, and finally all these varieties of steel is cut into desired widths or strips at the slitting operation.

2. Pyramid (Standard Tree or A Plant) The pyramid pattern reflects the fact that the product has few end items but a great many lower-level items. The end item level—the top end of the pyramid offers maximum control because everything is underneath the master-schedule in the product structure. Electric motors shipped from stock are an example of a product for which pyramid shape is applicable. Here, a relatively small number of finished products 'explode' downwards into a complexity of sub-assemblies, components, and raw materials. An A logical structure consists of a logistics flow where many raw materials, parts, and components are fabricated and assembled into a few finished products. A primary characteristic here is product convergence—scheduling decisions have to be made prior to convergent work centers to ensure that parts are available when needed at the convergence points. The fabrication of several parts and sub-assemblies and their assembly into a few products is another characteristic in A structures.

3. Hourglass (Modular, or T Plant) This pattern indicates many different end items, then a 'choke or neck point' then an increasing number of items. For example, an ATO or MTO product with many options. The hourglass pattern also applies to process-type industry for example, the choke point for pharmaceutical company manufacturing tablets is the unpackaged (or bulk) tablets. Below that level are a number of raw materials used to make the product. Above the choke point are a variety of different packaging configurations (such as, strips with 10 tablets, bottles with 50, 100, 250, 500, and 1,000 tablets). Master-scheduling the unpackaged tablets provides the greatest control with the minimum number of items. A T-logical structure consists of a logistics flow where many sub-assemblies and assemblies are combined to make many different finished products. A primary characteristic is commonality of parts. In a T structure, materials flow into common basic assemblies to provide dozens or even thousands of types of final products. Some of the characteristics of a T structure are:

(a) Misallocation of materials (assemblies) creates a major problem; Stealing common parts and sub-assemblies to complete a customer's order can have serious consequences.
(b) Poor due date performance.
(c) Overtime is used excessively to expedite missing parts.

4. Combinations Any or all of these shapes can be combined, and regardless of the combination, the objective is the same: to master-schedule the minimum number of items and get the maximum amount of control. Some companies use all these different shapes because they have different products, each of which may follow a different pattern.

The MTS and ATO environments both have a conical lower level structure indicating that a large number of raw materials and/or base components at the bottom of the structure go into making of a small number of end items in case of MTS, and a small number of standard modules in case of ATO environment. However, in case of ATO, this small number of modules can further be configured into a large number of end items each with a unique configuration based on customer choice, so that the structure diagram acquires an hourglass-like shape. In case of the MTO environment, a small number of raw materials or base items at the bottom of the structure can produce a large number of custom-engineered products at the top of the structure diagram.

Generally, the master-scheduler will schedule these items at the narrowest level of the BOM to reduce the complexity of his task. Doing so also provides greatest flexibility and control. The reason for this is that at this narrow point there are the fewest number of items to forecast and master-schedule, this makes it the easiest place to master-schedule. Similarly, the desired customer requirements can be configured from this point using the fewest number of master-scheduled items. These product structure diagrams also demonstrate the need for Final Assembly Schedule (FAS) which is described in a later section.

Following points may be noted about the MPS level:

1. In the make-to-stock (MTS) environment, the MPS and the FAS are at the same level. The MPS corresponds to the FAS, and thus, there is no need for the FAS. Sometimes, however, because the MPS is an *anticipated* build schedule, an FAS is used with MTS environments as the *actual* build schedule.
2. In the assemble-to-order (ATO) environments, the MPS is used to schedule up to the major assembly level, and the FAS is used to finally assemble those major assemblies into customer orders.
3. In the make-to-order (MTO) environment, the MPS schedules at the raw materials level, and the FAS drives the customer delivery process.

6.10 CERTAIN OTHER ITEMS THAT ARE LIKELY TO BE MASTER SCHEDULED: MULTILEVEL MASTER SCHEDULING

Certain items, because they are critical, or because their effect on the production process is substantial, would likely be master-scheduled. These might include:

1. Items which have a critical impact on lower level components.
2. Items with high value or long lead-times.
3. Items that have significant impact on capacity and/or material.
4. Items that need a high degree of human control in terms of evaluation of the need to change, and evaluation of the (production system's) ability to change. (Items that don't need human control can be handled by MRP I).

The point is: One can master-schedule anything that makes sense to the business. All of the factors and issues discussed above must be considered when a company is identifying what it plans to master-schedule. In addition, a company can elect to master-schedule at multiple levels in its product structure.

There are times when it may make sense to master-schedule also at levels other than the top, or final, product. A part that is expensive, difficult to obtain, difficult to manufacture, or subject to material or capacity constraints may need the kind of attention and human control that a master-scheduled item deserves. Although planning of such items normally occurs at MRP I levels, it may be necessary to state a schedule with firm planned orders in order to recognize the constraints. This schedule then places significant constraints on the higher-level MPS. Thus, since there are no hard-and-fast rules for deciding *what* to master-schedule, the management team and the master-scheduler may elect to master-schedule not only the end items, but other items one, two, or three levels deep in the product structure. This approach is known as 'master-scheduling at multiple levels,' or sometimes referred to as 'master-scheduling at two levels.' Multiple-level MPS processes are common, for example, when a manufacturing facility logically breaks down into different processes. Bulk pharmaceutical manufacturing may have batch-process characteristics requiring one set of MPS criteria. The packaging process with line orientation may be subject to another set of master-scheduling considerations. In such circumstances, two separate but related, person-controlled 'master-schedules' may be in use.

6.11 PLANNING HORIZON AND TIME PERIODS (BUCKETS) OF MPS

The time that the MPS spans is termed the planning horizon, while the time buckets refer to the units of time into which the planning horizon is divided.

They are related to each other as follows:

$$\text{Length of the Planning Horizon} = \Sigma \text{ No. of Periods of a Particular Size} \times \text{Size of the Particular Period}$$

At the master-schedule level, it is necessary to deal with two different types of lead-times. One is the lead-time required to produce the master-scheduled item itself when all items one level down in the item's bill-of-material are available. The second is the cumulative lead-time—the longest planned length of time required to produce the master-scheduled item from scratch. This takes into account *all* the lead-times of *all* the items that go into the master-scheduled item. In short, it is the critical path that recognizes all processes that can be done in parallel.

If MRP I is to provide valid information on the MPS items, the planning horizon of the MPS must be at least as long as the cumulative procurement and manufacturing lead-time for all components of the MPS items. The planning horizon normally exceeds this cumulative lead-time by significant amount, as operations managers are more comfortable with additional planning visibility. The MPS planning horizon should extend for some period beyond the cumulative lead-time, typically three to six months, so that purchasing can make informed decisions on the size of orders for raw materials, particularly where quantity discounts, or expected price increases, are involved. If management uses the MPS in simulation mode for long-term rough-cut capacity planning of personnel, machines, warehouse space, and so forth, the MPS horizon should be considerably longer in order to cover the longer procurement times of these long-range resources. In the past a year was commonly used, but attention to shortening lead-times has made 3 and 6 month horizon common.

The difficulty of making accurate forecasts for periods farther out on the planning horizon for individual items, coupled with the fact that the minimum planning horizon is determined by the longest cumulative lead-time, provides a strong motivation to reduce lead-times. Shorter lead-times in general will reduce the length of the period over which the schedule will need to be developed, which in turn means forecasts are likely to be more accurate. Hence manufacturing companies must try to be as responsive as possible. In this way they do not need to plan so far ahead thus reducing the associated uncertainty (Gibson *et al.* 1995).

Time periods of MPS (also called *Time Buckets*) must be identical to those of MRP I. The length of planning periods (time buckets) varies among environments and is dependent upon the software and the company's choice of planning horizon. The size of a time bucket is one of the major planning factors involved in designing MPS and MRP I systems. The possible options are monthly, weekly, or daily (that is, bucketless). Typically time buckets are one-week periods (Plossl, 1994).

1. Monthly Time Periods For the front end of the MPS the monthly increments are simply unsatisfactory as they make the plan too insensitive; attempting to manage the time blocks of this size increases the chances that the master-scheduler will miss the details necessary to convert production rates into specific part numbers, quantities, and due dates. These might, however, be used by the master-scheduler for further planning visibility in the mid-range periods. Aggregate planning, incidentally, is done with monthly time buckets.

2. Weekly Time Buckets These are widely used as the time accounting period for MPS and MRP I systems.

3. Daily Transaction (or Bucketless) Scheduling This is being increasingly used because of available computer support. The scheduling information is displayed according to the time/date of the requirement with no aggregation of time 'buckets.'

Most MPS software systems now on the market manage all supply and demand by dates. If the real dates are known, master-schedule data can be displayed in variety of ways. The number of time periods displayed on the MPS display matrix depends upon the software and company's choice of planning horizon.

By convention the first time period displayed is the 'current' period—the present—and *remains so as time passes.* Sometimes a column is displayed just to the left of the current period and is labeled 'past due.' The columns to the right represent future time periods and are used to display data by identified activities. The data in each column shifts to the left as time passes. The cells in the master-schedule matrix are for convenience of display when using a horizontal MPS format. Inside the computer, the quantities are stored by real dates, and, therefore, can be displayed using any time period arrangement that the master-scheduler requires.

6.11.1 Rolling the Plan: Adding to and Deleting from the Time Periods from the MPS

An MPS (or MRP I) represents a 'rolling' schedule. Period 1 is usually the current, or the next, period. When period 1 expires with the passage of time, period 2 will become period 1; and all other periods will adjust accordingly. As the time passes, the master-schedule needs to be updated. At the end of this week, the current week drops off the MPS. In addition, a new week is added to the MPS at the end of the planning horizon. The rolling process is illustrated graphically in Fig. 6.2.

Weeks >	*Weeks 6*	*Weeks 7*	*Weeks 8*	*Weeks 9*	*Weeks 10*	*Weeks 11*	*Weeks 12*
Bucket >	1	2	3	4	5		
MPS >	a	b	c	d	e		
After 1 Week	Bucket >	1	2	3	4	5	
	MPS >	b	c	d	e	f	
After 2 Weeks		Bucket >	1	2	3	4	5
		MPS >	c	d	e	f	g

Fig. 6.2 *Rolling the Schedule with a Rolling Frequency of 1 Week*

When the master-schedule reaches the lead-time to make the product, the master-schedule order is converted into a manufacturing (released) order. When the manufacturing order is completed, products are either put into stock or shipped to the customer. This closes out the manufacturing order and removes it from the system.

6.12 THE FREQUENCY OF MPS REVISION (Updating/Maintenance)

The frequency of MPS revision is usually *geared to the forecasting cycle*. This is often *weekly* when computer forecasting modules are used, but may be *monthly* when management reviews data. Between 'official' updating, however, MPS revisions may be needed to respond to significant changes in the mix of new customer orders and to serious unplanned developments in procurement and manufacturing.

Ideally, the master-schedule will be reviewed continuously or daily using an on-line computer system. At a minimum, it should be reviewed weekly. With today's technology it is possible to keep the master-schedule constantly on-line, where changes can be seen on the MPS screen as they are made. Particularly, the Available-to-Promise (ATP) row of the master-schedule screen needs to be analyzed in a real-time environment if customer orders are received on a regular basis.

6.13 MAJOR INPUTS TO MASTER SCHEDULING PROCESS

1. The Production Plan It provides a set of constraints on the MPS. The MPS must sum up to the production, inventory, and resource levels in the production plan.

2. Demand Data It includes sales forecast, customer orders, distribution warehouse requirements, interplant requirements, service demand forecasts, and safety stocks.

3. Inventory Status Information It includes inventory on-hand, allocated stock, released production/purchase orders (scheduled receipts), and firm planned orders.

4. Planning Data from Item Master File This includes lot-sizing rule to be used, shrinkage/safety factor, and lead-time.

5. Rough-Cut Capacity Planning (RCCP) It helps in determining the feasibility of the proposed MPS by ascertaining capacity requirements on critical resources.

6.14 FACTORS THAT GO INTO DEVELOPING AN MPS

The factors that go into developing an MPS can be categorized into five major groupings:

1. Sales and Production Forecast.
2. Customer Order.,
3. Branch Warehouse Demands and Interplant Orders.
4. Management Decisions.
5. Vendor, Capacity, or Material Limitations.

None of these are the MPS itself. They are combined in a process of human judgment and evaluation to produce MPS (Landvater and Gray, 1989).

6.14.1 Sales and Production Forecast

Most companies need to master-schedule prior to the receipt of customer orders. In other words, they must forecast demand and create a sales plan. Even in MTO and ETO situations, some sales planning is typically done beyond the backlog of customer contracts in order to calculate capacity. For these reasons, the sales plan is a key input to the MPS in almost every company.

The forecast, as one of the factors used to make the MPS, can be the sales forecast for the item or it can be a production forecast. A sales forecast is used whenever the MPS item is forecast individually. A production forecast is generally used whenever a number of related master production schedules have to be coordinated, for instance, for products offering variants and options.

In case of modular product line, the option packages are not individually forecast but are master-scheduled based on the MPS for the type of product. In other words, the type of product is master-scheduled and some forecast percentages are used to help breakdown the production for the product into estimated requirements for each of the master-schedule items.

Distinction between Forecasts and MPS

Although forecast is one of the most important factors that go into the development of the MPS, there are significant differences between forecast and MPS.

- MPS is a statement of production, not a statement of market demand although it is based on forecast of demand. Forecasts represent estimates of demand from outside sources, whereas MPS constitute plans of internal production. These are not necessarily the same. Products are often made at different times and in different quantities from those desired by the customer. For any number of reasons, the MPS may be set at levels above or below the sales forecast. For example, the MPS may be set higher than the sales forecast if stock is being built in anticipation of a peak selling period.
- A clear distinction exists between forecasts and MPS, forecasting attempts to predict when products will leave the finished product inventory; final assembly scheduling (FAS), supported by MPS, plans when they will enter the inventory.
- Unlike a forecast of demand the MPS represents a management commitment authorizing the procurement of raw material and the production of component parts.
- The MPS differs from the forecast in significant ways. The MPS takes into account capacity, vendor, and material limitations, and attempts to utilize capacity effectively. This means that some items may be built before they are needed for sale, and other items may not be built even though the marketplace could consume them.

6.14.2 Customer Orders

The total customer orders that have been booked and not yet shipped is called the *backlog of customer orders.* Customer orders are the second factor used to develop the MPS. Regardless of the forecast, actual customer orders never appear exactly as the forecast predicts. This means that the MPS may have to be changed to satisfy the customer orders. Sometimes the forecast is close and other times the errors are significant since forecast is, and will continue to be, an imprecise art. The forecast that is significantly in the error should be flagged and brought to the attention of the appropriate people.

In deciding whether or not to change the MPS when customer orders deviate from the forecast, the customer orders are used with the forecast, management decisions, branch warehouse demands, interplant orders, and vendor, capacity, or material limitations.

Deviations of Customer Orders from Forecast (and the Response of the Master-Scheduler to them) Fall into two Categories

1. Deviations in the Total Quantity of Customer Orders for a Family of Products In this case, the master-scheduler may decide to increase or decrease the quantities in the MPS to meet the customer objectives.

2. Deviations in the Mix of Products within a Family where the Total Forecast Quantity is Nearly Correct Faced with a situation like this, the master-scheduler will generally shift the production in the MPS from one item to another.

6.14.3 Branch Warehouse Demands and Interplant Orders

Where they exist, branch warehouse demands and interplant orders are planned orders or unshipped scheduled receipts either from another MRP I system or part of the same MRP I system. Distribution Resource Planning (DRP II) creates planned orders to supply the branch warehouses. These planned orders are the branch warehouse demands that appear as one of the factors in the MPS at the central supply location.

MRP I in a multiplant company creates planned orders for items needed at one plant and produced at another. These planned orders at the receiving plant produce the interplant orders which appear as one of the factors in the MPS at the supplying plant. In a similar way, allocations from unshipped orders are shown as a demand against the MPS in the supplying plant (Landvater and Gray, 1989).

6.14.4 Management Decisions

Management decisions of a nearly unlimited variety go into making up of the MPS. Following are some of the examples of management decisions that are typically made while master-scheduling:

1. Increasing or decreasing production as a way of providing stable employment.
2. Building stock in anticipation of a shutdown.
3. Putting on a third shift.
4. Beginning to master production schedule an item even though the customer order has not been received (as a way to reduce the delivery time).
5. Deciding among customer by changing the MPS when everyone cannot be supplied.

The need for responsible management decisions make it clear why master production scheduling be done by an intelligent human being and not by a computer. The factors and variables involved in these decisions are intangible and require judgments and assessments that cannot be made by a computer. Developing and managing the MPS is a difficult and challenging task that must be done by people.

6.14.5 Vendor, Material, and Capacity Limitations

Making MPS Realistic Managing inventories and production properly requires valid and realistic MPS. *Valid* means that they support management's strategies and policies. *Realistic* means that resources needed for execution will be available. Vendor, material, and capacity limitations are constraints on the MPS. MPS must be constrained by limits to available productive capacity of suppliers and company facilities over the short horizon. In the long term, with enough money and dedication, the material and capacity availability constraints can be eliminated. But in the short term, it can have a significant effect on the MPS and must therefore be considered. Both material and capacity constraints must be used as inputs to the process of master-scheduling. They are not, however, generally presented in the master-schedule report or display. Instead, they are usually handled as memos or just identified during the analysis of a particular plan.

- The MPS must not create a load on a work center, or a vendor that exceeds what that work center or vendor can achieve. To do so would create an MPS that is not attainable. An overstated MPS, which calls for more output than production can achieve, may lead to two problems:

    1. There will be failure to supply customers by the dates planned.
    2. If the recommendations of MRP I for buying components, raw materials, and making intermediates or sub-assemblies to support the inflated MPS are achieved, there will be a buildup

of raw materials and in-process stocks, since these will be generated faster than they are actually consumed (Kenworthy, 1997).

- Vendor, capacity, and/or material limitations are handled in the MPS by highlighting the problem and seeking resolution, and not by simply changing the MPS without attempting to solve the problem.
- An MPS that creates an overload on a work center must be identified and all attempts must be made to devise a solution before considering a change to the MPS. Overtime, extra shifts, subcontracting, and offloading work to another work center are possible solutions. If none of these provide a solution, the MPS must be reduced to a level that creates an attainable workload on the critical work center.
- Similarly, if a vendor cannot supply components or material at the rate called for in the MPS, a solution should be sought that does not require a change to the MPS. If no solution can be found, then MPS must be reduced to the level that is determined by the supply of components or material.
- When there is no real alternative to revising MPS, the tough question to be answered is, 'Exactly what in which MPS will be changed, and how?' 'Pegging' in the MRP I can help with the answer. Pegged requirements permit tracing upward through BOM from the specific problem component to the MPS of affected end-item or end-items. Pegging permits tracing to any higher level in the BOM as necessary, to pinpoint specific lots needing to be changed to restore harmony between MPS and reality of resources.
- MPS provide data for determining, over the long horizon, plant, workers, and machinery capacity requirements as well as estimates of other resources (tooling, energy, utilities, etc.) required to execute them.

6.15 ORDER MANAGEMENT (Process of Creating Orders)

The process of creating orders is a three stage process and involves moving through the following three stages of order management (Stonebraker, 1994) as shown in Table 6.1.

1. Computer Planned Orders (CPOs), Planned Orders (POs) or Suggested work Orders.
2. Firm Planned Orders (FPOs) or Work Orders.
3. Released Orders (Scheduled Receipts).

Table 6.1

Handling of Different Types of Orders

(Plossl, 1994)

Type of Order	*Exploded to Lower Levels?*	*Rescheduled Automatically?*	*Exception Messages Generated?*	*User Control of Quantity, Start and Need Dates?*
Computer-Planned Order	Yes	Yes	No	No
Firm Planned Order	Yes	No	Yes	Yes
Released Order (Scheduled Receipts)	No	No	Yes	Yes

6.15.1 Computer Planned Orders (CPOs)

Computer Planned Orders (CPOs) are also called Planned Orders (POs) or Suggested Work Orders. It is an order created by the computer software rather than by the master-scheduler. It is not a green light to manufacturing, but serves as a suggestion to the person doing the scheduling that a firm planned order of the indicated size will be needed if supply and demand are to balance. Generally, for a computer planned order to have a lasting effect on a master-scheduled item, the master-scheduler must convert the CPO into either a firm planned order or a released order (Proud, 1994).

- Planned orders are created by the computer and they exist only within the computer. CPOs may be changed (replanned) or deleted *by the computer* during subsequent MPS processing if conditions change.
- The computer bases its creation of CPOs on demand need dates, predetermined lot sizes, and lead-times to ensure that the items would be ready by the demand need date.
- Computer planned order creation rules are generally built into the MPS system's software and controlled by a 'Planning Time Fence (PTF).' This date-related boundary allows the computer to only generate CPOs with due dates *after* predefined PTF dates, and not before. Inside the planning time fence, the master-scheduler must control all the supply orders.
- It is important to note that CPOs are but computer's recommendations. The master-scheduler ultimately determines whether the computer's recommendations are valid for the particular situation at hand. There is a universal principle in both MPS and MRP I: *Machines make recommendations; people make decisions* (Proud, 1994).

- CPOs generate lower-level requirements via the BOM explosion process and thus plan matched sets of component parts.
- CPOs are included when capacity requirements are calculated.

6.15.2 Firm Planned Orders (FPOs)

A firm planned order can be defined as a planned order that can be frozen in quantity and time. Firm planned orders (FPOs) override computer-driven changes; they are substituted for computer planned orders (CPOs) when scheduler decides to freeze the quantity and/or timing of planned order releases.

- An FPO is an order that the master-scheduler places to take control away from the computer. The computer is *not* allowed to change the FPOs automatically. Instead only the planner in charge of the item is authorized to make changes in it and he is accountable for the changes.
- By 'reclassifying' an order from CPO to FPO, the scheduler identifies those orders whose order quantities and dates are to be controlled personally and directly by him.
- This technique can aid schedulers in planning materials and capacity by firming selected computer planned orders. Firm planned orders are the normal method of stating the master production schedules (MPS). In effect, the master-scheduler says, 'I am going to produce so many units of this product, which will be due on this date, but I am not yet ready to release the work or issue a work order.'
- Use of the FPO ensures a stable MPS that is not changed by the computer even if it differs from the forecasted demand for the item. This also gives the master-scheduler the opportunity to consider several important issues before actually changing either the quantity or the timing of the MPS.
- Generally, firm planned orders explode through the MRP I system to plan materials and other resources. This explosion process is the same as the one used on computer planned orders.
- The FPOs can, for example, be used to override the normal lead-times or calculated net requirements when the production manager is aware of an unusual situation. FPOs are used to effect *one-time* deviations from the standard item master data or system logic *without modifying the item master data,* such as:
  (a) To compress lead-times (expediting).
  (b) To change standard order quantities.
  (c) To change planned order due dates for specific orders.
  (d) To express the MPS.
  (e) To make one-time substitutions of standard routing or standard BOM with alternate routing or BOM.

- Although, computer logic cannot automatically change FPOs, the standard time-phased logic of master production scheduling and material requirements planning that calculates the projected available balance critiques the schedule of FPOs and recommends changes through action and exception messages and reports. Any change, however, is the responsibility of the master-scheduler for that item.
- There is a significant difference in the way FPOs are used in MRP I as opposed to MPS. Typically, all MPS planned orders are FPOs. In MRP I, FPOs are used on an exception basis. The difference in the way FPOs are used in MPS, as opposed to MRP I, shows that MRP I is carried out with greater reliance on the computer system than on human intervention. In part, this is because we assume that MRP I is a logical extension to the MPS plan. Furthermore, MRP I items are usually less costly and much more numerous than the MPS items, which means that heavy human intervention would be quite expensive in case of MRP I.

6.15.3 Released Orders (ROs)

ROs are also called Scheduled Receipts, Manufacturing Orders, Work Orders, Shop Orders, or Purchasing Orders. A released order is the one which has already been released to the shop floor for production or to the vendors for supplying. When a production order is classified as a released order, MRP I *allocates* the required resources for that order. These orders initiate the production process by authorizing material, labor, and equipment to be used to manufacture a specific item.

- Released orders are not exploded via MRP I because these orders create lower-level allocations for materials when they are placed, which are then treated as demand until the material is issued in accordance with the build plan.

These three stages coincide with the two time fences, the Planning Time Fence (PTF) and the Demand Time Fence (DTF) which divide the MPS into three areas namely unrestricted, flexible, and firm. In the unrestricted area of the schedule the orders are managed by the computer automatically as planned orders.

- At the PTF and during the flexible period of the schedule, the scheduler converts as many of the Computer Planned Orders (CPOs) into Firm Planned Orders (FPOs) as possible. During this stage the scheduler compares the required FPOs with the production rate. If the summed FPOs are less than the production rate, some of the CPOs are converted into FPOs so that the summed MPS equals the production rate. If the summed FPOs are greater than the production rate the management team that develops the production rate is advised accordingly (Stonebraker, 1994).
- At the DTF, the scheduler ascertains the balance between the

demands of the FPOs on the one hand, and the production rate, material and vendor limitations on the other hand, and then releases the order. This explains why DTF is generally made equal to the lead-time for moving from level 1 immediately below end-item level to the end item level (that is level 0). The end item is planned by the MPS and hence the production order for it should be released by MPS. Since backward scheduling is used, the order-release date would be the delivery date minus the lead-time of the end item, which means the DTF should made equal to lead-time so that as soon as the delivery date is within DTF, a message to release the order should be generated by the system. Thus, the order for the final assembly of the MPS item would also be released when the delivery date falls within this DTF. Changing the order at this stage would involve almost the total cost of the end item, hence, the MPS should be frozen within DTF.

Each type of supply order plays a unique role in the master-scheduling process as explained below (Proud, 1994):

- Computer planned orders in effect represent the computer software's own version of a firm planned order. Once the master-scheduler creates firm planned orders and balances supply to demand, the computer software will not attempt to create additional supply orders because they will not be needed.
- Firm planned orders are also essential; without them, the computer software logic would assume responsibility for balancing demand with supply, and would then attempt to rectify the situation by launching CPOs where needed, regardless of the actual material or capacity availability.
- A released order is a manufacturing order against which material, labor, overhead, and other resources can be charged.

Note that there can be many variations on the preceding supply order scheme; the movement *from CPO to FPO to RO* is only a generic approach. In fact, it is possible for the master-scheduler to take a computer-planned order and convert it directly into a released order without ever going through the firm planned order phase.

6.16 FALLACY OF FENCES

The arrangement of fences, according to George Plossl (1994), locks the execution of material and capacity plans to an MPS which was planned much earlier and thus blurs the distinction between planning and execution. In this type of system, the fences limit near term replanning for differences between customer orders and MPS schedules. Thus, when the implications of fences is considered this whole approach appears to be a fallacy. This approach leads implicitly to the following *two naive and untrue statements*:

1. 'If we haven't planned it, we can't make it even if we need it now.'
2. 'If we planned it, we must make it even if we don't need it now.'

Whereas the only true statement, according to George Plossl, is:

When enough customers' orders arrive to fill the front end (that is, the initial 'firm' period) of the MPS, stop replanning and execute.

A better approach, according to Plossl, is to treat MPS as schedules of planned order releases, subject to possible later change. MRP I will plan the needed components to produce the end items only if still desirable when the scheduled time comes. This views the MPS and MRP I as planning tools for components (and other resources), and *not* as part of end-item netting and scheduling. Netting and scheduling, according to this approach, should occur in the execution phase (Plossl, 1994).

6.17 PROVIDING SAFETY AND HEDGES

6.17.1 Safety Stock

In the context of Master Production Scheduling, *Safety Stock* (or *Buffer Stock*) can refer to additional inventory and/or capacity planned as protection against forecast errors and/or short-term changes in the backlog (sometimes referred to as 'overplanning' or a 'market hedge'). It is a cushion of inventory in excess of planned requirements to help meet unplanned needs. Order point stock-replenishment systems must include safety stock because of the fallacies in their basic assumptions: that demands during replenishment periods are uniform and can be predicted accurately. Safety stocks only provide protection against timing errors in demand or supply. Protection ceases once the stock has been consumed. Safety stock cannot protect a product from the effects of sales consistently exceeding forecast. They can however 'buy time' to enable production to be increased.

The list of contingencies against which people desire safety stock protection is incredibly long. It includes unexpected customer demand, late supplier deliveries, machine breakdown, tool trouble, power failures, scrap and rework, record errors, and many more surprises experienced in production. There are several ways to determine the amount of the safety stock. One popular method is to establish a safety stock that will meet a specified service level. The service level is the probability that the amount of inventory on hand during the lead-time is sufficient to meet expected demand—that is, the probability that a stockout will not occur and the customer will be served. The service level is typically a policy decision based on a number of factors, including carrying costs for the extra safety stock and lost sales if customer demand cannot be met.

There are four basic approaches to determining how much safety stock is needed on products and components: guesstimates, rules of thumb, statistical analyses, and inflating MPS. If safety stocks are planned in MRP I programs, inflating the MPS is the best means, but it is dangerous. It results in overstating the requirements for all resources unless precise adjustments are made to those in which cushions are not desired. There are a few strategic purposes to offset the extra cost of inventory (Proud, 1994):

- Items with long lead-times can add to the cumulative time needed to build the product. By safety stocking these items, the cumulative lead-time to build the product may be reduced.
- If a family of finished product is available in many options (colors, trim pieces, etc.), the forecast for the entire family is likely to be more accurate than the forecast for the specific configured product. Thus, the surprises in demand are likely to be most pronounced among the options; therefore, these semifinished options would be candidates for safety stocking.
- For many businesses, customers expect that the producer will have certain items on hand *all of the time,* other items *some of the time,* and other items *occasionally.* Thus, the customer requirements determine which items should be safety stocked and in what quantities.

The Mechanics of Using Safety Stocks

With modern MPS software, the desired stocking level for a master-scheduled item is entered into the system, and the system flags any situation in which the projected available balance (PAB) falls below the safety stock level. Some MPS and MRP I computer software systems handle safety stocks using an alternative display format. The MPS matrix for these systems does not reflect the safety stock in the projected available balance (PAB) line. It is there, but the computer software does not show it. This means that the master-scheduler can take those balances down to zero without violating safety stock policy. Cushions can be added to plans to help guard against unexpected shortages in two other ways: safety lead-time and safety capacity.

6.17.2 Safety Lead-Time

Safety lead-time is used to schedule orders earlier than otherwise required. This results in varying amounts of safety stock, depending on items' time-phased requirements. The amount of safety lead-time usually is guesstimated or set by rules of thumb. It should be noted that both the order release date and planned receipt date are changed by the amount of the safety lead-time. Hence the safety lead-time does not result in inflated lead-time.

Both safety stock and safety lead-time are used in practice, and there is no structural reason why both cannot be used simultaneously. Following are the thumb rules to decide which technique should be used:

(a) *Safety stock* is preferred in MPS/MRP I systems where uncertainty about quantities is the problem, (for example, where some small amount of scrap, spare part demand, or other unplanned usage is a frequent occurrence).

(b) *Safety lead-time*, on the other hand, is preferred when the major uncertainty is in timing rather than the quantity; for example, if a firm buys from a vendor who often misses delivery dates.

However, it should be noted that both, safety stock and safety lead-time, are hedges that to some extent result in indicating that orders should be released or that they need to be received when, in fact, this is not strictly true. To use safety stock and safety lead-times effectively, the influence of the techniques on plans must be well understood. If it is not well understood, the wrong orders can be sent to the factory, meaning that the workers are trying to get out part A because of safety lead-time or safety stock when, in fact, part B will be required to meet a customer order, so it would be better to have them working on that (Vollmann *et al.* 1997).

6.17.3 Safety Capacity

Safety capacity planning attempts to provide extra capacity to handle unplanned needs for additional components. It assumes that raw materials will be available to make the components needed; this is not unreasonable, since raw materials typically are ordered in large lots. When shortages occur, specific parts needed can be made quickly if capacity and materials are available.

6.17.4 Hedge or Overplanning

In Master Production Scheduling, a *Hedge* is defined as a *scheduled* quantity to protect against uncertainty in demand or supply. The hedge is similar to safety stock except that a hedge has the dimension of timing as well as amount.

Overplanning

When safety stocks appear to be necessary, the computational logic of MRP I includes a method called *overplanning* to create the effect of safety stock without actually carrying safety stocks. In overplanning, the random fluctuation of demand for an item and all of its components is buffered within the gross requirements instead of with the use of safety stocks. The effect of overplanning is analogous to maintaining safety stocks by creating a calibrated amount of buffer inventory, but the advantage of overplanning is that the buffer inventory is *dynamically* keyed to the master-schedule, which exploits the MRP I logic to rapidly consume the excess inventory.

Two Types of Hedges

There are two types of hedges
(a) Product Mix Hedge, and
(b) Volume (or Market) Hedge

(a) The product mix hedge is an approach where several interrelated optional items are *overplanned*. Sometimes, using a planning bill of material, the sum of the per cent mix can exceed 100% by a defined amount, thus triggering additional hedge planning. It provides for extra units of components, the proportions of which may be difficult to forecast. A typical example of a product mix hedge might be for a company that makes a box with either gold or silver handles (everything else being the same). If 40% of the boxes usually took gold handles and 60% silver, an MPS for 200 boxes would explode into 200 common boxes, 80 gold handles and 120 silver handles. The mix hedge might be to carry 50 *extra* gold and 50 *extra* silver handles. If at some point of time, 180 of the 200 boxes are sold, uncertainty only exists as to the remaining 20; any silver or gold handles above twenty can be rescheduled until the next uncertainty occurs (Stonebraker, 1994).

(b) The volume hedge sets increments for the total number of products by a certain amount or proportion. A volume or market hedge is carried at the master-schedule or production plan level. The master-scheduler plans excess quantities over and above the demand quantities in given periods beyond some time fence such that, if the hedge is not needed, it can be rolled forward before major resources must be committed to produce the hedge and put it in inventory. A volume hedge would be for the common boxes in the above example.

Summary of Various Ways of Achieving Safety

1. *Safety Stock* is held in the event that the demand is greater than the forecast.
2. *Safety Lead-times* ensure against delivery delays
3. *Safety Capacity* provides reserve capacity at key work centers.
4. *Product Mix Inventory Hedges* are held in the event that demand mix for component varies from the forecast.
5. *Volume Inventory Hedges* are held in the event that the demand is greater than the forecast or that the timing of demands is earlier than predicted by the forecast.

All the above ways of achieving safety are all different forms of 'overplanning' of requirements. Although overplanning provides greater safety for the master-scheduler, it also can overstate the schedule and, thus, should be used with caution and should always be visible (Stonebraker, 1994).

6.18 MASTER SCHEDULE, OR MASTER PRODUCTION SCHEDULE REPORT

The master production schedule report (also called *Master Schedule*) is a display of the information that will assist the master-scheduler to develop and manage the MPS. According to APICS Dictionary, master-schedule is a time-phased planning chart which is used primarily by the master-scheduler to help create the MPS.

- Several of the factors that go into making up the MPS which can be presented in the form of numbers appear on the MPS report. These include sales/production forecast, customer orders, branch warehouse and interplant demands.
- Several other factors which go into making up the MPS but are subjective type of information and hence cannot be represented numerically are not typically presented on the MPS report. These include management decisions, and vendor, capacity, or material limitations. These are typically handled in the form of a memo.

The master production schedule report can be thought of as having three segments. These three segments are listed below with the type of information that can appear on the report.

6.18.1 Demands

1. Sales Forecast.
2. Production Forecast.
3. Customer Order.
4. Branch Warehouse Demands.
5. Interplant Orders.

(All orders are displayed in the MPS report at their ship dates)

6.18.2 Production

Master Production Schedule (MPS) receipts are displayed at dates on which the master-scheduled items are due to be completed and will be available for shipment to the customer.

6.18.3 Calculations

1. Projected Available Balance (PAB).
2. Available to Promise (ATP).

- In case a category of information does not exist for a particular master-scheduled item, it should not appear as a line on the MPS report. The master-scheduler should only see a line of information in the report if it is meaningful. The best arrangement should be to allow each of these lines to be printed as options. This makes the MPS report clearer and easier for the master-scheduler to use.
- The calculations section is a way to assist the master-scheduler. The computer can do some calculations using the MPS and the demands. The purpose of these calculations is to provide the master-scheduler with information about the MPS.

- The Projected Available Balance (PAB) and the Available-to-Promise (ATP) are two key techniques used by the master-scheduler to project available inventory or capacity and to promise future production against customer orders.

6.19 PROJECTED AVAILABLE BALANCE (PAB)

The Projected Available Balance (PAB) calculation shows the projected stock build-up or depletion (that is, expected future inventory) to the master-scheduler. It *predicts what will be in inventory at a specific point in time.* There is a similar projected on-hand balance calculation in MRP I.

PAB is also the basis for the computer's critique of supply-and-demand balance. The outcome of the critique determines what actions will be recommended. Action recommendations are sent to the master-scheduler by the software in the form of 'action' or 'exception' messages. Thus PAB is used to generate action/exception messages for the master-scheduler.

The PAB considers the availability of goods based on consuming the forecast with orders. That is, in the distant future periods, the PAB uses the forecast while in the near future periods, customer orders are used as statement of demand.

Unless a company uses safety stock, the perfect supply-and-demand balance will be a projected available balance of zero (the supply of the product perfectly matches the demand for the product so that there is no projected inventory remaining). If a company uses safety stock, such as 100 units, then the perfect balance is 100 units. Rarely do we have a perfect balance in the imperfect world of manufacturing. Assuming that a company is not using any safety stock, a positive balance for any period of time in the projected available balance row suggests a potential surplus or excess stock condition. If there is a negative balance, the system is projecting a potential shortage. In the case of a potential surplus, the computer system may recommend that the master-scheduler move orders out or even cancel orders altogether. In the case of a potential shortfall, the master-scheduler will receive recommendations from the system to release new orders or move future orders in to cover the projected deficit.

6.19.1 Formulas for the Projected Available Balancc (PAB)

The formula of the PAB depends upon whether it is computed prior to or after the demand time fence.

Prior to the Demand Time Fence (DTF):
1st Period:

PAB = On-Hand Balance + MPS – Customer Orders

Subsequent Periods:

PAB = Prior Period PAB + MPS – Customer Orders

After the Demand Time Fence (DTF):

PABT = Prior Period PAB + MPS – Greater Value of Total Customer Orders or Forecast

6.19.2 Analysis of the Projected Available Balance Status

Analysis of the Projected Available Balance (PAB) depends on the period. Actions differ depending on whether the period under consideration is firm (frozen), flexible (slushy), or unrestricted (free/liquid) period of the MPS planning horizon.

1. Inside the Firm (Frozen) Area of the MPS Planning Horizon

- If the PAB becomes negative the scheduler tries to expedite the scheduled receipts as necessary and creates Firm Planned Orders (FPOs) outside the firm period to adjust orders outside of the demand time fence (DTF).
- If the PAB falls below safety stock but remains positive, no action is taken as safety stock is basically meant for taking care of such situations only.
- If the PAB becomes greater than the safety stock, some of the scheduled receipts are rescheduled, if possible, outside the firm period, so that the capacity can be made available for other items which might require expediting.

2. Inside the Flexible (Slushy) Period of the MPS Planning Horizon

- A negative PAB, or a PAB below safety stock, would cause the master-scheduler:
  (a) to create an FPO if there are no FPOs, or
  (b) if there is an FPO, to 'reschedule in' an existing FPO to meet the customer orders.
- A PAB greater than safety stock would be ignored by the master-scheduler if no FPOs are scheduled during the period. However, if there is any scheduled FPO, the master-scheduler would like to reschedule (defer) it out.

3. In the Unrestricted (Free/Liquid) Period of the MPS Planning Horizon

- In the unrestricted period, the master-scheduler lets the computer create or cancel Planned Orders (POs) as it encounters the net requirements. However, the master-scheduler keeps an eye on these actions particularly as they get closer to the planning time fence (PTF).
- If the Projected Available Balance (PAB) goes negative and then

returns positive, the system recognizes that a *timing problem* exists—that is, there is enough on order, but some orders are scheduled too late, and hence need to be rescheduled in.

- If the PAB goes negative and stays negative, the system recognizes a *volume problem* and calls for additional orders.

6.20 AVAILABLE TO PROMISE (ATP)

The available-to-promise calculation shows where new customer orders can be promised for delivery based on the *existing* master production schedule. The emphasis should be on slotting customer orders for shipment based on the existing MPS and not on attempting to change the MPS for each customer.

- ATP is uncommitted future inventory or capacity.
- It is calculated from the MPS. It is used as a tool for customer order promising (or commitment of the future capacity). The result of this calculation informs the sales and marketing department of the products that still can be sold *without* modifying the master-schedule. ATP identifies what can honestly be promised to a customer in terms of delivery.
- The ATP uses customer orders alone as the basis for its computation
- The process of consuming the forecast with orders normally yields some availability of either unconsumed capacity or resources in future periods. This unconsumed capacity or available resource is the basis for available to promise.
- In the make-to-order and assemble-to-order environments, there is a particular opportunity to evaluate future capacity or resource availability and to commit against those resources for future delivery, which is the logic of available-to-promise concept.

The available-to-promise calculation compares the MPS to customer order in either of two ways:

1. By master-schedule order (Discrete ATP).
2. Cumulatively (Cumulative ATP).

6.20.1 Discrete ATP without Lookahead (Discrete ATP-WOL)

In the first method, an available-to-promise is computed for the first time period and for each time period where there is an MPS order.

(a) *For the first period*:

ATP (1st period only) = On-Hand Balance + MPS – Sum of All Customer Orders Before Next MPS

(b) *For subsequent periods*:

ATP = MPS – Sum of Customer Orders Before Next MPS

(c) *For the periods in which no MPS quantity is scheduled*:

The ATP is zero, even if deliveries in the period have been promised.

This is called the regular method of computing ATP, and this type of ATP is termed as *Discrete ATP without Lookahead*. Discrete ATP-WOL is not useful for order promising as it does not include the inventory carried over from previous periods. For example, if the total actual order quantity in the periods from and including one MPS up to but excluding the next MPS supply is equal or more than the MPS supply then the discrete ATP-WOL would show zero or negative (shortage) ATP quantity, even if sufficient uncommitted inventory is available from previous MPS supplies. Secondly, if some orders are promised to be delivered in some future periods taking the balance units from this discrete ATP-WOL, that quantity is not deducted from this ATP and it continues to show the units that should be reserved to be delivered in the future period as available for promising to new orders. Thus, discrete ATP-WOL is a misleading figure. It is useful only for getting an idea of the discrepancies between an MPS supply and actual demand to be supplied during the period from that MPS to the next MPS.

6.20.2 Cumulative ATP without Lookahead (Cum. ATP-WOL)

In this second method, the ATP is a cumulative calculation. The salient difference between cumulative and discrete ATP described earlier is that cumulative ATP in any period is likely to include units already included in the ATP of other periods.

For the first period:

$$\text{Cum. ATP}_1 = \underset{\text{(1st period only)}}{\text{On-Hand Balance}} + \text{MPS} - \underset{\text{Before Next MPS}}{\text{Sum of Customer Orders}}$$

For later periods:

$$\text{Cum. ATP} = \text{Prior Period ATP} + \text{MPS} - \underset{\text{Before Next MPS}}{\text{Sum of Customer Orders}}$$

Cumulative ATP-WOL is better than the discrete ATP-WOL in that it includes the uncommitted inventory carried over from previous periods. Still it cannot be used for order promising as it includes inventory which might have already been reserved for future deliveries. However, it is useful to get an idea as to how much of inventory would be *physically* present in the store at the end of the particular period.

6.20.3 Discrete ATP with Lookahead (ATP-WL) (The Backward Method of Computing ATP)

In the regular methods of calculating ATP, where there is no 'lookahead' procedure, the ATP for a week may include units committed to fill requirements for a later period. For example, a customer order has been received for delivery in a period where the discrete ATP is either not available or the available quantity is less than the required order quantity; but the cumulative ATP may show that enough quantity is available if units available from previous periods are included. In such cases, the customer order can certainly be promised by reserving the deficit quantity from the closest previous period or periods where the discrete ATP (ATPs) is (are) available. Obviously, when such an order is promised the discrete and cumulative ATPs of previous periods will become misleading as they will show even the reserved items as available-to-promise. If new orders are promised based on misleading ATP, orders already confirmed cannot be honored. To resolve this problem, total order quantity must be subtracted from all discrete and cumulative ATPs of those periods from which the units have been reserved for fulfilling this order. In other words, the units produced in one period but committed to be delivered in some future period should be omitted in all periods from the production period to the delivery period. This method is called the backward method of calculating ATP, and the ATP calculated by this method is called 'ATP with Lookahead.' It can be either *Discrete ATP with Lookahead,* or *Cumulative ATP with Lookahead.*

Thus, it can be said that in the backward method, the ATP calculation includes some logic to prevent the available-to-promise quantity from dropping below zero. If the calculated available to promise is negative in a period, that ATP quantity is set to zero and the available to promise in an earlier period or periods is reduced by the over-promised amount. A negative discrete ATP-WOL indicates that the quantity of the item in question has already been promised to be shipped to the customer but it cannot be satisfied by the MPS supply *in that particular period* because the sum of customer orders till the next MPS is greater than the MPS supply. The cumulative ATP quantity should, however, be sufficient to fulfill the order, otherwise the promised delivery cannot be made. The only solution to this problem, other than changing the MPS, would be that the overpromised quantity be carried in inventory from some previous period where the MPS supply is in excess of the sum of customer orders. This excess quantity is represented by the discrete ATP-WOL of that period. To reserve the required quantity of the item for a subsequent period, we should, obviously, deduct that amount from the discrete ATP-WOL, implying that the quantity is committed to an already promised order in a subsequent period and hence is no more available to be promised against any new orders.

Computing the Discrete ATP with Lookahead (ATP-WL or Backward)

In the time bucket where the *discrete ATP-WOL* (*regular* or *without lookahead*) becomes negative, the discrete ATP-WL is shown as zero and the negative quantity shown as discrete ATP-WOL is subtracted from the nearest previous bucket where there is a positive discrete ATP-WOL, the reduced ATP-WOL becomes the ATP-WL for that period. If the ATP-WOL quantity is not sufficient to make good the shortage, we again go backwards and try to deduct the balance short quantity from the ATP-WOL. The procedure is continued until the total short quantity is reserved from previous periods. Once the entire short quantity is reserved, the ATP-WL in all previous buckets is the same as ATP-WOL for the respective periods. If, however, there is not enough discrete ATP-WOL right up to the beginning of the MPS planning horizon (that is, the current time bucket), then the balance short quantity is shown as a negative ATP-WL in the first period of the MPS report. This indicates to, rather warns, the master-scheduler right at the beginning of the planning horizon, that there is a shortage of supply and the orders already promised will not be delivered unless some corrective action is taken.

Discrete ATP-WL is obviously not useful for order promising as it does not include the uncommitted inventory carried over from previous period. However, it gives the idea as to how many units *from that particular MPS supply* are available for promising to new customer orders for the same period or for deliveries in future periods. Discrete ATP-WL can be summed up from the current period to any particular period to compute the *cumulative ATP with lookahead* for that period, which is described next.

6.20.4 Cumulative ATP with Lookahead (Cum. ATP-WL)

The cumulative ATP with lookahead (ATP-WL) is calculated as follows:

$$\text{Cum ATP (WL)}_i = \text{Cum ATP (WL)}_{i-1} + \text{MPS}_i - \text{B}_i - \Sigma(\text{B}_j - \text{MPS}_j)$$

$$\text{until all } \Sigma\, \text{MPS}_j > \Sigma \text{B}_j \text{ where } j > i$$

That is the cumulative ATP with lookahead for a period (i) equals the ATP with lookahead (ATP-WL) of the preceding period ($i - l$) plus the MPS of the period (i) minus the backlog (B_i) of the period (i) minus the sum of the differences between backlogs and master production schedules of all future periods until, but not including, the period from which point production exceeds the backlogs (Fogarty *et al.* 1991).

Computing the Cum. ATP with Lookahead (Cum. ATP-WL or Backward)

The discrete ATP-WL is summed up from the current period up to the particular bucket for which cum. ATP-WL is being calculated to give the cumulative ATP-WL.

Cumulative ATP-WL is the one ATP that is really useful for order promising as it gives the balance quantity *after* providing for all the committed orders—whether to be delivered in the current period or in the future periods; it also includes uncommitted units carried over from previous MPS supplies.

6.21 AN EXAMPLE OF MPS REPORT (With PAB and Regular/Backward ATP Calculations)

In the standard MPS Report as shown in Fig. 6.3 (Master Schedule) format, the header information contains *the following fields:*

1. *The Lead Time* It is the number of periods required to manufacture or purchase this item, and it, unlike *cumulative lead-time*, is exclusive of any time required to manufacture or purchase the lower level components.

LEAD TIME: 1	LOT SIZE: 20	DEMAND TIME FENCE (DTF)	: 2
ON HAND: 10	SAFETY STOCK: 0	PLANNING TIME FENCE (PTF)	: 4

WEEK	1	2	3	4	5	6
Forecast	10	10	10	10	10	10
Orders	12	5	20	5	0	0
PAB	18	13	13	3	–7	–17
Discrete ATP (Regular) or WOL	13		–5			
Cum. ATP (Regular) or ATP WOL	13	13	8	8		
MPS	20		20			
Discrete ATP (Backward) or WL	8		0			
Cum. ATP (Backward) or WL	8	8	8	8		

Fig. 6.3 *MPS Report (Master Schedule): Regular and Backward Methods* (Adapted from Stonebraker, 1994)

2. *On-Hand* It is the quantity of this item physically in stock.
3. *Lot Size* It is the quantity of this item normally ordered from the plant or supplier.
4. *Safety Stock* It is the additional stock of this item planned to be in inventory as protection against unplanned demands.
5. *Demand Time Fence (DTF)* It is a future master-schedule period inside of which changes to the MPS are, typically, not accepted due to the excessive cost caused by schedule disruption.
6. *Planning Time Fence (PTF)* It is a future master-schedule period inside of which changes to the MPS are evaluated to prevent costly schedule disruption.

6.21.1 Projected Available Balance (PAB Calculations)

Prior to the Demand Time Fence (DTF)

1st Period:

PAB = On-Hand Balance + MPS – Customer Orders

$PAB_1 = 10 + 20 - 12 = 18$

Subsequent Periods:

PAB = Prior Period PAB + MPS – Customer Orders

PAB_2 = 18 + 0 – 5 = 13

After the Demand Time Fence (DTF)

PAB = Prior Period PAB + MPS – Greater Value of Total Customer Orders or Forecast

PAB_3 = 13 + 20 – 20 (Orders) = 13

PAB_4 = 13 + 0 – 10 (Forecast) = 3

PAB_5 = 3 + 0 – 10 (Forecast) = –7

PAB_6 = –7 + 0 – 10 (Forecast) = –17

6.21.2 Available to Promise (ATP) Calculations

Discrete ATP-WOL, Prior to the Demand Time Fence (DTF)

1st Period:

Discrete ATP-WOL = On-Hand Balance + MPS – Sum of Customer Orders before Next MPS

Discrete $ATP\text{-}WOL_1$ = 10 + 20 – (12 + 5) = 13

Subsequent Periods:

Discrete ATP-WOL = MPS – Sum of Customer Orders before Next MPS

Discrete $ATP\text{-}WOL_3$ = 20 – (20 + 5) = –5

Note that although this ATP is showing a shortage of 5 units, actually 13 units from previous MPS supply are available and thus true ATP is (13 – 5 =) 8. Hence this ATP, being misleading, cannot be used for order promising.

Cumulative ATP-WOL

1st Period:

Cumulative ATP-WOL = On Hand + MPS – Sum of Customer Orders before Next MPS

Cumulative $ATP\text{-}WOL_1$ = 10 + 20 – (12 + 5) = 13

Subsequent Periods:

Cumulative ATP-WOL = Prior-Period Cumulative ATP-WOL + MPS – Sum of Customer Orders before Next MPS

Cumulative $ATP\text{-}WOL_3$ = 13 + 20 – (20 + 5) = 8

Cumulative ATP-WOL in periods 1 and 2 are shown as 13 units implying that 13 units are available and can be promised to new customer orders. This again is misleading, since if we promise this to new orders, there would not be sufficient quantity available for the already promised order for 5 units in period 4. This order was promised on the assumption that at least 5 units would be carried over from period 1 as quantity reserved for delivering against this order. Thus, cum. ATP-WOL is also not useful for order promising. It only indicates how many units would be *physically* present in the stores during that period.

Discrete ATP-With Lookahead (ATP-WL)

The discrete ATP-WOL in period 3 is –5; hence discrete ATP-WL is shown as zero; and 5 units are deducted from closest Discrete ATP-WOL (13) to get the discrete ATP-WL (13 – 5 = 8). This shows that out of 13 units physically present in the stores, only 8 are actually available for promising to new orders. 5 units are reserved for the already promised order for 5 units in period 4.

Cumulative ATP-With Lookahead (ATP-WL)

The cumulative ATP-WL for a period is calculated by adding the discrete ATP-WL for all the previous periods. It is 8 for period 1 and (8 + 0 =) 8 from period 3 onwards. This ATP figure is to be used for order promising as it takes care of units carried over from previous periods as well as units reserved for future orders. It should be noted that this ATP figure is cumulative, hence, although 8 units are shown as available for each bucket, it's not that a total of (8 × 4 =) 32 units are available for promising. As soon as 8 units are promised for any of the periods ATP figures in all four buckets will become zero.

It should be noted that:

1. Projected available balances (PAB) are stated beyond the PTF for information purposes only.
2. MPS and ATP are not planned beyond the planning time fence (PTF).
3. Most assemble-to-order (ATO) and make-to-order (MTO) companies disregard unconsumed forecast inside the DTF when calculating the MPS.
4. The MPS is usually stated as firm planned orders (FPOs) inside the PTF.
5. Outside the PTF, planned orders (POs) may be controlled by the computer logic.
6. PTF is often established at or just inside cumulative lead-time.
7. The MPS chart shows the data for each field at the end of a time period.

6.22 MASTER SCHEDULING IN MAKE-TO-ORDER ENVIRONMENT

Essentially in make-to-order environment, customer orders are loaded onto the MPS in the first slot at which adequate capacity is available. If the production capacity needed for the order can be easily determined, the MPS process and the associated giving of a delivery promise to the customer are relatively simple. An order clerk with access to the MPS can give a reliable delivery promise to a customer whilst answering the initial telephone enquiry. If there is more than one order clerk, it is important that both can access common data so that the same capacity slot is not promised to two customers. In many cases, a simple spreadsheet accessed via a network of PCs may be all that is required.

6.23 MPS FAMILY TOTALS PER MONTH

The MPS should also contain monthly family totals for each family of products in the MPS. The purpose of these totals is to verify that the MPS is in fact a reflection of the production plan. In the normal course of making changes to the MPS, it is possible for the master-scheduler to make an error so that the monthly totaled MPS for all the items belonging to a particular group of family is not within a specified tolerance of the production plan for the family. Thus, the weekly MPS quantities for specific items are totaled into monthly production figures for families of product. These totals can be compared to the quantities in the production plan for agreement.

The MPS family totals per month shown in Fig. 6.4 are compared to the production plan for this family that came out of the S&OP meeting.

Weeks / *Items*	1	2	3	4		...		49	50	51	52
A	050	–	050	–		...		060	–	060	–
B	610	610	610	610				610	610	610	610
C	340	360	380	400		...		300	310	320	330
D	180	180	180	180		...		150	150	150	150
E	15	–	–	–		...		–	–	20	–
F	205	205	205	205		...		225	225	225	225
Family Total per Month	5575							5340			

Fig. 6.4 *Typical MPS Format* (Adapted from Plossl, 1994)

If they are in agreement (within some agreed-upon tolerance, say ± 5%), no further work is necessary for that family of MPS items. If they are out of agreement, the master-scheduler is responsible for either changing the individual MPS, so that when added together they equal the production plan, or asking to have the production plan changed. What is not acceptable is for the master-scheduler to keep the top management 'unplugged' from the system.

6.24 MIXED-MODEL MASTER SCHEDULING

In a Just-In-Time (JIT) environment, companies are making plans to manufacture only what is needed, eliminating the 'waste' of excess inventory. Ideally, if customers were to buy a little of every product every day, then companies would manufacture a little of every product every day—an approach called mixed-model master-scheduling as shown in Fig. 6.5(b).

	Weeks					
	1	2	3	4	5	6
Product A	12			12		
Product B		10		10		10
Product C			5			

Fig. 6.5(a) *Batch Master Scheduling*

	Weeks					
	1	2	3	4	5	6
Product A	4	4	4	4	4	4
Product B	5	5	5	5	5	5
Product C	1	1	1	1	1	1

Fig. 6.5(b) *Mixed-Model Master Scheduling* (Landvater, 1997)

Following are the advantages of mixed-model master-scheduling:

1. It is more responsive to the need of the customer than is *batch* master-scheduling as shown in Fig. 6.5(a).
2. If you can build a little of everything every day, then you can make what the customers need and they don't have to wait.
3. Inventory drops because there is no need to stock each of the different products when you can quickly make what is needed.
4. It creates a more uniform demand for raw materials, components, and capacity. It's easier, simpler, and more efficient to schedule a

plant when the flow is uniform than to deal with large peaks and valleys.

In JIT environments, mixed-model master-scheduling is becoming the norm. The mixed-model master-schedule is used to plan the material and capacity, and kanbans are used to authorize production.

6.25 EXCEPTION MESSAGES

Master production scheduling leaves control in the hands of the planner, but the power of the computer can be used to draw his attention to possible changes. Should the stock of item fall below the safety stock level, a manufacturing order is not automatically created or moved, but a suitable message is generated for the master-scheduler, who can assess the significance of the change and above all determine whether it is feasible, before accepting or rejecting it. Thus, the master production schedule software includes exception reporting to direct the attention of the master-scheduler to the items that require his attention. The purpose of these exception messages is to allow the master-scheduler to go directly to the items that require evaluation. In some situations this is essential; for example, in a company where there are several thousand items in the MPS, the system should point out the specific items requiring attention. Even if there are fewer items in the MPS, exception messages are highly desirable since they relieve the master-scheduler of the responsibility for scanning the entire master-scheduling information for all the items in the MPS.

Following are the seven types of exception messages that are part of the MPS system:

1. Reschedule a master-schedule order to an earlier date (expedite).
2. Reschedule a master-schedule order to a later date (deexpedite).
3. Not enough in the MPS to cover demands.
4. Cancel a master-schedule order.
5. Master schedule order due for release.
6. Over-promised customer orders.
7. Past due master-schedule order.

(1–4) The First Four Exception Messages

These indicate that the MPS is out of agreement with the latest forecast, customer orders, branch warehouse demands, or interplant orders. The master-scheduler may or may not choose to change any of the MPS. These exception messages are based on the Projected Available Balance (PAB) calculations in the MPS report.

Forecast Consumption Logic Prevents Exception Messages Due to Minor Fluctuations. Having forecast consumption logic in the system helps to prevent exception messages being flagged due to routine minor sales fluctuations that need not be reviewed by the master-scheduler. Allowing the customer orders to reduce the sales forecast over a period of time has the effect of assuming the sales forecast to be correct over that period. Consequently rescheduling messages are produced only when significant variations occur.

It should, however, be noted that if abnormal demands are not identified and flagged, the master-scheduler may not be alerted to sales that are above or below forecast. When deciding how to set the rules for identifying abnormal demands, it is generally better to err on the side of tolerances that are too tight rather than too loose. In effect, this might force the master-scheduler to review marginal cases. The master-scheduler does not have to change the MPS each time he is alerted to sales that are above or below the forecast.

(5) Master Schedule Order Due for Release

The exception message for a master-schedule order due for release is generated when the master-schedule order is ready for release. This exception message is produced by checking the start date of the order against the current date plus a few days or a week. By producing this message a few days before the actual start date of the order, the master-scheduler can make a final check of the order quantity and date, check component availability, prepare any documentation or paper work, etc.

(6) Overpromised Customer Orders

The exception message for 'overpromised customer orders' is generated when even more customer orders have been promised than the MPS can support. Generally, this indicates that an error has occurred; either in customer order promising, or possibly a change to the MPS that was not made. The exception message is created whenever the available to promise in the MPS report is less than zero.

(7) Past Due Master Schedule Order

This exception message is generated when a scheduled receipt for the master-schedule item is past due. This is an extremely serious condition since it probably means that the company has been unable to ship the product to the customers or distribution centers to whom it has been promised. This exception message is produced by checking the due date of the order against the current date (Landvater and Gray, 1989).

6.25.1 Responding to Action Messages

Much of the master-scheduler's day is spent responding to action messages. This is because of the volatility of the marketplace and the ever-changing nature of manufacturing. Also, customer orders come in at rates different from the forecast, and problems inevitably occur in the plant. So what looked like a wonderful MPS last week looks like a total disaster this week. Most master-scheduling software points out exception conditions so the master-scheduler can manage more effectively. It's much easier to deal with the exceptions than to plow through tons of 'normal' data. The master-scheduler looks at the situation for the item, then works with people in the plant or with sales and marketing to resolve the problem (Landvater, 1997).

6.26 THE CHALLENGE OF MANAGING CHANGE AND MPS DECISIONS

The life of the master-scheduler is ruled by change. For that reason, 'master rescheduler' would actually be a more accurate job title. The master-scheduler will encounter two types of changes: those within the company and those outside it.

Examples of changes outside the company include:

1. Demand from Customers

Selling more or less than what is forecasted. The customer wants an order moved to an earlier date or to a later date. The customer goes on credit hold, wants the quantity changed, wants one item substituted for another, and so on.

2. Supplier Problems

A supplier is unable to deliver all or part of an order. The reason may be equipment breakdown, a quality problem, a wildcat strike, suppliers who fail to deliver on time, engineering changes, process problems, and so on.

The MPS must be stable and hence the MPS in the near future is kept under manual control to prevent spurious change being fed to MRP I. The master-scheduler will clearly not react to trivial changes, but there will frequently be perfectly good marketing reasons for changing the MPS. Following are some of the situations necessitating changes in the MPS:

1. An epidemic may result in large unforecast increase in demand for a particular drug.
2. Demand for roofing material may experience a dramatic but unexpected increase in business after a hurricane.
3. Withdrawal of a competitive product may present a marketing opportunity that must be seized.

All the above are good reasons for considering a change to the MPS, but the feasibility of the change must be assessed before it is accepted.

Given this steady stream of changes that bombard the master-scheduler, his or her task is to constantly ask, 'Do we have an attainable MPS?' Many general managers want an 'aggressive' MPS so they can go after as much business as possible. Another way of saying the same thing is that they want an ambitious MPS to stretch resources as much as possible. Although this seems like a reasonable goal, the plan must be *attainable* by the company. And must be one that people will be accountable for hitting (Landvater, 1997).

6.26.1 Providing Stability to MPS through Time Fences

The MPS provides a tremendous amount of control over other schedules. It truly is a 'master' of all lower-level schedules. A change in the MPS of a single item means changes in quantity and timings of a large number of component items at the MRP I and CRP levels. Thus minor changes in MPS have the potential of getting magnified and causing significant instability at lower levels, this phenomenon, referred to as 'nervousness,' is a characteristic in an MRP I system when minor changes in higher level (such as level 0 or 1 in the BOM) records or MPS cause significant timing or quantity changes in lower level (such as level 5 or 6) schedules or orders (Cox III and Blackstone Jr, 1998).

Changes in the MPS become more difficult, disruptive, and costly as they are made closer to product completion. To stabilize schedules and to make sure that changes are properly weighed before they are approved, MPS planning horizon can be divided into several zones, with different procedures governing schedule changes in each time zone. Time fences separate the time zones as shown in Fig. 6.6.

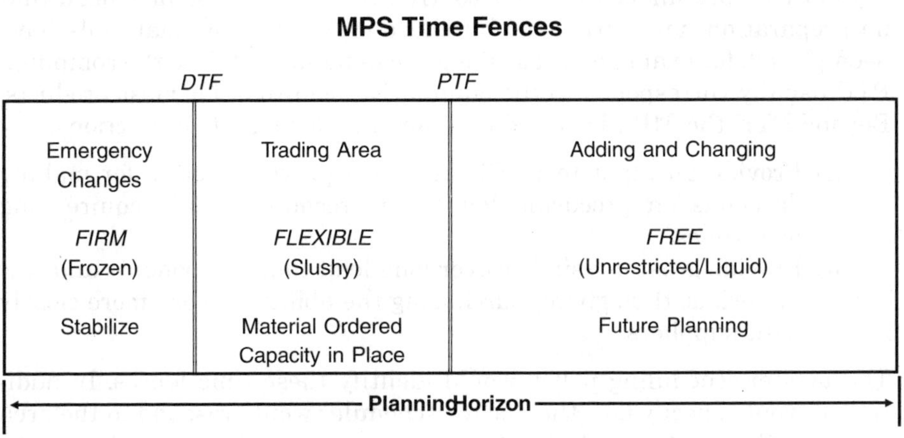

Fig. 6.6 *MPS Time Fences*

Most companies control the problem of instability of production plans through a master-scheduling policy that includes time fences. The master-scheduling policy spells out how the master-scheduler will respond to changes. The time fences determine the response.

1. Before Demand Time Fence, (DTF)

Before the first time fence, namely the DTF, few changes are made. The DTF is usually equal to the final assembly lead-time for assembled products or packaging time for soft goods. Quality or process problems may mean reducing the master-schedule. It's unlikely that anything could be added to the master-schedule this close to the current date. In this period, only an act of God or top management can change the MPS (Fogarty *et al.* 1991).

2. Between DTF and the Planning Time Fence (PTF)

Between the first time fence (DTF), and the second time fence, namely planning time fence (PTF), changes can be made but trade-offs come into play. You may be able to make a change but it will cost you, perhaps you will have to reduce production on another product, resulting in longer lead-times to customers. Particularly when operating at full capacity, any additions to the schedule must be counterbalanced by comparable deletions or increases in capacity. Changes in demand patterns, unusual orders, or equipment failures may warrant changes in the MPS. These changes are usually negotiated between marketing and manufacturing with the master-scheduler determining their feasibility before the final decision.

3. Beyond PTF

Beyond the second time fence, the situation is fairly flexible. Orders beyond PTF are under computer control. If an order is so far ahead that no preparation has started (no purchase orders for raw materials have been placed, for example), then changes can be safely left to the computer. PTF usually corresponds to the cumulative lead-time for most products. Beyond PTF, the MPS is stated only for the following two functions:

(i) Provide an input to RCCP, and thus provide a basis for making decisions on procuring long-range resources that require long lead-times.

(ii) Provide greater visibility over long lead-time components and raw materials, thus giving purchasing the ability to work more closely with suppliers.

The master-scheduling policy would identify these time fences. In addition, it would specify how the master-scheduler would respond in the area of tradeoffs (Landvater, 1997). In summary, as the time for order execution and manufacturing approaches, labour and material are committed. A change in the schedule can be disruptive and costly, and the costs must be compared with the benefits of change. Following time fence control guidelines described above will result in an MPS that promotes manufacturing stability and productivity while providing reasonable flexibility in meeting market demands. The guidelines for determining the length of periods from the current period at which the two time fences should be placed for different manufacturing environments, according to Stonebraker (1994), are shown graphically in Fig. 6.7.

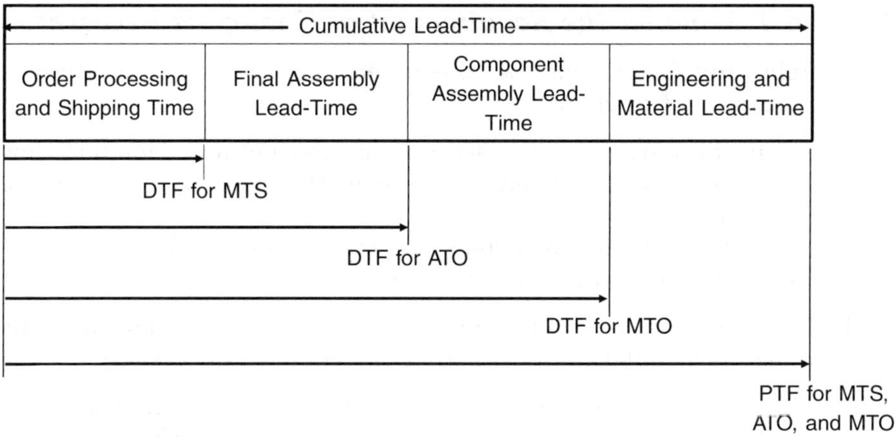

Fig. 6.7 *Time Fence Guidelines*

Note that the planning time fence (PTF) for all the manufacturing environments is equal to the cumulative lead-time of the MPS item. The demand time fence (DTF), however, varies for different manufacturing environments. DTF for make-to-stock (MTS) products is the shortest—equal to the order processing and shipping time. Whereas DTF for make- and engineer-to-order environments is the longest—equal to the sum of order processing, shipping, final assembly, and component assembly lead-times.

6.26.2 Responding to Sales Plan Changes

Most companies measure their sales plan accuracy by using some kind of tolerances. When a sales plan exceeds the tolerance one of two things happens:

(i) A new sales plan is developed.
(ii) It is reconfirmed that the old sales plan is still valid and the deviations represent a 'statistical fluctuation.'

In the event that a company reforecasts a particular item, a series of action messages is likely to appear the next time the master-schedule is run. These action messages will suggest moving the master-schedule orders up to earlier dates or out to later dates, depending on the type of change to the sales plan. The master-scheduler deals with a sales plan change in the same way as any other action or exception message (Landvater, 1997).

6.27 MPS DECISIONS

Following are the three decisions and the relevant sources of guidance to which a master-scheduler might turn:

Decision 1 Whether the Requirement or Change is Necessary

(a) Following factors may be considered to judge the necessity of a requirement or change:

1. Is the requirement based on a clearly defined demand? How stable has that demand been over the visibility period of the planning horizon?
2. Can the change be avoided by more timely action?
3. How important is the customer?

(b) Following are some of the information sources that might help the master-scheduler decide whether the change is necessary.

1. Master scheduling policies, as stated in a manual or a job book, are a formal way to state guidelines for scheduling. The policy should address such considerations as the following:

   - The procedures used to convert a planned order into a firm order.
   - The methods or procedures required to request changes.
   - The justification or rationale necessary for a change request.
   - The procedures for review of change requests.
   - The approval cycle for normal master production scheduling actions, as well as change actions.

2. The available-to-promise (ATP) or the projected available balance (PAB) computations may be used to evaluate how capacity and resources will be affected by responding to a requirement or change request.

Decisions 2 Whether the Requirement or Change is Feasible

(a) First Feasibility Consideration Initially, the change request should be considered in terms of the ability of the plant to respond, without consideration of all other loads on the system. Simulation, change budgets, and time fences are helpful in deciding this question; for example:

- Computer simulation may be used to review the planning impacts of the proposed change.
- Change budgets could confirm approval authority for the specific change action.
- Time fences would also indicate the feasibility of the project.

(b) The Second Feasibility Consideration The impact of the change on other requirements of the production system must be considered since,

though the change may be necessary and feasible when considered by itself, it may affect other jobs that, in the opinion of the master-scheduler, are equally or more important. Thus, the effect of the proposed change on all current requirements must also be evaluated.

Decision 3 Whether the Resources are Available

Capacity limitations and/or material and vendor limitations can be used to evaluate any key resources required as follows:

1. *Capacity load reports* are used to consider the key variables of labor or equipment availability. Work centers are often labour or machine constrained.
2. *The availability of materials*, either components or raw materials, may be assessed through on-hand inventories, supplier lead-time reports, and MRP I exception reports (which might identify overages of materials availability or underages of usage).

In summary, following are the two basic rules that need to be followed while making a change:

1. A change can only be made when it has been demonstrated that production capacity and materials are or can be made available.
2. Even if a change has proved to be feasible, it is still necessary to decide if making it is the best option. A change is made when the cost to the business of not making it exceeds the cost of making the change.

In practice, however, these costs can rarely be quantified and by default the decision is often left to the judgment of the master-scheduler who is probably being lobbied furiously by all interested parties.

It has been suggested that master-scheduler should be responsible directly to the managing director in order to maintain independence from the accountancy, sales, or production functions, but few, if any, companies have adopted this radical but enlightened suggestion. In practice, the master-scheduler needs support so that when necessary he can resist insistent demands for change either from senior sales managers who understandably put the interests of their customer first, or from senior production manager keen to maximize output, possibly at the expense of difficult to make but urgently needed products. The necessary support is achieved by having a formal change approval mechanism. The changes closer in time need approval from senior manager.

6.28 SUMMARY OF MASTER SCHEDULING ACTIVITIES

Master scheduling activities involve three stages, namely designing MPS, creating MPS, and controlling MPS.

A. Designing the MPS

It includes the following steps:

1. *Select the items to be master-scheduled* This involves selecting the levels in the BOM structure to be represented by the items scheduled (both components and final assemblies may be included).
2. *Organize the MPS by product groups.*
3. *Determine the planning horizon, the time fences, and the related operational guidelines.*
4. *Select the method for calculating and presenting the available-to-promise (ATP) information.*

B. Creating MPS

It includes the following steps:

1. *Obtain the necessary informational inputs* This includes the forecast, the backlog (customer commitments), and the inventory on hand.
2. *Prepare the initial draft of the MPS.*
3. *Develop the rough-cut capacity plan (RCCP).*
4. *If required, increase capacity or revise the initial draft* This process is repeated until the time a feasible MPS is obtained.

C. Controlling the MPS

It includes the following activities:

1. *Track actual production and compare it to planned production to determine if the planned MPS quantities and delivery promises are being met.*
2. *Calculate the ATP to determine if an incoming order can be promised in a specific period.*
3. *Calculate the projected on hand to determine if planned production is sufficient to fill expected future orders.*
4. *Use the results of the preceding activities to determine if the MPS or capacity should be revised.*

6.29 ROLE OF THE MASTER SCHEDULER

The role of master-schedulers, individuals who oversee the development and use of MPS in manufacturing planning and control, is of special importance. Master scheduler is the link between marketing, distribution, engineering, manufacturing, and planning. Master schedulers have the following duties:

1. Compare Actual and Forecasted Demand, and Suggest Revisions to Forecast and MPS

2. Convert Forecasts and Order-Entry Data into MPS They change general descriptions and generic model numbers on customer orders into specific end-item BOM numbers, dividing monthly into weekly quantities and applying forecasts of product options.

3. Correlate MPS with Shipping and Inventory Budgets, Marketing Programs, and Management Policies They check MPS data with budgets for shipment and inventory to detect significant differences for management review. They track actual demand against marketing programs and compare actual performance with management policies of customer service, inventory investment, and cost control.

4. Maintain MPS Data Files They maintain MPS data by tracking the use of safety stock provided at the MPS level, accounting for differences between quantities of end items produced and those consumed by the final assembly schedule, and enter and edit all changes to MPS files.

5. Participate in MPS Meetings, Preparing Agendas, Anticipating Problems, Providing Data for Possible Solutions, and Bringing Conflicts to the Surface Master schedulers play important roles in production meetings, preparing the agendas, evaluating problems brought to their attention by factory supervisors, inventory planners, and others detecting them, anticipating suggested solutions and having relevant data available to avoid wasted decision-making time, and bringing out conflicts among people to avoid their being ignored, concealed, and neglected. They also call the basic conflicts of demand and capacity to the attention of other members of management, especially marketing and manufacturing, who need to participate in resolving the problems.

6. Evaluate the Impact of Top-down Inputs For example, a request for introduction of a new product in much less time than the normal lead-time are evaluated by the master-scheduler.

7. Evaluate Suggested MPS Revisions They review suggested MPS changes, determine which end-item lot should be changed and what the effect will be, and initiate recommendations to management on acceptance or rejection.

8. Revise MPS when Necessary The MPS may have to be revised due to lack of material or capacity.

9. Develop and Monitor Customer Delivery Promises They provide delivery dates for incoming orders; evaluate the impact of new or changed orders on customer deliveries, and recommend and monitor promise dates.

10. Manage the Master Production Schedule This involves one or more of the following activities:

(a) Adding to and deleting from the MPS.
(b) Responding to action messages.

(c) Evaluating marketing requests for early delivery.
(d) Responding to sales plan changes.
(e) Responding to manufacturing or supplier problems by evaluating bottom-up inputs, such as anticipated delay reports from the shop or purchasing.
(f) Verifying the MPS against the production plan.

6.30 CHARACTERISTICS OF A MASTER SCHEDULER

MPS is a matter of judgment. Companies have discovered, sometimes painfully, that master-scheduling requires judgment, creativity, and product knowledge. Creating and maintaining the MPS is done by the master-scheduler and involves:

(a) identifying options (creativity),
(b) evaluating the options (judgment), and
(c) selecting the best of the options and implementing them (responsibility and accountability).

Computers can and should be used to help the master-scheduler in doing this job. But accountability must ultimately rest with a human being; the master-scheduler must be responsible for creating and maintaining a valid MPS (Landvater, 1997). Following are the characteristics of a good master-scheduler:

1. Experienced in production and inventory management.
2. A good communicator.
3. Good with numbers.
4. Cool under pressure.
5. Disciplined in maintaining data accuracy.
6. Assertive to make things happen.
7. Be able to command respect from all organizations within the company.
8. Farsighted so as to anticipate and prevent problems rather than fire-fighting them.
9. Flexible in approach.

6.31 MASTER PLANNING

The master planning activities of demand management, capacity management, and production planning culminate in the delivery of a Master Production Schedule that is feasible, cost efficient, and responsive to customer needs. The MPS is the visible output of all the master planning activities. It plans for and controls the impact of independent demand on resources.

Master planning balances customer demands and resource availability.

Thus, it may be viewed as a process of managing demand and planning for resource use. The objectives of master planning activity and of the master-scheduling are the following:

- A level, repeating schedule.
- Few scheduling changes.
- Close correspondence to the production plan.

6.32 SYMPTOMS OF A MISMANAGED MASTER PRODUCTION SCHEDULE

Following are symptoms of a mismanaged MPS

1. Overloaded or extensively front-loaded capacity plan. (The capacity plan should be neither overloaded nor extensively front-loaded. The load should be reasonably level over the entire planning horizon.)
2. Excessively short lead-times and frequent need for expediting.
3. Excessive past due shop orders and purchase orders.
4. Poor on-time delivery performance.
5. Increased component (mostly, work-in-process) inventory.

6.33 MEASURES OF PERFORMANCE

The measurement of how well the MPS is doing involves three issues (Landvater, 1997):

(a) Validity

Valid means *doable.* People who follow the plan will encounter no surprises and will achieve their goal, whether it's shipping product to customers on time or sending finished goods into inventory on time.

(b) On-Time Deliveries.

(c) Hot Lists versus Schedules

A company with a valid scheduling system has no need for a hot list because the system provides better information. If anything in the formal system does breakdown, the hot list will reappear. No hot list means that there are no breakdowns in the formal system and the MPS is valid (Landvater, 1997).

Following are the measures of performance of master-scheduling.

1. Capacity utilization (= Actual output ÷ Design capacity).
2. Accuracy of supplier evaluation and effectiveness of communication with them.
3. Inventory turnover ratio (= Cost of sales ÷ Average inventory investment).

4. Delivery performance (measured as a percentage of on-time, on-specification delivery).
5. Deviations of actual production from the MPS.
6. Conformance to the production plan.
7. Frequency of need for expediting (or the unplanned activities).

6.34 COMPUTER ASSISTANCE IN MASTER SCHEDULING

The complexity of most manufacturing environments requires a computerized planning and control system with human interfaces at appropriate *decision points*. The master-scheduler requires such an interface. The requirement for computer assisted planning is due to a combination of the number of items on the MPS, the large number of sub-assemblies and components, and the magnitude of recording and processing inventory transactions, material requirements, and capacity requirements. Following are the ways in which the computer can be a powerful tool for assisting the master-scheduler for reviewing the MPS for each item:

(a) To Display the Master Schedule

Here the master-scheduler frequently reviews the MPS for each item. The role of the computer is to present the big picture to the master-scheduler and let him or her decide what changes, if any, are needed. This usage is typical when the master-scheduler is managing a small number of items with high levels of activity, for example, 50 or 100 items that have customer order activity every day.

(b) To Critique the MPS

This activity is done through exception messages, such as, situations in which the projected available balance (PAB) will fall below the safety stock quantity or in which master-schedule orders can be rescheduled to later dates without affecting customer service. The computer directs the master-scheduler to situations that require his or her attention.

(c) To Perform What-If Analyses

The availability of an MRP II software allows the organization to perform what-if analyses to answer questions regarding effect of change in product volume/mix on capacity requirements. Answer to such queries from a computerized simulation run enables the management to prepare plans for such contingencies.

(d) To Maintain the MPS of a Company Whose Maintenance Load is Significant

This usage has become especially prevalent because more companies are implementing JIT approaches. Most high volume/repetitive manufacturing plants schedule in terms of daily production. For example the MPS may be to produce 10,000 units of 'product A' *per day for the next 45 days,* 7,500 units of product B, and so on. Computerized assistance of this type is called a *rate generator*.

Rate Generator It generates transactions based on a rate of production. A typical master-scheduling software system allows the master-scheduler to enter an MPS order for 10,000 units of 'product A' with a due date of today. The master-scheduler would then have to add another order for 10,000 for tomorrow, and so on for the remainder of the 45 days. A rate generator allows the master-scheduler to specify that the rate of production will be 10,000 units per day from September 1 until October 15. Instead of entering 45 transactions, the master-scheduler has to enter a single transaction, which in turn generates 45 transactions. The rate generator works the same way for changing the master-schedule from one daily quantity to another and for deleting master-schedule orders.

(e) Finite Scheduling Algorithm to Assist the Master Scheduler

This system is used in situations—such as high-volume food or consumer packaged goods business—in which the maintenance workload is high and the master-scheduler is adding little intelligence beyond what a scheduling algorithm can do. Additional help is either needed or beneficial in creating and maintaining the MPS. In these situations, a finite scheduling system is used to make a demand stream (customer order and forecast) and create an MPS. That is, if the logic in the finite scheduling is similar to what the master-scheduler would use to create a schedule, and if the master-scheduler is not adding significant additional judgment or creativity to the schedule, then this type of power assistance can be valuable. The finite scheduling systems can recalculate a schedule quickly and can include good logic, typical of how a scheduler would work. Using a tool like this does not change the responsibilities or accountabilities in master-scheduling. The master-scheduler must approve the MPS created in the finite scheduling system.

6.35 MASTER PRODUCTION SCHEDULING IN PROCESS INDUSTRIES

Following are the characteristics of the master production scheduling in process industries:

1. *Capacity-Driven Master Production Scheduling* Discrete item manufacturing industry follows the material-driven master production scheduling method, because materials are scheduled first and then capacities are checked. Because of the high capital invest-

ment, process industries concentrate on achieving high equipment utilization, and usually have a good estimate of the capacity of their line or plant. Thus, process industries tend to first schedule capacity and then materials. This is called capacity-driven master production scheduling.
2. There is more emphasis on Distribution Resource Planning (DRP II) and intraplant transfer planning.
3. There is closer coupling of MPS with forecasts.
4. MPS is less compliant to customer requests.
5. Lot sizes may be dictated by facilities design and manufacturing practices for ensuring product quality.
6. Schedules generally have smaller time intervals.
7. The MPS is the authority to produce and determines sequencing.

6.36 FINAL ASSEMBLY SCHEDULE (FAS) OR FINISHING SCHEDULE

Final Assembly Schedule (FAS) is a schedule of end items to finish the product for specific customer orders in a make-to-order (MTO) or assemble-to-order (ATO) environment. It is also referred to as *finishing schedule* because it may involve operations other than just the final assembly; also it may not involve 'assembly,' but simply final mixing, cutting, packaging, etc. The FAS is prepared after receipt of a customer order (which is accepted subject to material and capacity availability constraints). FAS schedules the operations required to complete the product from the level where it is stocked (or master-scheduled) to the end-item level (Sari, 1997). In short, the FAS controls that portion of the business from fabricated components and sub-assemblies planned on the basis of forecast to customer-ordered shippable products in ATO environment. In MTO businesses, it states the specific schedule for satisfying customer orders.

The term *final assembly schedule* is in common use because many of the earliest companies to apply MRP I techniques were assembled products manufacturers. The term *finishing schedule* is probably the better generic term and has gained acceptance in recent years. It better describes the many companies using MRP I that do not produce assembled products, for example, textiles, metals, and chemicals. Synonymous terms include *packaging schedules, blending schedules,* and *line schedules* (Sari, 1997).

Final assembly schedule (FAS) entries are needed when end products do not appear in the MPS. These end items are assembled to order or have several customer options that can be combined in various configurations. These products belong to the category of products with variants and options wherein many shippable end-item products are assembled from a few standard components (in modular construction). This category includes automobiles, machine tools, home appliances, electrical equipment, and many others. The MPS for such products are best stated in 'standard modules' rather than end-item models. Hence for these products, two different schedules are required: MPS for end-item *components*, and final assembly schedules for shippable products. Whereas the MPS may be based on planning bills, the FAS will require a single-level product bill (manufacturing BOM) for the purpose of component and sub-assembly order picking as well as product costing. The customer or assembly order itself serves as a one-time bill of materials. It is also the basis for historical records of what was produced, that is, warranty and necessary safety records for automobiles. The lead-time of the FAS should be as short as possible and expressed in lesser periods, such as shifts or days.

The traditional approach has too often been to try to predict the final configuration of each product and push it into the distribution network or to try to push manufacturing to develop the flexibility to produce what we need more flexibly and more quickly. The results are usually shortages at every level of what we need, excess inventory of configurations that are not selling, and half started products that were *hot* and now are *not*.

FAS in Make-to-Stock Businesses In MTS businesses, the FAS and MPS merge as the MPS itself plans the end item, hence, usually there is nothing like an FAS in MTS businesses. However, some MTS businesses use the term FAS to represent a refined and detailed MPS. Here, the distinction between the FAS and the MPS may be unclear because both schedules represent statements of end items to be produced. Where a formal FAS is in use, it usually represents a refined MPS plans that considers specific assembly lines or finishing areas to be used, more specific timings of production, and control of the material issue process. The MPS itself is usually not sufficiently detailed for these purposes (Sari, 1997).

The Final Assembly Schedule is the schedule that serves to plan and control final assembly and test operations. Following activities are generally included in the FAS:

(a) launching of final assembly orders,
(b) picking of component parts,
(c) sub-assembly,
(d) painting or other finishing operations,
(e) scheduling the fabrication or purchase of any component items not under MPS control but needed for final assembly, and
(f) packing.

The final assembly schedule is often used to plan for products that:

(a) have relatively low volume,
(b) are highly customized,
(c) have short procurement or manufacturing lead-time,

(d) do not get quantity discounts, or do not involve long setup times, and
(e) have relatively high-cost inventories.

These products are put into modular bills of material used in the execution phase and are manufactured or procured during execution of the Final Assembly Schedule (FAS). Assigning these to final assembly schedules permits fast machining of small quantities during final assembly of specific product configurations. This eliminates the need for forecasting these items, carrying them in planning and execution BOM, and committing expensive materials to specific components long before true requirements are known. Table 6.2 shows the comparison between FAS and MPS in a tabular format.

Table 6.2

Comparison of the FAS and the MPS

Aspect	*FAS*	*MPS*
Time Span	Short	Long
Product Level	Finished End Item	Semifinished Assemblies and Components
Constraints	MPS	Production Plan
Demand Source	Customer Orders	Forecast
Period	Days, shifts	Weeks or Months

- Whereas, the MPS is the *anticipated* build schedule based on forecast (with actual orders thereafter imperfectly consuming these forecasts), the FAS is the actual build schedule based on actual customer orders (backlog).
- The MPS disaggregates the production plan into end items, options, or group of items, whereas the FAS is the last disaggregation—into exact end-item definitions. The FAS represents the last possible adjustment that can be made to the MPS, therefore, it is advisable to make that adjustment as late as possible. Avoidance of firming up the FAS until the last possible moment means that the time horizon for the FAS is only as long as dictated by the final assembly lead-time (including document preparation and material release). Techniques that help to delay the FAS commitment include bill structuring, close coupling of order entry/promising systems, partial assembly, the stocking of sub-assemblies, and process/product designs with this objective.
- Because the required manufacturing lead-times are long but the customer-allowed delivery periods are short, MPS must be formulated and MRP I employed long before final assembly schedules are prepared.

- MPS usually extends several months into the future; the FAS usually covers only a few days or weeks.
- MPS may be stated in weekly priorities (buckets) whereas the finishing schedule may be stated by shifts within the week, by day, by hour, or in other increments. The automotive industry, for example, states final assembly schedules by the minute or less.
- MPS are often based on forecast customer demand; FAS usually contain actual customer orders and may be constrained by shortages of components. Thus, in case of assemble-to-order products, order entry, which is one of the drivers of the execution phase, normally does not interface with MPS but with FAS.
- In the assemble-to-order environment, the MPS is typically stated in super modular bills showing all the possible variants and options, whereas the FAS must be stated in terms of the exact end-item configurations.

Final Assembly Schedule (FAS) allows three things to change (Thorne, 1995):

1. It does not try to predict what final configuration or package, each customer will order when. It changes the focus to committing inventory to its final configuration only either just before shipment to customer or as it is pulled by demand in the distribution channel. It achieves this by breaking the MPS from the scheduling of the final configurations or packaged product.
2. By using a planning bill of material we maintain the flexibility of options for the final configurations or packaged product. This allows us to focus our planning on just those items that have a high level of variability. It drives them to agreed levels based on the projections of future demand. It tends to be self-correcting, since by exploding available to promise through the planning BOMs, it dynamically adjusts to recent demand while still maintaining the future profile of demands that you have established.
3. It is linked to the S&OP process. This keeps it from being either a manufacturing or marketing focus; it becomes both, but is driven by the realization that we need to adjust our capacity to meet our customer's needs as quickly and flexibly as possible. Finishing or packaging lead-time is relatively short, normally less than a day.

6.37 THE MECHANICS OF FAS

1. The first step is decoupling the MPS from the final configuration or packaged end product that is manufactured. This is achieved by master-scheduling *at* the unpackaged product level or 80% assembly, where the number of different styles or options is smallest and where the time to finish or package is relatively short, normally less than a day.

2. When the master-scheduled product (80% assembled) is completed it is put into inventory to be pulled by the FAS to its shippable configuration or package. This pull signal is driven by customer demand or demand from the distribution network, which is in turn created by customer demand (Thorne, 1995).

6.38 BENEFITS OF FAS

1. Shorter Cycle Times

An effectively linked final assembly scheduling allows companies with a mix of multi-product, multi-configuration items to shorten cycle times, with the resulting improvement in quality and costs.

2. Substantially Lower Investment in Inventories

By no longer trying to forecast each configuration or packaged product, the manufacturer can free up a substantial amount of unproductive inventory and reduce exposure to excess and obsolete inventory. To support the FAS approach there is an increase in the inventory carried of semi-finished product and packaging, but far less than the inventory that is removed in finished goods. Service expectations of customers and finished goods inventory have always been seen as two forces that must be traded off. FAS is one of the tools that allows us to mute these tradeoffs.

3. Significant Improvements in the Flexibility to Respond to Customer's Changing Needs

By decoupling the MPS from the FAS the manufacturers have provided a tool to react in hours to customer demand, as opposed to days or weeks, and in the process have dramatically reduced run sizes, adding to flexibility.

4. Better On-Time Performance

Use of FAS results in substantial improvement in customer fill rate and on-time performance.

5. Faster Response to Unanticipated Problems

With feedback from the marketplace in daily increments, changes are seen more quickly, as are the swings in demand that require more capacity than is available. As a result, problems are seen and reacted to faster.

6. Reduction in Waste Caused by Changes in the Plans

In many of the companies that have moved to FAS, by simplifying the process they have largely eliminated the waste caused by changing priorities, rush orders, stockouts, and special handling (Thorne, 1995).

References

Correll, James G, 'Reengineering the MRP II Environment' In *APICS: 37th International Conference Proceedings.* American Production and Inventory Control Society, Falls Church, V A, pp 1–4, 1994.

Cox III, James F, and John H Blackstone Jr, *APICS Dictionary* (ninth edition). American Production and Inventory Control Society, Inc, Falls Church, V A, 1998.

Dougherty, John R 'Master Planning: A Key to Satisfying Customers in the '90s.' In *APICS: 38th International Conference Proceedings.* American Production and Inventory Control Society, Falls Church, V A, pp 80–84, 1995.

Gibson, P, G Greenhalgh, and R Kerr, *Manufacturing Management: Principles and Concepts.* Chapman & Hall, London, 1995.

Kenworthy, John, *Planning and Control of Manufacturing Operations.* John Wiley & Sons, New York, Inc, 1997.

Landvater, Darryl V and Christopher D Gray, *MRP II Standard System: A Handbook for Manufacturing Software Survival.* Oliver Wight Limited Publishing Company, Essex Junction V T, 1989.

Landvater, Darryl V, *World Class Production and Inventory Management* (second edition). John Wiley and Sons, Inc, New York, 1997.

McMahon, Chris and Jimmie Browne, *CADCAM: From Principles to Practice.* Addison-Wesley Publishing Company, Wokingham, England 1993.

Narasimhan, S L, Dennis W McLwavey, and Peter J Billington, *Production Planning and Inventory Control* (second edition). Prentice-Hall, Inc, Englewood Cliffs, N J, 1995.

Oden, Howard W, Gray A Langenewalter, and Raymond A Lucier, *Handbook of Material and Capacity Requirements planning.* McGraw-Hill, Inc, New York, 1993.

Plossl, George, *Orlicky's Material Requirements Planning* (second edition). McGraw-Hill, Inc, New York, 1994.

Proud, John F, *Master Scheduling: A Practical Guide to Competitive Manufacturing.* John Wiley & Sons, Inc, New York, 1994.

Sari, F John, 'Master Production Schedule' In James H Greene, *Production and Inventory Control Handbook* (third edition). McGraw-Hill Inc, New York, 1997.

Shafer, Scott M and Jack R Meredith, *Operations Management: A Process Approach with Spreadsheets.* John Wiley & Sons, Inc, New York, 1998.

Stonebraker, Peter B, *Master Planning: (APICS) CPIM Certification Review Course, Student Guide* (Revision 3), American Production and Inventory Control Society, Falls Church, V A, 1994.

Thorne, Fred, 'Blending Quick Response and Effective Control: The Final

Assembly Schedule.' In the *38th International Conference Proceedings of APICS.* American Production and Inventory Control Society, Falls Church, V A, 1995.

Toomey, John W, *MRP II: Planning for Manufacturing Excellence.* Chapman & Hall, New York, 1996.

Vollmann, Thomas E, William L Berry, and D Clay Whybark, *Manufacturing Planning and Control Systems* (fourth edition). McGraw-Hill, Inc, New York, 1997.

Chapter 7

Bill of Material (BOM)

7.1 INTRODUCTION

A bill of materials is a list of all the components—items, ingredients, or materials—needed to produce an end item or product or one of its sub-assemblies. It lists all of the sub-assemblies, components, and raw materials that go into a parent assembly, showing the quantity of each required to make an assembly. It shows how much of materials are needed in what order to manufacture a product highlighting the child-parent relationships existing between them.

Bill of Material (BOM) is defined by APICS Dictionary as:

A listing of all the sub-assemblies, intermediates, parts, and raw materials that go into a parent assembly showing the quantity of each required to make an assembly.

- The bill of material is a list of all items that make an assembly and is structured level by level so as to show the parent-component relationship (Toomey, 1996). It contains information on each input to the product, such as part numbers, descriptions, quantity needed for each part, and the unit of measure.
- Unlike order point inventory planning, material requirements planning (MRP I) uses bills of material as its basis. BOM acquired a new function with MRP I, becoming a *framework on which the entire planning process depends* (Plossl, 1994).
- BOM is used by the MRP I system in conjunction with the master production schedule (MPS) to determine the items for which purchase requisitions and production orders must be released.
- The bill of material may also be called *product structure, part list, formula, formulation, recipe, ingredients list*, or *practice* in different types of industries (Cox III and Blackstone Jr, 1998).

- In production planning, a bill of material is a key input document for establishing a proper inventory control system.
- Because inventory records (status) are an important part of MRP I, an accurate BOM becomes a vital input to MRP I systems. A collection of these bills—one for each assembly and one for each of its components, which are sometimes assemblies themselves—describes the flow of material through the manufacturing process in the plant. Thus, a BOM also describes the relationships among parts.
- As a vital part of MRP I, BOM must be accurate and up-to-date if its outputs are to be valid. Parent and component items omitted from BOM will not be available when needed, and those in BOM not belonging there will add to excess inventory (Plossl, 1994).
- In addition, BOM must be unambiguous and structured properly for planning. The term 'BOM structure' refers to the arrangement of parents and components within the BOM file, not to the organization of this file in some storage medium (Plossl, 1994).
- The BOM shows the components that go into making the parent. It does not show the steps or process used to make parent or the components. That information is recorded in a *routing file* (Arnold, 1998).

Computerized BOMs

A complete and accurate bill of material is needed to summarize parts requirements if an MRP I system is to be used efficiently. The number and complexity of parts in most businesses today necessitates the use of computerized BOMs. It is important to note that the computations are only as accurate as the information you provide to the computer. The real value of computers in manufacturing BOMs, however, comes from three sources:

(a) its ability to store massive amounts of data,
(b) the speed with which information can be retrieved, and
(c) the availability of software packages that organize and retrieve this information.

Bill of Material or Product Structure Files

The BOM files are those computer records designed to provide desired output formats (Vollmann *et al.* 1997).

Bill of Material Structure

The BOM structure relates to the architecture or overall design for the arrangement of BOM files. The BOM structure must be such that all desired output formats or reports can be provided (Vollmann *et al.* 1997).

Bill of Material Processor

A BOM processor is a computer software package that organizes and maintains linkages in the BOM files as dictated by the overall architecture (BOM structure). Most BOM processors operate using the single-level BOMs and maintaining links or chains between single-level files. It is the BOM processor that is used in MRP I to pass the planned orders for a parent part to gross requirements for its components (Vollmann *et al.* 1997).

Different types of BOM representations are needed for various planning and manufacturing purposes. This does not mean that a number of different BOM computer files are necessary. The software requirement and prudent application of an integrated manufacturing system demands that a single BOM satisfy all users as opposed to individual bills for manufacturing, accounting, engineering, etc. BOM processor programs can code the basic data to link components and produce BOM for each specific purpose. It should be noted that BOM processor programs do not structure BOMs; they edit, organize, load, maintain, and retrieve BOM records already properly structured by knowledgeable people to serve the needs of MRP I (Plossl, 1994). A fundamental rule is:

> *A company should have one, and only one, set of bills of material or product structure records. This set should be maintained as an entity and be so designed that all legitimate company uses can be satisfied (Vollmann et al. 1997).*

Master Bill of Material

The master bill of material is the company database for a particular product or family of products. It is the only real bill of material. In fact, all the different bills should be nothing but the sorted versions of the master BOM. A single master bill is much easier to maintain. It is also much easier to handle configuration control from a single master bill (Garwood, 1997).

Items to be Included in the BOM

Every item that needs to be planned and scheduled should be on the BOM, because if it is not on the BOM you risk not having the item when you need it. Following is the list of items that should be considered for inclusion on the BOM:

1. Packaging material.
2. Printed material.
3. Instruction sheets.
4. Manufacturing sub-assemblies.
5. Hardware (bolts, washers, screws, company-name plates etc.).
6. Semi-finished materials.
7. Intermediates.

8. Raw materials.
9. Expendable tooling.
10. Reference materials and drawings.

7.2 BILL OF MATERIAL STRUCTURING

A bill of material structure represents more than just a parts list. It shows the assembly and sub-assembly breakdowns of a product 'as built' rather than 'as designed'—these two are often at variance (John, 1993). Following are the APICS definitions of the terms related to BOM structure:

Product Structure

It is the sequence of operations that components follow during their manufacture/assembly into a product. A typical product structure would show raw material converted into fabricated components, components put together to make sub-assemblies, sub-assemblies going into assemblies, etc. (Cox III and Blackstone Jr., 1998).

Product Structure Record

It is a computer record defining the relationship of one component to its immediate parent and containing fields for quantity required, engineering effectivity, scrap factor, application selection switches, etc. (Cox III and Blackstone Jr, 1998).

Figure 7.1 below shows a product structure in the form of a tree—a representation often used for learning purposes.

The topmost item on the tree structure represents the master scheduled product, most often an actual end item. At the next level down are shown the assemblies and components that directly make up the product. At successive levels, the assemblies are further broken down into sub-assemblies, components, and raw materials. The BOM tree structure in Fig. 7.1 adopts the following convention:

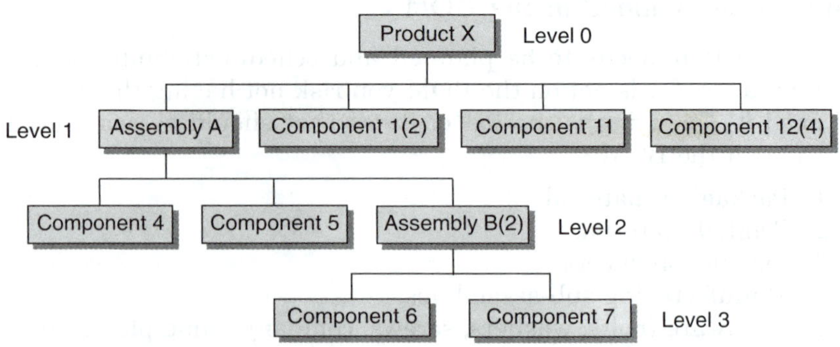

Fig. 7.1 *Bill of Material Tree Structure* (John, 1993)

1. Single letters are used to represent products and assemblies that contain further breakdowns.
2. Single numbers are used to represent raw materials or purchased components that are not further decomposed.
3. A usage quantity or 'quantity-per' of one is assumed *unless otherwise noted* by a usage number shown in parentheses, for example: 'Assembly B(2),' indicates the usage of 2 nunbers of assembly B in each unit of assembly A.
4. The topmost product number appears at level zero. Parts that directly enter into assembly of that product are considered level 1; and successive levels are numbered consecutively with the level number increasing as we move down in the tree.
5. A product or assembly at one level is often referred to as the *parent*, with its component parts at the next level down referred to as the *children* of that parent.

7.2.1 Bill of Material Restructuring

Different departments in a company use BOM for a variety of purposes. BOM issued and maintained by the engineering departments rarely meet the needs of all users; they have to be restructured without affecting the integrity of product specifications and in such a way as to permit modifications complying with engineering changes. The severity of the restructuring problem varies from company to company, depending on the complexity of its products and the nature of its business. Although each user has individual preferences for the way the bill should be structured, there must be only one structure, and it should be designed to satisfy most needs. However, there can be several formats, or ways, to present the bill (Arnold, 1998).

The creation of a product structure file can take a considerable amount of time. Accurate bills of material are essential to an effective MRP I system. The bill of material must specify how a product is actually manufactured rather than how it was designed to be manufactured. Redundant or obsolete part numbers must be purged from the system. This may not seem like a big task, but in some companies every time a part is purchased from a different supplier, it is assigned a different part number. One firm in the process of implementing MRP I was able to eliminate 6,000 extra part numbers from its database and dispose of thousands of dollars of obsolete inventory that had not previously been identified as such (Russell and Taylor III, 1998)!

7.2.2 Principles of BOM Structuring

1. Data in the Part Number and BOM Records should be Complete Enough to Satisfy the Needs of Every Internal Customer in the Company
The internal customers include engineering, manufacturing, planning and scheduling, order entry, finance, the shop floor and others.

2. Part Numbers should be Unique If an existing item is revised, and the new item is different in form, fit, or function from the old item, a new part number is required. Simply changing revision letter on a drawing and not assigning a part number is not adequate.

3. BOMs should Consist of Part Numbers, not Drawing Numbers.

4. BOM should Identify Material, not Labor.

5. all Parts in the BOM should be Additive.

6. The BOM must Include all Items to be Scheduled and Indicate Relationships between them Sub-assemblies, semi-finished items, components, raw materials, and any other item necessary to schedule in the production of the finished product and the relationships between them should be included.

7. BOM must Indicate Levels of Manufacture to Generate Low-level Codes; and the Number of Levels in the BOM should be Minimized—BOM should be as Shallow as Possible Each level in the bill creates extra part numbers, extra orders to be released, extra paperwork, extends the lead time, and increases the work necessary to maintain the bill. All these ultimately result in higher costs. Part number should not be assigned and used to identify work-in-process steps. Operation numbers on the routing serve that purpose. A set of guidelines, agreed upon by everyone, should be used to identify the legitimate conditions for assigning new part numbers and thus creating another level in the bill.

Every level in the BOM needs to be examined and challenged. If it is not there for planning and scheduling purposes, maybe it shouldn't be there at all. That challenge, in fact, needs to be extended to the manufacturing process itself. Can we change the process so we don't have to move items in and out of stockrooms or inventory points, thus doing away with the need to plan and schedule them? If the answer is yes, then there goes one more level in the bill (Garwood, 1997).

8. Requests to change the BOM should be Approved by Every User of the Bill Bill changes should be made only *after* total company approval.

7.3 SINGLE-STAGE AND MULTIPLE-STAGE BILLS OF MATERIAL

A bill of material can include single or multiple stages of production. For example, the paint manufacturing and packaging process can be accomplished by a manufacturing order to make paint from its ingredients. Subsequently, a number of orders are released to fill the paint into containers of various sizes. Here, each order shall have a single-stage bill of material. Alternatively, a single manufacturing order can be released along with one *multiple-stage bill of material* to manufacture filled container of a particular capacity of a particular type of paint from the ingredients. The single order covering multiple manufacturing stages through one multiple-stage BOM reduces the amount of effort, time, paperwork, and reporting required to accomplish the production. A precondition for this, however, is that all the product from one stage should go to the same next stage. If branching takes place at any stage, that is the output of one stage goes to more than one processes then separate orders with corresponding BOMs have to be released for each stage. For example, a separate order will be required for manufacturing the bulk paint. The bulk paint will be taken into stock and issued back to manufacturing in required quantities for various packing processes.

Advantages of treating the whole of the manufacturing as a single stage:

(a) Intermediates need not be recognized as materials but as Work-In-Process (WIP).

(b) The manufacturing and packaging activities simply become operations on the composite routing for the whole process.

(c) It increases control and reduces possibility that stocks of intermediates, such as bulk paint, will be left unconsumed in the stores over their shelf lives.

7.4 TYPES OF BILLS OF MATERIAL

Following are the different types of bills of materials used in production and inventory management:

1. Single-Level Bill of Material.
2. Multilevel or Indented Bill of Material.
3. Planning Bills of Material [such as, a) Modular, b) Pseudo, c) Super and d) Kit BOMs].
4. Manufacturing Bill of Material (M-Bills).
5. Transient/Phantom/Blowthrough Bill of Material.
6. Summarized Bill of Material.
7. Matrix Bill of Material.
8. Costed Bill of Material.

Nomenclature Unfortunately, the BOM area suffers from a lack of agreed international definitions. The terms phantom bill, planning bill, pseudo bill, and modular bill are used in different ways by different authors.

7.5 SINGLE-LEVEL BILL OF MATERIAL

Single-level BOM is a display of components that are directly used in a parent item. It shows only the relationships one level down (Cox III and Blackstone Jr, 1998). Figure 7.2 shows how the structure of a multilevel end item (product X) can be represented using multiple single-level bills of material. With single-level bills stored in modern computer databases, it is a simple matter to bring together multiple bills to prepare a complete multilevel display or printout. The single-level

BOM supports order launching by providing the data for component-availability checking, allocation, and picking.

Advantages of Using Single-Level Bills

1. Duplication Due to Common Sub-assembly Usage is Avoided

If a complete, multilevel BOM were constructed for each and every product in a large product line, there would be considerable duplication. For example, one particular assembly might be used at different levels in a single end item or it might also be used in different end items. Thus when complete multilevel BOMs are used, that assembly together with its components and any further breakdowns would appear at number of places in a single BOM or appear in a number of different BOMs. However, the breakdown of the assembly would always be the same in every product. Hence, it makes more sense to create single-level BOMs for each product, assembly, and sub-assembly so that each needs to be retained in data files only once. No matter where the sub-assembly is used, the same BOM will be retrieved by the computer. Thus, bills of material are necessary to guide the MRP I explosion process and are usually represented in database files as single-level structures to avoid duplication where common usage is evident. Storing the same information in multilevel form would not only result in needless duplication but also in the increased probability of obtaining differing information for the same part. Computer creates multilevel screens and reports by simply chaining together the appropriate sequences of single-level BOMs (Oden *et al.* 1993).

2. The Size of BOM File is Reduced.

3. Maintaining BOMs is Simplified

For example, if there is a change in one commonly used component, the change needs to be made in only one place.

7.6 MULTILEVEL OR INDENTED BILL OF MATERIAL

It is a form of multilevel bill of material. Multilevel bills are formed as logical groupings of parts into sub-assemblies based on the way the product is assembled. One convention used with multilevel BOMs is that the last items on the tree are all purchased items. Generally, a BOM is not complete until all branches of the product structure tree end in a purchased part (Arnold, 1998).

A multilevel BOM can also be shown as an indented BOM. This bill uses indentations as a way of identifying parents from components. The components are linked to their parents by identifying them as sub-entries and by listing them immediately below the parent (Arnold, 1998). It exhibits the highest-level parents closest to the left margin, and all the components going into these parents are shown indented toward the right. All subsequent levels of components are indented farther to the right. If a component is used in more than one parent within a given

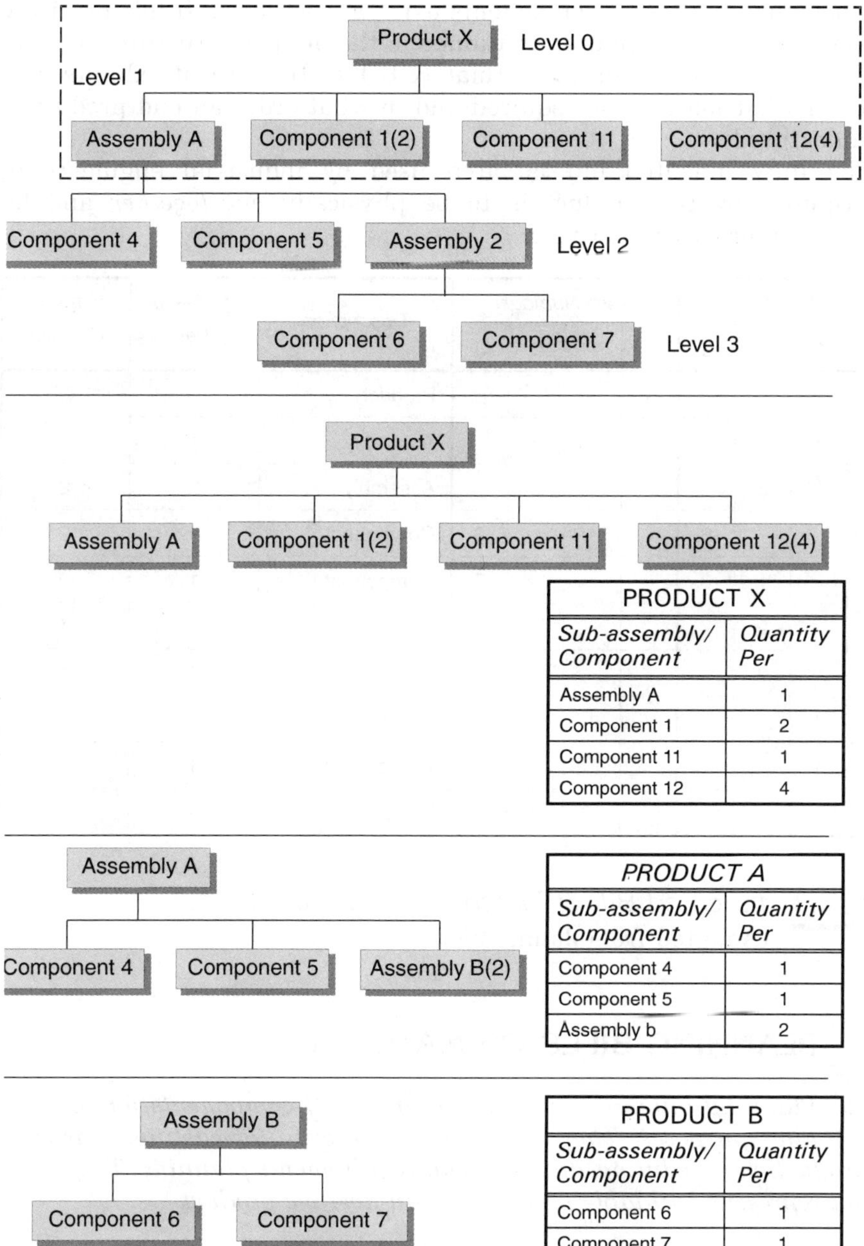

PRODUCT X	
Sub-assembly/ Component	*Quantity Per*
Assembly A	1
Component 1	2
Component 11	1
Component 12	4

PRODUCT A	
Sub-assembly/ Component	*Quantity Per*
Component 4	1
Component 5	1
Assembly b	2

PRODUCT B	
Sub-assembly/ Component	*Quantity Per*
Component 6	1
Component 7	1

Fig. 7.2 *Single-Level Bills of Material* (John, 1993)

product structure, it will appear more than once, under every sub-assembly in which it is used (Cox III and Blackstone Jr, 1998).

The indented BOM is the most convenient way to represent the BOM so that it is easy to retrieve for use in MRP I and in MPS computation. It shows concisely how many levels of components the end product has and how many different components exist on each level. It has precisely the same type of information contained in the product structure tree and in the cross-classification chart (matrix BOM). In short, it exhibits how much of what material is required and in what order an end product is manufactured.

The fully indented bill is often used by industrial engineers to determine how the product is to be physically *put together* and by accounting for cost implosions.

Level				*Part Number*				*Description*	*Quantity Per*	*Total Quantity*
0	1	2	3	0	1	2	3			
0				X				Product		
	1				A			Assembly	1	1
		2				B		Assembly	2	2
			3				6	Component	1	2
			3				7	Component	1	2
		2				4		Component	1	1
		2				5		Component	1	1
	1				1			Component	2	2
	1				11			Component	1	1
	1				12			Component	4	4

Fig. 7.3 *Indented Bill of Material for Product X*
(Adapted from John, 1993)

7.7 PLANNING BILLS OF MATERIAL

Planning bills are artificial groupings of components for planning purposes. They are used to simplify forecasting, master production scheduling, and material requirements planning. They do not represent buildable product but an average product

Arnold, 1998

The bill of material is normally a document written and maintained by design engineering. The structure of the bill is based on the perspective of the design engineer and is *after the fact* in that it is defining a product that has been designed. It can, therefore, be assumed that the BOM is not needed to design a product, but it is required to manufacture the product. Based on the needs of the manufacturing operation, it may require restructuring the BOM while not changing the product itself (Toomey, 1996). The restructuring of BOM required to better perform MPS activities has led many people to see the techniques and discipline of BOM approaches as having many applications. An example is the planning BOM, which is any use of BOM approaches for planning only, as opposed to use for building the products.

The Traditional Approach to Product Structure

The traditional way of viewing product structure is that the bill of material corresponds to the sequential build process or to how many lower-level items go into each higher level item. Following are some of the aspects of traditional product structure (Stonebraker, 1994):

- Component linkages are defined as 'gozinto,' or part list relationships.
- These linkages primarily facilitate engineering and manufacturing.
- The objectives are inventory planning and operations scheduling.
- The approach is most effective for stable environments, limited competition, and no demand pressure.

The Planning Approach to Product Structure

The traditional approach to product structuring is from an engineer's point of view, that is, the way the product is *built.* The key change required to achieve superior master production scheduling is to include BOM structures based on the way the product is *sold.* In this way, the BOM can support some critical planning and management activities. Planning methods of product structuring allow planners to meet other operations and non-operations objectives. In addition to accomplishing the traditional objectives of inventory control and operations scheduling, planning methods of product structuring also accomplish some or all of the following objectives (Stonebraker, 1994):

1. More accurate forecasting.
2. A reduced number of end items to be master scheduled.
3. A more accurate scheduling and planning process.
4. Simplified order entry.
5. Two-level scheduling.
6. A more accurate method of product costing; and
7. A more efficient and flexible data storage and maintenance system.

Planning or Pseudo BOM

It is defined by APICS Dictionary as an artificial grouping of items or events in bill-of-material format, used to facilitate master scheduling and master planning (Cox III and Blackstone Jr, 1998). A planning bill is an artificial grouping of components under a non-engineering part number to streamline planning for materials. It does not ordinarily reflect the way the product is actually built. It represents a group of parts it is convenient to plan together, even though they are never assembled into a recognizable sub-assembly. It is most often done to reduce the number of items that must be forecast and master scheduled. Planning bills are referred to as 'pseudo bills'—false or artificial bills. They cannot be used to actually build any configuration of the product. They are used for planning purposes only. To build the product, manufacturing must use an actual BOM and process instructions created by engineering.

Variations of Planning Bills

The growing interest in planning bills has led to the development of several types of planning bills. These bills are most often used in conjunction with a process called modularization of BOM data. Three variations of planning bills, namely modular BOM, super BOM, and kit BOM, are described here.

7.7.1 Modular Bill of Material

Modular BOM is a type of planning bill that is artificially arranged according to product modules or options (Stonebraker, 1994). It groups sub-assemblies and parts based on whether they are unique to a specific product option or common to all configurations of the product. Each group is called a module. It might be considered a major sub-assembly. It is often used in companies where the product has many optional features, for example, assemble-to-order (ATO) companies such as automobile manufacturers. Since there are many options and combinations of final end-product configurations in such environments, it becomes impractical to forecast and master-schedule what might amount to thousands of possible combinations. Hence, instead of end items, major sub-assemblies or options are master-scheduled independently and then combined into final products later using a final assembly schedule (FAS). When this is the case, the bill of material needs to be rearranged, or modularized, to accommodate the correct planning of components. Thus, basically, a modular BOM provides a company with a solution for dealing with many product options that create multiple end items.

The modular BOM structure has an architecture that links component parts to options, but it does not link either options or components to end-item configurations. If the options are simply buildable (assemblable) sub-assemblies, then all that is required for the new architecture is to treat the sub-assemblies as end items; that is, designate them as level zero, instead of level one.

In most cases, however, the options are not stated as buildable sub-assemblies but as options that provide some services to the customer, for example, the air-conditioning option for an automobile. The single-level BOM would show this option or module as consisting of a particular radiator, fan, hoses, compressor, and interior knobs and levers. These items are not, however, assembled together. They are assembled with still other parts as sub-assemblies, which eventually are assembled into the automobile. The use of the air-conditioning option as a BOM will pass demand from the customer who wants this option down to the necessary parts. It can also be used to forecast demand for air-conditioners. However, this BOM is not useful in the physical building process for air-conditioners. For example, the air conditioning knobs are planned by the BOM for the air-conditioning option. They are not planned by the bill of dashboard assembly where they are installed. Thus, the industrial engineer needs other means to say how the dashboard is to be assembled and from what components.

The MPS is stated in terms in which it is sold, rather than in terms in which it is built; that is, air-conditioning, two doors, automatic transmission, fancy trim and so on. This approach of stating MPS in terms in which it is sold is compatible with marketing perceptions of models, options, and trends in options (for example, more people buying cars with air-conditioning), which tends to improve forecasting (Vollmann *et al.* 1997).

- Modular bills are defined only once but may be incorporated into other bills as often as required.
- The modularization process begins by grouping parts in a product by the option they represent. Then a BOM is developed for each option, which provides the advantages of fewer bills of material, less maintenance, effective order entry, and a means of forecasting the options (Garwood and Rice, 1996).
- The modularization process, in effect, removes the end item itself from the bill of material and promotes level one sub-assemblies and components to level zero items. Often fake level-zero assembly part numbers are assigned, for planning purposes only, to show parts or assemblies that are common to all possible end items and for each option grouping.
- Eventually, of course, the actual end product to be delivered must be defined. Once the individual customer order (representing a unique collection of options) is entered, it serves the function of a one-time, unique single-level BOM, that is, it specifies which options or modules are to be included for the particular customer order. It is controlled by a separate Final Assembly Schedule (FAS). The FAS would use a special form of bill, sometimes called an M-bill or manufacturing bill. Thus, when a product BOM is being modularized, each assembly, sub-assembly, or module is assigned either:

(a) to the MPS, by making it part of the planning BOM, or
(b) to the FAS by putting it in the manufacturing BOM when needed for execution.

- A sub-assembly promoted to end-item status, as a result of modularization, must appear in the MPS in the proper time-bucket, earlier than the level-0 products they replace would have appeared, if necessary to allow time for final assembly of the finished product. Each time an item's original bill is modularized and it is transferred to the manufacturing BOM file, such lead time must be recognized. Final assembly and most sub-assembly lead times are usually short and require only small, if any, time shifts in MPS (Plossl, 1994).

7.7.2 Super Bill of Material (S-Bill, Top-Level Planning Bill or Master BOM)

There are three different types of super bills:

(a) Super Company Bill of Material,
(b) Super Family Bill of Material, and
(c) Super Modular Bill of Material.

(a) Super Company Bill of Material

Figure 7.4 shows a type of super bill that describes the individual products manufactured by a company. The bill is used to disaggregate the

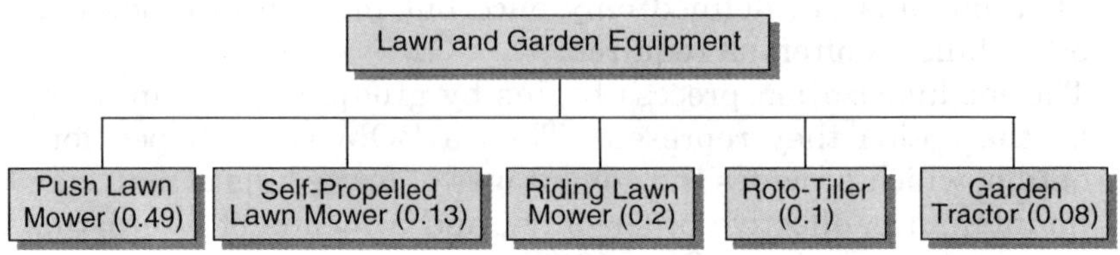

Fig. 7.4 *Super Company Bill of Material for a Lawn and Garden Equipment Manufacturer*
(Adapted from Oden et al. 1993)

total forecast quantity for a group of similar products into forecast for individual products. The master scheduler enters total demand for the super (and pseudo) assembly into the top level of the MPS. The MPS system then explodes the top level, using the fractions in the super bill, calculating production schedules for the actual end products in the lower level of the MPS.

(b) Super Family Bill of Material

To improve the accuracy of demand forecasting, many companies form families of products with similar demand patterns. The aggregate (family) forecast is normally much more accurate than the forecasts for any one individual product. To use the aggregate forecast they must develop a super family BOM as shown in Fig. 7.5, consisting of the family (pseudo) assembly as the parent and the individual family end products as the children.

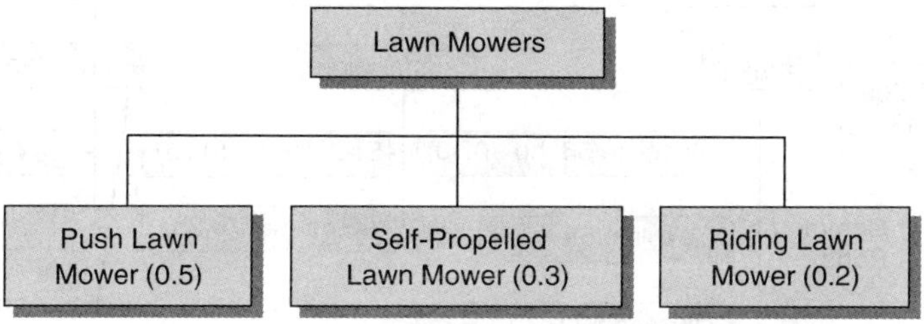

Fig. 7.5 *Super Family Bill for a Manufacturer of Different Models of Lawnmowers* (Adapted from Oden et al. 1993)

Super Modular Bill of Material

The most widely used planning BOM is the *super modular bill of material* also called *planning bill of material.* Super modular BOM is a type of planning bill, located at the top level in the structure, that ties together various modular bills (and possibly a common parts bill) to define an entire product or product family. In it, the parent is an *unbuildable* group of modules that is used for planning purposes only. The children are the modules that can appear in a final product. A super bill shows all the modules associated with each individual option by assigning an artificial parent identity to the option. The quantity per relationship of the super bill to its modules represents the forecasted percentage (or decimal values) of demand of each module. The master-scheduled quantities of the super bill explode to create requirements for the modules that also are master scheduled (Cox III and Blackstone Jr, 1998). Thus, the super bill describes the related options or modules that make up the *average* end item. Figure 7.6 shows a part of the super modular bill of material for the tractor family described in chapter one.

The major difference between a super modular bill and a modular bill is that whereas in a modular bill, the level zero end items are removed (because there could be so many combinations), and level one sub-assemblies and components are promoted to level zero, in a super bill an artificial part number is assigned to a standard representation of the group, or family, of possible end items. The components in the super bill constitute the option and common part groupings. The usage quantity for these 'components' would be a decimal representing the forecasted percentage of the total expected to be sold. Thus, the master scheduled demand for each option and common part grouping is calculated using normal MRP I logic based on the much-easier-to-forecast super bill product number.

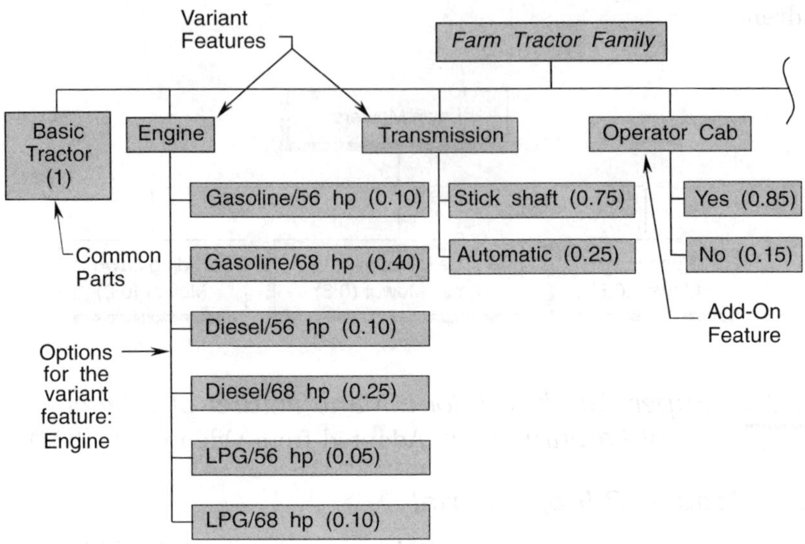

Fig. 7.6 *Planning, Super Modular, or Generic Bill of Material with no Options 'Overplanning' (Only a Part of the Bill is Shown)*

- The bottomline in planning bill accuracy is that sales and marketing must be accountable for creating and maintaining the planning bill percentages (Proud, 1994).
- Once the percentages are obtained, they are entered into the option-planning bill. But should they add up to 100 per cent? If the options are required (that is, it is a variant), the answer would be yes, however to have a safety stock of options, sometimes overplanning is used, in such cases, the total may be greater than 100 per cent. But if the options are 'add-ons,' then the percentages or probabilities of sales may be less than 100 per cent.
- As more options are added, the complexity of the end-item family increases, creating a more difficult situation for manufacturing and the master scheduler. One technique for managing this complexity is to group common parts (that is group all those parts that are always needed); and list them together on a common parts bill.
- The common parts bill is also a *pseudo*, since nothing can be build with just the common parts without the variants. The probability that common parts would be needed is 100 per cent.

7.7.3 Kit Bills (Kit Number or K-Bills) or Common Parts Bills

Many times, to build a product a large number of small loose parts such as nuts, bolts, washers, and other fasteners are frequently used. Rather than identifying each of the items individually, they are grouped under an artificial part number and name called 'kit.' Thus, with one name/number an entire group of components can be scheduled.

Sometimes, a sub-assembly or group of parts are commonly used together and appear in many assemblies. This group may consist of parts used together when a product option is ordered. The size of the BOM database can be reduced with a corresponding ease of BOM maintenance if the groups of common parts are identified with a part number, even if the components in the kit are never actually assembled together as a single sub-assembly. This is the 'bag of parts' concept. For example, if the same seal kit consisting of ten parts was used in each of 1000 unique diaphragms, a part number could be assigned for the seal kit and be shown as a single item on each of 1,000 BOMs for diaphragms. If a change were made in one of the common parts in the seal kit, only one bill would need to be changed—not each of the 1000 bills.

The common parts bill, or kit bill, groups common components for a product or family into one bill of material or one section of the bill of material. It is useful where there are many small, loose parts on level one in the product structure. Usually, the use of this type of planning bill occurs when several small component parts such as fasteners, nuts, bolts, other hardware and packaging materials used in assembling major product units are required in the same combinations, yet are called for in different structures. This is commonly the case when level 1 items are promoted to level 0 status in modularization. MRP I users find it impractical to deal with such items individually on the MPS level. This collection of logically similar parts can be put in an imaginary bag and treated as sub-assembly to which a non-engineering part number is assigned. Thus, individual nuts, bolts, etc., are not treated individually, the fundamental planning unit being the imaginary 'bag' of items.

It is convenient, from a planning standpoint, to treat such items as components of an imaginary assembly and group them under one 'pseudo item number.' This is called a *kit number*, or more often a *K-bill* as shown in Fig. 7.7. It is as though the small components are combined in an imaginary bag, with a part number to identify it, and then treated as an end item to be forecasted and master scheduled as a single unit. In this way, requirements for the items are processed only once (for the group), rather than for each individual item. K-bills reduce paperwork, processing time, and file space required in generating orders for small, inexpensive items that are usually ordered infrequently in large quantities (Russell and Taylor III, 1998).

- Kitting is a commonly used adaptation of the common parts bill of material. It is represented by the same diagram as the common

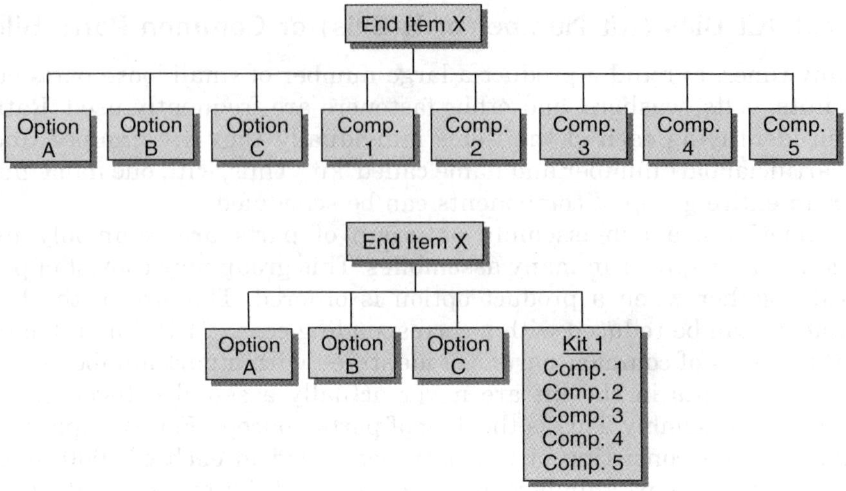

Fig. 7.7 *The Kit Bill*

parts bill (Stonebraker, 1994). The difference between common parts bills and Kit Bills is that whereas the former groups common parts that might include both major and minor components, the latter groups only small items like hardware and packaging materials.

- If the kit (that is the 'bag of parts') never really exists *physically* but exists only as a *logical* kit, the parent part number is coded as a phantom. If the kit physically exists and is inventoried, it's treated like any other scheduled part number and is not coded as a phantom.
- The objective with K-bills and S-bills is the same: to assign a single new identity code to sets of items that constitute a logical planning group and, employing the format of a BOM, to relate such items to an MPS requirements (Plossl, 1994).
- The planning BOMs like modular, super, and kit bills are intended for use only in the planning phase to improve forecasting, master-scheduling, and MRP I. They represent a superstructure in the BOM file that must be maintained along with the rest of this file. This increases the difficulty, labor, and cost of BOM file maintenance, but the extra effort and cost is very much worthwhile when they are used properly (Plossl, 1994).

7.8 MANUFACTURING BILLS OF MATERIAL (M-Bills)

In assemble-to-order environments, level 0 end items, that is the products that are sold to the customers, are rarely used in master scheduling. Instead, particularly for complex products with options, level 1 and level 2 modules are established, associated with individual options, and promoted to end-item status in MPS. However, BOM for level 0 products and other sub-assemblies excluded from planning BOM are needed in execution. These end items have to be ordered, scheduled, and assembled, and requisitions must be issued for their components. The end items are needed for the production control system so that the sales department can place orders, the industrial engineering department can use the information for standards and product costing, and the production department can schedule manufacturing properly. Therefore, we should keep their identity for manufacturing purposes.

Manufacturing BOMs, called *M-bills* meet these needs. An M-bill item can be a component only of another M-bill item or of an end product. Since the components of the M-bill are included in the planning bill, it is not necessary for the computer to access M-bills for generating material requirements. The M-bills are coded in such a way that the MRP I system bypasses them. Since the M-bills are not used by MRP I, they are segregated in a separate M-bill file and are used for final assembly, ordering, scheduling, and costing only.

7.9 PHANTOM, BLOWTHROUGH, OR TRANSIENT BILL OF MATERIAL

A phantom bill is a bill of material for an item which can be made from its components and which can be stocked, but in normal circumstances is not so made or stocked. It is a BOM coding and structuring technique used primarily for transient (nonstocked) sub-assemblies. A phantom BOM represents an item that is physically built but rarely stocked before being used in the next step or level of manufacturing. It is required in situations where several parts are put together into a sub-assembly and immediately passed along to the next stage of manufacture. These sub-assemblies can either be built and placed in stock or remain on the shop floor for immediate consumption by a parent assembly. In the latter case, the sub-assembly is commonly referred to as a 'phantom' assembly—that is, an assembly that is real, but not planned to be stocked. A phantom has a BOM just like a normal part but MRP I does not plan a phantom. It 'blows through' the requirements to the next level down until it finds a real part, that is MRP I ignores a phantom item and moves straight to its components.

The BOM for the parent includes the phantom as a child, indicating that it is a phantom. When the parent is being planned, the order-release program checks to determine if any units of the phantom are in stock. If so, the computer prints a pick list so they will be withdrawn and used. If not, the order-release system skips the phantom and directly requisitions the phantom's components. The process of skipping the phantom and calling out its components is called *blowing through.* Stocks can be recorded against a phantom should it arise for any reason. The MRP I system should take it into account and use the stock to find the net requirement of the phantom item before planning the requirements for its components. *MRP I will never plan to create stocks of phantom parts.* A phantom bill is real, in that the item exists but it is transparent to the inventory planning system.

7.9.1 Need for Phantom Assemblies

A sub-assembly that is assembled on a feeder assembly line and consumed within a few minutes or hours in the final assembly is an example of a transient sub-assembly. If the transient sub-assembly is not coded as a phantom, it has to be given a separate identity, and hence a separate part number, in the BOM, and MRP I will treat it the same as any other sub-assembly moving into and out of stock. This will require all receipts and disbursements to be reported for this assembly. Since this assembly does not *usually* go into stock—it is simply passed on to the next work center—artificial transactions will have to be entered to update time-phased inventory records. Such transactions, unrelated to actual movement of materials, are difficult to record accurately using conventional means. The question then becomes whether such a sub-assembly needs to be identified separately with a unique part number. The answer is 'no' as long as there are no service part requirements, customer returns, or overruns of the sub-assembly. The reason is that the sub-assemblies that are immediately consumed do not require a part number and thus a level in the bills. With fewer bills and fewer levels in the bills, BOM maintenance costs are reduced, and engineering changes are easier to manage and faster to implement. Each level in the bill creates paperwork, extends the lead-time, and increases the work necessary to maintain the bill. Hence, the sub-assembly components can simply be shown as additional components of the parent assembly.

However, it is sometimes necessary and appropriate to assign a part number to a transient sub-assembly (process industries often call these transient items *intermediates*); and put them in inventory. Following are the cases when transient assemblies have to be stored and hence assigned a part number:

1. An imbalance in inventory may occur due to overruns of the assembly or shortage of mating parts downstream due to quality rejections or some other causes. This will result in an excess quantity of the sub-assembly. In such situations, the excess quantity of the transient sub-assemblies may be left in the feeder assembly area at the end of a final assembly run. In such cases, the best thing to do is to return the sub-assemblies to the stockroom and inventory them, so the next time the end items are planned by MPS or MRP I, the sub-assemblies will be used. In order for that to happen, the sub-assemblies must be on the BOM; a part number must be assigned to them. This part should, however, be coded as a phantom so that MRP I can blow through it when sub-assemblies are not in stock.

2. Similarly, when a customer returns a product, it is disassembled and all good sub-assemblies and components are returned to the stockroom. If a transient assembly is sent to the store, a part number must be assigned to the sub-assembly to account for the inventory and consider its availability in replanning future inventory requirement for its components.
3. If a component part, apart from being used in the end item or end items manufactured by the company, is also planned, scheduled, and/or sold as a service part then it must obviously have a separate and unique part number.

In summary, where one of those conditions namely overruns, customer returns, or service part requirements occur, there these sub-assemblies must be identified separately in BOMs and their inventory status maintained. This means an artificial or dummy transaction has to be entered to add the quantity assembled to the on-hand quantity each time this item is assembled. Similarly, a dummy transaction has to be entered to 'issue' these assemblies to the line; although, in reality, this assembly is simply passed on to the next workstation. Obviously, these artificial transactions, although essential for the functioning of the MRP I system, are undesirable as they serve no productive purpose. An efficient system should treat the phantom assembly as a *real* part for the netting purposes when there is a positive on-hand balance quantity; but as soon as the on-hand quantity is used up, or, there is no on-hand quantity, then it should blow through the phantom assembly directly to its components to avoid entering of unnecessary dummy transactions and to save processing time. The technique used to achieve this is to bypass phantoms during an MRP I run when the on-hand balance is zero; that is, in this situation, we treat the parent assembly X to have the simplified structure as shown in Fig. 7.8(b). When the on-hand balance is greater than zero, the gross requirements should be netted against the inventory and then the net requirements exploded to components; that is, in this situation, we consider assembly X to have the actual product structure shown in Fig. 7.8(a). Thus, the on-hand balance quantity of the phantom assembly acts as a toggle to choose different treatments for the phantom assembly. Following are the guidelines to decide whether assembly P in Fig. 7.8(a) should be coded as phantom or not:

1. If assembly P is *always* taken into the stock, it should *not* be coded as phantom but should be made a regular assembly with a unique separate part number.
2. If assembly P is *never* taken into stock, then again it should not be coded as phantom; instead, assembly P should be ignored completely for the planning purposes. The assembly X should be regarded as having the simplified BOM structure as shown in Fig. 7.8(b).

3. If assembly P is *usually not* taken into stock, but *occasionally* may have to be taken into stock, then there are two options. First is to treat assembly P as a regular assembly as in 1 above. However, in this case, most of the time the dummy transactions will have to be entered unnecessarily. Second, and more efficient, option is to code assembly P as a phantom and use up the on-hand balance, if any, during netting process and still avoid the dummy transactions.

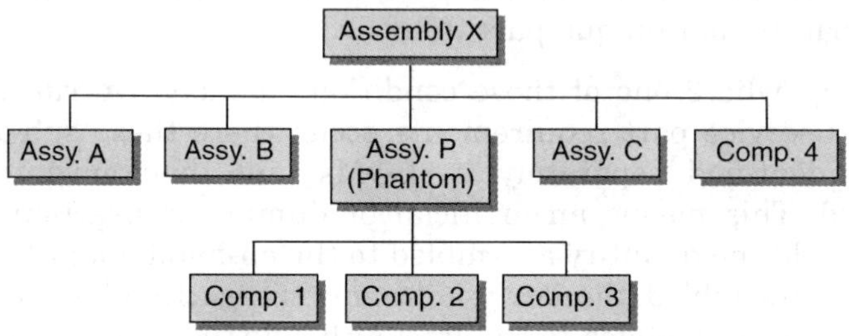

(a) The Actual or Manufacturing Bill of Material with Assembly Coded as Phantom

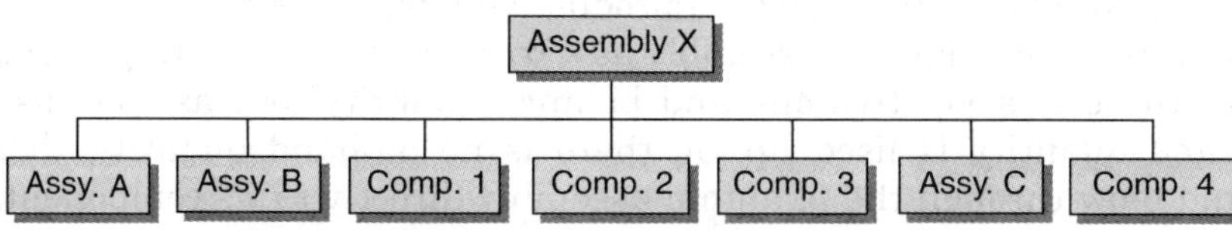

(b) The Simplified Bill of Material for Planning and Reporting Purposes

Fig. 7.8(a) and (b) *Phantom Bill of Material*

With the technique employing 'Phantom Bills,' transactions related to work-in-process orders of these assemblies do not have to be reported and posted. However, transactions involving stock room receipts and issues of these assemblies are reported and posted.

Thus, phantom or transient assemblies are usually not stocked. When these are not in stock, the MRP I software blows through this level bypassing the phantom item's record as though the phantom assembly does not exist and goes from its parent to its component directly. Occasionally they might have to be stocked because of overruns etc. In such situations, MRP I, using this technique, picks up and uses any transient sub-assemblies on hand, that is, before exploding the phantom assembly into its components, netting is done, and explosion process calculates the requirements only for the netted quantity. Service-part requirements also can be entered into the record and will be correctly handled. Such sub-assemblies are referred to as *transient, phantom, or blow-through assemblies.*

7.9.2 Ensuring the Equivalence of the Actual and Simplified BOMs

Since, for the parent assembly with a phantom, the MRP I logic might use either of the BOMs—actual or simplified—depending upon whether or not there is on-hand quantity of it, it is absolutely essential that these two BOMs should be equivalent *from the point of view of production planning.* They certainly are not, and need not be, equivalent from the manufacturing and assembling point of view. There are two requirements for the equivalence of the actual and the simplified bills:

1. The Total Lead-Time between the Parent X and the Components of the Phantom should be Equal in Both the Cases This can be achieved by applying the total lead time from the components of the phantom to the parent X (that is t_x) to the parent X itself and making the lead-time of the phantom equal to zero.

2. The Exploded Quantities of Components in Both the Cases should be Equal This can be ensured by making the lot-sizing policy for the phantom assembly as lot-for-lot; and multiplying the usage quantity of the components by the usage quantity of the phantom. This is illustrated in Fig. 7.9. It can be seen that in both the bills, the total quantity of the component Y will be equal to *(quantity of* $X \times a \times b$*).* Also, it is quite obvious that if a lot-sizing rule other than lot-for-lot is applied to the phantom, this equivalence cannot be assured.

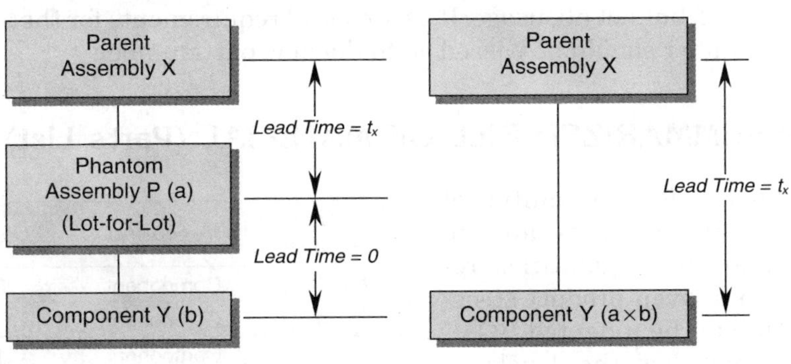

Fig. 7.9 *Equivalent of the Actual and Simplified Bills*

Thus, the technique used for the phantom assemblies is to give them a special code in the BOM or item data record, assume *no* (that is *zero*) *lead-time,* and perform *no lot-sizing* (that is use *lot-for-lot* ordering policy). In this way, the gross requirements for the sub-assembly simply pass through to become gross requirements for its components in corresponding time periods. This permits MRP I logic to drive requirements straight through the phantom item to its components, but the MRP I system usually retains its ability to net against any occasional inventories of the item.

7.9.3 Use of Phantoms to Phase-out Old Items and Phase-in New Items while Implementing Engineering Changes

The simplest method to phase out the old component while implementing an engineering change on 'when-existing-stock-of-the-old-item-gone' basis, is by making the new part a component of the old part and coding the old part as a phantom with a lead time equal to zero and lot-for-lot ordering policy as shown in Fig. 7.10. As long as there is a stock of the old item, the old item is planned and scheduled by MRP I. When the on-hand quantity of the item reaches zero, MRP I blows through it to the component (the new part) and begins planning and scheduling the new part. This method, however, has following disadvantages:

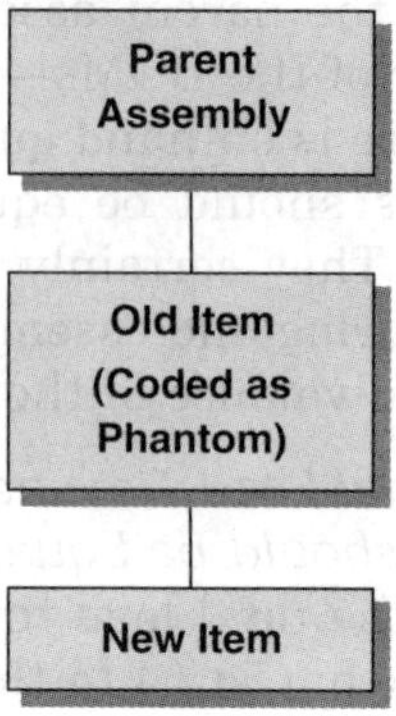

Fig. 7.10 *Phasing Out the Old Item Using Phantom*

1. BOM does not show how the part is actually made.
2. Costing of the parent item will include costs of both the new as well as the old part.
3. This method cannot be used where a component is being replaced in some, but not all, usage. It assumes all requirements for the old part number should be passed on to the new part number.

7.10 SUMMARIZED BILL OF MATERIAL (Parts List)

It is a form of multilevel BOM that lists all the parts and their quantities required in a given product structure. Unlike the indented BOM, it does not list the levels of manufacture and lists a component only once for the total quantity used (Cox III and Blackstone Jr, 1998). A summarized BOM is used for quick reference to the total number of components used in an assembly or product without showing the indented breakdowns. The summarized BOM is prepared

Part number	*Descrip- Tion*	*Usage*
1	Component	2
4	Component	1
5	Component	1
6	Component	2
7	Component	2
11	Component	1
12	Component	4
A	Assembly	1
B	Assembly	2

Fig. 7.11 *Summarized Bill of Material for Product X* (John, 1993)

by product design engineer and does not contain any information about the way the product is made or assembled. Figure 7.11 shows a summarized BOM for product X.

7.11 MATRIX BILL OF MATERIAL OR THE CROSS-CLASSIFICATION CHART

It is a chart made up from the BOMs for a number of products in the same or similar families. It is arranged in a matrix with components in columns and parents in rows (or vice versa) so that requirements for common components can be summarized conveniently (Cox III and Blackstone Jr, 1998).

The matrix BOM, also known as the cross-classification chart, is the earliest and simplest method of representing the relationships among parts. It exhibits the sub-assemblies, parts, and raw materials that are used in each of the primary products. For example:

- Row 1 in Fig. 7.12 represents product 1, which requires one unit of sub-assembly 4 and one unit of sub-assembly 6.
- One unit of sub-assembly 4, in turn, requires one unit of part 9 and two units of part 10.
- One unit of part 9 needs two units of raw material 12 and one unit of raw material 13.

Thus, the cross-classification chart (Fig. 7.12) shows the complete explosion of a product line (Narasimhan *et al.* 1995).

Item		*Sub-assembly (SA)*				*Part (P)*				*Raw Material (RM)*			
Type	*No.*	4	5	6	7	8	9	10	11	12	13	14	15
Finished Product	1	1		1									
	2		2										
	3		1		2								
Sub-assembly (SA)	4						1	2					
	5					3							
	6					1		1					
Part (P)	7						2		2				
	8									1			
	9									2	1		
	10										1	1	
	11										2		3

Fig. 7.12 *Matrix BOM or Cross-Classification Chart*
(Adapted from Narasimhan *et al.* 1995)

7.12 COSTED BILL OF MATERIAL

It is a form of BOM that extends the quantity-per of every component in the bill by the cost of the components (Cox III and Blackstone Jr, 1998).

7.13 FUNCTIONS OF BILLS OF MATERIAL

1. *Product Definition: Provides an Enterprisewide Common Description* The bill provides critical information on how to build a product. Engineering uses BOM as a means for setting up, maintaining, and communicating the information about components/ingredients contained in a company's product.

2. *Engineering Change Control* Implementation of engineering changes must consider phase out of the old product, introduction of the new, and could possibly include announcement to the marketplace. BOM provides the method for recording and controlling engineering changes. BOMs can be used to identify the items and components potentially affected by an engineering change.

3. *Lends Itself to Options Forecasting* Planning BOMs help improve forecasting of optional product features.

4. *Basis for Configuration Control during Order Entry* BOMs make order entry simple and easy by expressing customer orders either in model numbers or as a configuration of optional features. When a configurator is used to enter customer orders, sales order processing requires the BOM to properly configure products to customer-specified features and options.

5. *Basis for Stating the MPS in Fewest Numbers* BOMs help in stating the master production schedule (MPS) in the fewest possible number of items. These items will be products, major sub-assemblies, or standard end-item modules.

6. *Basis for MRP I Explosion Logic* Material planning, through the use of MRP I, depends on product structure records to plan and schedule items to maintain the integrity of the MPS. BOMs help in planning sub-assembly and component priorities so orders for them are released at the right time, with valid due dates kept up-to-date.

7. *Used for Planning and Scheduling* Because the bill tells the user the order in which the various components are needed (information found in the lead-time estimates), the information can help determine how to schedule the components supplied by the plants and outside suppliers. Assembly scheduling, including all components required to build parent items, is performed using BOM (Schlussel, 1995).

8. *Used for Material Availability Checking, Backflushing, and for Creating Pick Lists* The stockroom and inventory control relies on the BOM to facilitate backflushing or issuing material to support work orders for products (Schlussel, 1995).

9. *Basis for Preparing Routings and Capacity Planning* BOMs specify how products are built, including added sub-assemblies and semi-finished items for economical production and for replacement parts. Manufacturing engineers need BOM to develop routings to manufacture the product. Shop production departments use the BOM to build the product and ensure its conformance to design specifications. BOM plays a critical role in determining the impact of the new product on existing resources. Together with routings and work center data, it can be used to construct a *bill of resources, bill of labor,* etc., which can then help plan capacity requirements (Schlussel, 1995).

10. *Basis for Product Costing* Product cost is usually broken down into direct material, direct labor, and overhead. BOM provides not only a method of determining direct material but also a structure for recording direct labor and distributing overheads (Arnold, 1998). The BOM can be used to determine the cost of the end product by starting at the bottom of the bill, rolling up, and totaling the costs associated with each component. Accounting depends on the BOM for cost roll-ups for product costing, product pricing, and the calculation of work-in-process material variances (Wassweiler, 1996).

11. *Facilitates Efficient Computer File Storage and File Maintenance.*

12. *Facilitates Spare Parts Planning* The service department needs the BOM to determine which replacement parts to order when a customer complains about a product breakdown.

13. *Forms Part of Documentation Submitted to Regulatory Authorities* Many times the product definition must be approved by regulatory agencies such as standard setting boards, independent testing laboratories, or governmental agencies. BOM forms a key part of the documentation submitted to these agencies for approval before the product can be sold on the open market.

14. *Forms a Part of Defense Against Liability Suits* As bills of material define the product they have to be used as defense against liability suits or warranty claims.

15. *Can be Used to Trace Significant Escalation or Reduction in Warranty Claims to a Particular Engineering Change* Significant escalation in warranty claims for a product should be traceable directly to a revision level in the bills of material. The changes made at that time can be reevaluated to determine the problem and find a solution. Similarly, if a change has been implemented to reduce warranty costs, its success can

also be checked through the bill of material history file and claims by serial or model number from the field.

7.14 WHERE-USED BILLS OF MATERIAL

Another possible way in which the bill of material data might be displayed on a computer screen or in a printed report is the where-used format. It is a listing of every parent item that calls for a given component, and the respective quantities required, from a BOM file. It is thus a 'BOM in reverse' and indicates for each component part number, the part numbers of all sub-assemblies and/or end items on which the component part is used.

The process of determining the where-used relationship for a given component is called 'implosion.' Implosion can be single-level—showing only the parents on the next higher level—or multilevel—showing the ultimate top-level parent (Cox III and Blackstone Jr, 1998).

Pegging Report

The pegging report is similar to a where-used report, however, it shows only those parents for which there is an existing requirement (that is, parents which are in the MPS in given periods) whereas the where-used report shows *all* parents for a component. The pegging report shows the parents creating the demand for the components, the quantities needed, and when they are needed as shown in Fig. 7.13. Pegging keeps track of the origin of the demand (Arnold, 1998).

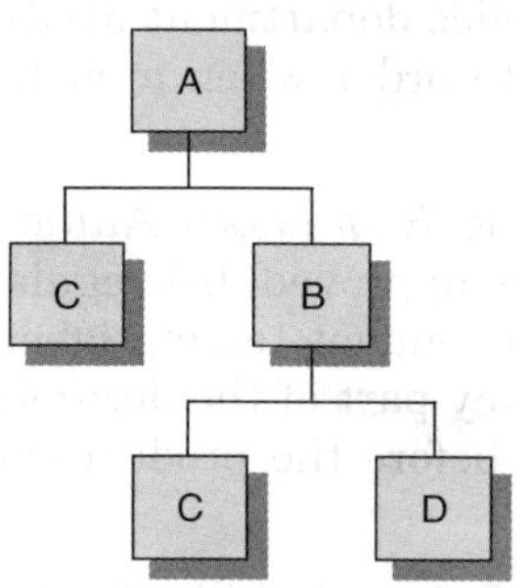

Pegged Requirements

Item	*Week*				
No.	1	2	3	4	5
C	50	124	25	50	150

Sources Requirements

A	1	2	3	4	5
C		100			100

Fig. 7.13 *Pegged Requirements* (After Arnold, 1998)

Although where-used formats are not directly used in the MRP II process, they are very useful in maintenance of the bills, and are used by purchasing, production control, and engineering in various decision-making situations, such as in implementing an engineering change, or when materials are scarce, or in costing a product. Where-used report, for example, can be used when considering an engineering change to an item;

all the parents that utilize this item must be evaluated to see if the change can apply in all places or only in some of them. The where-used bills can be displayed as single-level, indented, or summarized. Figure 7.14 depicts a Bill of Material tree structure.

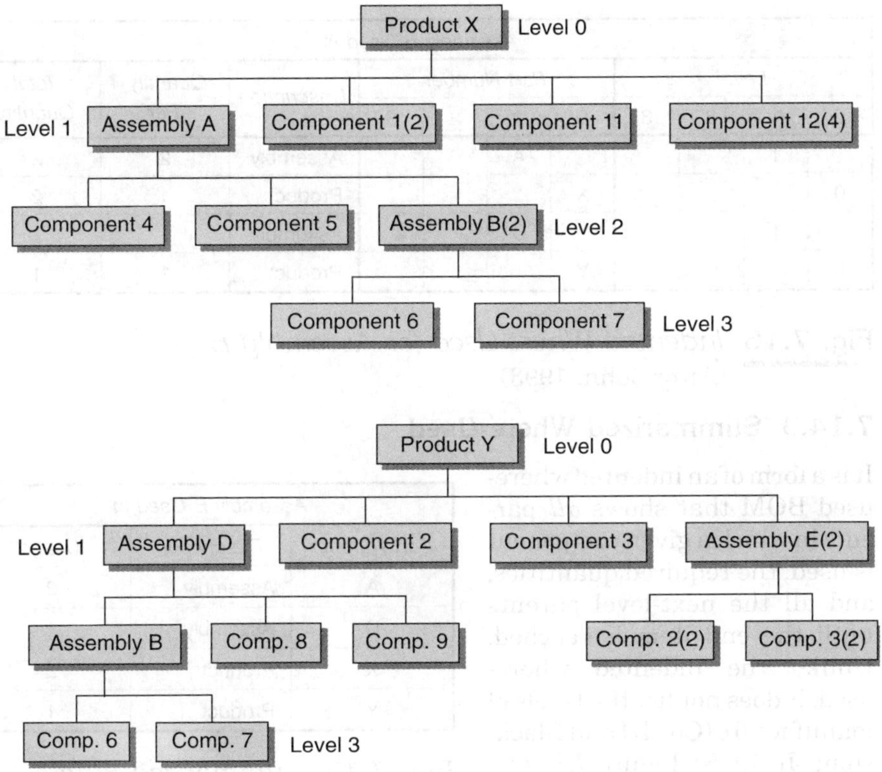

Fig. 7.14 *Bill of Material Tree Structure for Product X and Y* (After John, 1993)

7.14.1 Single-Level Where-Used (for a Component)

It lists each parent in which a particular component is directly used and in what quantity as Fig. 7.15 clearly shows. This information is usually made available through the technique known as implosion (Cox III and Blackstone Jr, 1998).

Assembly B Used in		
Part Number	*Description*	*Quantity Per*
A	Assembly	2
D	Assembly	1

Fig. 7.15 *Single-level Where-Used for Assembly*

7.14.2 Indented Where-Used

It is a listing of every parent item, and the respective quantities required, as well as each of their respective parent items, continuing until the ultimate end item or level 0 item is referenced. Each of these parent items calls for a given component item in a BOM file. The component item is shown closest to the left margin of the listing, with each parent indented to the right, and each of their respective parents indented even further to the right (Cox III and Blackstone Jr, 1998) as shown in Fig. 7.16.

Assembly B Used in										
Level				*Part Number*				*Description*	*Quantity-Per*	*Total Quantity*
0	1	2	3	0	1	2	3			
	1				A			Assembly	2	2
0				X				Product	1	2
	1				D			Assembly	1	1
0				Y				Product	1	1

Fig. 7.16 *Indented Where-Used for Assembly B* (After John, 1993)

7.14.3 Summarized Where-Used

It is a form of an indented where-used BOM that shows *all* parents in which a given component is used, the required quantities, and all the next-level parents until the end item is reached. Unlike the indented where-used, it does not list the levels of manufacture (Cox III and Blackstone Jr, 1998). Figure 7.17 is a clear representation of what a summarized where-used report looks like.

Assembly B Used in		
Part Number	*Description*	*Usage*
A	Assembly	2
D	Assembly	1
X	Product	2
Y	Product	1

Fig. 7.17 *Summarized Where-Used for Assembly B* (After John, 1993)

7.15 MAINTENANCE OF THE BILL OF MATERIAL

Maintenance of the BOM is generally the responsibility of engineering. In many companies, changes to product specifications are made frequently and therefore a good engineering-change system and a consistent method of updating the relevant data files are prerequisites to an effective MRP I system. Many times, an engineering change may be planned to become effective at some specific date or from some specific lot number in future or when all current supplies of the present part are exhausted. This gives rise to the need for effectivity information as part of the BOM data and must be specially handled by the maintenance procedures and programs (John, 1993).

Accuracy of the BOM data is a requirement for the use of MRP I to perform its job of planning the correct materials at the right time. Just as inventory data accuracy should be ensured before attempting to implement MRP I, accurate BOM data, coupled with an effective maintenance procedure, should also be in place (John, 1993). With multiple uses for the bill and multiple users relying on its data, poor BOM accuracy can have catastrophic impact on business operations.

7.15.1 Types of BOM Errors

1. Incorrect components (that is, incorrect part numbers listed),
2. Incorrect quantity and unit of measure, and
3. Incorrect structure.

BOM Accuracy
= (No. of single-level BOMs correct
• No. of single-level BOMs checked) • 100%

7.15.2 Methods for Checking BOM Accuracy

1. Review a sample BOM.
2. Pick and issue the parts according to the bill and pay attention to feedback from the plant floor.
3. Monitor any unplanned issues from and receipts to the stockroom. This might be indicative of a BOM inaccuracy. However, it could be due to wrong scrap or shrinkage factor as well.
4. Watch for production line excesses as it might be indicative of excess issues.
5. Break a product or component down and check the parts against the BOM (or the solution against the formula in chemical products).
6. Do a cost roll-up using the bill, then compare it to the real cost of producing the item.

The goal should be 100 per cent accuracy. Further, it should be noted that just because a bill is accurate doesn't mean it will stay accurate, an on-going audit is essential.

7.16 ITEM IDENTIFICATION (Part Numbers)

Every item, purchased or manufactured, that has to be scheduled to satisfy requirements or requires a work order, or if an inventory balance needs to be managed for it, must have a part number and be included in the BOM database. Semifinished materials, sub-assemblies, intermediates used in chemical manufacturing, and manufacturing sub-assemblies are all included. When a part is sold for service or replacement, a parent part number must be assigned and a structure created, even if it is not manufactured and issued from inventory when the end item is produced. An item without a unique part number cannot be carried in inventory.

Part number is a critical consideration. An incorrect part number affects everything and everyone at least twice. If the incorrect part number is used on a requisition you will get the incorrect part increasing your inventory, wasting space and resources. In addition, you would still be short of the correct part (Schlussel, 1995).

Part Number Significance

The part numbering system used to identify unique items can be significant, nonsignificant, or semisignificant.

Significant Part Number

It is defined by APICS Dictionary as a part number that is intended to convey certain information, such as the source of the part, the material in the part, or the shape of the part. Using numbers to represent this information usually makes these part numbers longer than corresponding *nonsignificant* part numbers.

A significant part number is one in which the digits and/or the placement of the digits in the part number carries a specific meaning. A significant part numbering system assigns a part number that will be used not only to identify and control the item but also to define or partially define the item. The advantage of a significant part numbering system is that users are often more comfortable with the descriptive feature of the part number, as it allows recognition of the item. The disadvantages are that a greater number of digits are required, the allocated grouping of unused numbers may be used up after a period of time, and there can be confusion in defining the grouping (Toomey, 1996).

Nonsignificant Part Number

It is defined by APICS Dictionary as a part number that is assigned to each part but does not convey any information about the part. Non-significant part numbers are identifiers, not descriptors.

Thus, a non-significant part number is just a number, an identifier, telling nothing about the item it is identifying. The advantages of a non-significant numbering system are that less digits (shorter part numbers) are required and there are no concerns about number allocation and descriptive categories. The disadvantages are the user's initial discomfort with a non-descriptive number and the increased possibility of more than one number being assigned to the same item (Toomey, 1996).

Semisignificant Part Number

A Semisignificant Part Number attempts to gain the advantages of a significant system but to avoid the problems. Here only a few digits of the identification number are significant, the remaining are insignificant.

7.17 PART NUMBERING PREREQUISITES AND PRINCIPLES

When assigning a part number or designing a numbering system the following rules should be followed (Schlussel, 1995):

1. Part numbers should be simple.
2. They should be easy to assign and control, and they should be capable of human processing.
3. They should have a minimum number of characters.
4. Those characters should be numeric only with no spaces, dashes, asterisks, hyphens, or other separators.
5. The system should require a minimum amount of documentation.
6. It must serve the purpose and function of a part number.

Following points should be noted regarding item identification.

Unique Identification for Each Planned Item Every item, purchased or manufactured, that has to be scheduled should have a unique part number and be included in the BOM. The single purpose of a part number is to act as a unique identifier. Computer systems in particular demand this. MRP I requires BOMs in which each item to be planned must be uniquely identified. One part number must not identify two or more different items. BOM accuracy is totally lost when the form, fit, or function of a part/sub-assembly in the BOM changes but the part number does not change. It should be noted that a part number identifies an item and when that item's form, fit, or function changes its part number must also change. The consequence of this type of error will affect all users of the BOM file. One part number that describes two or more substantially different configurations of a BOM will create problems with costing, material planning, quality, inventory control, the production process, etc.

Uniquely Identify Content Part number must also serve to uniquely identify the content of the item. If there are two sub-assemblies identical in all aspects except for one component, there must be a unique part number to identify the difference in content.

Additional Part Numbers Required for Various Purpose Product designers, industrial engineers, cost accountants, inventory planners, and production workers might each prefer to assign them differently; part numbers have many uses including identifying products to customers, raw materials to suppliers, service parts to field service people, and many more. Our concern is with MRP I needs. Design engineers are concerned primarily with functions of components and products in assigning part numbers; they view the BOMs as the end of their design task. Others in manufacturing view BOMs as the beginning of their work, and often need additional part numbers and different structuring. Examples are added part numbers for service parts made differently from components used in production, added sub-assemblies that make processing easier and less expensive, and modules that improve planning and control.

Separate Part Numbers for Commonly Used Subassemblies It is essential that item identification (part numbers) be unique and unambiguous in BOM data for use with MRP I. Sometimes, the design engineers depict a major assembly as put together all at once on a single engineering drawing. However, in manufacturing, perhaps because of common usage of certain parts on multiple products, a decision may be made to put together sub-assemblies as a separate process. These sub-assemblies must then be assigned unique part numbers to be planned and prioritized properly by the planning system.

New Part Numbers for Modified Parts/Products In case of an engineering change that appears to be very minor to a product, the engineering might prefer to record it by an engineering change notice (ECN) number or letter separately, from the old part number itself. This will result in two physically different parts with the same number. For MRP I to function properly, the change must be reflected in a new part number.

Identify State of Completion by Assigning Separate Part Numbers at Different Stages of Processing As a unique identifier, the part number may also identify a state of completion. This must not be confused with the simple completion of one of several steps on a routing or work order. Of course, when a routing or work order is completed the converted material or parts take on a new identity reflected by the new unique part number.

Where a single raw material goes through several stages of finishing, the same part number is sometimes retained at each stage of finishing since it is physically the same piece. Again, if MRP I is to properly perform its priority planning function, separate part numbers must be assigned. That is, planning BOMs must specify not only the composition of a parent but also the process stages in its manufacture. BOM product structure levels must define each step in the buildup of the product so that MRP I can establish, using planned lead times, the timing of order releases and completion.

A raw die casting that is first machined and then receives one of three

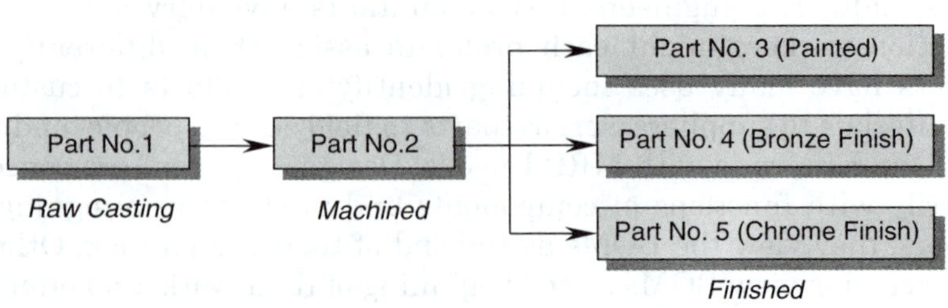

Fig. 7.18 *Unique Identity of Semifinished and Finished Items*

finishes—painted, bronze, or chrome—is shown in Fig. 7.18. If the part number for the original raw casting is used as the part number throughout the cycle described, there would be no way of determining the number of raw versus semifinished castings. Nor could we determine the true value of our inventory of raw materials, work in process, or finished. Here the semifinished and finished items should be assigned separate identities even though the finishing operations are minor. Further, the three finished items (Parts 3, 4, and 5) must have separate identities if their priorities are to be planned by MRP I and if they are to be identified in stock. By assigning a unique part number to identify state of completion all of this doubt, confusion, and inability to plan and control is eliminated.

'Collapsing' the BOM A separate identification ('part number 2' as shown in Fig. 7.18), for the machined but unfinished castings is required if the raw castings are machined in a single lot, stored in inventory (even for a short duration), and then taken from inventory and finished. If the castings are machined and finished as part of a single two-operation process, the part number for the unfinished machined casting would not be required. The simplified BOM structure, as shown in Fig. 7.19, has the

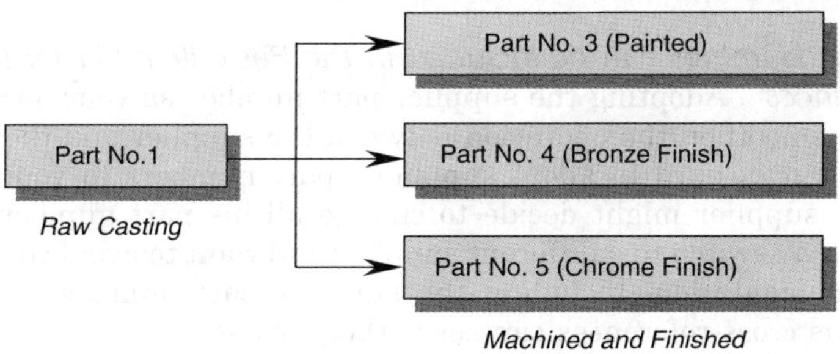

Fig. 7.19 *Unique Identity of Semifinished and Finished Items with Three Levels Collapsed to Two Levels by Combining Machining and Finishing Operations into a Single Routing*

advantage of one less part number to be maintained and one less level in the bill of material. However, the manufacturing process consistent with this simplified structure requires that the castings be machined and painted as part of a single routing. The process of eliminating part number and reducing the number of levels in a structure is called 'collapsing' the BOM.

Part Numbers with Fractional Usage for Consumable Tools Tooling information may also be included in BOM, in cases where tools such as dies or drill bits are used up in the production process. We can schedule tool replacement just like we can any other component. An expendable tool may be assigned a part number with a fractional usage in order to properly plan for tool availability based on the lot size to be produced. In this instance, the tools to be planned would have to be assigned unique part numbers and included in an item master file, in addition to the BOM. This will minimize down time on the line. The only other option is to stock an excessive inventory of expendable tooling, which again is money trapped in inventory.

An interesting example, reported by Garwood (1997) is that of *Connaught Laboratories* which had to include white mice in one of their BOMs. The mice were used for testing the product; without a requisite number of mice, the product couldn't be shipped. The decision was to put the white mice on the BOM and plan and schedule them like any other item.

Engineering Change and Part Numbering A universally accepted rule—*if either the form, fit, or function is affected by the change then a new part number should be assigned.* If neither the form, fit, nor function is affected then a revision letter is acceptable. A new part number at one level in the BOM can have an impact *all the way up the bill.* Bear in mind that by changing a part number at a lower level we have modified the content.

Supplier Part Numbers can be Included in the Basic Item Master File as Cross-References Adopting the supplier part number as your own part number can smoothen the operation between the supplier and the main company, but it is hard to adopt supplier's part numbers to your own. Further, the supplier might decide to change all his part numbers any time or you may switch to a different supplier and want to avoid changing all your documentation. Including the supplier part numbers in item master file as cross-references can serve the purpose.

The Old Part Number Should be Retained and Coded as Obsolete with a Description 'Obsolete, See Part ⟨New Partart Number⟩' Attached to It This protects against having to deal with a customer replenishment order where only the old part number is in the customer's records.

References

Arnold, J R Tony, *Introduction to Materials Management* (third edition), Prentice-Hall, Inc, New Jersey, 1998.

Cox III, James F and John H Blackstone Jr, *APICS Dictionary* (ninth edition), American Production and Inventory Control Society, Falls Church, V A, 1998.

Garwood, Dave and Don Rice, *'Making Your Bills of Material Work for You'*, In Thirty ninth International Conference Proceedings of APICS, American Production and Inventory Control Society, Falls Church, V A, 1996.

Garwood, Dave, *Bills of Material: Structured for Excellence.* Dogwood Publishing Company, Inc, Marietta, G A, 1997.

John, Ralph E St, *Material and Capacity Requirements Planning: (APICS) CPIM Certification Review Course, Student Guide* (Revision 3), American Production and Inventory Control Society, Falls Church, V A, 1993.

Landvater, Darryl V and Christopher D Gray, *MRP II Standard System: A Handbook for Manufacturing Software Survival,* Oliver Wight Limited Publishing Company, Essex Junction V T, 1989.

Narasimhan, S L, Dennis W McLwavey, and Peter J Billington, *Production Planning and Inventory Control* (second edition), Prentice-Hall, Inc, Englewood Cliffs, N J, 1995.

Oden, Howard W, Gray A Langenewalter, and Raymond A Lucier, *Handbook of Material and Capacity Requirements Planning,* McGraw-Hill, Inc, New York, 1993.

Plossl, George, *Orlicky's Material Requirements Planning* (second edition), McGraw-Hill, Inc, New York, 1994.

Proud, John F, *Master Scheduling: A Practical Guide to Competitive Manufacturing*, John Wiley & Sons, Inc, New York, 1994.

Russell, Roberta S and Bernard W Taylor III, *Operations Management: Focusing on Quality and Competitiveness* (second edition), Prentice-Hall International, Inc, New Jersey, 1998.

Schlussel, Ben, 'Principles of Product Structuring: How to get the Most out of Your Bills of Material', In *The Thirty-eighth International Conference Proceedings of APICS,* American Production and Inventory Control Society, Falls Church, V A, 1995.

Stonebraker, Peter B, *Master Planning: (APICS) CPIM Certification Review Course, Student Guide* (Revision 3), American Production and Inventory Control Society, Falls Church, V A, 1994.

Toomey, John W, *MRP II: Planning for Manufacturing Excellence,* Chapman & Hall, New York, 1996.

Vollmann, Thomas E, William L. Berry, and D Clay Whybark, *Manufacturing Planning and Control Systems* (fourth edition), McGraw-Hill, Inc, New York, 1997.

Wassweiler, Bill, 'How to Achieve and Maintain Bill of Material Accuracy', In *Thirty-ninth International Conference Proceedings of APICS,* American Production and Inventory Control Society, Falls Church V A, 1996.

Chapter 8

Capacity Management

8.1 CAPACITY

Capacity is the amount of work that can be done in a specified time period. Capacity is defined by the APICS Dictionary as 'the capability of a worker, machine, work center, plant, or organization to produce output per period of time'.

- Capacity is a *rate* of doing work, not the *quantity* of work done. Capacity is the rate at which work is withdrawn from the system.
- Manufacturing resources have historically included direct manufacturing labor and machine hours, but can also include tools, fixtures, material handling equipment, warehouse space, NC tapes, quality measurements equipment, and indirect labor for both shop (setup, engineering assistance, and quality) and office (design engineering). Currently most CRP systems only plan for the availability of the two resources: direct labor and machine hours. Some CRP systems can only analyze the most critical of these two resources. However, CRP can potentially be used to analyze any one, several, or all of the manufacturing resources.

8.1.1 Capacity Management

It is defined by APICS Dictionary as the function of establishing, measuring, monitoring, and adjusting limits or levels of capacity in order to execute all manufacturing schedules.

Capacity management is basically concerned with supplying the necessary resources. It is responsible for determining the capacity needed to achieve the priority plans as well as providing, monitoring, and controlling that capacity so that the priority plans can be met. As with all management processes, it consists of planning and control functions.

Capacity decisions affect product lead times, customer responsiveness, operating costs, and a company's ability to compete. Inadequate capacity can lose customers and limit growth. Excess capacity can drain a company's resources and prevent investments in more lucrative ventures. When to increase capacity and where to add capacity are critical decisions (Russell and Taylor III, 1998).

8.1.2 Capacity Planning

It is the process of determining the resources required to meet the priority plan and the methods needed to make that capacity available. It takes place at each level of the priority planning process. Production planning, master production scheduling, and material requirements planning determine priorities: What is wanted and when? These priority plans cannot be implemented, however, unless the firm has sufficient capacity to fulfill the demand. Capacity planning, thus, links the various production priority schedules to manufacturing resources (Arnold, 1998).

8.1.3 Capacity Control

It is the process of monitoring production output, comparing it with capacity plans, and taking corrective action when needed.

8.1.4 What-If Analysis

One important need for managing capacity of the company is the ability to perform 'what-if analysis'. This implies the ability to look at the capacity implications of alternative material plans. This can be done in detail with Capacity Requirements Planning (CRP) after performing MRP I, but at a substantial computation time and cost. Alternatively, what-if analysis can be accomplished using rough-cut methods directly from the MPS at a lower computation time and cost. This means that the MPS can be changed and the capacity implications reviewed for several alternatives within a much shorter period of time.

8.2 CAPACITY EXPANSION STRATEGIES

Figure 8.1 shows three basic strategies for the timing of capacity expansion in relation to a steady growth in demand.

1. Capacity Lead Strategy

Capacity is expanded in anticipation of demand growth. This aggressive strategy is used to lure customer from competitors who are capacity constrained or to gain a foothold in a rapidly expanding market.

2. Capacity Lag Strategy

Capacity is increased *after* an increase in demand has been documented. This conservative strategy produces a higher return on investment but

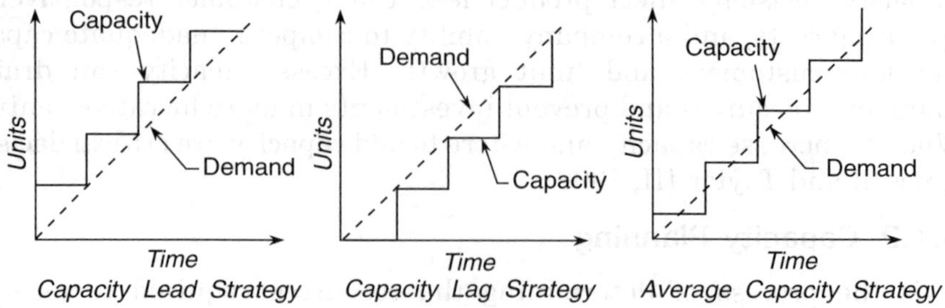

Fig. 8.1 *Capacity Expansion Strategies*

may loose customers in the process. It is used in industries with standard products and cost-based or weak competition. The strategy assumes that lost customers will return from competitors after capacity has expanded.

3. *Average Capacity Strategy*

Capacity is expanded to coincide with average expected demand. This is a moderate strategy in which managers are certain they will be able to sell at least some portion of the additional output.

How much to increase capacity depends on:

(i) the volume and certainty of anticipated demand,
(ii) strategic objectives in terms of growth, customer service, and competition, and
(iii) the costs of expansion and operation.

Capacity can be increased incrementally or in one large step. Incremental expansion is less risky but more costly. An attractive alternative to expanding capacity is *outsourcing,* in which suppliers absorb the risk of demand uncertainty (Russell and Taylor III, 1998).

8.3 LOAD

The APICS Dictionary defines load as the amount of planned work scheduled for and actual work released to a manufacturing facility, work center, or operation for a specific span of time, usually expressed in terms of standard hours of work, or in units of production when items consume similar resources at the same rate. Simply stated, load is the amount of work there is to be done by a work center or work cell.

- Load is the rate at which work is being imposed upon the system. Load is an input rate, capacity is an output rate of a system.
- Load is initiated by the sales and operations plan, which sets the rate of production for the long term. A supporting MPS will then drive MRP I, which will plan future orders for both purchased and manufactured items.

- These planned orders are released for execution when their planned start dates are neared. Progress is reported against orders to keep 'open' amounts (quantities and remaining operations) up-to-date.
- CRP will calculate the hours of load for each work center by time periods, using information from the routing file to calculate the hours required for all planned and released manufacturing order's operations.

8.3.1 Load Types

Following are the two types of loads:

1. Planned Load (Scheduled Receipts and Planned Order Releases)
2. Unplanned Load

Planned Load

It is generated by the MRP I system. This type of load is known some time in advance, and capacity for meeting this planned load can be arranged long before it is required. Following are the two types of planned loads:

(a) Scheduled Receipts (Released Load) The released load consists of those orders that have been released to the shop by the MRP I planner. These released orders appear on the MRP I display as scheduled receipts, and on the shop dispatch lists as shop orders.

(b) Planned Order Releases (Unreleased Load) The unreleased orders, or planned order releases, are orders in the MRP I system for planning purposes only, they do not show up in the shop at all until actually released by the planner.

(c) Unplanned Load It includes the total load that was not generated by MRP I. This can include emergencies, engineering requests, unplanned response to very important customers, personal work, and various other situations. The unplanned load should obviously be minimized.

To obtain an accurate picture of capacity requirements in the future, both released and unreleased orders *must* be included when running CRP, and the unplanned load should be also entered into the MRP I system as soon as it is known.

Standard Hours Required at a Particular Work Center
= Setup Time (S/U) in Hours
+ [No. of Hours of Run Time per Piece • No. of Pieces]

However, the actual hours required, which would be different than the standard hours required because of efficiency and utilization factor of the work center, can be calculated as follows:

Actual Hours Required
= Standard Hours Required • (Efficiency • Utilization)

Of course, the loads would have to be accumulated for all open and planned orders requiring working at the particular work center. Operation start and finish dates are also established and adjusted during this process. These dates will be used to determine the period (week) in which the load will be accumulated. Thus, the load (work to be done) can be compared for each time period to the capacity available (how much can be done). Imbalances between capacity and load will require corrective action. Capacity must be adjusted to handle the load, or schedules will have to be revised to fit within capacity constraints.

8.3.2 Long-Range to Short-Range Balancing

- Significant changes to capacity, such as a new plant or major pieces of equipment, require long lead-times. Major changes must be anticipated in long-range planning (business planning, sales and operations planning).
- Some capacity changes, for example, adding a third shift, will take a reasonable amount of time. These changes are considered in the MRP I and CRP phases of planning and control.
- In PAC we are primarily concerned with short-range capacity and load adjustments: such as working overtime on a work center, or temporarily assigning people to a different work center, or using an alternate source to relieve an overloaded area, for example, by routing work to a different work center or sending to an outside subcontractor.

Loads and capacities for each work center must be brought into balance, by changing either or both. Capacities may be adjusted by changing work hours, people assigned, adding equipment, improving productivity, and so on. Load may be scheduled in or out, out-sourced, moved to alternate work centers, etc. The primary objective is to balance needs and resources. As a last resort, higher level plans (MPS, Production Plan) may have to be revised to fit within capacities (Crepeau, 1995).

8.4 CAPACITY PLANNING PROCESS

Priority plans are usually stated in units of product or some standard unit of output. Capacity can sometimes be stated in the same units as that of the priority, for example, tons of steel or meters of cloth. If there is no common unit, capacity must be stated as the hours available. The priority plan must then be translated into hours of work required and compared to the hours available. The process of capacity planning is as follows:

1. Determine the capacity available at each work center in each time period.

2. Determine the load at each work center in each time period. This consists of the following two steps:
   (a) Translate the priority plan into the hours of work required at each work center in each time period.
   (b) Sum up the capacities required for each item on each work center to determine the total load on each work center in each time period.
3. Resolve differences between available capacity and required capacity. If possible, available capacity should be adjusted to match the load. Otherwise, the priority plans must be changed to match the available capacity.

This process occurs at each level in the priority planning process, varying only in the level of detail and time spans involved as shown in Fig. 8.2.

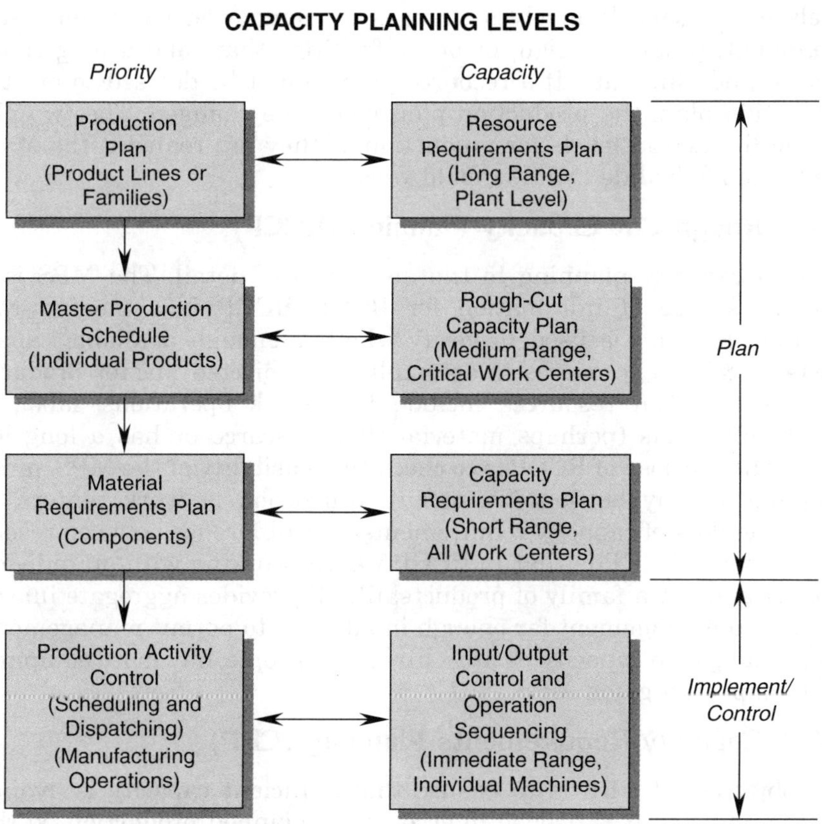

Fig. 8.2 *Planning Levels*

The master schedule must not list more quantities of products than the factory is capable of producing with its given number of machines and workers each month. The production quantity that the factory is capable of producing is referred to as the plant capacity. *Capacity planning* is concerned with planning the manpower and machine resources of the firm. This process may be performed at an aggregate or product-line level for long-term (*resource planning*), at the master scheduling level for intermediate term *(rough-cut capacity planning)*, and at the detailed or work-center level for short-term *(capacity requirements planning)*.

8.4.1 Resource Requirements Planning

It involves long-range capacity resource requirements and is directly linked to production planning. Typically, it involves translating monthly, quarterly, or annual product priorities from the production plan into some total measure of capacity, such as gross labor hours. Resource planning is a top management responsibility that determines long-range, overall levels of capacity. It involves changes in manpower, plant and capital equipment, product design, or other facilities that take a long time to acquire and eliminate. If a resource plan cannot be devised to meet the production plan, the production plan has to be changed. The two plans set the limits and levels for production, if they are realistic, the Master Production Schedule (MPS) should work.

8.4.2 Rough-Cut Capacity Planning (RCCP)

It takes capacity planning to the next level of detail. The MPS is the primary source of information for RCCP. RCCP is a medium-range planning aid and is used to verify whether enough available capacity exists at critical resources to accomplish a projected master production schedule. Critical resources include bottleneck operations, labor, and critical materials (perhaps material that is scarce or has a long lead-time). The purpose of RCCP is to check the feasibility of the MPS, provide warnings of any bottlenecks, ensure utilization of work centers, and advise vendors of capacity requirements. The difference between resource planning and RCCP is that in RCCP we are working with an individual product and not a family of products. RCCP provides aggregate information to top management far enough in advance to permit management to make changes in capacity (that is, hire more people, buy more equipment) to accomplish a given MPS.

8.4.3 Capacity Requirements Planning (CRP)

The objective of CRP is to ensure that sufficient capacity is available *where* and *when* it is needed to accomplish planned production. Without materials, production capacity is idle, and without productive capacity, materials are idle. To plan capacity, an operations manager needs to coordinate CRP with MRP I. Hence CRP is directly linked to the material requirements plan. With an MRP I schedule of planned order releases calculated, it now becomes necessary to test this tentative schedule against practical realities of capacity availability to validate the schedule and provide some assurance that it is workable. Since this type of planning focuses on component parts, greater detail is involved than in rough-cut capacity planning. CRP is concerned with individual orders at individual work centers and calculates work center loads and labor requirements for each time period at each work center.

8.4.4 Input/Output Control

It provides detailed assessments of the planned and actual inputs and outputs of the shop floor. It is a capacity control process used to control the input and output at a workstation. Production Activity Control (PAC) is usually the last stage of planning conducted at the shop floor level. The planned workstation input and output are defined in the CRP plan. The monitoring techniques used to perform this control process may be computer-based or manual.

8.4.5 Operations Sequencing

These techniques are used for planning job orders through workstations. The techniques can consider the capacity requirements planning and existing human and machine resources to devise a job-sequencing plan.

Figure 8.2 shows the relationship between the different levels of priority planning and capacity planning. It should be noted that, although the upper levels of priority planning are input to lower levels, the various capacity plans relate only to their level in the priority plan, not to subsequent capacity planning levels. Resource planning, for example, relates to production planning but is not an input to rough-cut capacity planning.

After the plans are completed, production activity control (PAC) and purchasing must be authorized to process, or implement, shop orders and purchase orders. Capacity must still be considered for shop orders. PAC includes operations sequencing and input/output control methods that are covered in shop floor control.

8.4.6 Supplier Capacity Planning

MRP I generates planned order releases for both purchased and factory-produced items. These need to be separated, using some sort of code in the item data records, for the next stage of processing. The factory orders are processed by CRP to evaluate load versus capacity in the plant. The purchase planned orders would likely be sent to a purchasing department. Although not considered part of MRP I or CRP, the subject of supplier capacity planning should parallel the CRP process of considering capacity constraints. Through regular contact with suppliers, purchasers should determine that sufficient capacity exists with the supplier base to fulfill the planned order release schedule. Multi-sourcing from more than one supplier or evaluating alternative suppliers would be logical alternatives. If sufficient supplier capacity is not available to meet the schedule of planned purchase orders, then this information must be fed back to the material planning and master scheduling functions so that appropriate action may be taken (John, 1993).

8.5 MEASURING CAPACITY

8.5.1 Unit of Measure

The unit of measure selected for capacity planning should be common to the mix of products encountered, such as, pieces, tons, feet, or standard work hours per time period. It must be a common denominator for all products measured. One of the primary purposes of capacity planning is to identify bottlenecks; the unit of measure selected should be meaningful in terms of such identification. For an operation producing only one product, a measure such as pieces produced is meaningful. But for an operation producing variety of products, it would be far more meaningful to convert from pieces produced to the standard time required to produce these pieces (Blackstone Jr, 1989).

1. Units of Output

If the variety of products produced at a work center or in a plant is not large, it is often possible to use a unit common to all products. Paper mills measure capacity in tons of paper, breweries in barrels of beer, and automobile manufacturers in numbers of cars. However, if a variety of products is made, a good common unit may not exist. In this case, the unit common to all products is time.

2. Standard Time/Hours

The work content of a product is expressed as the time required to make the product using a given method of manufacture. Standard hours data specify the number of hours that it *should* take to produce some volume of work (pieces, gallons, etc.). Using time-study techniques, the standard time for a job can be determined—that is, the time it would take a qualified operator working at a normal pace to do the job. It provides a yardstick for measuring work content and a unit for stating capacity. It is also used in loading and scheduling.

8.6 DETERMINING CAPACITY AVAILABLE

There are three ways of determining the capacity available:

(i) Theoretical Capacity.
(ii) Rated (Calculated) Capacity.
(iii) Demonstrated (Measured) Capacity.

8.6.1 Theoretical Capacity

Theoretical capacity is based on simple clock hours available for work, without any allowances for rest breaks, machine downtime, or any other disruptions. It is the maximum amount of time, or possible capacity that could be made available, working all people and/or machines all available hours, each day of the week, with no rest breaks or machine downtime.

Theoretical Capacity per Week
= No. of Machines or People × No. of Shifts per Week
× No. of Hours per Shift

This can be translated into work production through the use of standard hours. Theoretical capacity can never really be achieved and is not normally used in capacity determination. It can serve, however, as an upper bound to test the reasonableness of other capacity calculations.

8.6.2 Rated or Calculated Capacity

Rated capacity, also called calculated capacity, is the most frequently used version of capacity in CRP, particularly when dealing with a jobshop. It is determined by adjusting the theoretical capacity by productivity adjustment factors that have been determined by previously demonstrated capacity. These factors are 'utilization' and 'efficiency'.

Rated Capacity = Total clock Hours Available
× (Efficiency × Utilization Factor)

= Total clock Hours Available × Productivity Factor

where Total Clock Hours Available (Theoretical Capacity) per Period
= No. of People or Machines × No. of Shifts per Period × No. of Hours per Shift

Utilization is defined by APICS Dictionary as a measure (usually expressed as a percentage) of how intensively a resource is being used to produce a good or service. Utilization compares actual time used to available time. Traditionally, utilization is the ratio of direct time charged (run time plus setup time) to the clock time available. Utilization is a percentage between 0% to 100% that is equal to 100% minus the percentage of time lost due to unavailability of machine, tool, worker, etc. Thus, machine utilization is an estimate of the percent of the available scheduled hours the machine actually will be running.

Utilization Factor = Production Time Available
÷ Total Clock Hours Available

= Actual Hours Charged
÷ Scheduled Available Hours

where Production Time Available = Total Clock Hours
– Non-working Time (such as, Breaks, Routine Maintenance, etc.)

Efficiency is defined by APICS Dictionary as a measure (usually expressed as a percentage) of the actual output to the standard output expected. Efficiency measures how well something is being performed relative to existing standards, in contrast, productivity measures output relative to a specific input, such as, tons per labor hour.

Efficiency = Standard Hours of Work Earned (Produced)
÷ Setup and Run Hours

= Standard Hours of Work Earned (Produced)
÷ Actual Hours Charged

Rated Capacity (Standard Hours) per Week
= Shifts per Week × Hours per Shift
× No. of Work Stations × Utilization × Efficiency

John H Blackstone, Jr (1989) has proposed that the rated capacity be defined in the following fashion:

Rated Capacity = Total Clock Hours Available
× Efficiency × (Availability × Activation)

That is, he has split utilization into availability and activation.

Availability = (1 – fraction of time down due to machine breakdown and/or absenteeism)

Activation = (1 – fraction of time down due to lack of work)

From a management perspective the change is significant because both availability and activation can be used as performance measures. The downtime of a machine can be due to machine breakdown, absenteeism, or lack of work or material. Absenteeism and machine breakdowns are negatives and the goal should be to eliminate them completely. Clearly, then *the ideal availability is* 1. However, lack of work in some situations may be planned to allow for the random arrival of work from other work centers. It is not realistic to set a goal of 100% utilization of a machine or work center unless it is a bottleneck operation. Thus, ideal activation is harder to define. In an effort to achieve high utilization, queues (work-in-process) have often been carried at unreasonably high levels. Unless the manufacturing environment is continuous flow, it is not reasonable to expect or plan for a completely balanced plant.

Productivity Factor Some companies combine the *utilization factor* and *efficiency* into a single percentage called productivity. This calculation represents the standard hours produced divided by the total clock hours available. However, it should be noted that the productivity percentage can also be distorted by the amount of work available to run. Differences between the actual standard hours produced and total clock hours available may result from errors in the standard rates used for planning or from machine breakdowns, shortages of materials, shortage of labor, unauthorized extension to meal brakes, etc. Planned idle time, for which no work is planned because it is not required, is not counted as lost time.

Productivity Factor = Efficiency × Utilization Factor

= Standard Hours Earned
÷ Total Clock Hours Available

8.6.3 Demonstrated or Actual Capacity

One way to establish how much *can* be done is to measure how much *was* done in recent periods. This measurement is also called '*actual capacity*'. This calculation is a simple average of historical information—usually an average based on *normal* workloads. Enough history is provided so that the data is representative, but not going back so far that the average is out-of-date. Demonstrated capacity can sometimes be misleading. For example, If a work center should run out of work to do for a period of time, then its demonstrated capacity for that period would be less than the normal average. Hence, the historical data should be reviewed for abnormal conditions (extensive overtime/undertime, changes in personnel levels, etc.). Further, to estimate capacities for future periods, numeric averages should be adjusted to reflect planned changes to capacity such as more people or additional equipment (Crepeau, 1991). Practitioners normally use demonstrated capacity only if the data needed to determine calculated capacity is not available. Following are the relevant APICS definitions:

Demonstrated Capacity is the proven capacity calculated from actual performance data, usually expressed as:

Average number of items produced × Standard hours per item

It should be noted that demonstrated capacity is average, not maximum, output. It also depends on the utilization and efficiency of the work center, although these are not included in the calculation. Efficiency and utilization can be obtained from the historical data if a record is maintained of the hours available, hours actually worked, and the standard hours produced by a work center.

Maximum Demonstrated Capacity is the highest amount of actual output produced in the past *when* all efforts have been made to 'optimize' the resources, for instance, overtime, additional personnel, extra hours, extra shifts, reassignment of personnel, or use of any related equipment. Maximum demonstrated capacity is the most one could ever expect to produce in a short period of time but represents a rate that cannot be maintained over a long period of time.

Although the effectiveness and wisdom of planning to demonstrated capacity only has been proven many times, it is not always supported by production managers who quite reasonably have a desire to increase output and beat targets. The argument that if we do not plan for improvement, we are unlikely to achieve it, appears compelling (Kenworthy, 1997).

8.6.4 Comparing and Reconciling Rated and Demonstrated Capacities

It makes good sense to calculate the rated capacity, and compare it to the demonstrated capacity. This provides a test of reasonableness when establishing work center capacities. If there are significant differences, the reasons must be determined and reconciled to determine a useful measure of capacity. Some reasons include:

1. Demonstrated capacity hours include such variables as holiday weeks, overtime worked, manning changes, and amount of work available to be done. Rated capacity is a logical estimate of how much can be done in a standard work week with given staffing levels, utilization factors, and efficiencies.
2. Historical data used in either method could be incorrect.
3. Different periods of historical data may have been used, for example, 13 weeks for demonstrated and the previous month for rated.

It must be noted that capacity will change as the factors that affect capacity change (number of people and equipment, length of work week, break policies, etc.). Therefore, capacity determinations must be updated to reflect such changes and reviewed periodically to assure that they reflect current conditions (Crepeau, 1991).

8.7 RESOURCE REQUIREMENTS PLANNING

(a) The critical resources must first be identified.
(b) The following four-step resource-planning process should be followed:
   (i) Obtain planned production data,
   (ii) Identify product structures,
   (iii) Determine the bill of resources, and
   (iv) Calculate the total resource requirements.
(c) Per-period resource requirements must be evaluated against available resources, and shortfalls must be identified.
(d) The resource plan is used to validate the production plan or to highlight shortfalls, and then to identify and provide the necessary resources.

Resource requirements planning decisions are among the most difficult and potentially disastrous decisions a company can make. Whenever a company experiences a substantial increase in unit sales, it must answer the question, 'Is this increase permanent?' If the increase is not permanent, the company should provide additional output by subcontracting, thereby avoiding a permanent addition to capacity. If, on the other hand, the increase is permanent, the company will make more money by adding capacity (Blackstone Jr, 1989).

8.8 ROUGH-CUT CAPACITY PLANNING (RCCP)

Rough-cut capacity planning is defined by APICS Dictionary as the process of converting the master production schedule into requirements for key resources, often including labor, machinery, warehouse space, supplier's capabilities, and, in some cases, money. Comparison to available or demonstrated capacity is usually done for each key resource. This comparison assists the master scheduler in establishing a feasible MPS.

The intent of rough-cut capacity planning is to convert high-level plans (production plans or MPS) into the impact on resources needed to carry out those plans. Rough-cut capacity planning is an approximate type of capacity planning using some representative routings for product families or for items in the MPS. Normally companies validate the sales and operations plan against major resources using rough-cut capacity planning. With RCCP, managers and master schedulers can quickly identify obstacles to the plan without wading through all the detail. Simply put, RCCP attempts to identify 80 to 90% of the issues or potential problems.

In some instances, rough-cut capacity planning can't be done by product family, and has to be done using the items in the MPS. This can happen when resource requirements vary significantly based on the mix of items in the MPS. In this situation, rough-cut capacity planning should be run directly from the MPS. It can be an invaluable tool when assessing changes to the MPS, when the product mix is changing. In short, RCCP makes it possible:

(a) to test the validity of a production plan and MPS before doing any detailed material or capacity planning, and
(b) to be able to initiate action for making mid- to long-range capacity adjustments.

In some companies, spare parts account for a significant part of the total business and consequently the total resources. As a result, spare parts demands should be included in rough-cut capacity planning when assessing the effect on MPS changes.

8.8.1 RCCP versus Resource Planning

The rough-cut capacity plan is similar to the resource plan, but is likely to be more detailed than the resource plan in several regards. Following are some of the differences (Stonebraker, 1994) also shown in a tabular format in Table 8.1.

Table 8.1

Resource Planning versus RCCP

	Resource Planning	*Rough-cut Capacity Planning*
Inputs	Production Plan.	Master Production Schedule (MPS).
	Product Groups or Items.	Products and Components.
Period	Quarterly/Monthly.	Weekly/Daily.
	No offset.	Lead-Time Offset.
Outputs	Resource Plan.	Rough-cut Capacity Plan.
Methods	Bill of Resources.	Detailed Product Load Profile.

1. The rough-cut plan is disaggregated to individual products; the resource plan is stated in product groups.
2. The rough-cut plan is disaggregated to days or weeks; the resource plan is stated in months.
3. The rough-cut plan considers more production resources and uses a more accurate statement of those resources. The bill of resources for the rough-cut plan is stated in product-defined resources; the resource plan uses a family average resource requirement.
4. In the rough-cut plan, lead-times are offset; the resource plan is not lead-time offset.

8.8.2 RCCP versus CRP

- RCCP is a medium range planning aid whereas CRP is a short-term planing aid.
- RCCP is not as accurate as CRP because it uses estimated values like averages whereas CRP uses the actual routings.
- Further, RCCP is not as specifically detailed as CRP because it focuses on capacity at the department rather than at a specific workstation. RCCP considers critical resources only. CRP is very detailed and time consuming and in many MRP II systems a major revision to the capacity plan requires a batch run on the computer, often run overnight. Master schedulers need a quick way of assessing the feasibility of any proposed changes to the plan and RCCP was developed for this purpose.

Table 8.2

Differences between RCCP and CRP (Adapted from Proud, 1994)

	RCCP	*CRP*
What	■ Projected gross capacity. ■ Requirements for key resources.	■ Projected net capacity. ■ Requirements for each work center.
How	Explode production plan or MPS through resource profiles.	Explode MPS and MRP I planned orders through detailed routings; combine with current WIP status from shop floor control.
When	As required for simulation.	Weekly, monthly.
Why	■ Pre-MRP I evaluation of MPS. ■ Intermediate- to long-range planning.	■ Post-MRP I detailed analysis. ■ Periodic check of all work centers.
Precision	Aggregate or gross key resources only	Detailed—considers inventory, lot sizing WIP completions, work center lead times.
Complexity	Much less than CRP.	Usually exceeds MRP I.
Planning Horizon	Production plan limits.	MRP I Horizon.
Implementation	Short	Requires work centers, routings, MPS, MRP I, and WIP status of SFC.

Table 8.2 demonstrates the differences between RCCP and CRP.

8.8.3 Benefits of Rough-Cut Capacity Planning

1. A Simple and Quick Tool to Use For performing rough-cut capacity planning (RCCP) master schedulers do not need a detail routing for every item in the plan. This is what makes RCCP a simple and quick tool to use. In contrast, CRP requires MPS, MRP I, inventory control, bills-of-material, detail routings, and shop floor control. In addition, CRP requires a high degree of accuracy in bills of material, detail routings, and inventory records. Only then can accurate detail capacity planning be done.

2. Fast Implementation According to Proud (1994), on the average, RCCP can be productively used in as little as thirty to ninety days after implementing a RCCP system. In the standard implementation scheme, detail CRP is not generally effective until twelve to fifteen months after the MRP II implementation is commenced.

3. Better Simulation Tool Than CRP Because RCCP requires minimal computer time (relative to CRP), it is a better simulation tool than CRP, though its output is more of a shadow of reality, owing to its use of only key resources. Simulations such as the following are possible with RCCP: What is the effect of changing the mix of expected demand, the booking of a large order, or the shifting of replenishment orders either in or out? RCCP also allows the master scheduler to test the proposed MPS if necessary before obtaining more detail via MRP I and CRP.

8.8.4 Limitations of Rough-Cut Capacity Planning

1. It does not calculate capacity requirements by work centers.
2. It does not take into account Work-in-Process (WIP), completed parts and assemblies. This negates its value as a short-term planning tool and limits it to an intermediate- and long-range planning tool. The WIP netting does not take place until material requirements planning, capacity requirements planning, and shop floor control systems are run. Thus, RCCP data is often invalid in the short-term. It is not useful in planning for this week's production; its eyes are on the future.
3. It assumes lot-for-lot ordering in the calculation of resource requirements.
4. RCCP is based on representative products for an entire family. Incoming orders, however, may not exactly fit the predicted mix, causing discrepancies between aggregate and detail planning. Also, the manner in which setup time is handled may affect load predictions in various ways.
5. RCCP is further limited by the fact that it considers only *key* or *critical* resources. Actual building of a product, however, requires the resources in *all* work centers. Thus, execution of the production plan and MPS is always vulnerable to contingencies not highlighted during the rough-cut process (Proud, 1994).

In spite of these clear limitations, rough-cut capacity planning does provide a way to assess the capacity requirements created by an S&OP or MPS in a rough-cut way. A plan that is grossly overstated or a change to the plan that generates a significant overload on a resource or key piece of equipment will be clear from an RCCP run.

8.8.5 RCCP Techniques

The capacity requirements of a particular MPS can be estimated by any of the following techniques:

1. Capacity Planning using Overall Factors (CPOF).
2. Capacity Bills (Bills of Resources) .

3. Resource Profiles.

These techniques provide information with which to modify the resource levels or material plan in the medium range to ensure efficient execution of the MPS.

1. Capacity Planning Using Overall Factors (CPOF)

In this method, the MPS items and quantities are multiplied by the *total time* required to build each item to provide the *total* number of hours on *all* the required work centers to produce the schedule. Historical work-center percentages are then applied to the total number of hours to provide an estimate of the hours per individual work center to support the master schedule. This technique eliminates the need for engineered time standards. CPOF is the simplest of all the four techniques and is based only on accounting data.

Capacity Bills or Bills of Resources

It is a listing of the required capacity and key (bottleneck) resources needed to manufacture one unit of a selected item or family. RCCP uses these bills to calculate the approximate capacity requirements of the MPS. Resource planning may use a form of this bill. The capacity bill procedure provides a much more direct link between individual end products in the MPS and the capacity required for individual work centers. BOM and routing data are required, and direct labor or machine-hour data must be available for each operation.

3. Resource Profiles

'Resource profiles' is, perhaps, the most commonly used technique in RCCP. Here the top-level products are extended by a summarized resource profile for each product. Both the CPOF and the basic bill of resources calculate the resource requirements in the same time period as the master-scheduled family or item; that is, there is no time phasing. Consideration of time phasing is accomplished with the resource profile approach, a refinement of the bill of resources. Each key resource is time phased through the use of a resource profile. The resource profile identifies:

(a) the key resources required,
(b) the number of hours of the key resources required to make the product, and
(c) approximately when those resources are required relative to the completion of the product.

- Rough-cut capacity plan can be calculated from either the production plan or the MPS. The production plan would be used when the items in an S&OP family are similar (that is when they use about the same amount of key resources and hence one resource profile will work for the entire family). If, however, the number of hours of key resources required for the different items of the S&OP family are significantly different, then a resource profile is needed for each MPS item.

- Most companies with flow-type manufacturing use RCCP *only*. This includes most process manufacturers and Just-in-Time (JIT) companies with extensive use of kanban to control the movement of material.
- Almost all companies start with RCCP. Generally, this is because they implement MRP II in stages with the first stage consisting of RCCP, MPS, and MRP I modules.
- If a company finds that RCCP is sufficient to meet the needs of the business, it has no reason to implement detailed CRP. However, if RCCP doesn't show a capacity shortage problem but capacity bottlenecks appear on the manufacturing floor, then CRP is needed.

8.9 CAPACITY REQUIREMENTS PLANNING (CRP)

The Capacity Requirements Planning (CRP) is done at the level of material requirements planning (MRP I). It is the process of determining *in detail* the amount of labor and machine resources needed to achieve the required production. It answers the questions, what re-usable resources—equipment, manpower, cold storage, tooling, etc.—do you need and when do you need it?

Capacity Requirements Planning (CRP) is defined by APICS as the function of establishing, measuring, and adjusting limits or levels of capacity. It refers to the process of determining in detail the amount of labor and machine resources required to accomplish the tasks of production. Open shop orders and planned orders in the MRP I system are input to CRP, which through the use of parts routings and time standards translates these orders into hours of work by work center by time period. Even though RCCP may indicate that sufficient capacity exists to execute the MPS, CRP may show that capacity is insufficient during specific time periods.

The process involves each work center and covers the same horizon and time periods as the MRP I. Both open (released) shop orders as well as planned orders are considered in the determination of the required capacity (the workload).

The primary objective of CRP is to show a comparison between the load imposed on work centers by existing and proposed work orders and the capacity of each work center, distributed by time period. By identifying overloads and underloads, if any, the replanning actions may be taken to relieve those situations to achieve a balanced load pattern (John, 1993).

The capacity requirements planning calculations are nearly identical to the rough-cut capacity planning calculations. However, instead of using the production plan or MPS, as is done in RCCP, CRP uses the routings to identify the resources that are needed to make an item. CRP is the most detailed, complete, and accurate of the capacity planning techniques as Fig. 8.3 aptly shows. This accuracy is most important in the immediate

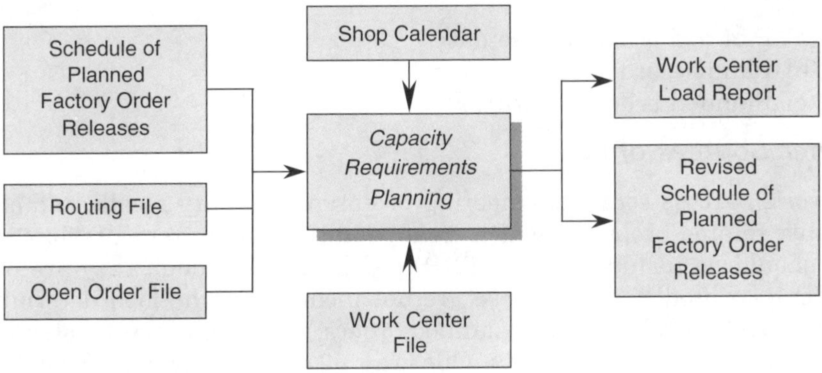

Fig. 8.3 *CRP Input/Output Model*

time periods. Because of the detail, a great amount of data and computation are required. Detailed capacity requirements are then compared with the available capacity.

8.10 INPUTS TO CRP

Capacity requirements are based on released and planned orders generated by the MRP I system. Work-load calculations are based on relating these orders to routing and work center files. Thus, the inputs needed for a CRP are open (released) shop orders, planned order releases, routings, time standards, lead-times, and work center capacities. The information can be obtained from the following files:

8.10.1 Open Order (Scheduled Receipts) File

An open shop order appears as a scheduled receipt on the material requirements plan. It is a released order for a quantity of a part to be manufactured and completed on a specific date. It shows all relevant information such as quantities, due dates, and operations. The open order file is a record of all the active shop orders and includes the following data by part number and work order:

(a) Order due date.
(b) Order quantity.
(c) Operations completed.
(d) Operations remaining.

8.10.2 Planned Order Releases

Planned orders are determined by the computer's MRP I logic based upon the gross requirements for a particular part. The planned orders include the firm planned orders. They are direct inputs to the CRP process in assessing the total capacity required in future time periods. The data include:

(a) Planned order release date.
(b) Planned order receipt date.
(c) Planned order quantity.

Other Sources of Load

Reworks, excess scrap, engineering prototypes, quality recalls and other quality related problems, which create additional requirements for capacity, should be included in the CRP calculations although they are often difficult to predict. Since these are not included in the load determined on the basis of released and planned orders, separate provision should be made for these requirements. One way of providing capacity for these unpredictable requirements is to provide certain amount of 'safety capacity' based on the statistical analysis of past performance.

8.10.3 Routing File

The necessary details of how to manufacture an item are found in the routing file. Routing is the path that work follows from work center to work center as it is completed. Routings are typically used as *shop travellers,* which means that they accompany the work from start to finish. Routing is specified on a route sheet or, in a computer-based system, in a route file. A routing file should exist for every component and assembly manufactured and contain the following information:

- Operations to be performed (Operation identification code and description),
- Sequence of operations,
- Work centers in which the operations should be performed,
- Possible alternate work centers,
- Tooling needed at each operation,
- Standard times (associated with each operation): setup times, and run times per piece,
- Operator skill levels, and
- Inspection operations and testing requirements.

Routing data are used in CRP and SFC systems. The key questions, 'Which resources?' and 'How much resource is needed?' are answered by the routing or scheduled operations file. The third key question, 'When is the resource required?' is usually answered by a process known as backward scheduling. The routing information, together with standard time estimates, completes the basic input data necessary to perform CRP. As with bills of material, there have been efforts to simplify the process routings, but again the routing is merely a recording of the required process. The best way to collapse the routing is to standardize and simplify the process. If the part were produced in a dedicated work cell, the routing would be much less detailed. The more detail in a routing is an indication of the greater amount of scheduling control required for production (Toomey, 1996). The format of a Routing File is shown in Fig. 8.4.

Part Name: Gear shaft
Drawing Number: D 123 X
Part Number: SG 123

Operation No.	*Planned Work Center*	*Setup Time (Standard Hours)*	*Run Time/Piece (Standard Hours)*	*Operation Description*	*Tooling Required*
10	12	1.50	0.20	Turn shaft	
20	14	0.50	0.25	Mill slot	
30	17	0.30	0.05	Drill 2 holes	
40	03	0.45	0.10	Grind	
50	Stores			Inventory	

Fig. 8.4 *Routing File*
(Adapted from Arnold, 1998)

Alternate Routing

The standard routing for an item is its *primary routing*. An item can also have one or more alternate routings, which describe additional methods of manufacturing the same item. Alternate routings are used when the primary routing is unavailable, owing to capacity constraints or equipment unavailability. Thus, they serve as a way to expedite a single job. A primary routing is chosen for one of two reasons: either the work center produces the parts better than other work centers or it produces them more cheaply. If using an alternative work center will lead to significant quality problems, then the original work center must be used. When a work center is overloaded, however, there are often some jobs that can be performed elsewhere with no loss of quality. Even when the manufacturing database indicates that the primary station should be chosen because it produces the part more cheaply, alternate routing should be strongly considered if the primary station is overloaded and the secondary station is underloaded.

Alternate Operations

A routing can include one or more alternate operations, which provide options for manufacturing the item when a given operation has capacity constraint or is otherwise unavailable. The standard operations are called *primary operations.*

Routing File Organization

The bill-of-material file and the routing file have historically been two distinct data files as they have been created, maintained, and controlled separately by design engineering and manufacturing engineering respectively. It has been suggested that the routing file be eliminated and the routing data become part of the bill of material file because in the last few years:

(i) the routings have become less detailed because of the increasing use of cellular manufacturing,
(ii) more teamwork has brought design and manufacturing functions together, and
(iii) their has been an enormous increase in the computer storage capacities.

The CRP system explodes each manufacturing order through its routing, rather like MRP I did with its bills of material.

Total Time on a Resource = Setup Time + (Run Time per Unit Required on the Resource × Order Quantity)

These times are stored on a scheduled operations file. Scheduled operations can be attached to a firm planned manufacturing order in the same way as materials requirements. Operation details attached to an FPO can be changed *manually* and will not be changed back to the routing values by a subsequent run of CRP. For a CPO, scheduled operations are recalculated from the latest version of the routing each time CRP is run. The PAC system maintains the data, operation by operation, in the order-specific routing in the scheduled operations file. From this file, CRP learns the status of each open order. So that CRP can compute the amount of work left. For planned shop orders, CRP uses standard routing file for the part.

8.10.4 Work Center File

A work center is defined by APICS Dictionary as a specific production area, consisting of one or more people and/or machines with identical capabilities, that can be considered as one unit for purposes of capacity requirements planning and detailed scheduling. A work center is composed of a number of machines or workers capable of doing the same work. The machinery will normally be similar so there are no differences in the kind of work the machines can do or the capacity of each. Several sewing machines of similar capacity could be considered a work center. A work center does not have to be a machine, *per se.* It can be a work area, such as an assembly bench. A work center can also be any of the following:

(i) *A Functional Cell.* In this a group of similar machines are physically located together. These machines have a range of slightly varying capacities and capabilities.

(ii) *An Entire Assembly Line or Flow Line.* This can be considered as a single work center having a single capacity, because the entire sequence is run as a unit, or not at all.

(iii) *A Group Technology Cell.* Here, several dissimilar machines are arranged in a cell to produce a group of products with similar manufacturing requirements in a flow.

(iv) *Groups of People Working on Similar, or Closely Related Tasks or Products.*

One of the major inputs to the CRP process is the data that pertains to each work center which is stored in the work center file. The purpose of the work center file is to assemble in one place all the relevant data related to a work center, such as information concerning cost, capacity, lead-time, crew size, and function. Each work center has a unique identity number, or name. Much of the work center data, other than descriptive information, is related to capacity and to lead-time and are used in CRP calculations. Production scheduling, shop loading, and capacity planning all use the work-center capacity data that resides in the work center file.

Work Center Data Elements

1. Identification and descriptions (may also include cost center and department ID).
2. Number of shifts scheduled.
3. Number of machines or work stations.
4. Hours scheduled per shift.
5. Workdays per period.
6. Utilization factor.
7. Efficiency factor.
8. Default/average setup time.
9. Average queue time allowance.
10. Average wait and move times.

A work center file also contains information on the capacity and move, wait, and queue times associated with the center.

- *The queue time* is the (planned) time a job waits at a work center before being handled. It is a timing factor that becomes very significant in the calculation of lead-time and is generally associated with a work center rather than with routings. A queue of orders is needed to allow for the random variation of arrival of other orders.

If a high machine utilization is the goal, there must be a large queue time allowance so as to ensure job availability at all times. Reducing the queue will reduce lead-time and work in process and increase flexibility. Assuming random variation of arrivals and lot sizes, the only way to reduce the queue is to have capacity available for all arrivals, which, in turn, calls for reduced machine utilization (Toomey, 1996). Queue time is usually an average time associated with a work center, and resides in the work center file.

- *The wait time* is the time a job is at a work center after completion and before being moved. It is a data element that can be maintained either with the routing or with the work center data, whichever makes more sense in a specific case.
- *The move time* is the time normally taken to move material from one workstation to another. Move time might be a factor included in routing data, work center data, or it might be maintained separately in a move-time matrix. The move-time matrix shown in Fig. 8.5 shows the standard (or planned) move times, in hours, for movement of material, regardless of quantity, from one work center to any other work center. Also shown is the time to move material from any work center to stores, which would follow the last operation in a routing.

Move-Time Matrix (Hours)		*To Work Center*				
		100	200	300	400	Stores
From Work Center	100	0	4.0	2.0	4.0	2.0
	200	2.0	0	2.0	2.0	2.0
	300	2.0	4.0	0	4.0	2.0
	400	4.0	2.0	2.0	0	2.0

Fig. 8.5 *Move-Time Matrix*
(Adapted from John, 1993)

Lead-time is the sum of queue, setup, run, wait, and move times as shown in Fig. 8.6. The lead-time data are required in order to place the work load in the proper time period .

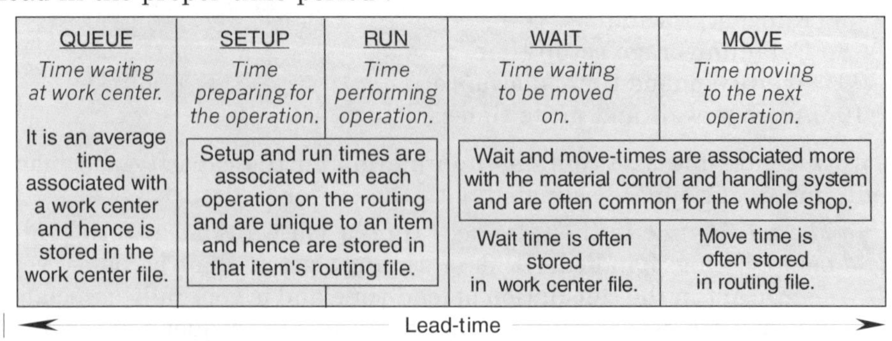

Fig. 8.6 *Manufacturing Lead-Time Elements*

The actual work load in the time period is only the setup and run calculations as these elements are the ones that use work center capacity; although they are the value-added functions, they are normally a small part (5 to 20%) of the total lead-time. The queue, wait, and move elements do not use capacity; they just use up space and time. In planning calculation, the queue time is an estimate, but it is the key to lead-time control (Toomey, 1996).

The Manufacturing Cycle It consists of all the activities that occur to an item between the release of an order to the shop and the receipt into stores, or the shipment to the final customer. The manufacturing cycle in a jobshop typically consists of the following activities: order preparation, queue, setup, run, wait, move, inspection, and store.

Load Activity Any activity that requires the resources or capacity of a work center is called a load activity. Only setup and run are load activities; others are nonload activities, because they occur without using work-center resources. Time elements associated with load activities are called *operation times* whereas times associated with nonload activities are sometimes called *interoperation times.*

Internal and External Setup Setup time can be divided into two types: internal, which absorbs machine capacity, and external, which allows the machine to perform other work.

8.10.5 Shop Calendar

Another piece of information needed is the number of working days available. The Gregorian calendar (which is the one we use every day) has some serious drawbacks for manufacturing planning and control, hence, it is desirable to develop a shop calendar. Shop calendar has been described in detail in Chapter 2.

- Open shop orders and planned orders in the MRP I system are input to CRP, which through the use of parts routings and time standards translates these orders into hours of work by work center by time period.
- Even though rough-cut capacity planning may indicate that sufficient capacity exists to execute the MPS, CRP may show that capacity is insufficient during specific time periods. Detailed capacity planning begins with the MPS stating the quantity of each item to be produced by time period (usually weekly). The MPS drives Material Requirements Planning (MRP I), which calculates the quantity of all components (purchased and manufactured) that will be needed, and when. Order releases are planned (and kept up to date as requirements change) for all parts, covering requirements throughout the planning horizon.
- CRP uses the planned order releases, along with orders already

released to the shop, to calculate the load (how much work needs to be done) at each work center, usually in standard hours by week. Work centers are one or more machines with the same capabilities.

- These loads are time-phased calculating when each operation should start, considering order completion dates and queue, setup, run, wait for move, and move times through all operations in the production sequences.
- Available capacity is determined based on hours of people and equipment available, efficiencies (ratios of standard hours produced per actual hour of production) and utilization factors (percentages of time work centers could be producing, considering breaks, daily maintenance, slack time, etc.).

8.11 SCHEDULING STRATEGIES

8.11.1 Backward Scheduling

CRP most often uses the technique of backward scheduling to place load hours into work centers. With backward scheduling, we begin with the requested date (a scheduled or planned receipt date) and work backward accounting for all lead-time elements in order to place the loads (operation times) in the correct time periods. It is a form of simulation using standard and estimated times. Backward scheduling reveals, in addition to work center load profiles, the latest possible start date for each operation in a work order to achieve on-time completion. The backed-off start time for the *first* operation (at what is sometimes referred to as the 'gateway' work center) indicates the latest date that an order could be released to meet the scheduled due date (John, 1993).

8.11.2 Forward Scheduling

This begins with the planned order release date from MRP I, or the completion date of the previous operation in line for open orders, and works forward through the routing to predict the earliest completion date for the order. In addition to a projected earliest completion date for the order, it also shows the earliest completion date for each operation.

When CRP uses backward scheduling, employing *detailed scheduling* (using the precision of actual routing data), it assists the planners to release orders to start just in time to meet their due dates, and therefore minimize the WIP inventory. However, if this backward scheduling step results in some orders with past due start dates, these orders must be forward scheduled, starting on the current date.

8.12 INFINITE VERSUS FINITE LOADING

8.12.1 Infinite Loading

Infinite loading means calculating the work center loads by time period without regard to capacity. It does not consider the existence of other shop orders competing for capacity at these work centers. Because CRP plans an operation when it is needed, irrespective of capacity, it is sometimes called an infinite capacity scheduling system, that is, it schedules jobs as though capacity is infinite—it takes no account of *scheduling interferences*. For example, if four jobs each requiring one day of working are planned in a week of 5 working days, then CRP will show the week as 80% loaded even if all the four jobs had to be completed on the Monday. The queuing allowance at each operation is the key to resolving such scheduling conflicts. The flexibility which the queue provides permits the supervisor of a work center to adjust the actual start dates to devise a workable schedule and still meet the operation due dates. Dispatch list is used to devise a practicable schedule which should aim to meet all operation due dates and be feasible. In a dispatch list, all scheduled operations on a file are sorted by supervisor and operations due for completion in the next few days or weeks are printed daily.

It assumes infinite capacity will be available. Planning to infinite capacity has the effect of revealing potential overloads and underloads in various periods. According to Stonebraker (1994), the term 'infinite loading' is obsolete now although the specific computer programs used to do infinite loading can now be used to perform the technique called CRP. Infinite loading was a gross misnomer to start with, implying that a load could be put into a factory regardless of the factory's ability to perform. The poor terminology obscured the fact that it is necessary to generate capacity requirements and compare these with available capacity before trying to adjust requirements to capacity. The success of infinite scheduling is dependent on the initial capacity management, the flow of orders, and shop floor prioritizing at each work center (Toomey, 1996).

8.12.2 Finite Loading

Finite loading is a technique that will not allow work to be loaded beyond the stated capacity of a work center. It assumes there is a defined limit to available capacity at any workstation. If there is not enough capacity available at a workstation because of other shop orders, the order has to be rescheduled in a different time period. The prioritizing of work is based on predetermined rules such as work center completion date or next operation availability.

CRP uses infinite loading, in conjunction with backward scheduling, to generate the capacity requirements plan, or work-center load reports. Operations sequencing, which is part of Production Activity Control

(PAC) module, uses finite loading and various forms of scheduling. Before finite loading can begin, however, priorities must be established for individual orders, generally by sales and marketing prioritizing customer orders. The highest priority orders get first claim on available capacity in each work center. The jobs are then loaded into the work centers in priority sequence. As soon as a work center is filled to its limiting capacity, the 'overflow' jobs are scheduled where unused capacity exists, either earlier or later, depending on the rescheduling algorithm and the planning parameters.

- Finite loading generally utilizes backward, forward, and other alternative scheduling methods, to distribute the load evenly among the periods.
- Finite loading also utilizes alternate work centers and routings; infinite loading only uses the primary work centers and routings; because it would not know when to switch to the alternate work center or routing.
- Finite loading does not solve the undercapacity problem. Finite loading will, however, determine which jobs will come out first, based on priorities.

8.13 CRP OUTPUT

Work Center Load Profile and Report

The work center load profile is developed by listing both open order and planned order loads by work centers and time periods and then comparing to available capacity. Because MRP I assumes infinite capacity, overloads and underloads can be expected. A load report is a comparative analysis of the expected versus actual capacity usage in a department or at a work center. The work center load report shows future capacity requirements based on released and planned orders for each time period on the plan as shown in Fig. 8.7. Figure 8.8 shows the same data in graphical form. Load reports can be very detailed and prescribe units of

	Week					Total
	20	21	22	23	24	
Released Load	51.5	45	30	30	25	181.5
Planned Load	100.5	120	100	90	100	510.5
Total Load	152	165	130	120	125	692
Rated Capacity	140	140	140	140	140	700
Overload	+ 12	+ 25	– 10	– 20	– 15	– 8

Fig. 8.7 *Work Center Load Report*
(After Arnold, 1998)

output required and labor hours by skill level or even by employee required.

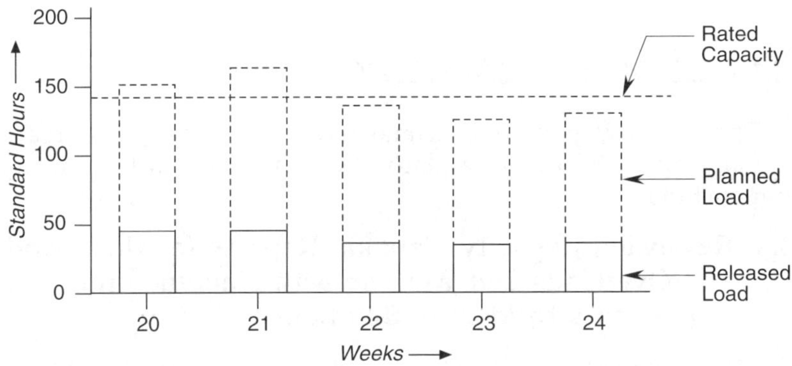

Fig. 8.8 *Graph of a Work Centre Load Profile* (After Arnold, 1998)

The graphic representation is very useful to see the relationships between projected load and capacity and to readily identify overloads and underloads. CRP would ordinarily produce a work center load profile such as this for each work center identified in the factory. This type of display gives information used to adjust available capacity or to adjust the load by changing the priority plan. In this example, week 1 and week 2 are overloaded, the balance are underloaded, and cumulative load is less than the available capacity. For the planner this shows there is enough total capacity over the planning horizon, and available capacity or priority can be juggled to meet the plan. Reporting cumulative loads versus cumula-

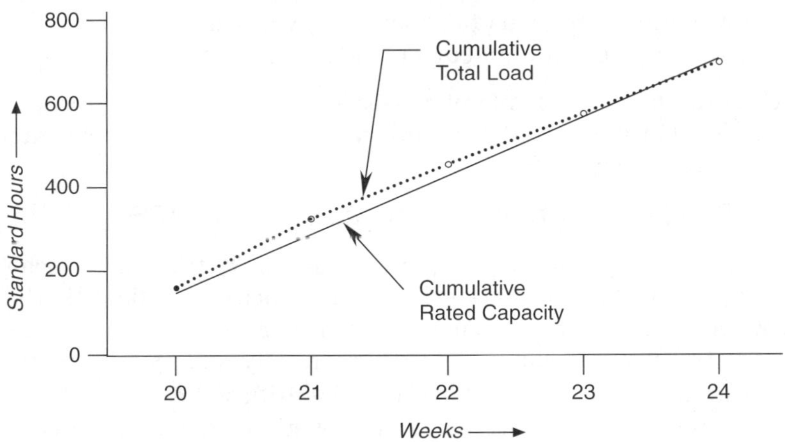

Fig. 8.9 *Graph of Cumulative Rated Capacity and Cumulative Load Profiles.*

tive capacity shows whether adequate capacity exists, in general, across the planning horizon. Figure 8.9 shows the cumulative load and cumulative capacity graphically.

8.14 ROLE OF THE CAPACITY PLANNER

The capacity planner's primary responsibility is to create and maintain valid capacity plans. Typical activities for the capacity planner include:

8.14.1 Reviewing Capacity Planning Reports for Underloads and Overloads and Working with Manufacturing Supervisors to Identify Solutions

The capacity planner uses the capacity requirements planning summary report to identify overloads and underloads. Using the capacity requirements planning detail report, the planner identifies possible solutions. These solutions include such options as:

(a) Increasing capacity by using overtime, weekends, and/or extra shifts.
(b) Off-loading work to another work center in the plant.
(c) Subcontracting work to an outside supplier.
(d) Changing the material plan to reduce the capacity requirements and relieve an overload.
(e) Changing the material plan (increasing quantities or doing work early) to relieve an underload.
(f) Changing the MPS to reduce the capacity requirements and relieve an overload.
(g) Working with engineering or development to redesign, retool, or develop new methods to relieve an overload.
(h) Acquiring new equipment and/or hire new people.

In each case, the capacity planner works with the manufacturing supervisor to develop plans that the supervisor can be held accountable for meeting (Landvater, 1997).

8.14.2 Dealing with Requests to Change the MPS

When investigating a request to change the MPS, the master scheduler must ask, 'Do we have the materials and capacity to make this change?' The capacity planner must provide an answer to the capacity question. There are two ways to check capacity: manually and by computer.

(a) Manually, the material planner identifies the items that would change if the master schedule were changed and feeds this information to the capacity planner. The capacity planner uses this information and routings to calculate the additional capacity requirements from these changes. Finally, the planner marks up the capacity planning reports, noting where underloads and overloads appear.

(b) The capacity planner can also evaluate the change using the computer. This is done through simulation. Both the material plan and the capacity plan would be rerun, and the planner would evaluate the results. The savings from computerized capacity simulations, like material plan simulations, are in the calculation of the data. The problem resolution, which is usually most of the work, takes the same amount of time as the manual method (Landvater, 1997).

8.14.3 Maintaining the Work Center Data (on Planned and Demonstrated Capacity, Scheduling Rules)

Like the material planner, the capacity planner is responsible for the basic planning data. The material planner is responsible for the item data, and the capacity planner is responsible for the work center data. This includes the data on demonstrated and planned capacities, and in companies with shop floor control systems, it includes certain scheduling rules, such as move and queue rules for the work center and rules on the number of shifts the work center is operated. The capacity planner works with the manufacturing supervisor to develop these data and the capacity planner is responsible for maintaining them (Landvater, 1997).

8.15 ROLE OF THE MANUFACTURING SUPERVISOR

The manufacturing supervisor's responsibility is to meet both order due dates and the capacity plan. With respect to capacity planning, the manufacturing supervisor would:

1. Work with the capacity planner to develop capacity solutions.
2. Meet the planned capacity numbers.

1. Work with the Capacity Planner to Develop Capacity Solutions

In most companies, the capacity planner meets weekly with the manufacturing supervisor to review the CRP reports and identify possible solutions for work centers that have underloads or overloads. If they cannot agree on a solution, they can appeal to their respective managers for a decision, although this is rarely necessary.

The different options for solving these capacity problems have already been covered earlier in the section on the role of the capacity planner. In a number of these cases, the capacity planner will be responsible for doing follow-up work, such as meeting with the material planner to see if an order quantity can be reduced or meeting with engineering to discuss off-loading or new methods.

2. Meet the Planned Capacity Numbers

In the same way that the manufacturing supervisor is responsible for meeting the due dates, he or she is responsible for meeting the planned capacity numbers. Planned capacity is typically developed by looking at the demonstrated capacity (history) of a work center and including any changes that would either increase or decrease the capacity.

What happens if a work center is not achieving the planned capacity number? If the problem is an equipment breakdown, the manufacturing supervisor should work with the maintenance people. If there is a process problem, the supervisor would work with engineering. The problem could also be a lack of work coming into the work center. Regardless of the reason, the manufacturing supervisor is responsible for bringing the work center up to its planned capacity or getting the capacity changed to a more realistic number.

Machine Loading

In general, the capacity of production facilities is limited; its limit is the effective work time multiplied by the number of machines, expressed in terms of machine-hours, or, in the case of operative workers, man hours. Hence, it is necessary to adjust and optimally determine the workload for manufacturing products in every time period so as not to exceed the production capacity, in connection with due dates for those products. This decision is called machine loading.

Machine loading is the accumulation by work centers, machine, or machine group of the hours generated from the scheduling of operations for *released* orders by time period. Machine loading differs from CRP in that it does not use the planned orders from MRP I but operates solely from released orders. It may be of limited value because of its limited visibility of resources (Cox III and Blackstone Jr, 1998).

We have seen that machine loading provides capacity-planning information based only on released orders. However, for most companies, only the work to be done in the next few weeks exists in the form of released orders. But as you go further out in time, increasingly the workload is in the form of planned orders, hence the machine loading report gives the picture of diminishing workload overtime. That, of course, is incorrect and hence machine loading is not an effective way to do capacity planning, and few companies use it now, making it an outdated technique.

8.16 PRODUCTION SMOOTHING

It quite often happens that in some periods workload exceeds the production capacity to meet due dates established by market demands. The function to level down the workload so that it is within the production capacity and to respond to fluctuations of the external environment of the manufacturing system is production smoothing. In the short run, capacity can be adjusted by following certain management policies. There are two ways of balancing capacity available and load: alter the load, or change the capacity available. Altering the load means shifting orders ahead or back so the load is leveled. If orders are processed on other workstations, the schedule and load on the other work centers have to be changed as well. It may also mean that other components should be rescheduled and the MPS changed.

Policies for Production Smoothing

Following are some of the management policies for production smoothing:

1. Time Adaptation

It is a method of absorbing production fluctuation by increasing or decreasing production time, thus increasing the current production capacity; for example, by setting work overtime or undertime. The most commonly used method is to schedule overtime or weekends, followed by addition of extra shifts. This will provide temporary and quick relief for cases where the load/capacity imbalance is not too large.

2. Inventory Adaptation

It is a method of absorbing production fluctuation by controlling inventory level, that is, by producing and storing excessive products at periods of low demand and shipping them at periods of high demand.

3. Intensive Adaptation

It is a method of absorbing production fluctuation by production intensity; for example, intensive utilization of production facilities by increasing or decreasing the production speed or the speed of rotation of the main spindle of the machine tool.

4. Quantitative Adaptation

It is a method of absorbing production fluctuation by increasing or decreasing the number of usable production facilities. It might involve adjusting the level of the workforce by hiring or laying off workers. The ability to do so will depend on the availability of the skills required and the training needed. The higher the skill level and the longer the training needed, the more difficult it becomes to change quickly the level of the workforce.

5. Subcontract or Outsource Adaptation

It is a method of absorbing production fluctuation by subcontracting production of the amounts unable to be manufactured by the existing production facilities of the manufacturing system. It may be more costly to subcontract rather than make the item in-house, but again it is important to maintain schedule. In case the available capacity is more than the required capacity the items normally purchased or subcontracted are manufactured in-house.

6. Load/Workforce Shifting Adaptation

This method involves the shifting of the workforce or the load to another work center. Workforce can be shifted from an overloaded work center to an underloaded work center. This requires a flexible cross-trained workforce. Or, alternate routings can be used to shift some load to another work center. Often the other work center is not as efficient as the original. Nevertheless, the important thing is to meet the schedule, and this is a valid way of doing so. Another possibility is to adjust operation dates forward or backward in time.

7. Load Adjustment Adaptation

This method involves reducing or increasing the load for balancing the load and capacity. The possible options are:

(a) Reducing or increasing lot-sizes.
(b) Holding work in production control, increasing delivery lead-time, or releasing orders early.
(c) Reducing or increasing MPS.

- Obviously, planners have far more options available when they can look farther into the future. For example, they can deal with a problem far out on the horizon by subcontracting, using new tooling, acquiring new equipment, or implementing a methods change. If they are trying to solve a problem this afternoon, their options list is reduced to overtime, extra shifts, or changing the MPS, none of which may be desirable (Landvater, 1997). Revising MPS is the last resort. Only after retesting, considering outsourcing, overtime, and so forth, should there be an adjustment to the master schedule.
- The demand-frequency adjustment due to lot size can be automatically handled through the MPS system logic, but load leveling and capacity adjustments require management intervention (Toomey, 1996).
- CRP is important because it allows people to deal with problems quantitatively. Too often, companies stop at the fact that they have an overload or underload. What's the size of the underload? Five hours? Five hundred hours? It's important to quantify problems in order to develop practical, cost-effective solutions that can be implemented in the necessary time frame (Landvater, 1997).
- In the context of MRP II, Capacity Requirements Planning (CRP) is medium range. Long-term adjustments to capacity (new facilities, procurement and installation of long-lead equipment, etc.) must be evaluated and planned at higher levels in the process (business

planning, S&OP, and MPS). The broader term 'capacity management' thus encompasses long-range capacity planning and control as well as intermediate-range and short-term capacity planning (Crepeau, 1991).

8.17 ADVANTAGES AND LIMITATIONS OF CRP

8.17.1 Advantages of CRP

1. Provides time-phased visibility of load and capacity imbalances.
2. Confirms that adequate capacity exists on a cumulative basis across the planning horizon.
3. Considers specific lot sizes and routings.
4. Uses a more precise estimate of lead-times than MRP I.
5. Contributes to the elimination of erratic lead-times by providing data for the smoothing of loads across work centers (John, 1993).

8.17.2 Limitations of CRP

1. Applies primarily to jobshop production.
2. Requires a great deal of computation, making computer processing almost mandatory.
3. Usually performs only a backward schedule and does not show where slack times might be utilized for better balance.
4. Requires extensive amount of input data.
5. Often results in misleading computational details, especially planned queue times.
6. Unable to give precise level of detail for day-to-day, short-term decisions.
7. Does not clearly show the effect of master schedule revisions on achieving balance, which might make the situation even worse (John, 1993).

8.17.3 Effects of Not Doing CRP

Much of the effort devoted to the master production scheduling (MPS) and material requirements planning (MRP I) systems may be almost wasted if capacity needs are not considered. Failure to perform a capacity analysis could result in non-realization of the benefits of otherwise sound MPS and MRP I analysis. Following are the possible consequences of bypassing the capacity planning process:

1. Bottlenecks,
2. Low productivity,
3. Higher work-in-process inventory,
4. Late deliveries and shortages,
5. Extended lead-times, and
6. Higher labor costs.

8.18 CAPACITY PLANNING IN REPETITIVE MANUFACTURING

Repetitive manufacturing is often characterized as an assembly line although there are many variations in practice that are much different from the image of a never-stopping conveyor system such as those typically used in automobile assembly. Still, the concept of the assembly line does offer a good basis for comparison between types of manufacture. In general, the emphasis in repetitive production is on the flow of materials. Machines and workstations are organized with a product orientation rather than the functional orientation (grouping of similar machines) that is prevalent with job shops. There is usually no job lot control based on a manufacturing order for some specific batch quantity. Instead, an assembly line is setup to produce a high volume of a certain product mix that will remain constant for some period of time, days or even weeks, before the line is changed over to produce a different mix of products.

The principles of capacity planning with repetitive manufacturing is basically the same as with intermittent processing. There must be a balance between the load imposed by the MPS and the capacity resources available to produce that load. The determination of capacity is done much more frequently at the MPS level rather than as a follow-up to MRP I. The master schedule indicates the rate at which a given product mix will be produced and this determines the load. Then capacity must be adjusted along the assembly line to maintain that rate. Detailed CRP may not need to be performed at all. All capacity planning might be done at the rough-cut level, and various mathematical methods for line balancing might be substituted for the load profiles that are characteristic of CRP.

Primary Differences between the CRP Principles as Applied to Repetitive and Intermittent Manufacturing

1. All routings tend to be the same for all products so that the database organization for repetitive manufacturing might be different from that of intermittent manufacturing.
2. Standard run times at each work station may differ for various products so that capacity balancing needs to be performed whenever the line is changed over for a different product mix.
3. Queue time and wait time are not significant factors of lead-time. Once the production process is initiated, work flows along the line without intentional pileups in the form of queues.
4. Setup time is not a factor once a line is operational. Most of the setup occurs during line changeover, when the line is effectively shut down anyway. Run time is the key component of lead-time.
5. The rate of production is constrained by the slowest operation.

Therefore, capacity planning generally attempts to assign equipment and personnel to work stations to produce at the same rate as the slowest component.

6. Feedback on work in progress is collected at key points along the line and at the end of the repetitive process rather than at every work station.

References

Arnold, J R Tony, *Introduction to Materials Management* (third edition). Prentice Hall, Inc, New Jersey, 1998.

Blackstone Jr, John H, *Capacity Management*. Cincinnati, South-Western Publishing Company, Ohio, 1989.

Cox III, James F, and John H Blackstone Jr, *APICS Dictionary* (ninth edition). American Production and Inventory Control Society, Falls Church, V A, 1998.

John, Ralph E S., *Material and Capacity Requirements Planning: (APICS CPIM) Certification Review Course, Student Guide* (Revision 3). American Production and Inventory Control Society, Falls Church, V A, 1993.

Kenworthy, John *Planning and Control of Manufacturing Operations.* John Wiley and Sons Inc, New York, 1997.

Landvater, Darryl V, *World Class Production and Inventory Management* (second edition). John Wiley and Sons Inc, New York, 1997.

Proud, John F, *Master Scheduling: A Practical Guide to Competitive Manufacturing.* John Wiley & Sons, Inc, New York, 1994.

Russell, Roberta S and Bernard W Taylor III, *Operations Management: Focusing on Quality and Competitiveness* (second edition). Prentice-Hall Inc, New Jersey, 1998.

Stonebraker, Peter B, *Master Planning: (APICS CPIM) Certification Review Course, Student Guide* (Revision 3). American Production and Inventory Control Society, Falls Church, V A, 1994.

Toomey, John W, *MRP II: Planning for Manufacturing Excellence.* Chapman & Hall, New York, 1996.

Vollmann, Thomas E, William L Berry, and D Clay Whybark, Manufacturing Planning and Control Systems (fourth edition). McGraw-Hill, Inc, New York, 1997.

Chapter 9

Shop Floor Control

9.1 SHOP FLOOR CONTROL (Also called Production Activity Control)

Shop floor control is that element of an MRP II system that is directly responsible for the activities required to execute the planned production for the company. Its responsibilities include:

1. Reviewing and releasing the work to the shop.
2. Monitoring and controlling lead-times.
3. Establishing and receiving order priorities.
4. Making detailed schedules.
5. Planning capacity at each work center.
6. Controlling queue time and work-in-process.
7. Receiving feedback on the status of production activity.
8. Disposing orders.
9. Recording performance of actual results and comparing them with the standards and targets (Wassweiler, 1997).

Shop floor control is defined by APICS Dictionary as a system for using data from the shop floor to maintain and communicate status information on shop orders (manufacturing orders) and on work centers. The major subfunctions of shop floor control are:

1. Assigning priority of each shop order.
2. Maintaining work-in-process quantity information.
3. Conveying shop order status information to the office.
4. Providing actual output data for capacity control purposes.
5. Providing quantity by location by shop order for work-in-process inventory and accounting purposes.
6. Providing measurement of efficiency, utilization, and productivity of the workforce and machines.

Shop floor control can use order control or flow control to monitor material movement through the facility (Cox III and Blackstone Jr, 1998). SFC brings the entire manufacturing facility, including people, machines, materials, and tools together in an optimum relationship to execute the master production schedule. It controls shop priorities and work-in-process while providing acceptable customer service.

The people on the factory floor are responsible for meeting the due dates—both the operation due dates and the order due dates. Shop floor Control (SFC) is simply a tool to help make that happen. The scheduling aspect of SFC indicates when each operation should be started and completed. The tracking and monitoring aspect of shop floor control shows the location and status of a job. In jobshop fabrication operations, SFC maintains the detailed operation status and quantities of each open shop order. The data are used for prioritizing dispatch lists, output for CRP control, cost system management, and shop performance measurements, such as labor performance, machine utilization, and quality. SFC is responsible for executing the master production schedule (MPS) and material requirements plan. At the same time, it must make good use of labor and machines, minimize work-in-process inventory, and maintain customer service.

The MPS and MRP I authorize SFC:

1. To release work orders to the shop for manufacturing,
2. To take control of work orders and make sure they are completed on time,
3. To be responsible for the immediate detailed planning of the flow of orders through manufacturing, carrying out the plan, and controlling the work as it progresses to completion, and
4. To manage day-to-day activity and provide the necessary support.

The activities of the SFC system can be classified into planning, implementation, and control functions.

9.1.1 Planning

The flow of work through each of the work centers must be planned to meet delivery dates, which means shop floor control must do the following:

1. Ensure that the required materials, tooling, personnel, and information are available to manufacture the components when needed.
2. Schedule start and completion dates for each shop order at each work center so the scheduled completion date of the order can be met. This will involve the planner in developing a load profile for the work centers.

9.1.2 Implementation

Once the plans are made, SFC must put them into action by advising the shop floor what must be done. Usually instructions are given by issuing a shop order. SFC will:

1. Gather the information needed by the shop floor to make the product.
2. Release orders to the shop floor as authorized by the material requirements plan. This is called dispatching.

9.1.3 Control

Once plans are made and shop orders are released, the process must be monitored to learn what is actually happening. The results are compared to the plan to decide whether corrective action is necessary. SFC will do the following:

1. Rank the shop orders in desired priority sequence by work center and establish a dispatch list based on this information.
2. Track the actual performance of work orders and compare it to planned schedules. Where necessary, PAC must take corrective action by replanning, rescheduling, or adjusting capacity to meet final delivery requirements.
3. Monitor and control work-in-process, lead-times, and work center queues.
4. Report work center efficiency, operation times, order quantities, and scrap.

SFC for Effective Utilization of Capacity

If more than enough capacity exists to produce the peak loads, almost *any* SFC system will achieve material flow objectives. It is in the cases where there are some bottleneck areas and where effective utilization of capacity is important, that we see the utility of good SFC systems. A related issue is the extent to which capacity planning is done. If the detailed CRP activity provides sufficient capacity, with relatively level loading, SFC is straightforward. On the other hand, if peaks and valleys in CRP are passed down to the back end, execution becomes more complex and difficult. SFC criteria focus on shop orders; capacity criteria focus on the work centers through which the shop orders pass. The material plan provides information to the SFC system and sets performance objectives. The essential objective of SFC is to execute the material plan—to build the right part at the right time. This will result in being able to hit the MPS and to satisfy customer service objectives (Vollmann *et al.* 1997).

SFC Executes the Dictates of the Computerized MRP I

In a computerized MRP I system, each shop order would be a scheduled receipt for the part. As such, the scheduled receipt has a due date. From this due date, operational due dates could be established by backing off expected operation times, if these data are needed to establish priority sequence. The great advantage of this computer-based system is that, whenever the due date for a scheduled receipt changes, the operation due dates can be changed accordingly. These changes, in turn, lead to priority changes for shop floor control. The result is an execution system that works on the most-needed shop orders first. The objective is for high priority jobs to move through the shop very quickly, while low priority jobs are set aside. In this way, the SFC system can indeed execute the dictates of MRP I.

Computerized SFC Keeps Order Due Dates Current

In a computer-based SFC system, due dates would not be printed on any shop paper that travels with the work-in-process inventory. The shop paper would show routing or sequence of operations (static data), but not any due dates. The changing (dynamic) due date information would be printed daily in the form of work center schedule or dispatch list for each work center. It is the dispatch list, not the traveling paper, that shows the priority sequence. The dispatch list can be updated as rapidly as transactions are processed to the MRP I database. The daily dispatch list is, in fact, only one option. It is also possible to keep the data even more current, using on-line inquiry for which jobs to run next.

The Loop-Closing Aspect of SFC

There is an important feedback from the shop floor to the engine (MRP I and CRP). The feedback is of two types: *status information* and *warning signals*. Status information includes where things are, notification of operational completions, count verifications, accounting data, and so on. The warning signals are to help flag inadequacies in material and capacity plans (Vollmann *et al.* 1997).

9.2 FACTORY COORDINATION

The function of the factory coordination system is to manage the implementation of the MPS throughout the factory and to ensure that the MPS, which was exploded at the material requirements planning (MRP I) stage is realized across the various work cells at the operational level of the factory.

Factory coordination organizes the flow of products throughout a factory and ensures that the production in each cell is synchronized with overall production goals of the factory. It is therefore a set of procedures concerned with the planning and controlling of the flow of products at a plant level. Factory coordination consists of the following two tasks as shown in Fig. 9.1:

1. Production environment design (PED) task, and
2. Control task.

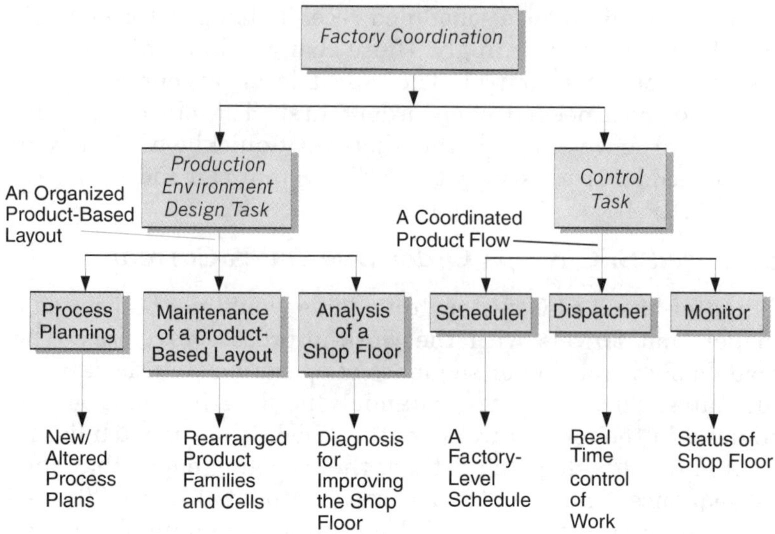

Fig. 9.1 *Factory Coordination*
(McMahon and Browne, 1993)

1. Production Environment Design (PED)

The PED module helps to reduce the variety of possible production related problems by organizing the manufacturing system in so far as possible into a product-based layout and allocating new products to existing product families.

The key role of PED task is to reorganize the manufacturing system to simplify it and to accommodate new products coming into production. The initial layout of the manufacturing system should be, where possible, product-based. This can be accomplished using a methodology such as Group Technology (GT).

2. Control

The control module provides guidelines and goals with which each workshop can manage its activities and deal with any problems that occur within its area of responsibility.

9.3 PRODUCTION ENVIRONMENT DESIGN (PED)

The PED task uses the following:

(a) Static Data

This includes the following documents for each new product:

(i) *Bill of Materials (BOM)* which defines the structure of a product by listing the names and quantities of each component of each assembly and subassembly in a product's structure.

(ii) *Routing (Bill of Processes, BOP)* describes the process steps involved in the production of a product in terms of the required resources, operation procedures, process times, and setup procedures.

(b) The Future Production Requirements Generated by an MRP I Type System

Using an initial product-based layout, the PED module integrates new products into the production environment with minimal disruption and reorganization to the existing product-based structure.

The various procedures involved in PED may include the following:

1. The Detailing of Process Plans for New Products

The inclusion of an element of the process planning activity within the factory coordination function helps to standardize and reduce the complexity in the planning task. It is clear that there may be more than one possible process plans for an individual component and that perhaps different workstations might be used to process a particular part depending on the availability of resources at any particular point in time. At the factory coordination stage, the production planning and control people should have the facility to select from alternative process plans, wherever feasible, in order to make best use of available resources. The process-planning task also allows existing process plans to be altered, if such a need arises.

2. Maintenance of the Product-Based Layout

The product-based layout must be maintained whenever a group of new products is introduced into production. Using a set of procedures based on GT principles, the new products are integrated into the existing range of product families and any necessary reorganization is carried out on the product-based layout with minimal disruption to the daily production activities.

3. Analysis of a Manufacturing System (Shop Floor)

The main purpose of this analysis is to pinpoint particular areas of the production process where there is room for potential improvement. This ensures continuous improvement to a production process, to simplify the control task.

9.4 THE CONTROL TASK WITHIN FACTORY COORDINATION

Product-based manufacturing facilitates the distribution of responsibility for the production of a family of products to each work group. The main purpose of the control task is to coordinate the activities of each Production Activity Control (PAC) system through the provision of schedule and real time control guidelines, while recognizing that each PAC system is responsible for the activities within its own cell.

The control task involves the following activities:

1. The development of schedule guidelines using a factory-level scheduler
2. Implementing the above schedule guidelines and providing real time guidelines for each of the PAC systems using a factory-level dispatcher
3. Monitoring the progress of the schedule using a factory-level monitor

The factory coordination control task, as we shall see, is in many ways a higher-level recursion of a Production Activity Control (PAC) system.

9.5 BUILDING BLOCKS ASSOCIATED WITH THE FACTORY COORDINATION CONTROL TASK

1. A Factory-Level Scheduler The task of the factory-level scheduler is to develop a factory-level schedule, which each PAC system uses as a guide when developing its own schedule.

2. A Factory-Level Dispatcher The task of the factory-level dispatcher is to control the movement of materials between cells on a real time basis and to communicate with each PAC dispatcher.

3. A Factory-Level Monitor The task of the factory-level monitor is to observe the status of the entire factory based on information coming from each PAC monitor.

Each cell on the shop floor may be regarded as a virtual producer, because each cell receives guidelines on its production activities from the factory-level dispatcher. In relation to the movement of materials between cells there may be two different types of factory-level mover.

1. The PAC mover himself can also coordinate the materials movement between cells provided that the material handling between cells is a simple task. Information on the next cell on a product's routing is given to the mover by the particular PAC dispatcher of the cell which the batch is leaving.
2. A factory-level mover is concerned with organizing the materials flow between the cells, and receives all of its instructions from the factory-level dispatcher.

Figure 9.2 is a representation of the above points.

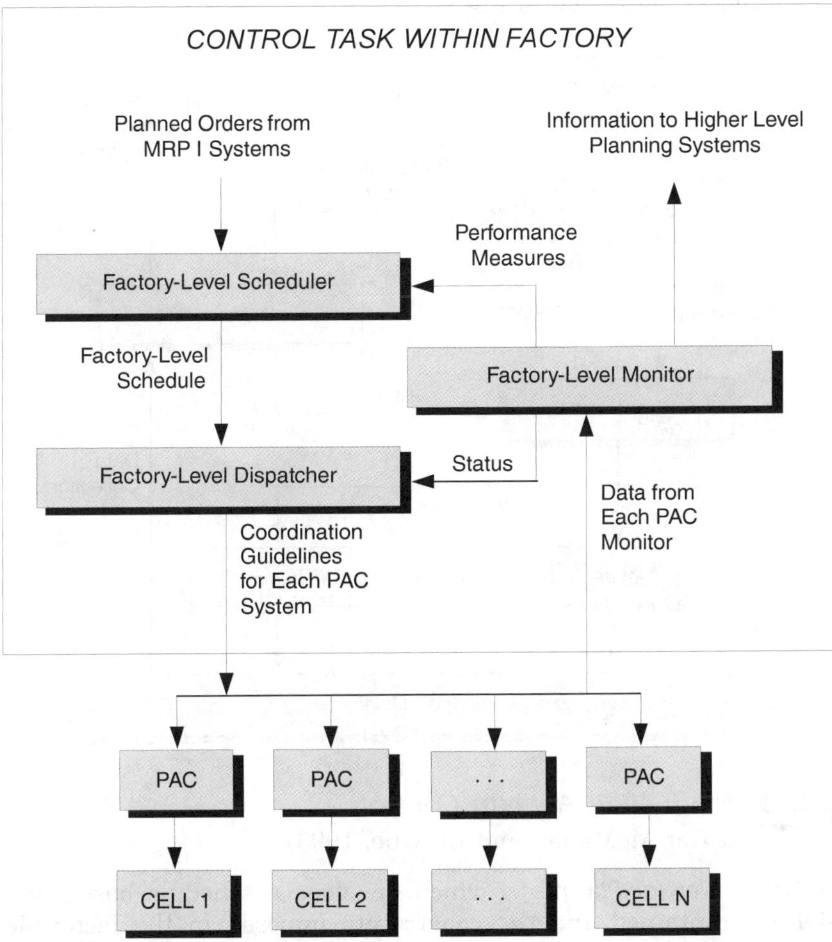

Fig. 9.2 *Data Exchange Between the Control Task of Factory Coordination and a Number of PAC Systems* (McMahon and Browne, 1993)

9.6 OVERVIEW OF PRODUCTION ACTIVITY CONTROL (PAC)

Production Activity Control (PAC) exists at the lowest level of the Production Management System (PMS) architecture. PAC describes the principles and techniques used by management to plan in the short-term (typically between one week and quasi real time), to control and to evaluate the production activities of the manufacturing organization. A PAC system provides the necessary functions to control the flow of products within a cell, through the interaction of the following five building blocks, as shown in Fig. 9.3.

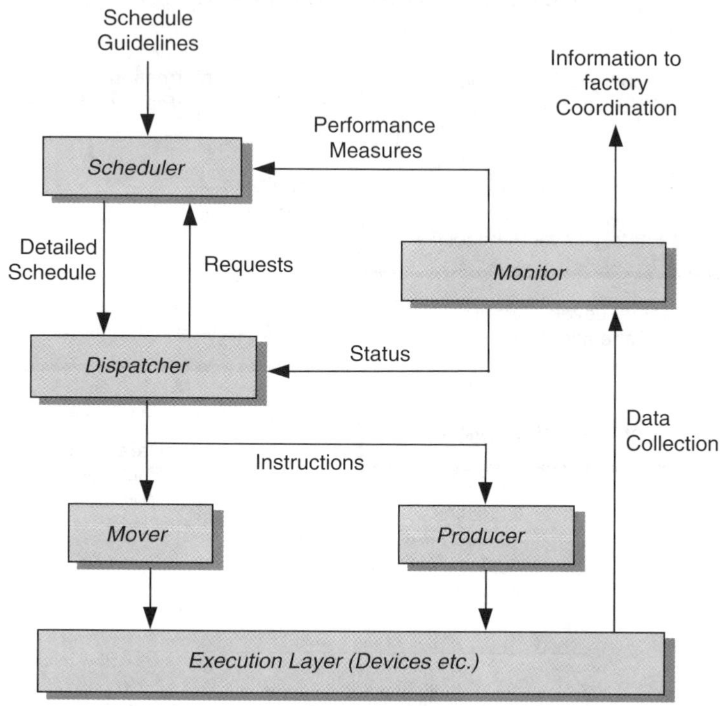

Fig. 9.3 *Production Activity Control*
(After McMahon and Browne, 1993)

1. A Scheduler (or Planner) which develops a schedule based on the guidelines contained and the constraints imposed in the factory-level schedule.

2. A Dispatcher (Supervisor or Foreman) which controls the flow of work within the cell on a real time basis. The term dispatching refers to deciding which job to do next. The dispatcher takes the schedule and issues appropriate commands to the movers and producers. The dispatcher ensures that the schedule is adhered to in so far as possible. It works in quasi real time by receiving data from the monitor on the current state of the system, and it issues instructions to the moving and producing devices so that the required tasks are performed. The term sequencing refers to the process of determining the order in which several jobs should be performed. Thus, sequencing involves a larger problem than dispatching. Although most dispatching rules can be used to sequence an entire queue, it is important to remember that the dispatching decision is always made one job at a time. In manufacturing, it is more important to make dispatching decisions than sequencing decisions since by the time the next job has been completed the set of choices may have changed (or their priorities may have changed).

3. A Monitor (or Expediter) which observes the status of the cell and passes any relevant information back to the scheduler, dispatcher, and the higher-level factory coordination system.

4. A Mover (or Material Handler) which manages the movement of materials and semifinished products between workstations.

5. A Producer (or Machine Operator) which controls the sequence of operations at each workstation.

Frequently the term shop floor control (SFC) is used to describe the operational control of the workflow through the factory floor and incorporates the combination of the factory coordination system and the production activity control (PAC).

9.7 DATA REQUIREMENTS

To plan the processing of materials through manufacturing, PAC must have the following information:

1. What and how much to produce (MPS).
2. When parts are needed so the completion date can be met (BOM).
3. What operations are required to make the product and how long the operations will take (routings).
4. What the available capacities of the work centers are (work center file).

Shop floor control must have an information system from which to work. Usually the data needed to answer these questions are organized into databases. The files contained in the databases are of two types: planning and control.

9.8 PLANNING FILES

Four planning files are needed: item master file, product structure (BOM) file, routing file, and work center file.

9.8.1 Item Master File

There is one record in this file for each part number. The file contains, in one place, all of the pertinent data related to the part. For PAC, this includes the following:

(a) Part number (a unique identification assigned to a component).

(b) Part description.
(c) Manufacturing lead-time (the normal time needed to make this part).
(d) Quantity on-hand.
(e) Allocations (quantities assigned to specific work orders but not yet withdrawn from inventory).
(f) Quantity available (on-hand—allocations).
(g) Scheduled receipts (on-order quantities, the balance due on all open/outstanding orders).
(h) Lot-size quantity (the quantity normally ordered at one time).

9.8.2 Product Structure (Bill of Material) File

This file contains a list of the single-level components and quantities needed to assemble a parent. It shows how a product is made and forms a basis for a 'pick list' to be used by storeroom personnel to collect the parts required to make the assembly. It also provides where-used and pegging information.

9.8.3 Routing File

There is a record in this file for each part manufactured. The routing consists of a series of operations required to make the item. For each product, this file contains a step-by-step set of instructions describing how the product is made. It gives details of the following:

(a) Part number.
(b) The operations required to make the product and the sequence in which those operations are performed.
(c) A brief description of each operation.
(d) Work Center(s) to perform the step.
(e) Equipment, tools, gauges, and accessories needed for each operation.
(f) Setup times, the standard time required for setting up the equipment for each operation.
(g) Run times, the standard time to process one unit through each operation.
(h) Lead-times for each operation.

9.8.4 Work Center File

The purpose of this file is to collect all of the relevant data on a work center. For each work center, it gives details on the following:

(a) Work center number and name,
(b) Capacity,
(c) Number of shifts worked per week,
(d) Number of machine hours per shift,
(e) Number of labor hours per shift,
(f) Efficiency,

(g) Utilization,
(h) Queue time (the average time that a job waits at the work center before work is begun),
(i) Alternate work centers (work centers that may be used as alternatives), and
(j) Costing information (average labor rate, overhead rate).

9.9 CONTROL FILES

Control in intermittent (batch) manufacturing is exercised through shop orders and control files that contain data on these orders. There are generally two kinds of files: the shop order master file and the shop order detail file.

9.9.1 Shop Order Master File/Open Orders File

Each active manufacturing order has a record in this file. The purpose is to provide summarized data on each shop order such as the following information:

(a) Shop order number, a unique number identifying the shop order,
(b) Order quantity,
(c) Quantity completed,
(d) Quantity scrapped,
(e) Quantity of material issued to the order,
(f) Due date (the date the order is expected to be finished),
(g) Priority (a value used to rank the order in relation to others),
(h) Balance due (the quantity not yet completed), and
(i) Cost information.

9.9.2 Shop Order Detail File

Each shop order has a detail file that contains a record for each operation needed to make the item. Each record contains the following information:

(a) Operation number,
(b) Setup hours—planned and actual,
(c) Run hours—planned and actual,
(d) Quantity reported complete at that operation,
(e) Quantity reported scrapped at that operation, and
(f) Due date or lead-time remaining.

9.10 ORDER PREPARATION

Once authorization to process an order has been received, SFC/PAC is responsible for planning and preparing its release to the shop floor. The order should be reviewed to be sure that the necessary tooling, material, and capacity are available. If they are not, the order cannot be completed and should not be released.

9.10.1 Material/Tooling Check

If MRP I software is used, it will have checked the availability of material and allocated it to a shop order so no further checking is necessary. If MRP I software is not used, SFC must manually check material availability. Tooling is not generally considered in the material requirements planning (MRP I) program, so at this stage, the availability of tooling etc. must also be checked.

9.10.2 Capacity Check

If a capacity requirements planning (CRP) system has been used, necessary capacity should be available. However, at this stage, there may be some differences between planned capacity and what is actually available. When CRP is not used, it is necessary to determine if adequate capacity is available.

Checking capacity availability is a two-step process. First, the order must be scheduled to see when the capacity is needed, and second, the load on work centers must be checked in that period. If there is not enough capacity, the release of the order can be delayed until the time that the necessary capacity becomes available. Because if the order is released to the transformation process when there is not enough capacity, then all that would have been accomplished would simply be the overloading of the operations management system. The result would be longer lead-time, greater lead-time variation, and greater confusion among the conversion-process operators trying to identify which orders they can or cannot work on because of a lack of capacity.

It is necessary to check actual capacity levels before releasing orders even though CRP is supposed to assure the SFC system access of adequate capacity. The reason is that the capacity requirements planning system sets capacity levels *in advance.* It tries to set expected capacity levels available. However, at any time, the amount of capacity that is available may vary greatly from what is expected. There are a number of factors that may account for this:

1. Absenteeism due to illness, vacation, etc.
2. Equipment unavailability due to breakdown or unplanned maintenance.
3. Inaccurate processing times.
4. Changes in priorities.

9.11 SCHEDULING

In scheduling, we are concerned with ensuring that the *right* tasks are conducted at the *right* time on the *right* items to produce the output—which is, in a sense, a matter of 'orchestration.' Problems with productivity are often attributable, in large part, to poor management of the schedule: *not* having the right items when they are needed, not having equipment available when it is needed, and so on. Scheduling addresses issues managers must face on a daily basis: where each input (material, machine, worker) must be, when it must be there, what form it must be in, how many must be available, and other such details (Shafer and Meredith, 1998).

The objective of scheduling is to meet delivery dates, minimize the lead-time, and to make the best use of manufacturing resources. It involves establishing start and finish dates for each operation required to complete an item. To develop a reliable schedule, the planner must have information on routing, required and available capacity, competing jobs, and manufacturing lead-times (MLT) at each work center involved. There are many techniques to schedule shop orders through a plant, but all of them require an understanding of forward and backward scheduling as well as finite and infinite loading.

Scheduling for continuous process and flow organizations is not anywhere near the problem that it is for jobshops. This is because scheduling is largely *built into* the transformation process when the facility is designed and therefore need not, and in fact *cannot*, be constantly changed. To reschedule these facilities, beyond just increasing or decreasing the rate of input, requires a rebalancing of the entire flow through the production system. At the other extreme, the scheduling of project operations is probably the most important single planning activity in the successful management of projects (Shafer and Meredith, 1998).

9.11.1 Backward Scheduling

CRP most often uses the technique of backward scheduling to place load hours into work centers. With backward scheduling, we begin with the requested date (a scheduled or planned receipt date) and work backward accounting for all lead-time elements in order to place the loads (operation times) in the correct time periods as shown in Fig. 9.4. That is, the last operation on the routing is scheduled first and is scheduled for

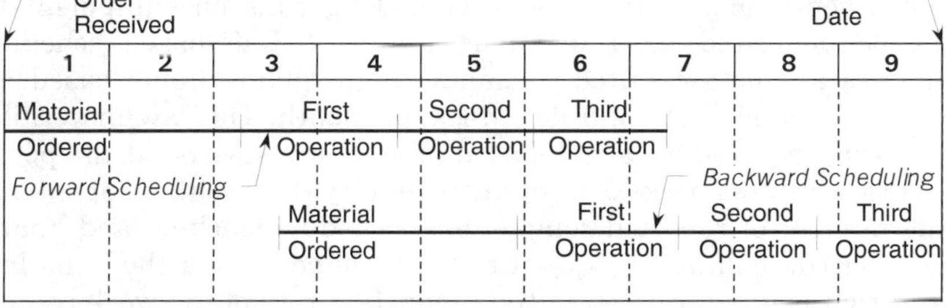

Fig. 9.4 *Forward and Backward Scheduling: Infinite Loading*

completion at the due date. Previous operations are scheduled back from the last operation. It is a form of simulation using standard and estimated times.

Backward scheduling reveals, in addition to work center load profiles, the latest possible start date for each operation in a work order to achieve on-time completion. The backed-off start time for the *first* operation (at what is sometimes referred to as the 'gateway' work center) indicates the latest date that an order could be released to meet the scheduled due date (John, 1993). Backward scheduling schedules item to be available as needed and is the same logic as used in MRP I system. Work-in-process inventory is reduced, but because there is little slack time in the system, customer service may suffer. It is not unusual when backward-scheduling to find operation start dates falling into past-due time periods. This condition must be rectified through lead-time compression tactics to keep the scheduling information in the system valid.

9.11.2 Forward Scheduling

It assumes that material procurement and operation scheduling for a component start when the order is received, whatever the due date, and the operation are scheduled forward from this date. This begins with the planned order release date from MRP I, or the completion date of the previous operation in line for open orders, and works forward through the routing to predict the earliest completion date for the order as shown in Fig. 9.4. In addition to a projected earliest completion date for the order, it also shows the earliest completion date for each operation. Often, the result is completion before the due date, which usually results in a buildup of inventory. Forward scheduling is used to calculate how long it will take to complete a task. The technique is used for purposes such as developing promise dates for customers or figuring out whether an order behind schedule can be caught up.

9.11.3 Rate-based Scheduling

The use of work orders, along with the creation of shop paper and a pick list, is not always appropriate for repetitive or continuous production environments. Work orders are not a practical means of scheduling items that run continuously or are subject to long runs on equipment that might be dedicated to a few part numbers. Rate-based scheduling techniques are better suited to handle high-volume items based on a periodic production rate of a day, week, or month. The raw materials or components required for a rate-based product are stocked at the point of use. This facilitates backflushing the inventory that was consumed based on the number of reported completions. Schedule tracking and analysis consists of the cumulative quantities completed versus the cumulative quantities scheduled related to the production rate of the work center or assembly line (Wassweiler, 1997).

9.11.4 Infinite Scheduling/Loading

Infinite scheduling means calculating the work center loads by time period without regard to capacity. It does not consider the existence of other shop orders competing for capacity at these work centers. It assumes that the work centers on which operations are to be performed would be available whenever required, that is, it assumes infinite capacity will be available. Planning to infinite capacity has the effect of revealing potential overloads and underloads in various periods.

Infinite scheduling involves loading work into work centers, leveling the load, sequencing the work, and monitoring its progress. Here leveling and sequencing decisions are made after overloads or underloads have been identified. This iterative process is time-consuming and not very efficient (Russell and Taylor III, 1998).

According to Stonebraker (1994), the tem 'infinite loading' is obsolete today, although the specific computer programs used to do infinite loading can now be used to perform the technique called CRP. Infinite loading was a gross misnomer to start with, implying that a load could be put into a factory regardless of the factory's ability to perform. The poor terminology obscured the fact that it is necessary to generate capacity requirements and compare these with available capacity before trying to adjust requirements to capacity. The success of infinite scheduling is dependent on the initial capacity management, the flow of orders, and shop floor prioritizing at each work center. The prioritizing of work is based on predetermined rules such as work center completion date or next operation availability.

The Dispatch List

The dispatch list displays all the jobs at a work center that are either in process or waiting to be processed in priority sequence. The dispatch list is a priority schedule for each work center that is normally published daily (as a printed document) or in real time (through CRT display) and lists the workload with a visibility of 2–5 days. Dispatch list is defined by APICS Dictionary as a listing of manufacturing orders in priority sequence. The dispatch list, which is usually communicated to the manufacturing floor via hard copy or CRT display, contains detailed information on priority, location, quantity, and the capacity requirements of the manufacturing order by operation. Dispatch lists are normally generated daily and oriented by work center (Cox III and Blackstone Jr, 1998).

Dispatching involves the selecting and sequencing of available jobs to be run at individual work centers and the assignment of those jobs to workers. The dispatch list is the starting point for the shop supervisor to assign shop orders to the operators.

The due dates on the dispatch lists arc important, as most hard-copy shop orders no longer carry due dates. Experience has shown that for a variety of reasons, the due dates at the time of order release cease to be valid during the lifetime of the shop order. The daily published dispatch lists have at times been confusing to the shop floor due to updating problems such as the job that completes 20 minutes after publishing or the priority that changed from yesterday and the job now being set up is no longer needed the next six days. On-line CRT display of the dispatch list has reduced much of the confusion (Toomey, 1996).

Dispatching or Priority Rules

Operation priorities are determined using decision rules commonly referred to as dispatching or priority rules. The resulting priorities are essentially the orders ranked—the most important order is at the top while the least is located at the end of the list. Priorities are communicated to people in the system through the dispatch list. Following are some of the commonly used priority rules:

1. Earliest operation due date or start dates. If all work is available as scheduled and there are no capacity problems at the work center.
2. Least total slack time remaining.
3. Jobs with the shortest processing time first. This rule causes lower Work-in-Process (WIP) but is only valid when all jobs are running late.
4. Job with the lowest critical ratio first. Critical ratio is (order time remaining to due date completion the expected manufacturing lead-time remaining to completion). A priority less than one indicates a job behind schedule (hence high priority), whereas a priority greater than one indicates a job that is ahead of schedule (hence low priority).

There are other variations of priority rules, but some such as first in, first out or longest processing time conflict with the first goal of the system—completing the fabrication by the due date. Judgment is required in the execution of the system. There is no point in running a job first, no matter what the priority rule, if it is going to sit in queue for 2 weeks at the next operation (Toomey, 1996).

Controlling the Schedule

A factor to be resolved is who is going to control the schedule. The foreman knows the people, the machines, and the tooling. The planner in production control understands the requirements of the order. The most reasonable compromise, according to Toomey (1996) is for production control to do the scheduling, but for the foreman to be allowed adjustments within predetermined limits. An advantage of an infinite oriented dispatch list is that the foremen can easily understand the logic. A common situation is when setup times can be substantially reduced (or in some cases eliminated) by combining jobs. The combining of jobs may cause running an order out of sequence. The desired result can happen when the foreman has enough system knowledge and understanding to know when such variation of sequencing will not have an adverse effect on the on-time completion of the part. In other words, will there be adequate slack time in the balance of the run schedule and will the system accommodate it?

Operations Overlapping

In some situations, the setup times for a number of operations may be excessive and therefore require large lot-sizes. In this situation, the shop order may not have to be completed for each operation for the whole lot before scheduling the next operation. The items completed can be sent to the next work center even though the whole lot is not complete. This method of scheduling is called overlapping. Most software systems allow for overlap allowances in the routing file and will schedule based on the allowances. Table 9.5 illustrates the effect of overlapping.

		Conventional Jobshop Schedule		*Overlapping Jobshop Schedule*		
Operation No.	Queue, Setup, and Run Time	Scheduled Start	Scheduled Completion	Overlap Allowance	Scheduled Start	Scheduled Completion
10	40 Hours	Day 1	Day 5	80%	Day 1	Day 5
20	48 Hours	Day 6	Day 11	50%	Day 2	Day 7
30	32 Hours	Day 12	Day 15		Day 5	Day 8

Fig. 9.5 *Conventional and Overlapping Schedule*
(Adapted from Toomey, 1996)

9.11.5 Finite Scheduling/Loading

Finite scheduling is a method that considers capacity and does not permit overloads. It assumes there is a defined limit to available capacity at any workstation. If there is not enough capacity available at a workstation because of other shop orders, the order has to be rescheduled in a different time period. Real-time forward finite loading may not only consider the capacity of the work center but also the availability of the shop order, manpower, and tooling. The system will collect and monitor the data and then load based on predetermined priority rules. In this approach, loading and sequencing decisions are made at the same time, so that the first jobs loaded onto a work center are of highest priority. Any jobs remaining after the capacity of the work center or resource has been reached are of lower priority and are scheduled for later time periods. Finite scheduling systems use a variety of methods to develop their schedules, including mathematical programming, network analysis, simulation, and expert systems or other forms of artificial intelligence. Finite schedulers are becoming more popular as software systems become more adaptable and easier to use and as manufacturing environments are simplified and are better understood.

In computerized finite loading, the computer is allowed to make decisions and solve problems. For instance, if an overload exists in a particular week, the computer identifies the situation, evaluates some alternatives, and recommends scheduling of an extra shift or proposes some other solution. Finite scheduling algorithms are being used successfully to schedule production and have been a significant help in maintaining the MPS. However, in the situations where finite loading has been used successfully calculation is being done for a single level in the BOM, not for multiple levels. The finite loading calculations are also being done for high-volume, repetitive, or flow-type of manufacturing, not for jobshop, disconnected-flow manufacturing environments.

Over the years, however, a number of finite loading systems have been aimed at multiple-level, jobshop, disconnected-flow manufacturing. But most finite loading systems of this type (as distinguished from those for single-level loading of high-volume production or packing lines) have been discontinued. The reason being that computers, although good at handling and manipulating massive amounts of data, don't fare well when given the opportunity to evaluate options and creatively make decisions. Some companies have even achieved operating improvements from using finite loading, but these improvements have come from the basic capacity planning that is a part of the technique, rather than from the finite loading logic.

An approach that does seem to work well is to use a finite scheduling system as a simulator. The planner allows the finite loading algorithm to run and then studies the results. If the planner and the manufacturing supervisor like the solution then they implement this solution. The finite loading software does the arithmetic, but people evaluate the solutions, make the decisions, implement them, and are accountable for the results.

Loading to finite capacity, also called operations sequencing, falls within the scope of the production activity control (PAC) in the shop floor control (SFC) module. CRP normally uses backward scheduling only and loads without regard to capacity (infinite loading). Priority sequencing rules used in CRP do not consider the work center capacities or the other jobs to be scheduled.

Finite loading systems simulate actual job order starting and stopping to produce a detailed schedule for each shop order and each work center; that is finite scheduling does, in fact, *load* all jobs in all necessary work centers for the length of the planning horizon. The result is a set of start and finish dates for each operation at each work center. One output of finite loading is a simulation of how each machine center will operate on a minute-by-minute basis for whatever time horizon is planned (Vollmann *et al.* 1997).

The selection of the next job to schedule is not just based on those jobs that are physically waiting at the work center. Most finite loading systems look at the jobs expected to come to the work center, when they will be completed at the prior centers, and what priorities these jobs have to decide whether to leave the work center idle for the anticipated arrival of a particular high-priority job so that the work center would be immediately available. Also, some systems would overlap the operations, so a job might start at a work center before the completion of all work at the prior center.

Vertical Loading

The approach of filling a work center job by job is called vertical loading. It is consistent with the way most jobshop scheduling research is conducted, as well as with the priority scheduling viewpoint; that is, one looks at a work center and decides which of a set of jobs (both those at the work center and those scheduled to have had their prior operations processed by the time of loading) to load next.

Horizontal Loading

With this approach, one entire shop order or job (that job with the highest priority) is loaded for all of its operations on the required work centers based on the sequence of operations specified in the routing, then the next highest priority job, and so on. This horizontal approach may well be at odds with a criterion of using a machine center to its maximum capacity. By creating detailed schedules with horizontal loading, the net result can be 'holes' in the capacity for a work center. What this implies is that at times a machine is to sit idle, even if a job is available, because a more important job is coming.

If the finite loading system were run on a Sunday night, the schedules set for a particular work center for Monday should be reasonably accurate. However, the schedules for Tuesday must deal with Tuesday's expected starting conditions. Moreover, the jobs scheduled for Tuesday are more likely to come from another work center, rather than already being at the work center. Differences between the simulated operations and actual operations in those work centers will be reflected in the actual job arrivals at the work center. These uncertainties on the shop floor mean that there will be a decay in information validity that grows as the time horizon for the schedule is lengthened. One way to overcome this decay is to redo the planning (finite loading) more often. If it is done each night, the next day's plan will be more accurate. The problem with doing finite loading every night is the computer processing time and cost. Finite loading can easily take 10 times as much computer run time as shop floor control by priority scheduling.

Because finite loading only schedules a work center up to its capacity (and tries to schedule it up to the limit), it will often be true that matching parts have inconsistent due dates. For example, if assembly A is made up of parts B and C. The due dates for both B and C would be identical under MRP I. If part B is delayed for two weeks at some work center, there is no reason to rush completion of part C. Finite loading cannot take explicit account of these product structure relationship in the scheduling process. The use of priority information, other than MRP I due dates, can compound this problem.

Problems with Finite Loading The results of finite loading may conflict with MRP I requirements. In other words, infinite loading is compatible with MRP I requirements, but may not be realistic while finite loading will be realistic, but may not be compatible with MRP I requirements. Secondly, a finite loaded order based on the priority rules may make sense based on the scheduled work center, but not if workload down the line is considered. For example, if a high priority job at a work center has to wait for a long time at the next work center and a lower priority job's next scheduled operation is on an open work center, it makes greater sense to schedule the lower priority job first since the high priority job if completed early on this work center would, in any case, just wait in the long queue in the subsequent work center (Toomey, 1996).

Finite scheduling is accomplished with stand-alone systems calling for sophisticated data control and calculations. It is most important that these systems be integrated with the planning system (that is, MRP I requirements). Finite loading must be continually compared to planning requirements in order to make expedient adjustments to either planning or execution (Toomey, 1996).

Operation Sequencing

It is a simulation approach in which the initial calculation generates a short-term plan at each work center based on finite scheduling priorities. These plans are then projected to show completion times and queues. The simulation continues by moving the job to the next work center and then second day's run is simulated. Simulation can predict both potential overloaded and idle work centers, and therefore assist the planner (dispatcher) in making priority adjustments to short-term plans prior to the extension of the simulation-based schedule into real-time dispatching. Operation sequencing should be generated on at least daily or on a shift-by-shift basis.

9.12 INPUT-OUTPUT CONTROL

Input-output control is the means of verifying that the capacity plan is being executed as planned. Input-output control is seen most often in jobshop, disconnected-flow environments in conjunction with capacity requirements planning and shop floor control. It can, however, be applied to almost any situation in which capacity planning is being done. Input-output must operate even if it is based on only rough-cut capacity data.

Input/output (I/O) control monitors the input to and output from each work center. Prior to such analysis, it was common to examine only the output from a work center and to compare the actual output with the output planned in the shop schedule. Using that approach in a jobshop environment in which the performance of different work centers is interrelated may result in erroneous conclusions about the source of a problem. Reduced output at one point in the production process may be caused by problems at the current work center, but it may also be caused by problems at previous work centers that *feed* the current work center. Thus, to identify more clearly the source of a problem, the *actual input* to a work center must be compared to the *planned input,* and the *actual output* must be compared to the *planned output.* Deviations between planned and actual values are calculated, and their cumulative effects are observed. The resulting backlog or queue size is monitored to ensure that it stays within a manageable range (Russell and Taylor III, 1998).

Figure 9.6 shows the output side of input-output control for a particular work center, in this case, a group of milling machines. Following formulas are used for both input and output:

$$\text{Deviation} = \text{Actual} - \text{Planned}$$

$$\text{Cumulative Deviation} = \text{Previous Deviation} + \text{Actual} - \text{Planned}$$

Work Center: Vertical Milling

	Weeks							
	– 4	– 3	– 2	– 1	1	2	3	4
Planned Output	112	112	112	112	112	112	112	112
Actual Output	90	97	95	90				
Cumulative Deviation	–22	–37	–54	–76				

Today (between weeks –1 and 1)

Fig. 9.6 *Input-Output Control—Output Side*

Work Center: Vertical Milling

	Weeks							
	– 4	– 3	– 2	– 1	1	2	3	4
Planned Input	112	112	112	112	112	112	112	112
Actual Input	114	115	116	107				
Cumulative Deviation	+ 2	+ 5	+ 9	+ 4				

Today (between weeks –1 and 1)

Fig. 9.7 *Input-Output Control—Input Side*

It can be seen in Fig. 9.6 that the planned output is 112 standard hours per week, but the actual output is less than this amount, yielding a cumulative deviation over four weeks of minus 76 standard hours. In other words, the work center is not producing the expected hours of output. The first question to ask would be 'Is the work coming into the work center?' The input side of the input-output report provides this information to a planner (Fig. 9.7).

This report shows that the planned input is 112 standard hours per week. The actual input is close to that, as seen by the cumulative deviation of 4 standard hours. It's clear from these numbers that the work is flowing into the work center but not coming out. Therefore, this work center is the problem. If the work weren't coming into the work center, then the problem would be somewhere 'upstream'.

In a jobshop environment with jobs moving in a network fashion, it's sometimes difficult to identify the problem, especially when multiple work centers contribute to the situation. By looking at the input-output control reports, planners can clearly see which are the problem work centers, even in a complex network of interconnected work centers.

To identify a problem work center and not respond to minor fluctuations in the numbers, companies establish tolerances. Whenever the cumulative deviation for either input or output exceeds a predetermined number of hours, action must be taken to correct the situation. Experience has shown that it's much easier to solve a problem if you know what the boundaries are ahead of time. In the case of the milling work center described above, the number is 30 hours. Since the cumulative deviation is out of tolerance, the manufacturing supervisor responsible for this work center must take action to correct the problem. *A planning and control system does not solve problems—it simply identifies them so people can do something about them.* This particular problem with the milling work center may be a quality problem, an equipment breakdown problem, a manpower problem, and so on.

9.12.1 The Data for Input-Output Control

The data for input-output report comes from a number of different sources. Shop floor control generates the actual input and output numbers, because it records the movement and completion of each operation (Landvater, 1997).

1. Actual Output

This line is based on the operation completion transactions. If an operation is finished and it required *n* standard hours, then *n* standard hours are shown as the actual output in the specified time period (most people use time periods of weeks, although it is possible to use days).

2. Planned Output

This line comes from the manufacturing supervisor. He commits to a number of hours that can be produced from the work center. This is the planned capacity.

3. Actual Input

This line of the input-output report results from the move transactions in the shop floor control system. If an operation is worth n standard hours and it gets moved into the work center today, then n standard hours get added to the actual input for the time period. (Some environments do not use move transactions. In some pharmaceutical manufacturing operations, for example, the completion of an operation has the same effect within the system as a move transaction to the next operation. When the previous operation is completed, the n standard hours of work are added to the actual input for the time period.) The input rate to a work center can be controlled only for the initial operations of a job. These first work centers are often called *gateway* work centers, because the majority of jobs must pass through them before subsequent operations are performed. Input to later operations, performed at downstream work centers, is difficult to control because it is a function of how well the rest of the shop is operating—that is, where queues are forming and how smoothly jobs are progressing through the system.

4. Planned Input

This line is taken from CRP. If CRP summary report for the work center shows n standard hours per week as capacity requirements, then n standard hours per week are shown as planned input for the time period.

5. Cumulative Deviations

These are based on simple math: the calculation takes the planned input and the actual input and compares them. The deviation of planned to actual input for downstream work centers can be minimized by controlling the output rates of feeding work centers. The same calculation for cumulative deviations is done with the output side. The input difference will be due to the unplanned variations in arrival patterns due to previous operations being completed either early or late against schedule. Output less than planned can be due to production problems such as machine breakdowns, quality considerations, absenteeism, and so forth, or lack of available work (input). Output greater than planned is usually due to an effort to reduce the queue of available work (Toomey, 1996).

6. Planned Queue/Backlog

It is a target set by the capacity planner and manufacturing supervisors. It's the amount of work they want to put in front of the work center. The larger the queue, the longer the lead-time and the larger the work-in-process. The objective should be to reduce the size of the queue. On the other hand, queues serve a useful purpose by evening out the uneven flow of work into work centers.

Planned Backlog = Planned Input + Previous Planned Backlog – Planned Output

7. *Actual Queue/Backlog*

It is the actual queue or backlog at a work center and can be calculated as follows:

Actual Backlog = Actual Input + Previous Actual Backlog – Actual Output

8. *Tolerances*

These are numbers agreed to by the capacity planner and the manufacturing supervisor. It is their estimation of the level of cumulative deviation which they consider as a problem situation and which might result in missed schedules. Most companies pick a number that seems reasonable—for example, 50% of the planned queue—and revise it if necessary. Input/output control is a form of *management by exception* because no action is taken unless the cumulative deviation exceeds some threshold. Whenever the threshold is exceeded, the computer creates a high-priority action message to be sent to the production manager, or whoever supervises order releasing. The threshold should be high enough to permit helpful deviations (Blackstone Jr, 1989).

9.12.2 Who Does Input-Output Control

The capacity planner meets with the manufacturing supervisors to review the input-output control report. This is done in the same meeting in which the CRP reports are reviewed. In the same way that overloads and underloads are discussed, work centers that have exceeded the tolerance on either input or output are identified. Working together, the manufacturing supervisors identify possible solutions. Again, if they cannot agree on a solution, they can appeal to their respective managers for a decision.

Input/output control provides the information necessary to regulate the flow of work to and from a network of work centers. Increasing the capacity of a work center that is processing all the work available to it will not increase output. The source of the problem needs to be identified. Excessive queues, or *backlogs,* are one indication of that *bottlenecks* exist. To alleviate bottleneck work centers, the problem causing the backlog can be worked on, the capacity of the work center can be adjusted, or input to the work center can be reduced. Increasing the input to a bottleneck work center will not increase the center's output. It will merely clog the system further and create longer queues of work-in-process, resulting in longer lead-times.

While trying to control the queues via input/output control, the first option, according to John Blackstone, Jr, the planner should consider is *deexpediting*, that is, moving a due date further out into the future. In most companies *expediting*—moving work up to be performed sooner—is frequently performed, but deexpediting is considered low priority. Nevertheless, jobs frequently are not needed as early as originally scheduled. Since failure to deexpedite is a common cause of queue buildup, appropriate rescheduling should be the first option considered in reducing queues. Deexpediting can be performed by using the pegging capabilities built into the MRP I system. Through pegging, a planner can trace a particular job to the order which created it. He or she can check to see if the end item is still desired on the original date. If a later delivery date is acceptable, then the end item may be rescheduled and a net change MRP I run to determine the new required date of the component under consideration. If the planner cannot achieve necessary reduction in input via de-expediting, other options are alternate routing (use of another machine within the same department or shop) and subcontracting (sending work out of the shop) (Blackstone Jr, 1989).

The input/output control technique is important because it is used to maintain actual queue lengths near planned queue lengths. Since queue time represents most of the time a typical job spends in a shop, queue control is tantamount to lead-time control. It is reasonable to say that MRP I will work properly only if lead-times are met by a technique such as input/output control (Blackstone Jr, 1989). The basic principle of input/output control, according to Oliver Wight is 'never put into a manufacturing facility or to a vendor's facility more than you believe that he can produce.'

9.12.3 Input/Output Control Example with Actual Backlog

Figure 9.8A shows the information required for preparing and Input/Output Report for a particular work centre.

- The Input/Output Report has planned a level production of 75 units per period for this work centre.
- This is to be accomplished by working off the backlog of work and steadily increasing the input of work.

A. Input/Output Report

		Weeks				
		1	2	3	4	Total
Planned Input		60	65	70	75	
Actual Input		60	60	65	65	
Deviation						
Planned Output		75	75	75	75	
Actual Output		70	70	65	65	
Deviation						
Actual Backlog	30					

B. Completed Input/Output Report

		Weeks				
		1	2	3	4	Total
Planned Input		60	65	70	75	270
Actual Input		60	60	65	65	250
Deviation		0	– 5	– 5	– 10	– 20
Planned Output		75	75	75	75	300
Actual Output		70	70	65	65	270
Deviation		– 5	– 5	– 10	– 10	– 30
Actual Backlog	30	20	10	10	10	

Fig. 9.8 *Input output Example*
(After Russell and Taylor III, 1998)

- The report is completed by calculating the deviation of the actual input/output from the planned input/output. Then summing the values in the respective planned, actual, and deviation rows.
- The initial actual backlog (at the beginning of period 1) is 30 units.
- Subsequent actual backlogs are calculated using the formula (Actual Input + Previous Backlog – Actual Output).

The completed input/output report (Fig. 9.8B). shows that the work center did not process all the jobs that were available during the four periods, therefore, the desired output rate was not achieved. Similarly the work center received less input than the planned input. Hence, the input/output reports from those work centers need to be examined to locate the source of the problem.

9.13 PLANT SCHEDULING

1. Production schedules, kanban, and operation scheduling (shop floor control) all work. This is because each is suited to the manufacturing environment in which it is deployed. The objective is the same: to effectively communicate the schedules and to provide a sufficient degree of control to assist people in meeting their delivery commitments. The methods differ because the requirements are different.
2. The production schedule is a simple and effective way to communicate the schedule in a manufacturing environment in which scheduling is done by item, not by operation. Production schedules are used in industries such as chemical manufacturing and food processing, and in high-volume repetitive manufacturing, in which the need to track every operation is unnecessary and the cost is prohibitive.

3. Where JIT is being implemented or used, kanbans are likely to be the primary tools for controlling the movement of work on the plant floor. The increased manufacturing velocity and the focus on flow eliminate the need to track every operation. Brand-name and generic kanbans provide the degree of control needed to run the business.
4. Operation scheduling (or shop floor control) is generally used in complex jobshop environments with disconnected flow. The level of control needed in this type of manufacturing requires scheduling by operation rather than just by item. Operation schedule dates are used to assess where work is relative to its completion date, and status reporting on the different operations is used to track the progress of the work.
5. The primary measurement in plant scheduling is 95% or better on-time performance to the order due date within a specified tolerance. The tolerance will vary by company and industry, ranging from zero to several days.

9.14 KANBAN

Kanban is a visual control system used to govern the movement of material on the plant floor. In its most basic form, kanban is a work authorization that restricts the amount of work between two work centers by setting limits. Two of the most popularly used kanban signals in manufacturing are empty containers and cards.

1. Empty Containers In manufacturing an empty container can be an authorization to produce. There are no work orders for these items. (The simplest example of kanban is the milkman. If you wanted two bottles of milk, you left two empty bottles in the milkbox.)

2. Cards Cards are also popular kanban signals. When a card returns to an operation, it is an authorization to produce more product. The card itself indicates the quantity. When production of an item is complete, the card is sent along with the work to the next operation. When these recently manufactured item are used, the card is returned, authorizing a replacement of the items.

The number of ways to signal an operation to make more is only limited by the imagination. Anything that communicates the information from one work center to another can be used as a kanban. Kanbans can actually take on an amazing variety of forms. Some examples include: squares on the floor, carts, boxes, colored flags, parts boards, clips, spaces, golf balls, and cards. Whatever the specific mechanism kanban limits the work-in-process.

9.15 PRODUCTION REPORTING AND STATUS

Information in the SFC system acts as the major link between the planning system and shop floor control system. The information flows in two directions. The MRP I system keeps the SFC system informed of any changes in the planned requirements, and in return the SFC system keeps the MRP I aware of the progress of all orders through the system.

Following pieces of information are collected by the data collection and monitoring system:

- Current location of the order (including scrap replacements required),
- Current state of an order's completion,
- Actual resources being used at each operation, and
- Actual material being consumed at each operation.

This information is often collected in terms of hours, units, weight, volume, and costs. The progress of order in terms of actual costs incurred is more easily understood by people outside the system. Labor and material variances are created when the standards are not achieved. Overhead variances occur when labor standards are missed because overhead is calculated as a per cent of labor. The usual causes for variances relate to scrap, bill of material errors, invalid routing times, rework, and the use of alternate materials and methods (Wassweiler, 1997).

Once the information has been collected, it must be reviewed. This is done by comparing the actual progress with the expected or planned progress. Progress is measured along several different dimensions:

- State of completion,
- Cost incurred to date,
- Amount of scrap produced, and
- How near to the due date.

Ultimately, the purpose of monitoring is to identify those orders with the largest gaps between actual and desired levels of performance so corrective action can be taken (Wassweiler, 1997).

9.16 MEASUREMENTS FOR CONTROL

The following performance measures can be used to measure the effectiveness of production:

1. Scrap It is a valid measurement of the necessary operator skill, the production process, and/or the design for manufacturing.

2. *Rework.*

3. *Queue Time* It represents the needless work-in-process that increases a company's inventory investment and inflates lead-times.

4. *Labor and Material Variances* These may indicate the accuracy of the standard times on a routing and the accuracy of the bills of material.

5. *Schedule Performance* It compares the number of shop orders completed by the due date to those that are past due for a time period.

6. *Statistical Process Control* It can control product quality by monitoring and analyzing process outputs by statistical techniques. Feedback can be used to maintain and improve process capability.

7. *Efficiency* It is the ratio of the actual units produced to the standard number of products expected in a time period.

8. *Productivity* It is the actual output of production compared to the actual input of resources.

9. *Setup Reduction* Single Minute Exchange of Die (SMED) is an excellent program for reducing setup time and cost.

9.17 ORDER DISPOSITION

When manufacturing is complete, the shop order is transferred out of work-in-process, and costs are accounted for. Completion of this stage indicates that SFC system is no longer responsible for the order. However, before the order is closed out, it is useful to study it and see what problems it encountered and why. This is an important step in such programs as total quality management and just in time (Wassweiler, 1997).

9.17.1 Objectives of Order Disposition

- To relieve the SFC system of the responsibility of the order.
- To provide information for dispersing the product either by shipping or storing.
- To provide the rest of the firm with order completion information on which to evaluate performance of the process.

9.17.2 Two Ways of Disposing of the Order

1. *Acceptable Items become Part of the Firm's Inventory Stocks* In case of acceptable orders, the items become part of the firm's inventory stocks. Such items are now treated assets by the accounting system, and they are available for filling customer orders.

2. *Scrapped Items are Charged off to a Salvage Account* In case the material is scrap, the items in the order can be disposed of to one or more

'profit-and-loss' accounts. If there is no recoverable salvage value, the cost of scrapped items can be charged off to a salvage account such as scrap expenses (Wassweiler, 1997).

9.17.3 Evaluation against Standards

After the completion of order disposition, the management can evaluate the performance of both operations and the conversion process against standards.

Factors to be Considered for Evaluation against Standards

- The amount of resources used (such as, the number of labor-hours used, amount of machine-hours used),
- The materials used for the orders,
- The number of setup hours used,
- The cost of tooling used,
- The date on which the order was completed,
- The amount of rework or scrap generated by the order, and
- The number of units completed compared with the number started (Wassweiler, 1997).

9.18 IMPORTANCE OF SHOP FLOOR CONTROL

You can use shop floor control to evaluate a job relative to the schedule. If it's behind schedule, this information triggers the manufacturing supervisor to look at the situation and think about how to get the job back on track and complete by its due date. Shop floor control makes the problem visible so the manufacturing supervisors can go about choosing the best solution out of various alternatives, such as:

- Adding capacity,
- Running a partial lot,
- Speeding up the process,
- Overlapping operations,
- Eliminating the move and queue time by moving it quickly from one operation to the next and working on it immediately.

Data integrity is most important and has not always been easily achieved in SFC. Recent improvements have been brought about by increased operator training, technological innovations such as bar coding, and the JIT approach of simplification by only collecting what is really important. Real time data collection in a distributed data processing mode, collects information on site which will be utilized for local control and scheduling (Toomey, 1996).

References

Arnold, J R Tony, *Introduction to Materials Management* (third edition). Prentice Hall, Inc, New Jersey, 1998.

Blackstone Jr, John H, *Capacity Management.* South-Western Publishing Company, Cincinnati, Ohio, 1989.

Cox III, James F, and John H Blackstone Jr. (Editors), *APICS Dictionary* (ninth edition). American Production and Inventory Control Society, Inc., Falls Church, VA, 1998.

Crepeau, Eugene A, *Production Activity Control: (APICS) CPIM Certification Review Course, Student Guide.* American Production and Inventory Control Society, Inc, Falls Church, V A, 1991.

Landvater, Darryl V and Christopher D Gray, *MRP II Standard System: A Handbook for Manufacturing Software Survival.* Oliver Wight Limited Publishing Company, Essex Junction, V T, 1989.

Landvater, Darryl V, *World Class Production and Inventory Management* (second edition). John Wiley and Sons Inc, New York, 1997.

McMahon, Chris and Jimmie Browne, *CADCAM: From Principles to Practice.* Addison-Wesley Publishing Company, Wokingham, England, 1993.

Plossl, George, *Orlicky's Material Requirements Planning* (second edition). McGraw-Hill, Inc, New York, 1994.

Russell, Roberta S and Bernard W Taylor III, *Operations Management: Focusing on Quality and Competitiveness* (second edition). Prentice-Hall Inc, New Jersey, 1998.

Toomey, John W, *MRP II: Planning for Manufacturing Excellence.* Chapman & Hall, New York, 1996.

Vollmann, Thomas E, William L Berry, and D Clay Whybark, *Manufacturing Planning and Control Systems* (fourth edition). Irwin Professional Publishing, Chicago, 1997.

Wassweiler, William R (Editor), 'Shop Floor Control' In James H Greene (Editor), *Production and Inventory Control Handbook* (third edition). McGraw-Hill, Inc, New York, 1997.

Chapter 10

An Introduction to Enterprise Resource Planning: Architecture, Implementation, and Cases

10.1 INTRODUCTION

Enterprise Resource Planning (ERP) is defined by APICS Dictionary as an accounting-oriented information system for identifying and planning the enterprisewide resources needed to take, make, ship, and account for customer orders.

- More generally, ERP is a method for the effective planning and control of all resources needed to take, make, ship, and account for customer orders in a manufacturing, distribution, or service company. An integrated ERP system manages the business processes required for each customer order and assesses the impact of executing that order on inventory and production. Manufacturing module of an ERP solution, for example, helps in reducing planning/replanning time and thus makes the system more flexible enabling the company to produce the right product quickly and cost-effectively. The distribution module keeps track of everything from raw material procurement to the shipping of finished goods.
- An ERP system differs from the typical MRP II system in technical requirements such as:
  (a) Graphical User Interface (GUI),
  (b) relational database,
  (c) use of fourth-Generation Languages (4GL), and Computer-Assisted Software Engineering (CASE) tools in development,
  (d) client/server architecture, and
  (e) open-system portability

- ERP is a comprehensive planning and control framework that has evolved over a thirty-year period.
- ERP is a package software solution—not a custom-made software for a specific firm—that addresses the enterprise needs of an organization by tightly integrating the various functions of an organization using a process view of the organization (instead of the functional view). The specific needs of a firm are met through a process of customization (Sadagopan, 1999-b).
- It is a computerized system which is used by manufacturing industries to derive competitive advantage in the production, distribution, and financial areas.
- ERP is the latest enhancement of MRP II with the added functionality of finance, distribution, and Human Resource Management (HRM) integrated to handle the global business needs of an integrated and networked enterprise.
- ERP provides relevant, timely, and accurate production-oriented information for operational planning and control. This capability to collect, collate, analyze and present high quality information facilitates better decision making. ERP forms the *transactional backbone* of an enterprise and works like an information lubricant, facilitating the exchange of data among corporate divisions through the unification of key processes.
- ERP solutions use technology to address business issues, at the same time striving to keep technology transparent to the users. Users do not need to learn more about bits and bytes but they need to know-how operational and long-term business issues could be effectively addressed with technology, albeit using a user-friendly interface (Ushasri, 1999).

10.2 IT AS ENABLER

There are many computer programs designed to convert data into information, usually for a specific purpose or function. Shop floor workers need one level of information, in real-time, about the process or production; executives need a different view of that data for long range planning and short-term deployment—same data, different information. In many situations, data are available but uncollected. This could require manual entry, bar code scanning or sensor monitoring to correct. In other cases, the data are collected but stored awaiting a request. Data mining and data warehousing applications involve this aspect of information development. Databases are becoming larger and larger as companies find more and more data to store. But all those bytes are still just data.

Answering that call, software developers have looked for ways to bring many divergent requirements under a single, if huge, umbrella program called enterprise resource planning or ERP. What ERP strives to do is combine the needed functions of every application a company requires to do its job and integrate them all together.

ERP allows organizations to balance customer demands impacted by multiple interrelated items and multiple plant locations. ERP adds applications for financials, supply chain, and distribution and requirements planning for multiple sites. It facilitates intelligent resource planning in the face of rapidly changing constraints such as materials availability, market readiness, plant capacities, personnel certification and business costs per location. When operations transactions are computerized they can be fed directly to the accounting and finance system, which means transactions are put in monetary terms. ERP systems also track transactions from their origin at the customer, to order entry, through operations and accounting until the transaction is completed. Also, all decisions made in one function will be apparent to other functions through the ERP system. In short, ERP systems will allow decision-makers to champion an intelligent, agile and responsive organization in the global marketplace.

In modern manufacturing operations, information starts at the top and the bottom simultaneously. Data originates at the machines, the process and the workers. It is collected by sensors, controls, and operators. It indicates what is being made, how and where, when it will be done, and why it won't be on time. Programs such as Manufacturing Execution Systems (MES) and shop floor control take the data and create the information, passing it up the chain to the planners and product developers. Here, the data are integrated with other bytes from advanced scheduling and planning systems to produce a report and forecast.

Data flows throughout the ERP system, entered once and available everywhere. But until it is needed, it is not information. Once it is called for, the application creates information from the data and presents it in the best possible manner for the user—operator, executive, order taker, maintenance worker, systems analyst or whomever—or the equipment—machine, control, monitor, sensor, computer, EDI system or whatever.

We are now seeing the advantages of using this concept of data flowing to produce additional productivity in other areas through the use of the technique called workflow. Workflow could be thought of as the movement of work, like the water in a river, from one place to another in a continuous flow. However, like many rivers, there are rocks along the way, representing points of transaction that disrupt the flow and slow the pace of our information river. In some extreme cases, they might totally upset the ship—that is, our company. If we can smooth the flow by removing the bumps and rocks in our way, we can create an even movement of work or, in the digital age, an even flow of information. The technique to do this is workflow, and the method of achieving it comes packaged in the best software systems. The Workflow Management Coalition (WfMC)—an international group in the U.K.—says, 'Workflow management consists of the automation of business procedures during which documents, information or tasks are passed from one participant to another in a way that is governed by rules or procedures'. Logically, work flow can be embedded in data flow so the computer can determine the best flow of both. The computer augments the system designer by allowing immediate simulation of workflows to prove, or disprove, the concepts. Workflow and demand flow work together, one from the inside out, the other from the outside in.

ERP is not confined within the walls of a single entity. Its data flows go out to suppliers and vendors, customers and prospects. Information is presented on Internet sites and transferred to other computers on a network at the speed of light, or around the world via cable and radio beams bounced off satellites. Nothing equals the computer for making the world a much smaller place. What you find out can be shared with another person or even millions of people at the same time you learn it yourself. Design and product development can be a collaborative process without barriers of space and time. Groupware, data management and communications combine to allow everyone to sit in the same office, regardless of where they are physically located.

E-mail and the World Wide Web have exploded onto the scene as mandatory business requirements. They have helped even the playing field to a point where the size of a company is far less apparent and in some cases unimportant. The Web offers a great resource and network capability that allows us to put more product information and customer service information in the hands of our customer base. We can also offer direct ordering of products and services and traditional order management services through the Web and improve our customer interaction. This should ultimately lessen our cost per order and allow us to focus more on customer 'wow' and less on traditional tasks.

10.3 NEED FOR ERP

1. As an Enabler to Quickly Access Internal Information and to Integrate Different Functions

In today's competitive environment, organizations need to introduce new products rapidly to survive in the market. Rapid new product development requires continuous design improvement, flexible manufacturing systems, fast and efficient inbound logistics, operations, and outbound logistics, and efficient management of the entire supply chain. This calls for fast access to accurate information about all the internal functions in the organization and also about the entire supply chain from outside. ERP with its centralized database provides tight integration of different functions, such as finance, marketing, production, distribution, and HRD, without loosing flexibility. Many companies found that their legacy systems were inadequate as they only automated the transactions without integrating the business functions.

2. As an Enabler to Achieve Improved Business Performance

ERP facilitates cycle time reduction, reduces inventory while simultaneously improving customer service levels, enables faster response to customer queries, provides capability to handle multiple locations, currencies, and languages, and provides flexible, integrated, real-time decision support.

3. As a Replacement for Outdated Legacy Systems

ERP has been used in recent past to solve the Y2K problem and to replace the legacy systems which formed islands of information unable to communicate with each other with an integrated system. Further, e-commerce and e-business drivers are also compelling companies to replace their legacy systems with web-enabled ERP solutions.

4. As an Alternative to Developing In-House MIS System

Installing a Commercial Off-the-Shelf (COTS) ERP solution from third-party vendors rather than developing an in-house MIS system has the following advantages:

(a) In-house development requires qualified manpower and long development, testing, and debugging times; a thoroughly-tested, readily available ERP package can be implemented relatively easily and quickly.
(b) ERP makes use of best-in-the-breed business practices used by some of the most successful companies.
(c) Training the users on the standard ERP system is lot easier than on an in-house system.
(d) Improving and updating the COTS ERP system is as easy as buying the upgrade.

COTS ERP solutions, however, require the organization to reengineer established business practices to accommodate application constraints, or to customize the software. Besides, in some cases, COTS ERP solution might result in an initial higher total cost. Finally, COTS solutions do not provide a competitive edge for long—any technology your company can buy today your competitor can buy tomorrow.

Following are the reasons, according to Travis (1999), that drive a company to pursue a new ERP solution:

(a) The company wants to achieve performance improvements, such as, reducing operational costs, gaining competitive advantage, improving customer service, and improving or reengineering business processes.
(b) The existing system in the company is not able to support its

needs and requires significant Information System (IS) resources for maintenance and support.

(c) The system uses multiple points of input, often with duplication of effort.

(d) Staff members are unable to answer questions easily or respond to information requests by key customers or suppliers.

(e) The enterprise has grown through mergers and acquisitions and contains a variety of incompatible systems.

(f) Key information is updated on a batch basis instead of in real time.

10.4 ARCHITECTURE OF SAP R/3 ERP SOLUTION

A SAP product is a suite of standard software made up of individual programs that have been written to carry out computing tasks in the most efficient manner possible. SAP R/2 is a system for mainframes. The SAP R/3 has an open system architecture and applies the client/server concept across multiple levels as shown in Fig. 10.1.

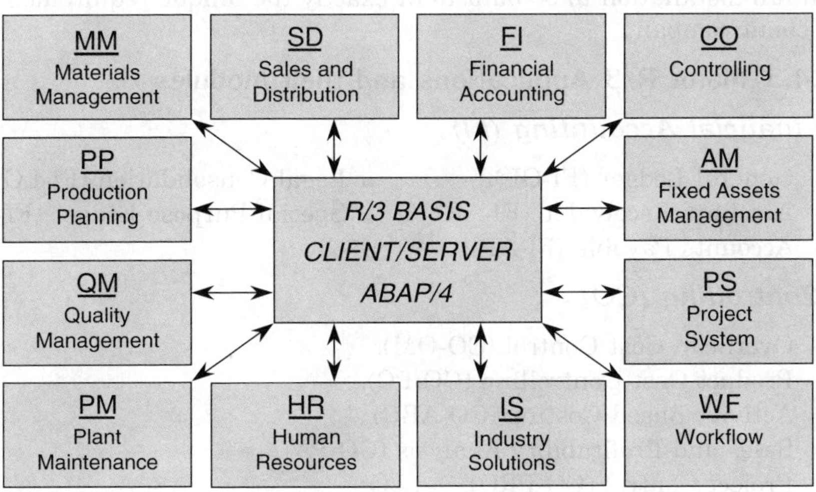

Fig. 10.1 *Major SAP Applications*

10.4.1 The SAP R/3 Basis (R/3 Standard System)

All R/3 installations include a set of components that form the core of the system and is referred to as R/3 Basis. It provides the users with a set of tools to build a suite of integrated programs that can be fitted exactly to the requirements of the company and modified as the company develops. Every implementation will need a SAP R/3 Basis module that provides the elements of the SAP R/3 runtime system. It includes the fundamental tools and functions of the R/3 Data Dictionary, the SAP R/3 Reference Model, the ABAP/4 Development Workbench, and R/3 Customizing Component. When designing an implementation, the R/3 Reference Model is used to select which module components will be needed in the target system.

10.4.2 The SAP R/3 Applications

A SAP R/3 application is a set of programs that has been designed for a specific type of business data processing. Each application addresses a main sector of business activity, ranging from financial accounting to human resources. Under each application are grouped the modules most likely to be associated with the title of the application. However, the fully integrated design of all SAP standard business programs allows great flexibility in the assembly of modules to form a specific implementation.

Each application is fully integrated with the R/3 Basis. This allows each application to communicate with any other application. Some application modules depend on other applications. For example, the Controlling (CO) module depends on the Financial Accounting (FI) module. Some of the components of a module may be optional. Some of the functions within a component may be optional. This flexibility allows each R/3 installation to be built to fit exactly the unique requirements of the client company.

10.4.3 Major R/3 Applications and their Modules

1. Financial Accounting (FI)

- General Ledger (FI-GL).
- Accounts Receivable (FI-AR).
- Accounts Payable (FI-AP).
- Legal Consolidation (FI-LC).
- Special Purpose Ledger (FI-SL).

2. Controlling (CO)

- Overhead Cost Control (CO-OM).
- Product Cost Controlling (CO-PC).
- Activity-Based Costing (CO-ABC).
- Sales and Profitability Analysis (CO-PA).
- Project Control (CO-PRO).

3. Fixed Asset Management (AM)

4. Project System (PS)

- Basic Data (PS-BD).
- Operational Structure (PS-OS).
- Project Planning (PS-PLN).
- Approval (PS-APP).
- Project Execution/Integration (PS-EXE).
- Information System (PS-IS).

5. Workflow

Workflow refers to the movement of work items through a series of operations that add value to these items. It also refers to the flows of information that must take place if this added value is to be optimized. A workflow management system is intended to manage business processes automatically or semi-automatically by controlling the sequence of activities. It should ensure that the appropriate steps are carried out at the right moment by specific people or groups, or by particular data processing programs and the machinery they control. SAP Business Workflow is devoted to developing and managing the flows of work and information that will make a business as effective and efficient as possible.

6. Industry Solutions (IS)

An Industry Solution is an enhancement of the standard R/3 system that may include some or all of the components of any R/3 applications, according to the sector of industry for which it has been designed. Some of the industry solutions offered by SAP are for the following sectors:

- Public Sector (IS-PS).
- Oil (IS-OIL).
- Hospital (IS-H).
- Telecom (IS-T).
- Banks (IS-B).
- Real Estate Management (IS-IS).

7. Human Resources (HR)

This application provides an integrated human resource management system through the use of the components of the Personnel Planning and Development (PD) module and Personnel Administration (PA) module.

Personnel Planning and Development (HR-PD).

- Organizational Management (PD-OM).
- Seminar and Convention Management (PD-SCM).
- Personnel Development (PD-PD).
- Workforce Planning (PD-WFP).
- Room Reservations Planning (PD-RPL).

Personnel Administration (HR-PA)

- Employee Management (PA-EMP).
- Benefits (PA-BEN).
- Compensation Administration (PA-COM).
- Applicant Manger (PA-APP).
- Time Management (PA-TIM).
- Incentive Wages (PA-INW).
- Travel Expenses (PA-TRV).
- Payroll (PA-PAY).

8. Plant Maintenance (PM)

- Equipment and Technical Objects (PM-EQM).
- Preventive Maintenance (PM-PRM).
- Maintenance Order Management (PM-WOC).
- Maintenance Projects (PM-PRO).
- Service Management (PM-SMA).
- Plant Maintenance Information System (PM-IS).

9. Quality Management (QM)

- Planning Tools (QM-PT).
- Inspection Processing (QM-IM).
- Quality Control (QM-QC).
- Quality Certificates (QM-CA).
- Quality Notifications (QM-QN).

10. Production Planning (PP)

- Basic Data (PP-BD).
- Sales and Operations Planning (PP-SOP).
- Master Planning (PP-MP).
- Capacity Requirements Planning (PP-CRP).
- Material Requirements Planning (PP-MRP).
- Production Orders (PP-SFC).
- Product Costing (PP-PC).
- Kanban/JIT Production (PP-KAB).
- Repetitive Manufacturing (PP-REM).
- Assembly Orders (PP-ATO).
- Production Planning for Process Industries (PP-PI).
- Plant Data Collection (PP-PDC).
- Information System (PP-IS).

11. Materials Management (MM)

- Material Requirements Planning (MM-MRP).
- Purchasing (MM-PUR).
- Inventory Management (MM-IM).
- Warehouse Management (MM-WM).
- Invoice Verification (MM-IV).
- Information System (MM-IS).
- Electronic Data Interchange (MM-EDI).

12. Sales and Distribution (SD)

- Master Data (SD-MD).
- Basic Functions (SD-GF).
- Sales (SD-SLS).
- Shipping (SD-SHP).
- Billing (SD-BIL).
- Sales Support (SD-CAS).
- Information System (SD-IS).
- Electronic Data Interchange (SD-EDI).

10.4.4 SAP Configuration and Customization by Applications Programmers

By itself the SAP program will not be very friendly to an individual user. It will not know what sort of business is to be conducted or exactly how the user company wants to invoice and conduct other interactions with its customers. If the SAP implementation is to look and behave as if it really understands the company it is working for, it will have to be configured and customized. Naturally, the SAP software expects to be told how to behave in a specific company and has standard routines which help the company experts to set out what has to be done in a format that SAP can accept. The experts who do this are referred to as Applications Programmers or Applications Developers (ASAP World Consultancy, 1997).

10.4.5 Development of SAP Product Range in Three Directions

1. A function within a SAP component may be elaborated so as to become a complex module. For example, the requirements of enterprise controlling can be met by the components of the Controlling (CO) module, but in addition, the Enterprise Controlling (EC) product is available as a separate module and includes functions not available in the Controlling (CO) module.
2. Extra integrating functions can be provided to interact with a group of SAP application modules. For example, the Logistics General (LO) module is designed to provide integrating functions for the following modules:
   - Sales and Distribution (SD).
   - Production Planning (PP).
   - Materials Management (MM).
   - Plant Maintenance (PM).
   - Quality Management (QM).
3. Where there are many companies in a particular sector of business that share a specialized requirement, a SAP partner may develop a specialized enhancement of the R/3 system that can be marketed as an 'Industry Solution.'

10.5 ARCHITECTURE OF BAAN ERP SOLUTIONS

Baan company's ERP solutions are available as BaaN IV (the older version) and BaanERP (the latest version). The architecture of both versions is described here.

10.5.1 BaaN IV

The Baan IV software can run on many platforms, for example, it can run on various UNIX platforms supplied by HP, IBM, Sun, Digital, etc. It is also available on Windows NT. The software can use database provided by the software manufacturer; or, third-party databases such as Oracle or Ingres can also be used. Figure 10.2 shows the menu browser displayed to the user having access to all the packages of Baan IV.

Menu browser [User:]

- BAAN IV Common
- BAAN IV Finance
- BAAN IV Project
- BAAN IV Manufacturing
- BAAN IV Distribution
- BAAN IV Process
- BAAN IV Transportation
- BAAN IV Service
- BAAN IV Enterprise Performance Manager
- BAAN IV Enterprise Modeler
- BAAN IV Constraint Planning
- BAAN IV Tools
- BAAN IV Utilities
- Distributed Data Collection

Fig. 10.2 *BaaN IV Menu Browser*

BAAN IV Common

The Baan IV common package allows you to maintain the common master data. This data is used by all the Baan IV packages. All the files that are used in more than one module are stored in this package. The Baan IV common contains the following tables:

1. Logistic Tables.
2. Financial Tables.
3. Employee Master.
4. Customer Master.
5. Supplier Master.

BAAN IV Finance

The finance package allows users to extract financial transactions from the sales and manufacturing areas and post them to the general ledger without having to key any transaction. It also has a budget system, and an activity base module. Following modules are included in the finance package:

1. General Ledger.
2. Accounts Receivable.
3. Accounts Payable.
4. Cash Management.
5. Fixed Assets,
6. Financial Statement.
7. Financial Budget System.
8. Cost Allocation.
9. Electronic Data Interchange. (EDI).

BAAN IV Project

Baan IV project is designed to support the management of projects through all stages, from estimating tenders to delivery and throughout the guarantee period. It is especially suited to project-driven companies for the coordination of multiple projects. The goal is cost-effective management of each project according to the time schedule, within the specified budget and to the required quality. Furthermore, allocation of personnel and equipment to projects is critical in cost-effective operation which will maximize company profits. The project package provides all the tools necessary to control project accounting and planning. A planning requirements process accurately tracks costs for the project-related industries. This package is linked with all the software's other functions to provide the information necessary to successfully manage the project within the enterprise environment. It includes following modules:

1. Project Estimating.
2. Project Definition.
3. Project Budget
4. Project Planning.
5. Project Requirements.
6. Hours Accounting.
7. Project Progress.
8. Project Monitoring.
9. Project Invoicing.

BAAN IV Manufacturing

The manufacturing package provides all the manufacturing planning functions such as, MPS, MRP I, CRP. It can also provide control for all operations related to product's fabrication, labor management, and capacity required. The following modules are included:

1. Engineering Data Management.
2. Item Control.
3. Bill of Material Control.
4. Routing.
5. Cost Accounting.
6. Master Production Schedule.
7. Material Requirements Planning.
8. Capacity Requirements Planning.
9. Repetitive Manufacturing.
10. Shop Floor Control.
11. Hours Accounting.
12. Project Budget.
13. Product Configuration.
14. Product Classification.
15. Project Control.
16. Quality Management System.

BAAN IV Distribution

The distribution package is designed to take care of day-to-day logistical management in production and trade companies. The package is fully integrated with all other packages of Baan IV. This package contains all the programs to create and manage the sales orders. It is a reliable source of information on market trends and developments. It also includes all the inventory related functions such as inventory control, location control, distribution requirements planning (DRP I), and replenishment control.

Baan IV Distribution package contains the following modules:

1. Item Control.
2. Cost Accounting.
3. Purchase Control.
4. Sales Control.
5. Sales and Marketing Information.
6. Electronic Data Interchange (EDI).
7. Replenishment Order Control.
8. Inventory Control.
9. Lot Control.
10. Location Control.
11. Distribution Requirements Planning.
12. Tables.
13. Common Data.
14. Distribution Parameters.

Characteristics of BAAN IV Distribution

1. Simple and fast data input and control.

2. Extensive purchase and sales statistics.
3. Comprehensive forecasts and planning techniques.
4. Interface with the graphical module enterprise performance manager.
5. Interface with EDI.
6. Use of multiple currencies.
7. Use of multiple warehouses..
8. Multi-company solution including supply chain management.

BAAN IV Process

The process package is designed to help manage the entire supply chain of any company operating in a process environment, such as the chemical industry. It helps manufacturers of identical product in different containers. It also helps to keep track of the various batches processed. It is able to account for the potency, the acidity, and the grade of items. Following are the modules provided in the process package:

1. Item Control.
2. Formula Management.
3. Routing.
4. Cost Accounting.
5. Capacity Requirements Planning.
6. Production Management.
7. Hours Accounting.
8. Quality Management System.

BAAN IV Transportation

Transportation package helps in managing transportation orders and to maximize equipment use. It can handle all types of transportation modes. It has powerful modules for managing warehousing and packaging. It keeps track of transportation costs and determines cost trends. It includes the following modules:

1. Employee Control.
2. Address Control.
3. Transport Fleet Management.
4. Transport Fuel Control.
5. Hours and Expense Control.
6. Central Data Entry.
7. Transport Control.
8. Invoicing Control.
9. Packing Control.
10. Electronic Data Interchange.
11. Warehouse Control.
12. Distribution Requirements Planning.

BAAN IV Service

Service package can be used to manage all the repair and warranty information for supporting installations in the field. With this package:

1. You can register at which customers and locations specific installations are situated.
2. An installation bill of material that lists the components requiring servicing including their serial numbers and service history can be linked to each installation.

3. This data helps in analyzing the causes of malfunction in the fastest possible way.
4. You can create maintenance contracts and record warranties together with associated warranty terms for each customer.
5. Calls reporting malfunctions can be recorded even as you are on the phone.
6. Based on periodic maintenance obligations and calls, a service plan is drawn and service job sheets are printed.
7. Finally, the service activities can be invoiced and a detailed service history is built up.

The modules included in this package are:

1. Service Tables.
2. Installation Control.
3. Contract Control.
4. Service Order Control.
5. Service Analysis Control.
6. Item Control.
7. Cost Accounting.
8. Inventory Control.

BAAN IV Enterprise Performance Manager

The Enterprise Performance Manager (EIS) incorporates tools that are designed to give various levels of management access to the data in the Baan IV tables. The data of the distribution, finance, and manufacturing modules are available in this module and they can be displayed using various formats.

- The tool can be used interactively to get an overview of the overall business performance by using Ishikawa fishbone diagrams.
- Via a flexible user interface the enterprise performance manager can drill down to the basic figures. These figures can be fetched from the integrated Baan IV repository and can be linked to the persons responsible in the organization. A set of predefined performance indicators is available and new indicators can be defined very easily. The EIS module is fully integrated with the manufacturing, finance, and distribution modules.
- The enterprise performance manager is especially meant for middle and top management. It can be used as a business benchmarking tool during Baan IV implementation and optimization cycles (business process reengineering) and as a management and reporting tool at tactical and strategic level.

BAAN IV Enterprise Modeler (Formerly Known as Orgware)

Enterprise modeling is a process in which customers can map all the processes used, and then develop an accurate and complete implementation plan.

BAAN IV Constraint Planning

The constraint planning package provides planning functionality that takes into account capacity and materials constraints. This package currently provides MPS functionality with both finite and infinite planning methods.

BAAN IV Tools

The Tools package consists of all the programs designed to maintain and customize the application. The form manager, report writer, and sessions manager are the options that allow the developers to tailor Baan IV to user's needs. It also includes programs to manage the database, devices, and user profiles. Following modules are included in the Tools package:

1. Software Installation.
2. Application Configuration.
3. User Management.
4. Device Management.
5. Job Management.
6. Database Management.
7. Audit Management.
8. Text Management.
9. Menu Management.
10. SQL Queries.
11. Business Objects.
12. Application Customization.
13. Application Development.
14. Terms and Definitions.
15. Translation.
16. Documentation.
17. Conversion.
18. Software Distribution.
19. Desktop Management.

BAAN IV Utilities

The existing utilities allow users to easily import or export information between Baan IV and any other system. This package facilitates the implementation of the software by helping in creation of master files imported from other software. This package has tools to facilitate the communication between Baan IV and other databases, spreadsheets, etc. This module can also be used to convert data of older versions of the application to new versions. The exchange module can be very useful for multisite applications as it facilitates the communication between two sites. In short this package provides the needed bridge between all other sources and Baan IV and includes following modules:

1. Import Module.
2. Export Module.
3. Generate Exchange Scheme.

Distributed Data Collection

This package allows the users to interface between Baan IV and third party data collection, vendor's data collection systems. This allows for the collection of data through the use of devices such as laser scanners etc. with the real time update of the interfacing Baan IV module. The data collection vendor must supply an interface to Baan IV to create a functional solution.

10.5.2 BaanERP

BaanERP, the successor to Baan IV, has the following enhancements, as claimed by Baan Co, over the older version:

1. Order to Cash foundation for all Baan business solutions.
2. Open Component Architecture.
3. Fully integrated allowing for consistency and visibility across the enterprise.
4. Comprehensive international capabilities, supporting multiple languages, tax structures and currencies including the Euro.
5. Modular components that allow for incremental implementation and migration.

BaanERP includes the following five components:

1. Manufacturing.
2. Finance.
3. Project.
4. Distribution.
5. Tools.

1. Manufacturing

BaanERP manufacturing module includes:

1. Bills of Material.
2. Cost Price Calculation.
3. Engineering Change Control.
4. Engineering Data Management.
5. Hours Accounting.
6. Product Classification.
7. Product Configuration.
8. Production Control.
9. Production Planning.
10. Project Budgeting.
11. Project Control.
12. Repetitive Manufacturing.
13. Routings.
14. Shop Floor Control.
15. Tool Requirements, Planning, and Control.
16. Capacity Requirements Planning*.
17. Master Production Scheduling*.
18. Material Requirements Planning*.

Benefits

1. Open architecture design allows for a seamless and simplified integration with popular CAD packages via 'BaanEngineering' elements.
2. Graphical simulations help analyze a 'what if' impact on financial requirements, capacity and inventory.
3. The system's object orientated configurator supports different production strategies.
4. Planning is integrated at every level and across multiple sites allowing smooth and consistent operational activity.
5. Within a dynamic environment, enterprise planning simulates alternative plans and reactive planning.

---

*These modules come with extensive enterprise planning capabilities.

6. Planning and tracking capabilities are extended to improve production resource management issues such as inventory.
7. The integrated quality management tool enables a wide range of statistics (from raw material to finished goods) to be monitored resulting in continuous improvement in manufacturing quality.
8. Multiple valuation methods help the company identify cost drivers and reduce product costs.

2. Finance

BaanERP Finance Module is designed to meet dynamic financial management and reporting requirements around the world. It includes:

1. Accounts Payable.
2. Accounts Receivable.
3. Financial Budgets System.
4. Cash Management.
5. Financial Reporting System.
6. Fixed Assets.
7. General Ledger.
8. Cost Accounting.
9. Sales Invoicing.

Benefits

1. This independent system allows for easy solution configuration to meet changing business strategies.
2. Integration with Hyperion financial software provides advanced budgeting, consolidation, reporting and analysis.
3. Accounting operations are simplified and duplicate data entry is eliminated with parameter-driven posting and updating tools.
4. Baan Accounts Payable streamlines vendor payments. It supports checks, electronic banking and payment on consumption.
5. Superior visibility enables the company to immediately focus and act on financial information to help increase margins, revenue and cash flows.
6. International business requirements are met with the use of multidimensional ledger and dual sets of books.
7. Provides cost analysis and cost allocation functionality on both a detailed and summarized level.
8. Costs can be proactively tracked via budget links.
9. Multicurrency functionality allows the company to hold up to 3 home currencies therefore complementing and complying with the Euro regulations.
10. Central point invoicing.

3. Project

The project module provides the control and visibility the company needs for profitable operations from estimates, and bids through site installation and maintenance. In addition the system supports project invoicing for all the different contractual agreements found in project environments. Baan Project Module includes:

1. Project Budget.
2. Project Definition.
3. Project Estimating.
4. Project Invoicing.
5. Project Monitoring.
6. Project Planning.
7. Project Progress.
8. Project Requirements Planning.

Benefits

1. Real time control of all aspects of project manufacturing.
2. Integration of project management and manufacturing resource enhancing visibility and timely consolidated reporting.
3. The link with Baan Manufacturing allows all the relevant, cross-functional information about each project to be easily accessible for effective enterprise resource management.
4. An integrated planning and scheduling system environment results in activity networks to be defined, resources to be allocated, and 'what if' analysis to be conducted.
5. Organizational, logistical, and contractual structures can be modeled and the resulting project activities reported which results in effective cross-functional management.

4. Distribution

To help develop the best solution for meeting customer requirements and balancing business constraints, this component manages the entire spectrum of distribution, sales, and logistics for manufacturers and distributors. BaanERP Distribution Modules include:

1. Sales Management.
2. Purchase Management.
3. Warehouse Management.

Benefits

1. Extensive simulation capabilities optimize purchasing and internal inventory decision making.
2. Top-down planning supports any distribution strategy.
3. With integrated workflow management and order templates, order processing is speeded up.
4. Shipping constraints, order blocking algorithms, and multilevel ATP component checks are supported by the system.
5. Integration with the Aurum Front-Office suite enhances the capabilities of Baan Sales solution.
6. Purchasing is simplified with online requisitioning.
7. Sophisticated supplier contract and release management enables your company to take advantage of economies of scale.
8. EDI is key in enhancing the speeds of communication with trading partners as well as providing a solid link between distribution operations and manufacturing planning.

5. Tools

All Baan applications are built using flexible BaanERP Tools to handle business needs that require software or configuration changes. BaanERP Tools include:

1. Open System Tools.
2. Client/Server Tools.
3. End User Tools.
4. Developer Tools.
5. Documentation Tools.
6. Translation Tools.
7. Software Distribution.
8. Implementation Tools.

Benefits

1. BaanERP enables quick reaction to new trends in the marketplace that require software or software configuration changes.
2. Helps in developing the Baan applications in such a way that they are kept independent of third party products.
3. Helps create tailored applications to meet special requirements.
4. Facilitates integration of Baan applications with third-party products.

10.6 SELECTING THE RIGHT ERP PACKAGE

1. The selection and implementation of ERP is primarily an operations management initiative and decision, not an IT or MIS project; hence it is critical that the Chief Executive Officer (CEO), the vice-president of manufacturing, and other key players be involved and support the Chief Information Officer (CIO) in the ERP selection and implementation process. The selection team must have top-level representation from all major functional areas including production, distribution, finance and accounting, human resources, sales, marketing, customer service, and information systems (Travis, 1999).
2. A project leader must be selected from among the team members.
3. The team must develop consensus on several critical issues that will shape the entire project: budget, time frames, goals, and deliverables.
4. A tentative schedule should be prepared for the selection process and implementation.
5. The team should determine what critical business needs/problems the company is trying to address and what benefits are to be gained from ERP; some possible needs, problems, and benefits are:
   - Reducing inventory investment.
   - New product planning and introduction.
   - Increasing fulfillment rate.
   - Customer service programs.
   - Lowering transportation costs.

- Company downsizing.
- Simplifying the manufacturing process.
- Company expansions into new areas or markets.
- Gaining market share.
- Potential acquisitions.

6. The team should then develop a document containing:
   - Total number of customers.
   - Company goals.
   - Company's most significant business process—areas that set it apart.
   - A narrative on how the company conducts its business.
7. By putting these businesses processes in the form of key (transaction) scenarios, potential vendors can prepare scripted demonstrations. This document helps in focusing the internal team's efforts and also becomes a valuable tool for potential ERP providers to understand the business and its needs.
8. The project team must determine its differentiating points to ensure that a vendor's product plays to those strengths.
9. The team should also find out which ERP packages the competitors are using or implementing.
10. After narrowing down the choices, the top two or three vendors should be invited to demonstrate how their products could be tailored to the specific work environment. The vendor should be asked to build and demonstrate business processes and transaction scenarios based upon the company needs.
11. Once the scripted demos narrow down the alternatives, team members must conduct site visits of the top ERP candidates' solutions to see the vendors' working environments, gauge the vendors' commitment to training and customer services, and get a sense of their overall business philosophies.
12. Finally, the project team must visit the sites of other companies that are using the particular ERP product to see how the software system functions in actual applications and assess vendor support, both during and after initial implementation.

The ERP package should be evaluated on three major criteria: a) product evaluation, b) ease of implementation including support and service provided by the vendor or the implementation partner, and c) the future roadmap (Rajani and Padmanabhan, 1999).

10.6.1 Product Evaluation

1. Functionality

- How the product functionality fits with the industry—the selected package should support the best-of-the-breed business practices followed in the company's line of business. Major gaps between the required and available functionalities can lead to major problems in the implementation.

- The package should satisfy the functional requirements of the whole organization and all the departments.
- Given the rapid pace of change in business, the future requirements of the organization should also be considered while selecting the ERP package.
- The functionality and report formats and country-specific features, such as treatment of various taxes, statutory reporting, payroll, should also be critically examined.

2. Technical Architecture (Software and Technical Aspects)

- Whether the package is scalable and is based on an open architecture supporting open, non-proprietary technology standards.
- Whether Relational Database Management System (RDBMS) has been used.
- What development tools and utilities are available.
- What kind of user interface is used, whether it has a graphical user interface.
- Whether the package is web-enabled enabling the organization to communicate with others in the supply chain through the Internet.
- Whether the package provides parametric customization, modular implementation, and plug-in upgrade (Chaterji, 1999).
- What type of administrative functions, such as security, archival, backup/recovery, auditing, import/export and EDI features, are included in the software.

3. Cost-Benefit Analysis

The Total Cost of Ownership (TCO) of an ERP implementation consists of the cost of software, hardware, additional software/interfaces such as EDI, consultant's fees, customization, networking and telecommunications, training, implementation, and maintenance. Important issues to be considered are: What kind of licensing structure the vendor offers? Whether licensing is based on modules opted or the entire suite has to be licensed. Is there a limitation on number of users? What is the warranty period and cost involved? It is estimated that the cost of the ERP package is generally only 30 per cent of the total cost of ERP implementation. It is generally agreed that the payback period of an ERP implementation should be less than 36 months. Finally, for the cost-benefit analysis to be unbiased, it is recommended that the project manager himself should not do it.

The top five underestimated implementation costs (Slater, 1998) are as follows:

1. *User Training* Training expenses are high because workers almost invariably have to learn a new set of processes, not just a new software interface. Employees at all levels have to accept different responsibilities, which means change, and change management appears in the budget on the training line. Expect a minimum of 10 per cent, and more likely 15 per cent, of the total project budget.

2. *Integration and Testing* Add-on applications for things like logistics, taxes, production planning and bar coding are required. As with training, testing ERP integration has to be done from a process-oriented perspective. Instead of plugging in dummy data and moving it from one application to the next, veterans recommend running a real purchase order through the system, from order entry through shipping and receipt of payment—the whole order-to-cash process—preferably with the participation of the employees who will eventually do those jobs.

3. *Data Conversion* The majority of companies seem to deny their data is 'dirty' until they actually have to move it to the new client/server setups that popular ERP packages require. But even clean data may demand some overhaul to match process modifications necessitated—or inspired—by the ERP implementation. Such work takes time and manpower, which obviously costs money.

4. *Data Analysis* Users with heavy data analysis needs should include the cost of a data warehouse in the ERP budget—and should expect to do quite a bit of work to make it run smoothly.

5. *Getting Rid of Your Consultants* When users fail to plan for disengagement, consulting fees run wild. To avoid this, companies should identify objectives for which its consulting partners must aim when training internal staff.

10.6.2 Ease of Implementation

1. Implementation Cycle Time

One of the most important factors to be considered while selecting an ERP package is the anticipated total implementation time. It has a twofold effect on the implementation process. First, longer implementation time means more expenses in terms of the cost of the implementation team's time and efforts. Secondly, longer implementation time would also mean a delayed start to the payback in terms of benefits of implementing ERP and return on investment. Hence shorter the anticipated implementation time the more attractive the package. The availability of well-defined implementatiom methodology and the actual implementation time at previous sites are the factors to be considered.

2. System/User/Reference/On-Line Documentation

The actual users of the system must have thorough understanding of and be familiar with the ERP system they are implementing and which they are expected to run. This mandates that the complete documentation of the ERP system in various forms namely systems manual, user manual, reference manual, and on-line help be readily available to the users during and after the training. Thus, a careful look at the quality and accessibility of the documentation is essential.

3. Ease of Setup and Configuration (Setting Up Parameters, Designing Menus and Authorizations, Role Definition, etc.), Adaptation, and Use of the Software

After the hardware—the computers and networking—is in place, the very first step in the implementation process is the setting up and configuration of the software on the hardware platform. Unless these activities are completed the software cannot be used at all. Hence the ease with which these activities can be completed is of utmost importance to the implementation process.

4. Support and Service Provided by the Vendor or Implementation Partner

The support and service provided by various vendors should be carefully compared on the following points:

- Availability of vendors and competent implementation partners.
- Vendor's willingness to provide single point of contact during implementation to eliminate finger pointing among various parties involved.
- Warranty and post-warranty support.

5. Ability of the Vendor or Implementation Partner and the Internal Team to Execute

- Organization, size, financial strength, number of installations and experience, of the vendor in executing similar projects as assessed through reference sights.
- The internal team's ability to support and manage the project.
- Availability of implementation partners and their industry experience, financial strength, and number of sites implemented.

6. An Assessment of Degree of Business Process Reengineering (BPR) and/or Package Customization Required

When there is a mismatch between processes embedded in the ERP package and the processes of the user company, either the user processes have to be reengineered (BPR) or the package has to be modified (customization) to match the user requirements. The scope of customization may range from simple 'cosmetic' changes mostly involving modification of the input and output report formats to major changes in the 'functionality' or the embedded processes of the ERP package. To the extent possible, the implementers should try to satisfy user requirements and business objectives within the existing framework of the selected package. Customization involving changes to any of the processes embedded in the ERP package should be resorted to only if it becomes impossible to proceed with the ERP project without the change. ERP being a totally integrated system, any change made to one process may have an adverse effect on some other processes or functions. Besides, during the upgradation of the system, the customization exercise has to be repeated and maintenance of consistency in changes becomes a problem. This makes it more difficult to implement the package and considerably increases the risks and throws the costs and schedules out of gear. Following are the two possible approaches to implementing BPR and ERP:

Pure BPR Approach The organization undertakes a BPR exercise first and then implements ERP. This, however, involves the risk that a particular ERP software package selected later may not be able to implement the reengineered processes.

ERP-Led or Channelized BPR Known as 'Package-Enabled Process Reengineering (PEPR)' The organization selects an ERP package and uses its capabilities to achieve business process improvement by changing the existing process to suit the ERP package. PEPR has been preferred by most organizations over pure BPR as it involves less drastic changes in the organization and reduces implementation time.

7. Change Management

- An assessment of readiness for or resistance to change; design of appropriate change management program to suit the environment.
- Gaining appropriate sponsorship for the project from the top management.
- Proper and timely communication.

10.6.3 Future Road Map

- Enhancements planned by the vendor over the next three to five years.
- Vendor's capability to use and support advanced technologies such as EDI, e-commerce, Internet, data warehousing and data mining.
- Capability to extend various functionalities to other supply chain partners.

10.7 ERP IMPLEMENTATION

In a typical ERP implementation project multiple agencies such as the hardware vendor, network vendor, ERP package vendor, implementer, and consultant need to be involved. Top management commitment is necessary for the success of ERP implementation as it involves major changes in the organization, huge costs, and long implementation time. Top management should ensure that key indispensable persons (rather than people who can be spared) are released for the ERP core committee. They should also have the political will to implement changes that radically alter the existing business practices. A typical ERP implementation project involves following stages:

1. ERP readiness assessment.
2. Preparing the organization for ERP.
3. ERP feasibility study.
4. Process modeling and documenting the 'as is' processes and 'to be' processes along with BPR.
5. Detailed plan for ERP implementation:
   (a) ERP software selection.
   (b) Selection of ERP implementation consultant (the company has to make sure that the consultants are unbiased as many consulting companies have links with some or the other ERP vendor).
   (c) Implementation methodology.
   (d) Final and precise extent of implementation.
6. Detailed Implementation including development, quality assurance, and production systems.
7. Preparing to go live including data migration.
8. Going 'live', which means shifting from the legacy system to the ERP system. ERP development, testing and quality assurance are the activities that *precede* this stage.
9. Performance audit, documentation, training, and future plans.

10.7.1 Implementation Teams

According to Alexis Leon (2000), the ERP implementation project needs people who can grasp new ideas quickly, who have an open mind to new technologies and concepts and who love challenges. Management should ensure that the best people from each of the functions are assigned to the implementation teams. Figure 10.3 gives the structure of the Implementation Project Organization.

10.7.2 Implementation Methodology

1. Formation of implementation team, which includes:
   (a) Formation of *Steering Committee* for monitoring and controlling the implementation process, resolving problems as they occur, and allocating the resources of implementation as required.
   (b) Selection of knowledgeable and experienced people from all functions including IT personnel to work in the implementa-

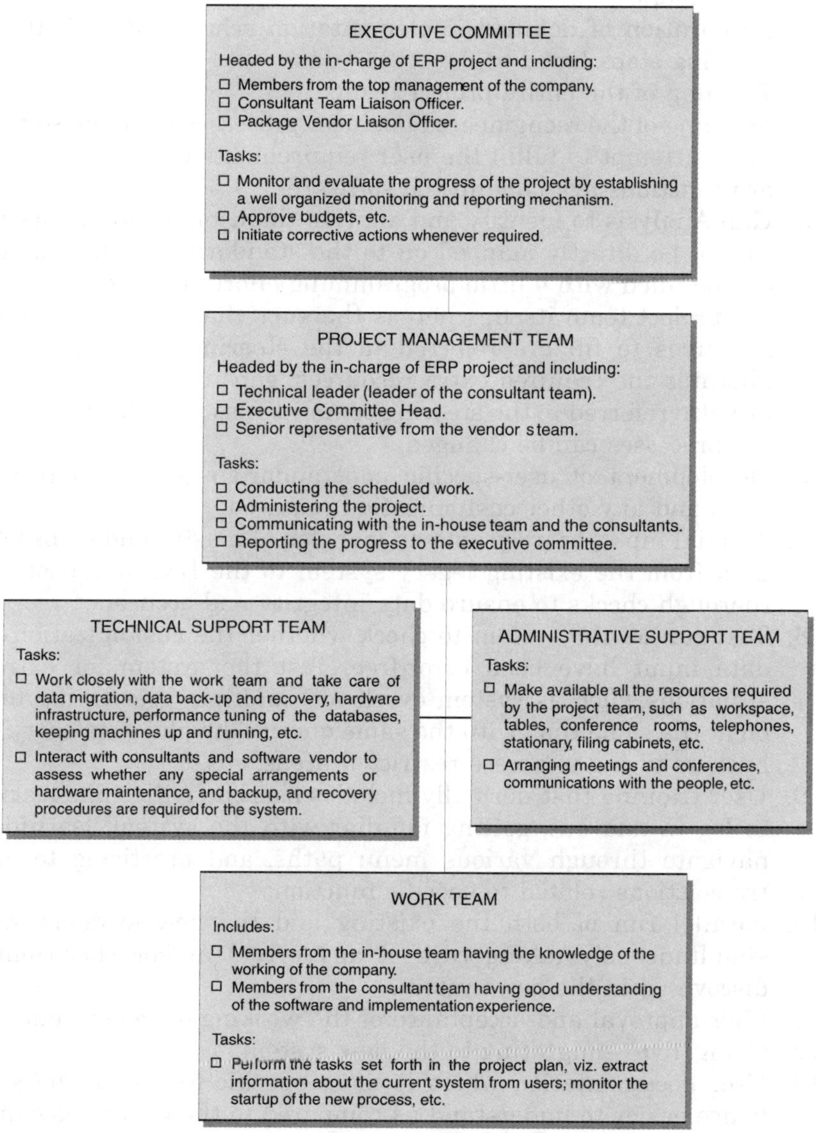

Fig. 10.3 *Implementation Project Organization*
(Adapted from Leon, 2000)

tion team along with the people deputed by the implementation partners.

(c) Identification of project manager, project leaders, and module leaders.

2. Preparation of detailed implementation schedule for all the remaining steps listed below.
3. Training of the entire project team.
4. Mapping of the reengineered business processes on to the software in an attempt to fulfill the user requirements through the standard functions available in the software.
5. Gap Analysis to identify and analyze the user requirements that cannot be directly mapped on to the standard system; gaps that can be filled with a little programming effort are taken care of by the project team itself; whereas the gaps that require additional resources to fill are referred to the steering committee which allocates the required extra resources; gaps that cannot be filled are also referred to the steering committee so that the corresponding processes can be changed.
6. Development of user-specific nonstandard reports and transactions and any other customization required.
7. Manual input, or programmed transfer, of master and transaction data from the existing legacy system to the ERP database with thorough checks to ensure data integrity and accuracy.
8. Test run on the system to check whether the customization and data input have been error-free. Test the system for extreme situations, such as system overloads, multiple user trying to log on at the same time with the same queries, invalid data entry, and hackers trying to access restricted areas, etc.
9. User training that normally includes basic activities like learning to log in and out, getting familiar with the system, learning to navigate through various menu paths, and practicing to enter transactions related to specific functions.
10. Parallel run of both the existing and the new systems, while simultaneously taking care of any errors and/or shortcomings discovered in the new system.
11. User approval and acceptance of the working of the system.
12. Going 'live'—migrating to the new system.
13. User documentation which is more specific to the business and hence easier to understand as compared to the general documentation supplied with the ERP package.
14. Post-implementation support, such as answering the queries from the users, or making minor modifications in the report and other formats.
15. System monitoring and fine tuning.

10.7.3 Major Difficulties in ERP Implementation

1. Long implementation times.
2. Communications gap between business users and IT professionals.
3. Getting all departments to agree on a single centralized system.
4. High turnover among ERP implementers.
5. Inadequate training and documentation.
6. Resistance to change.
7. Presence of large IT group in the firm which perceives ERP as a threat to their own existence.

Successfully implementing ERP the first time requires a structured methodology that is strategy-, people-, and process-focused. Starting an implementation with an undocumented, skimpy, or untailored implementation methodology is an open invitation to disaster or, best case, a long, drawn-out implementation. Everyone from the boardroom to the stockroom needs to understand his or her role and responsibilities for implementation. Implementation leaders should encourage dialogue to get people focused on business objectives and early identification and correction of any problems. Who will be accountable for results, and when, must be an integral part of this understanding. The idea that this is strictly a technology project because software is involved is wrong and, in fact, is one of the leading causes of ERP failure. The IT function is not well-positioned to evaluate the business implications of various tradeoffs or to determine their impact on day-to-day operating results versus strategic intent (Donovan, 1999).

10.7.4 Preconfigured Industry-Specific Best-Practice Templates

The high demand for rapid ERP implementation led to the development of off-the-shelf templates designed to speed up and simplify the software personalization process. These preconfigured templates incorporate industry-specific best practices that support cross-functional business processes and allow faster system deployment and faster benefits. Processes can be refined at a later date. Selected template processes must, however, be verified for appropriateness, at least for the near term, before going forward. The alternative 'clean-sheet-of-paper' approach involves business process redesign and ERP system configuration, which are complex, costly, and time-consuming. Consequently, most companies have come to accept the compromises and tradeoffs that industry-specific, best-practice templates require. However, by taking the easy way out, these companies end up with generic, albeit industry specific, functionality (Donovan, 1999).

10.8 BENEFITS OF ERP

1. ERP Software Provides Almost Instant Access to all the Transactional Information Across the whole Organization ERP provides an enterprise database where all significant transactions are entered, recorded, processed, monitored, and reported. This database enables the unified view and acts as a general corporate resource rich in information and knowledge about how the company is doing business, its customer relations, its performance over time, and many other areas.

2. ERP Eliminates Data Duplication and Inaccuracies In the initial period, a successful ERP implementation helps in streamlining the operational areas like purchase, production, inventory control, finance and accounts, maintenance, quality control, and sales and distribution mostly in the form of automating the day-to-day transactions. The main benefit is that the duplication of data is eliminated while accuracy, reliability, availability, and consistency of the data are enhanced.

3. ERP brings Cohesion and Vigor even to a Multi-plant, Multi-location Organization Since ERP addresses the entire organization's needs, and not selected islands of the organization, the introduction of ERP brings a new culture, cohesion, and vigor to the organization. The line managers would then no longer have to chase information, check compliance to rules or conformance to budgets (Sadagopan, 1999).

4. ERP Helps in Breaking down the Bureaucratic and Political Stronghold on Information Ushering in a Culture of Interdependence and Transparency ERP provides a unified enterprise view of the business common to all functions and departments, which increases the visibility of the interdepartmental cooperation and coordination.

5. ERP Systems Relieve Operating Managers of Routine Decisions and Fire Fighting and Leave them with Lots of Time to Think, Plan, and Execute Vital Long-term Decisions of an Organization and to Monitor the Overall Progress of the Organization.

6. In the Medium-term, ERP Helps in Reducing Overall Cycle Time for Order Fulfillment In the process, it helps in reducing working capital (inventory and receivables). Further ERP improves coordination between various processes in the organization.

7. ERP Enhances Competitive Edge as it Helps in Cost Reduction and makes the Organization more Responsive to Customers and Markets ERP brings significant cost savings while simultaneously enhancing competitive edge of the organization by providing it increased customer service level and the ability to change easily (flexibility). Increased communications and coordination between units within the company enables the companies to become more responsive to customers and markets, thus, ERP makes the organization agile. This can be exploited to improve profitability, market share, or customer service.

8. ERP can Provide the Necessary Enterprise Extension to embrace the Entire Supply Chain.

10.9 FUTURE OF ERP

The ERP Solutions of the Future will be More Tailor-Made and Industry-Specific

ERP is Now Evolving Toward Object-Oriented Design Architecture Object orientation promises to create extraordinarily flexible ERP tools and will enable vendors to update the system without upsetting client-made customizations. Clients, ERP vendors, and third-party contractors would be able to add functionality without affecting the current system. The resilience and robustness that object-orientation promises will eventually be considered a requirement in ERP systems (Dillon, 1999).

ERP Solutions would be Component-Based or Modular Some visionaries believe that the present monolithic structure of ERP systems will be replaced with a flexible, modular system based on standard interfaces—middleware—between closely coupled modules produced by different vendors. That is, best-of-the-breed ERP solutions for each specific business need (that is modules) picked from various vendors can be combined together in an open system and an ERP solution can be developed. For instance, it would be possible to run SAP financials with Baan Manufacturing modules, an Oracle Applications database, and MS Office and have each one behave as if it were operating in its native environment.

- Modularization of ERP systems would enable vendors and third parties to revise individual applications or bring new ones to market quickly.
- Modularization also means that smaller companies buying their ERP in pieces would not need the capital to purchase the entire expensive package at once. This is very significant since, with the large companies' market for ERP getting saturated, ERP vendors are now targeting smaller companies with lower prices and easier and faster implementation methodologies. Many smaller companies are already under pressure to adopt ERP as larger manufacturers are beginning to demand that their suppliers be ERP compliant.
- Supply chain partners having different ERP solutions would be able to exchange information using *transaction translation engines* that convert business transactions from one ERP tool to another in real time.

Third Party Bolt-on Applications will Become Routine Additions to ERP Software Suites This will enable users to exchange application packages easily as better ones reach the market (Latamore, 1999).

ERP Packages will be 'Even More Web-Enabled' to Support Closer Coordination Along the Supply Chain This would eventually lead to creation of 'virtual manufacturing enterprise' in which the manufacturing planning of customer and supplier are closely meshed.

ERP will have Capability to Create and Disseminate Business Intelligence (BI) The process of capturing, understanding, and analyzing corporate data from a variety of sources within a company is called business intelligence. It is designed to help companies focus on analysis as they solve their business problems. It gets information to the decision-makers—mostly the middle and senior managers. In that sense, the data mining, On-Line Analytical Processing (OLAP), and data warehousing activities of recent years help create BI. It gives a variety of company personnel access to the enterprise database, which has all the real-time operational information. Following are two of the most important BI tools.

- *Dashboards* These are specifically created for each manager and executive and have the measures and metrics necessary to inform him of any emergency, important task, or issues to be dealt with. With this on-the-spot, targeted information, companies can more readily achieve the flexibility their customers require.
- *Balanced Scorecards* Robert S Kaplan developed the concept of balanced scorecards, which presents managers with four different perspectives from which to choose measures: financial (past and present), internal, customer (external), and growth (future). These perspectives helps managers to identify and focus on important issues with a feedback mechanism to measure progress.

ERP may eventually provide a Variety of Advanced Planning Capabilities based on Artificial Intelligence (AI). Companies will be able to redesign the whole processes throughout the organization using the information and analysis ERP systems provide (Latamore, 1999).

10.10 ERP IMPLEMENTATION CASE STUDIES

10.10.1 Implementation of SAP R/3 at Mahindra & Mahindra, Mumbai, India*

Core Business Activity

Manufacturing utility vehicles, light commercial vehicles (LCVs), and agricultural tractors.

Facilities

Seven factories; 33 sales offices, and 500 dealers throughout India.

Pure BPR Approach

The company undertook a massive business process reengineering exercise before implementing ERP. The BPR exercise began sometime in 1994

*(Adapted from Garg, 1999).

whereas SAP implementation started in 1996. The company appointed Lucas Engineering Company as consultants who would do the hand-holding, but not carry out the actual implementation. The major objectives of the BPR exercise according to the ERP chief were:

1. The company was to be structured to be sector-based instead of being on a functional basis; each sector was to be evolved into a strategic business unit.
2. A design team was constituted for the factories, and a supplies model was made for their supply chain management. The entire manufacturing operation was reorganized and reoriented to a process-based structure.
3. The BPR exercise allowed the company to build ownership around the cells. From each cell activity the ownership was built around each department to each functional area to the organization and so on.
4. Top management's continued interest and commitment, communication to the employees through seminars and discussions convinced the employees that the management was earnest about the activity.
5. By the end of the year 1995 management realized that all the business functions and processes, such as sourcing, purchase, manufacturing, R&D, finance, marketing, distribution, and service needed to be integrated with each other to improve the customer service levels and to decrease costs. Secondly, management also realized that fast and accurate information must be available for the decision making to be effective. This naturally led to the realization that a solution like ERP was not only a need for the organization to be competitive but it was also a logical extension of the BPR exercise at Mahindra and Mahindra. Besides momentum had already been built up since change had been taking place for quite some time.

A Wholly Owned Subsidiary as Outsourced ERP Implementation Consultant

A remarkable feature of ERP implementation at Mahindra & Mahindra was that the ERP implementation expertise was outsourced to a wholly owned subsidiary—Mahindra Applied Systems Technology Ltd. (Mahindra ASYST)—which was especially floated by the parent company for the purpose. The team members required for the implementation task were shifted from the parent company to Mahindra ASYST; the team included people involved in the BPR exercise. A wholly owned 100 per cent subsidiary combined the benefits of both—the expertise, exposure, and competitive skills of an outsourced implementation consultant as well as the control that can be exercised only on a group company.

Selection of ERP Package

Initially some 13 ERP packages were shortlisted and were subjected to a comprehensive evaluation process; this process narrowed down the choice to four packages namely SAP, Baan IV, JD Edwards, and Ramco Marshall. These packages were reevaluated through a script test keeping in mind the specific requirements of the company. The major considerations were:

(i) the functionalities offered by and the cost of the solution, and
(ii) the skill, expertise, experience, and the strength of and the services offered by the solution provider.

SAP was chosen as it was found to be very rich in functionalities and secondly, it was felt that SAP matched with the needs of the company. However, before making the final selection, all the processes of manufacturing, sales and distribution, asset management, plant maintenance, finance, accounting practices were mapped and reengineered. SAP was finally chosen as the software package for enterprisewide implementation.

Formation of Implementation Team

Some 50-odd users were included in the team. About a third of the team members belonged to various functional areas while the remaining people were from the technical side. Experts from SAP and also a few experienced consultants were taken into the team to ensure that the pitfalls could be avoided. The team was divided into modules of Production Planning and Control (PP), Materials Management (MM), Sales and Distribution (SD), Finance and Controlling (FI & CO), Plant Maintenance (PM), and Quality Management (QM). By June 3, 1996, the implementation team was already in place, and by September 1996, training of the team, data collection, and mapping of business processes were over. Table 10.1 gives the format of the three phases of the implementation (ERP) at Mahindra & Mahindra.

User Training

The implementation team trained a total of about 2500 users on the system. The type of training depended on the level at which the particular user would be working; the levels ranged from those of the decision-makers to that of the data entry personnel.

- The implementation team faced integration problems only in implementing certain areas in finance which involved various taxes, such as excise and other levies. SAP people developed a solution to address these issues. This solution eventually became the 'India Version.'
- Some two million entry fields were filled in; these fields also needed to have interfaces for proper sequencing and dataflow.

Table 10.1

Three Phases of ERP Implementation at Mahindra & Mahindra

Phase	*Sites*	*Milestones*
I (Pilot Sites)	1. Nasik Automotive Plant. 2. Nagpur Farm Equipment Plant. 3. Sales Head Quarter at Worli.	▪ Configuration and testing phase was over by February 1997. ▪ All the three sites went live with FI & CO, MM, SD, PP, QM, and PM modules on April 1, 1997.
II	4. Two plants at Kandivali. 5. Head Office.	▪ Kandivali site went live on April 1, 1998.
III	6. Igatpuri Plant. 7. Zaheerabad Plant.	▪ Implementation at Igatpuri plant started in June 1998 and it went live on October 1, 1998. ▪ Zaheerabad plant was projected to go live on April 1, 1999.

10.10.2 Baan IV Implementation At Calcom Vision, New Delhi, India*

Company Characteristics

Calcom Vision is a manufacturer of color and black-&-white TV sets, fly-back transformers, deflection yokes, and electronic components. It has seven factories spread across three states in India. The management wanted to link them all to improve inventory management, manufacturing planning, distribution, and financial systems.

How the Need for Implementing ERP was Identified

Calcom Vision was using a customized UNIX-based Manufacturing Resource Planning (MRP II) solution running in batch mode; the solution had been developed by Wipro in 1989. The company engaged Price Waterhouse to study the structure of the organization and suggest changes in the processes employed to realize the vision of the top management. Price Waterhouse while studying their IT infrastructure suggested the implementation of ERP. Since the existing MRP II system was not giving the desired results the management decided to go for ERP as per the recommendations of the consultants.

Selection of the Right ERP Package

The ERP vendors were contacted by Calcom directly. Three packages were shortlisted based on a one-day-each demonstration of the packages

*(Adapted from Upadhyaya, 1999).

in a simulated environment by the vendors. The management felt that all the packages offered more or less similar features, hence the selection was done based on certain other considerations that are described below:

1. Quite a few electronic manufacturing companies having similar products like those of Calcom had implemented Baan IV abroad; this convinced the company that Baan IV would be the right solution for them.
2. Baan IV had already addressed the localization issues such as taxes etc.
3. Baan IV offered good prices—the total package cost was 10 million rupees, which included software from Baan, and implementation including entire business process mapping and project management by Price Waterhouse.

ERP Implementation Options

Option 1: Implement ERP unitwise—testing it, fine-tuning it and then moving on to the other sites.

Option 2: Implement ERP simultaneously at all sites.

Management decided to go for the second option to cut short the implementation time and to allow all users to get started on the ERP together.

Implementation Methodology

1. The hardware specifications were prepared to satisfy the company's requirements; hardware was procured and was put in place.
2. The software was loaded.
3. A key user group of 20 people was identified and trained on the software for three months.
4. *Simulation 1.* Price Waterhouse defined the entire business processes in detail and created a feedback loop, ensuring that there were no open-ended processes.
5. *Simulation 2.* Master data was structured and coded.
6. Process of setting WAN was initiated, VSATs for distant units and ISDN for factories inside and around Delhi city were suggested.
7. Software was set up and was put on the network.

Implementation Problems

- Calcom was the first Indian site implementing Baan IV on Windows NT and according to the company management the vendors lacked experience and also did not properly draw on the huge knowledge base that Baan being a company with an international standing must have had. The ERP vendor also admitted that Calcom had some unique problems and Baan, which was still in the process of setting up business in India, struggled to address these problems.

- Process mapping done by Price Waterhouse in Sim-1 had to be modified in Sim-2.
- According to the company the vendor had not done the hardware sizing properly; the server choked within the first six months of going live when it had actually been sized for two years. The vendor, on the other hand, blamed that the company had defined 11 companies within a company, this has resulted in many databases being created and this had changed the hardware sizing.
- Networking the various units was another major problem owing to the very poor communication infrastructure in India.
- According to Baan, a customer needs to do a lot of work as far as training is concerned; the vendor can at the most transfer about 60% of the knowledge existing in the software as the vendor does not have the knowledge of the business processes of the customer. Calcom and Price Waterhouse, according to Baan, fell short in addressing the training requirements which also contributed to implementation problems.

10.10.3 SAP R/3 Implementation at IBM*

In 1993, IBM operated as 20 near-autonomous units each with its own processes, systems and IT structure. It was then that the company adopted the business strategy to go to the market as one IBM—a single integrated global organization—to better serve the customer and present a consistent face to the customer. To achieve this the company had to undertake the following activities:

1. Streamline and integrate major businesses across the enterprise.
2. Redesign IT systems and architectures to help in integrating the whole of the organization.
3. Get people working as a team across IBM.

IBM transformed processes in three waves:

1. The first wave, which was completed in 1995, focused on finance and human resources with the intent to stabilize financial performance and address redundancies to show some immediate returns. Here SAP R/3 was implemented to support corporate financial processes. IBM implemented SAP R/3 financial modules in its Storage Systems Division to support a new and independent OEM business.
2. The second wave involved redesigning and integrating six key value-added business processes. Here SAP R/3 was implemented across their primary business processes, namely procurement, production, fulfillment, and finance.
3. The third wave comprised of web-enablement—leveraging im-

*(Adapted from www.ibm.com).

provements and linking internal systems to the web. Here e-business solution was used to tap into R/3 for critical data and transactions.

The Storage Products Company

- IBM's R/3 implementation is one of the largest implementations in the world. It encompasses 5 business units and 4 global corporate-wide functions. The Storage Products Company is part of IBM's Storage Systems Divisions (SSD), which is headquartered in San Jose, California. It was aiming to achieve an annual growth rate of 50 to 100 per cent. This company set out to develop a worldwide business model to fulfill its need for better systems.
- Storage products selected SAP R/3 as its application functions met their requirements very well. The hardware selected was IBM's RS/6000 with AIX—IBM's UNIX-based system.
- After the platform was selected, a handful of knowledgeable users from within the organization were selected and made full-time members of the SAP implementation team. Their task was to work with the outside systems integrator and the onsite consultant from IBM/SAP competency center to map R/3 applications to the reengineered business processes.
- As a result of process improvements achieved during the implementation, every hard-disk drive coming out of the warehouse today is bar-coded with its serial number and IBM part number. Once it is scanned into the order entry system, the serial number becomes the unique identifier that allows every drive to be tracked throughout the fulfillment cycle.
- Storage Products Company's future plans include supporting sales offices in Europe and Asia and extending R/3 functionality in materials management and production planning for the company's new manufacturing facility in Singapore and, eventually, for all plants worldwide.

IBM's Microelectronics Division (IMD)

The second installation to go live was IMD headquartered in Fishkill, New York. IMD has long made semiconductors for IBM products, but in 1992 the division started selling them to other companies. The division's order processing, quoting and fulfillment system was a home-grown solution that was never intended to serve the needs of outside customers, nor was it well-equipped to handle the volume of business the company was now handling. Further, lack of integration between the functions of the systems was leading to errors (as high as 20 per cent) between the quoting, manufacturing, invoicing and shipping stages. The implementation team found that the right way to implement SAP was to combine SAP implementation with reengineering of the processes.

10.10.4 Implementation of Mfg/Pro at Glaxo India Limited, Mumbai*

Company Characteristics

Glaxo India is one of India's largest pharmaceutical companies. It produces a wide range of pharmaceutical, veterinary, and laboratory products. It has a 5.4 per cent share in the domestic market. It was established in 1924 under the name H J Foster & Co. Now, it is a public limited enterprise with a 51 per cent stake being held by its parent company.

Need for ERP

Apart from the usual reasons for which a company goes for ERP, Galxo wanted to ensure that they could get reports from all their field staff on a daily basis. To eliminate the high cost of setting up and maintaining a network, Glaxo turned to the Internet to build a network which was cheaper and just as efficient as some of the other on-line networks implemented in India.

Selection of ERP Package

Glaxo's key criteria for the selection of ERP software were:

1. Stability of the product worldwide.
2. Experience of the vendor in pharmaceutical industry.
3. Availability of support from vendor or implementation partners.
4. Conformance of the product towards Indian environment and laws.
5. Cost.

Glaxo had already implemented BPCS in the early 1990s. But because there were serious limitations in the product functionality and also because proper support for the product was not available, the company decided to discontinue BPCS and started looking for their second ERP package. The company didn't go for SAP as the cost of implementing it could not be justified. Glaxo needed a slimmer, meaner package that could meet their functional requirements; MFG/Pro fitted the bill and was selected. Top management at Glaxo was not in favor of outsourcing the entire implementation to an external systems integrator without internal full-time specialists on the team. The option favored by Glaxo was to have internal managers as implementers having the combination of business knowledge as well as IT, technical, and ERP software knowledge. Lower level hands, such as technical writers, systems analysts, or programmers, were to be hired from the body shopping agencies.

The management at Glaxo believed that phased ERP implementation was the most appropriate methodology particularly since they had a

*(Adapted from Fernandes, 1999-a).

small team to work with. The company established a dial-up network across over 100 Glaxo-Wellcome India (GW) business locations. It was a conscious decision to have a distributed IT environment and integrate data periodically at a central key business decision location. The ERP team established the ERP environment with state-of-the-art technology and trained users in the use of software. The user requirements were fine-tuned and more and more locations were connected and greater bandwidth was provided to the users.

10.10.5 Implementation of Mfg/Pro at Raychem RPG Limited, Mumbai, India*

Raychem RPG is a 50:50 joint venture company between US-based $ 2 billion Raychem Corporation and RPG Enterprises, one of India's largest industrial houses. Its products include electrical, telecom, electronics and polyswitch products. It has a turnover of Rs 500 million with a net worth of Rs 100 million in the fiscal year ending March 1999.

Till 1995, the company was using a software package called PeachTree which allowed the company to do order entry, invoicing, inventory control, and purchasing. The company already had a staff/PC ratio of 1:1 and all employees were connected over an e-mail network. The company felt the need for a process control system that spanned the enterprise.

ERP Selection Process

Company wanted to implement something that was accepted globally and had been tested extensively. Their decision were in no way influenced by their principals as they did not have to send any reports to their principals in the US. However, their principals were prepared to send in support staff to help with the implementation. The company also looked for a package that would be able to deliver all that the company required with the least amount of customization. Costs and local support were the other major criteria for selecting the ERP package. The company invited a couple of implementers to demonstrate the chosen ERP's capabilities in a conference room pilot.

Benefits the Management Expected to Achieve

1. Reduction in inventories from 120 to 90 days and a corresponding reduction in inventory obsolescence by 50 per cent.
2. Reduction in air shipments as a result of better planning down to 20 per cent.
3. Reduction in manpower requirement by 10 per cent (by avoiding data entry duplication).
4. Reduction in sales outstanding by 20 per cent due to proper follow up.

---

*(Adapted from Fernandes, 1999-b).

5. Reduction in order cycle lead time from 42 days to 28 days.
6. Increase on-time deliveries from 75 per cent to 90 per cent.
7. Complete order acknowledgements/invoicing in 24 hours.
8. Decrease production cycle time from 18 to 14 days.
9. Closing the books in a single day.
10. Intangible benefits such as enhanced decision making capability, and ability to manage inventories, receivables, and costs better.

Implementation

The implementers worked over weekends to ensure that business was not interrupted as a result of moving to an ERP installation. Knowing that the IT team was working round the clock generated positive attitude and created a sense of ownership among the employees. The implementers consciously avoided customization of the package.

The ERP implementation yielded a return on investment of more than 47 per cent and by a conservative estimate the entire system has paid back in less than a year and a half.

10.10.6 Implementation of Baan IV at Godrej Telecom Ltd (At their Mumbai and Hyderabad Plants in India) in 122 Days!*

The Company Characteristics

Godrej Telecom Limited (GTL) is a small Indian company manufacturing analog/digital telephones, cordless telephones and payphones. Some of these products were made to stock (MTS) while others were made to order (MTO). GTL is a joint venture between Godrej & Boyce Manufacturing Company Limited and Singapore based Intraco. The company manufactures 20 different end products from more than 2,500 components procured from 400 business partners, comprising mostly of overseas component suppliers and local subcontractors who primarily assemble printed circuit boards from components like chips, diodes, resistors, etc. supplied by GTL. High obsolescence due to rapidly changing product design and highly volatile customer demands were the major problems faced by the company.

Objectives of ERP Implementation

1. To achieve flexibility and agility to respond to fast obsolescence of product designs and fluctuations in demand.
2. To manage a growth rate of over 50%.
3. To properly manage manufacturing processes, which were becoming more and more complex as more and more of outsourced components were manufactured in-house.
4. To replace the existing software solutions like Focus and Tally

*(Adapted from Thathachary, 2000).

which were becoming bottlenecks in terms of operations and also required a lot of manual intervention.

Package Selection

After evaluating a number of ERP packages including Makess, Scala, Magic, and Baan, GTL finally selected Baan IV (with a 16 concurrent user license). The package was installed on HP Netserver LC3 running Windows NT 4.6 operating system, the server was located in their Hyderabad office. The database used was Informix 7.2 TE1. Number of users was limited to 20 from their overall employee count of 260. Company's Mumbai and Hyderabad sites were connected using a 64 kbps leased line from Department of Telecommunications (DOT).

Implementation Teams

In-House (GTL) Team

- Senior Manager—HRD (Head).
- Functional Head—Finance.
- Functional Head—Logistics.
- Functional Head—Manufacturing.
- Functional Head—Materials.
- Functional Head—R&D.

Implementation Partner (GIL) Team

- Project Manager.
- Baan Consultant—Manufacturing.
- Baan Consultant—Distribution.
- Baan Consultant—Finance.
- Support personnel for Baan Tools, administration and customization.

Stage 1. Preparation (Activities Preparatory to Simulation 1)

(a) Business interviews were conducted and all business process details were collected from key users.
(b) The implementation teams visited the factory.
(c) Conceptual training for all the modules viz. manufacturing, distribution, and finance was conducted for the team.
(d) A discussion was held on item coding schemes, and a coding scheme acceptable to all functions was developed.

Stage 2. Simulation 1 (Mapping)

(a) Simulation 1 was done using a normal push-button phone, which was a high-volume make-to-stock product.
(b) All key users went through the simulations together and were able to easily identify with and relate to the outputs from the system as the item selected for simulation 1 was a common running product of GTL.
(c) The benefits as perceived by the individual departments and gaps between Baan processes and the current business processes of GTL were identified and documented.
(d) A document listing the 'as is' and 'to be' business processes was prepared.

Stage 3. Simulation 2 (Piloting)

(a) Advanced training for key users was conducted.
(b) The most complicated and expensive product of GTL, the digital payphone, which covered almost all the different types of business processes and situations in the company, was chosen for simulation 2.
(c) Blank data formats were distributed and data was collected through them familiarizing the people with the process of data collection. The data collected at this stage greatly facilitated the migration process in stage 3.
(d) Simulation 2 was carried out with full Finance integration setup and the results were analyzed.
(e) End user manuals with work instructions, data coding schemes, business processes, system parameters, customizations and authorizations were all prepared in this stage.

Stage 4. Simulation 3 (End User Training and Data Migration from Legacy System to the ERP System)

(a) End users were trained.
(b) Data was collected and entered for item data, bills of material, routing, customers, suppliers, etc.
(c) Stocktaking was carried out within the factory premises and with subcontractors.
(d) The data entry of products was done in the sequence of the production plan of the subsequent month so that the process of static and dynamic data migration did not affect production of GTL after going live.

Activities Performed after Going Live

(a) Bills of material, inventory status, and opening balances were continuously monitored for errors. Corrections were made immediately and files were kept updated.
(b) Every transaction was monitored for accuracy of data and setup.
(c) Technical problems were resolved by GIL's technical team.

References

ASAP World Consultancy, *Using SAP R/3* (second edition). Prentice-Hall of India Private Limited, New Delhi 1997.

Blanchard, David, *ERP: the Great Equalizer,* In Evolving Enterprise Volume 1, Number 1, Spring, 1998.

Chaterji, N G, 'ERP: Imperatives for Success'. In S Sadagopan, (Editor) *ERP: A Managerial Perspective.* Tata McGraw-Hill Publishing Company Limited, New Delhi, 1999.

Cox III, James F, and John H Blackstone Jr. (Editors), *APICS Dictionary* (ninth edition). American Production and Inventory Control Society, Inc, Falls Church, VA, 1998.

Dillon, Craig K, 'Stretching toward Enterprise Flexibility with ERP' in *APICS: The Performance Advantage.* October 1999, Vol. 9, No. 10, pp 38–43.

Donovan, R Michael, 'Successful ERP Implementation the First Time' in *Midrange ERP*, August 1999 (article downloaded from www.midrangeERP.com).

Fernandes, Louis, 'Connectivity Solutions'. In *MIS South Asia,* April 1999-a, pp 52–54.

Fernandes, Louis, 'Ray of Business'. In *MIS South Asia,* April 1999-b, pp. 41–43.

Garg, Pooja, 'A Patient Wait'. In *MIS South Asia,* April 1999, pp 26–29.

Hiquet, Bradley D, Anthony F Kelly, Kelly-Levey & Associates, *SAP R/3 Implementation Guide: A Manager's Guide to Understanding SAP.* Techmedia, New Delhi, 1998.

Latamore, G Burton, 'ERP in the New Millennium'. In *APICS: The Performance Advantage,* June 1999, Vol. 9, No. 6, pp 29–32.

Leon, Alexis, *ERP Demystified.* Tata McGraw-Hill Publishing Company Limited, New Delhi, 2000.

Rajani, Renu L and Sridhar Padmanabhan, 'ERP Evaluation Criteria: A Consultant's Approach'. In S Sadagopan (Editor) *ERP: A Managerial Perspective.* Tata McGraw-Hill Publishing Company Limited, New Delhi, 1999.

Russell, Roberta S and Bernard W Taylor III, *Operations Management: Focusing on Quality and Competitiveness.* Prentice-Hall, Upper Saddle River, NJ, 1998.

Sadagopan, S, 'The World of ERP'. In Sadagopan, S (Editor) *ERP: A Managerial Perspective.* Tata McGraw-Hill Publishing Company Limited, New Delhi, 1999-a.

Sadagopan, S, 'Frequently Asked Questions (FAQs) on ERP'. In Sadagopan, S (Editor), *ERP: A Managerial Perspective.* Tata McGraw-Hill Publishing Company Limited, New Delhi, 1999-b.

Slater, Derek, 'The Hidden Costs of Enterprise Software'. *CIO Enterprise,* Section 2, January 15, 1998, pp 48–55.

Schroeder, Roger G, *Operations Management: Contemporary Concepts and Cases.* Irwin McGraw-Hill, Boston, 2000.

Thathachary, Shridhar, 'The 122 Days Countdown' in *ERP Journal* Vol. 2, No. 1, January 2000, pp 14–19.

Travis, D Michael, 'ERP Selection'. In *APICS: The Performance Advantage;* June 1999, Vol. 9, No. 6, pp 37–39.

Upadhyaya, Aparna, 'Calculated Risk'. In *MIS South Asia;* April 1999, pp 30–32.

Ushasri, T S, 'ERP: Bridging Business and Technology'. In S Sadagopan (Editor), *ERP: A Managerial Perspective*. Tata McGraw-Hill Publishing Company Limited, New Delhi, 1999.

Chapter 11

Supply Chain Management

11.1 SUPPLY CHAIN (LOGISTICS NETWORK)

Supply chain is defined as the sequence of business processes and information that provides a product or service from suppliers through manufacturing and distribution to the ultimate customer (Schroeder, 2000). It is a term that is being increasingly used to emphasize the interactions among marketing, logistics, and production.

In a typical supply chain, which is also referred to as the *logistics network,* raw materials are procured, items are produced at one or more factories, shipped to warehouses for intermediate storage, and then shipped to retailers or customers, thus the supply chain is a 'process umbrella' under which products are created and delivered to customers. The cycle of buy-make-move-store-sell is the supply chain.

The facilities in a supply chain typically include suppliers, manufacturing centers, warehouses, distribution centers, and retail outlets. These facilities are connected by transportation and communication links, along which raw materials, work-in-process inventory, finished products, and information flow. Ideally, the supply chain consists of multiple companies that function as efficiently and as effectively as a single company, with full information visibility and accountability.

The Supply Chains are Identified by Product Families There are several supply chains in a large company. The supply chain for a particular product family is identified by tracing the flow of materials and information from the final customer backward through the distribution system, the manufacturer, suppliers, and the sources of raw material.

Supply Chains Offer Great Opportunity to Improve Operations The total time for materials to travel through the entire supply chain can be six months to a year or more. Since materials spend so much time waiting in inventory, there is a great opportunity to reduce the total supply chain cycle time, leading to a corresponding reduction in inventory, increased flexibility, reduced costs, and better deliveries (Schroeder, 2000).

Virtual Supply Chains

Virtual Company

A virtual company produces a product or provides a service without owning the facilities like offices, factories, warehouses, etc. and without traditional employees. It does business by coordinating other companies that do the design, production, and distribution work. Virtual company is being made feasible by use of computers and Internet. Rapid communications and access to well-organized information about suppliers and contractors make it possible to quickly put together groups of people who are not in the same physical location.

A virtual company contracts for all or most of its major functions. More commonly, virtual companies use traditional employees and assets for their *core competencies* so as to protect their intellectual property and their knowledge of technology in those areas where they plan to create a competitive advantage for themselves. All or most of the other parts of the business are contracted out to outside businesses, such as, subcontractor manufacturing firms and third-party logistics firms, etc. This gives the virtual company greater flexibility and a very high Return On Net Assets (RONA) making them highly competitive in rapidly changing markets. However, virtual companies are more expensive to operate and are unable to compete with traditional companies in stable markets.

Virtual Supply Chain

It consists of at least one virtual company that coordinates all of the activities of the supply chain. Virtual company does not own the entire supply chain, but it can control the design of products and the movement of material through the supply chain by use of contracts and partnerships with other companies.

11.2 CHARACTERISTICS OF SUPPLY CHAINS*

1. Decisions in Each Part of the Supply Chain Affect the Other Parts.

2. Accelerator or Bullwhip Effect Demand changes by the end-user create an accelerator effect in the supply chain, which magnifies the size of demand changes on upstream supply chain elements (wholesalers, warehouses, and factories). Each upstream element in the supply chain not only aggregates the inflated demand changes of all the elements fed

*(Schroeder, 2000).

by it but also overreacts to these demand changes and adds its own safety margins to take care of perceived fluctuations in future demand from downstream elements. Inaccurate forecasts, information time lags, and replenishment time lags account for the dynamics observed. Upstream elements must be careful to not overreact to inflated orders from downstream elements, rather they should consider only the real demand changes at the end-user level.

3. The Best Way to Improve the Supply Chain is to Reduce the tOtal Replenishment Time Reducing replenishment time allows the supply chain to react rapidly to real demand changes and reduces the inventory needs.

4. Accurate Forecasting of Demand Changes and Making Available Actual Demand Information to All Levels can dampen the Effects of the Demand Change Effective demand management can smooth the fluctuation in the demand.

11.3 SUPPLY CHAIN MANAGEMENT

Supply chain management is defined as the planning, design, and control of the flow of information, materials, and money along the supply chain in order to meet customer requirements in an efficient manner, now and in the future. Supply chain management is the function of coordinating the material, information, and financial flows between and among all the participating enterprises. What these flows typically involve is described below:

1. *Material flows* involve physical product flows from suppliers to customers through the chain, as well as the reverse flows via product returns, servicing, recycling, and disposal.
2. *Information flows* involve demand forecasts, order transmissions, and delivery status reports.
3. *Financial flows* involve credit card information, credit terms, payment schedules, and consignment and title ownership arrangements (Kalakota and Robinson, 2000).

- Supply chain management consists of a set of approaches utilized to efficiently integrate suppliers, manufacturers, warehouses, and stores, so that merchandise is produced and distributed in the right quantities, to the right locations, and at the right time, in order to minimize systemwide costs while satisfying service level requirements (Simchi-Levi *et al.* 2000).
- Effective supply chain management enables the company to make informed decisions along the entire supply chain, from acquiring raw materials to manufacturing products to distributing finished goods to the consumer. Fast and effective flow of both information and materials is essential for improving the supply chain.

- SCM packages are back-end applications designed to link suppliers, manufacturers, distributors, and resellers in a cohesive production and distribution network and thus allow an enterprise to track and streamline the flow of material and data through the process of manufacturing and distribution to customers. By enabling greater data sharing between supply chain partners, SCM applications improve production efficiency and flexibility. The objective of SCM software is to plan and optimize the supply chain as a continuous, seamless activity that integrates all planning functions across the supply chain.
- SCM is directing people's thinking toward opportunities of reducing costs and increasing revenues by properly managing at the cross-functional and cross-company boundaries. SCM requires looking at the business as one continuous process. A reason for increased interest in SCM is that now the companies are thinking in terms of *total delivered cost to the consumer,* which includes not only the initial cost of the product but also transportation costs; warranty, repair, and return costs; inventory carrying charges for the total time the product spends in the supply chain, and the carrying costs for accounts receivable until the company is actually paid.
- The SCM process absorbs such traditionally distinct functions as forecasting, purchasing, manufacturing, distribution, and sales and marketing into a continuous flow of business interaction.
- Although SCM theory is more than twenty years old, it has, until recently, been difficult for companies to put it into practice because of lack of information and communications technology support to achieve the cross-company integration.

11.4 OBJECTIVES OF A SUPPLY CHAIN MANAGEMENT SYSTEM

1. Decrease iNventory Costs by More Accurately Predicting Demand and Scheduling Production to Match it SCM forecasting applications utilize extremely complex planning algorithms to predict demand based upon information stored in the company database. These applications also facilitate incorporation of any changes in supply chain data into the forecast much faster than previous modes of calculation.

2. Reduce Overall Production Costs by Streamlining the Flow of Goods through the Production Process and by Improving Information Flow between an Enterprise, its Suppliers, and its Distributors SCM applications such as manufacturing scheduling, transportation management, and warehouse management all contribute to reduced production costs. By ensuring real-time connectivity between the various parties in a supply chain, these applications decrease idle time, reduce the need to store inventory, and prevent bottlenecks in the production process. SCM helps suppliers, producers, and shippers work in an integrated manner and, thus, enables a continuous flow of material through the supply chain with almost no stoppages and no inventory.

3. Improve Customer Satisfaction by Offering Better Quality, Higher Product Variety, and Fast Response SCM applications allow enterprise to reduce lead times, increase quality, and offer greater customization, enhancing the customer relationship and improving retention.

Thus, the objective of supply chain management is to be efficient and cost-effective across the entire system, and to accomplish these objectives:

1. Every facility that has an impact on cost and/or plays a role in making the product conform to customer requirements is taken into consideration.
2. An attempt is made to minimize total systemwide costs while simultaneously improving the service levels. All costs, for example, costs associated with inventories of raw materials, work-in-process, finished goods, transportation and distribution, etc. are included.
3. Activities at all the levels of the firm, namely strategic, tactical, and operational levels, are addressed.

SCM is a multidimensional management activity having three dimensions involving:

1. Coordination of intra-functional activities.
2. Coordination of inter-functional activities.
3. Coordination of inter-organizational activities in the supply chain.

The value proposition of SCM system is its capacity to integrate all the participant in the supply chain into a dynamic Internet-, intranet-, and/or extranet-enabled system that plans the operations of the entire chain taking into account all the relevant factors and constraints.

11.5 SUPPLY CHAIN MANAGEMENT AND LOGISTICS MANAGEMENT

Logistics management has been defined by the Council of Logistics Management (CLM) as the process of planning, implementing and controlling the efficient, cost effective flow and storage of raw materials, in-process inventory, finished goods, and related information from point-of-origin to point-of-consumption for the purpose of conforming to customer requirements.

Simchi-Levi *et al.* (2000) does not distinguish between the two terms and their definition of SCM given above is similar to CLM's definition of logistics management. Both definitions emphasize the integration of the different components in the supply chain as the only means for significantly reducing the costs and improving service levels. Thus, when logistics management is broadly defined, it is identical to supply chain management.

Some managers, however, define logistics narrowly as only concerned with inbound transportation and outbound distribution, in which case logistics management is a subset of supply chain management (Schroeder, 2000). The term 'supply chain management (SCM)' seems to be replacing more of the traditional terms of management of material and service flows.

11.6 SUPPLY CHAIN MANAGEMENT AND ENTERPRISE RESOURCE PLANNING

ERP systems are built for recording what already happened, rather than for medium- and long-term planning for what will be. They are capable only of short-term planning and recording of operations. This

Table 11.1

Comparison of ERP and SCM Systems

Sr. No.	*ERP*	*SCM*
1.	ERP systems are chiefly concerned with transaction processing and simulation.	SCM systems have the ability to optimize.
2.	ERP considers various demand, material, and capacity constraints separately, in isolation of each other.	SCM systems are able to consider all the relevant constraints, such as demand, capacity, and material constraints, simultaneously and to perform real-time simulations of adjustments in the constraints.
3.	ERP systems perform many functions—getting answers from an overloaded ERP system may take hours.	SCM systems can provide plans in minutes or even seconds.
4.	ERP systems provide only limited graphical capabilities.	Leading SCM products provide enhancements such as visible maps of the entire supply chain. The graphical user interface gives the users complete visibility into the inner workings of the supply chain through demand, supply, manufacturing, scheduling, ware-housing and transportation. It greatly helps in identifying the problem areas.
5.	The static sourcing tables in ERP cause high costs.	SCM systems are much more dynamic.

lack of advanced planning capability hinders the company's ability to make the right decisions at the right time. SCM system derives raw data from ERP system or any other transaction processing system. Next, it engages an integrated set of planning engines to produce an optimal solution based on a complete view of the enterprise and its trading partners. Then it feeds the optimal solution data back into the ERP (or any other transaction) system for execution. SCM is the complete decision-intelligence system that turns ERP data into actionable business intelligence. Table 11.1 compares ERP and SCM systems.

ERP-SCM Convergence Most ERP and SCM vendors are bringing about a rapid convergence between ERP and SCM systems. The ERP vendors are adding more sophisticated SCM functionality to their ERP products; and SCM vendors are also expanding their functionality to include many ERP functions.

11.7 DIFFICULTIES AND ISSUES IN SUPPLY CHAIN INTEGRATION

A typical supply chain is a complex network of facilities and organizations with different, conflicting objectives as shown in Table 11.2.

Table 11.2

Conflicting Objectives of Different Players in a Supply Chain

Supplier Objectives	*Manufacturer Objectives*	*Warehouse / Retailer Objectives*	*Customer Objectives*
■ Large stable de mand for less variety of materials.	■ Desire low product variety and high production volumes, but to satisfy the customers must have high product variety and low volume.	■ Reduce inventory and transportation costs.	■ Low price.
■ Flexible delivery schedules.	■ Have to be flexible to accommodate customer's changing requirements.		■ High product quality and variety.
■ Low cost of production; high price from the customer (manufacturer).	■ Reduce raw material and production cost, increase profit margins.		■ Short lead time.

Predicting and Satisfying Fluctuating Demand is the Most Difficult Task Predicting demand and then maintaining a match between fluctuating demand and supply is a major challenge.

System Variations in the Supply Chain and the Environment over Time Pose Major Challenge in Determining the Most Effective SCM Strategy The system variations might be in the form of variations in needs and bargaining power of customers, capabilities of suppliers, supply chain relationships, and nature of competition.

High-Tech Products with Ever Shrinking Life Cycles Pose New Problems These problems might, for example, include lack of opportunity to understand and learn the related issues, lack of historical data and consequent difficulty in predicting demand, and significant price declines. The proliferation of products that address the same market segment makes it even more difficult to predict the demand for a specific model.

11.8 EFFECTIVE SUPPLY CHAIN MANAGEMENT: THE NEXT FRONTIER

In the late nineties, it became clear that many companies have drastically improved their internal operations and reduced the manufacturing costs as much as is practically possible. Hence, they find it necessary to consider relationships with the external customers and suppliers in the supply chain to gain further improvements in operations. Thus, supply chain is the area where there is a great scope to significantly reduce costs and achieve increase in market share by reducing the time to market, reducing distribution costs, and getting the right products to the right place at the right time, cost, and price. In some industries, for example computer and peripherals industries, where most manufacturers use the same suppliers and identical technologies, companies ultimately compete on cost and service levels making supply chain management the single most important factor determining the success of the firm.

Companies spend a significant amount of money in supply related activities, for example, as cost of movement, storage, and control of products across the supply chain, both within manufacturing plants and warehouses and between different components of supply chain. Many opportunities exist to cut unnecessary costs and time arising due to excess stock, inefficient transportation strategies, and many other wasteful practices in the supply chain. Not surprisingly, a number of companies have been able to substantially increase revenue, decrease costs, and/or decrease delivery times through effective supply chain management (Simchi-Levi *et al.* 2000).

11.9 FUNCTIONS OF IT SYSTEMS IN SCM

1. *In Modern Supply Chains, Information Replaces Inventory*

The SCM experts often use this phrase, and rightly so, as databases, ERP systems, decision support systems (DSS), electronic data interchange

(EDI), intranets, and the Internet have become essential components of any complex supply chain. Although, the customers finally need products not just information, nevertheless this abundance of information can effect changes in the supply chains that can lead to more accurate forecasts, reduced lead-times and inventories, better coordination between manufacturing and distribution systems, and better customer service. Further, the capabilities of SCM applications are rapidly growing to include capabilities to manage the inventories that companies can't see and don't own.

2. Allow Collection of Information on Each Product from Purchase through Production to Delivery, and Provide Complete Visibility for all Parties Involved

The idea is to have an information trail that follows the product's physical trail and links the point of production seamlessly with the point of delivery or purchase. This allows planning, tracking, and estimating lead-times based on real data. Further, any party that has an interest in the whereabouts of the product should be able to have access to this information.

3. Allow Access to Any Data in the Supply Chain from a Single Point of Contact

Many times the components in the supply chain need to access the data that resides in other companies' information systems, or across functions and geographic locations inside a company. For example, a supplier wants to access the manufacturer's MRP I system to anticipate an incoming order. A retailer might want to know the current status of its orders. Furthermore, the participants need to see data in their own terms; that is, the quantities converted from the total quantity of the parent item assembled by the manufacturer to the quantities of component items to be supplied by the vendors. Therefore, translation tables, such as bills of material, are required throughout the system. To enable this, the system must have a single point of contact for all the available information so that the information would be the same regardless of who is making the enquiry.

4. Provide Decision-Support Tools in the Form of Analysis of Total Supply Chain Information

Provide means to analyze, plan activities, and make trade-offs based on information from the entire supply chain to find the most efficient ways to produce, assemble, warehouse, and distribute products.

11.10 SUPPLY CHAIN PERFORMANCE MEASURES

1. On-Time Delivery Performance

It is the percentage of orders delivered complete and on the date requested by the customers. Orders partially fulfilled or delayed are to be excluded.

2. *Quality* (measured in terms of customer satisfaction and customer loyalty).

3. *Business Cycle Time*

It is the sum of number of days in inventory and the number of days in accounts receivable.

4. *Cost*

This can be measured in two ways:

(a) Total Delivered Cost to the Consumer, and

(b) Cost Efficiency = (Value Added • Total Cost of Adding the Value)
= (Sales – Cost of Material) • (Labor Cost + Overhead)

11.11 IMPROVING THE SUPPLY CHAIN STRUCTURE

Supply chain structure includes capacity, facilities, process technology, and vertical integration. Changes in structure are long-term, capital intensive, and bricks-and-mortar related. These changes usually rearrange the elements of the supply chain in a major or dramatic way. The structural changes are aimed at removing sources of uncertainty and time delays from the supply chain.

Five Forms of Structural Changes of the Supply Chain

(Schroeder, 2000)

1. *Forward and Backward Integration*

This refers to the ownership within the supply chain. If a manufacturer buys a wholesale firm to distribute its products only through it, then the integration is *forward toward the market.* On the other hand, if the manufacturer buys a supplier company, the integration is *backward in the supply chain.* If one firm owns the entire supply chain, there is total vertical integration. The benefits of vertical integration are better control of the supply chain and reaping the profits which otherwise would have gone to the suppliers or distributors. The drawbacks are loss of flexibility and loss of economies of scale (Schroeder, 2000).

2. *Major Process Simplification*

When the processes are very complex, or hopelessly out of date, it is better to redesign the process from scratch without regard for the existing processes. This could include major conceptual changes in how business is conducted and major information system changes.

3. Changing the Number and Configuration of Factories, Warehouses, or Retail Locations

4. Major Product Redesign

Many times supply chain operations can be improved by a major change or a complete redesign of the product, for example, a possible solution to too many product variations and types is the modularization of the product design.

5. Outsourcing Logistics to a Third Party

For some companies, the best approach is to outsource all inventory management, distribution, and logistics to a third party.

11.12 IMPROVING THE SUPPLY CHAIN INFRASTRUCTURE

Supply chain infrastructure includes people, information systems, organization, production and inventory control, and quality control systems. Changes in infrastructure are related to people and systems and thus, are in the 'soft' side of the supply chain. Infrastructure changes involve changing the way the supply chain operates within the given structural arrangements. Whether in the structure or infrastructure, the improvements should either reduce uncertainty or replenishment time. Uncertainty can be reduced in demand or in supply times along the chain. This in turn reduces the need for inventory. For example, in the extreme case where demand uncertainty is zero and supply is completely reliable, no inventory is needed except in transit (Schroeder, 2000).

There are five possible ways to improve supply chain infrastructure:

1. Cross-Functional Teams These teams provide coordination that is lacking across the various departments.

2. Partnerships with Suppliers and Customers Partnerships start with a commitment from both firms to establish a long-term business relationship that will be mutually beneficial. The partners should develop trust for each other and establish teams of employees from the two firms to work together on important improvement projects.

3. Setup Time Reduction This makes making or purchasing of smaller lots economical, which in turn drastically reduces inventories.

4. Improving Information Systems to Enable Faster Information Sharing throughout the Supply Chain.

5. Cross-Docking In cross-docking system, which is a feature of Wal-Mart's supply chain design, products are delivered to Wal-Mart's warehouse on a continual basis where they are sorted, repackaged, and distributed to retail stores *without* sitting in inventory. Goods cross from one loading dock to another in forty-eight hours or less. This system allows Wal-Mart to purchase full truckloads of goods while avoiding the inventory and handling costs (Russell and Taylor III, 1998). Cross-docking is now being widely used wherever there is sufficient volume to make it possible.

11.13 PROCESS OPTIMIZATION

The integration of ad hoc and fragmented processes in the supply chain into a cohesive system requires process optimization to minimize the total cost of the order-to-delivery process by trading off the costs of inventory, transportation, and handling. The traditional optimization tools were not capable of optimizing multiple variables simultaneously and, thus, were unable to handle the complex interdependencies of real-life situations. Secondly the suppliers, manufacturers, transporters, distributors, and retailers were using separate business systems trying to optimize the costs under their direct control rather than optimizing the combined costs of end-to-end operations. Thirdly, the lack of effective information systems meant that no one player has had the information visibility needed to synchronize and optimize the entire supply chain. The advent of large-scale optimization techniques has enabled the optimization of multiple variables simultaneously. The new generation of supply chain optimization tools provide an integrated approach through which demand prediction, inventory stocking, and transportation decisions are made together to optimize cost, quality, service, and time factors, while simultaneously considering constraints, such as production and storage capacity, capital resources, time constraints, and also profitability objectives.

11.14 SUPPLY CHAIN APPLICATION MODULES

Supply chain modules help make better operating decisions, for example, decisions about product volumes in various periods, inventory levels, finished goods storage locations, transportation planning, etc. For most of their data, SCM applications rely on the ERP backbone. Following are some of the SCM application modules:

11.14.1 Supply Chain Planning Applications

Supply chain planning applications are tools used to access and analyze information stored in the company database in order to forecast product demand and plan manufacturing accordingly. Sophisticated computing platforms run data through extremely complex planning algorithms and optimization routines in a comparatively short time frame to produce anticipated demand figures and corresponding procurement and scheduling requirements, the calculation of which used to take days. These systems may be applied to both operational decisions (such as shop floor scheduling) and long-term strategies (such as factory construction and quarterly forecasting) (Cherry Tree & Co, 2000).

1. Demand Planning It identifies and analyses patterns and fluctuations in demand, and creates accurate dynamic demand forecasts. Actually, the SCM process begins in this application with forecasting and data mining applications analyzing information consolidated in the enterprise's database. Planning algorithms are used to generate a demand forecast, which is then used as a basis for planning procurement orders and production schedules.

2. Supply Network Planning It matches purchasing, production, and transportation processes to demand, and balances and optimizes the entire supply network.

3. Collaborative Planning It helps the enterprise to carry out collaborative supply chain planning activities with business partners.

4. Supply Planning This is a planning-level subprocess that spans the strategic and tactical supply-planning processes. It involves long-term planning, and collaborative procurement.

5. Manufacturing Planning and Detailed Scheduling This devises a production schedule based on real time analysis of constraints. It optimizes the use of resources and creates accurate plant-by-plant production schedules in order to shorten production life cycles and respond rapidly to changes in market demand forecasts. It can adjust manufacturing plans if certain materials or workers are not available.

6. Distribution Planning.

11.14.2 Supply Chain Execution Applications

These applications use the information generated by supply chain planning tools to guide the physical production, storage, and movement of raw materials, components, and finished products.

1. Order Promising (also called Available-to-Promise, Demand/Order Fulfillment, or Order Commitment) It is used to provide fast, accurate, and reliable delivery-date responses to customer orders by means of both real-time checks and sophisticated simulation methods. It is an execution-level subprocess that includes order capturing, customer verification, order promising, backlog management, and order fulfillment.

2. Inventory Management.

3. Transportation Management It helps to optimize the existing transportation network and select the best routes in response to changes in demand or costs. It supports all activities necessary to manage, execute, and optimize the entire transportation process. These activities include inbound and outbound, intra- and inter-company moves of materials and products. It considers such variables as loading dock space, trailer availability, load consolidation, and the best mix of available transportation modes and common carriers.

4. *Warehouse Management* It provides capabilities for managing inbound, outbound and storage processes.

5. *Supply Chain Modeler (also called Supply Chain Navigator, Supply Chain Cockpit)* It provides users with a bird's eye view of all activities and applications in the supply chain through a graphical user interface.

11.15 MARKET LEADERS

The top vendors of SCM software include i2 (part of their RHYTHM solution), Manugistics (part of their NetWORKS Applications); SAP (part of their mySAP.com applications), Baan (through their company CAPS Logistics), PeopleSoft, and Oracle.

References

Cherry Tree & Co, Extended Enterprise Applications: Spotlight Report, January 2000. A white paper prepared by Cherry Tree & Co. available at the *ITtoolbox Portal for SCM* at the site http://scm.ittoolbox.com

Kalakota, Ravi and Marcia Robinson, *e-Business: Roadmap for Success.* Reading, Addison-Wesley, Massachusetts, 2000.

Russell, Roberta S and Bernard W Taylor III, *Operations Management: Focusing on Quality and Competitiveness.* Prentice Hall, Upper Saddle River, N J, 1998.

Schroeder, Roger G, *Operations Management: Contemporary Concepts and Cases.* Irwin McGraw-Hill, Boston, 2000.

Simchi-Levi, David, Philip Kaminsky, and Edith Simchi-Levi, *Designing and Managing the Supply Chain: Concepts, Strategies, and Case Studies.* Irwin McGraw-Hill, Boston, Burr Ridge, I L, 2000.

Chapter 12

Customer Relationship Management and the Extended Enterprise

12.1 INTRODUCTION

Customer Relationship Management (CRM) is an information industry term for methodologies, software, and Internet capabilities that help an enterprise manage customer relationships in an organized way. For example, an enterprise might build a database about its customers that describes relationships in sufficient detail so that management, salespeople, service-people, and the customer can directly access information, match customer needs with product plans and offerings, remind customers of service requirements, and know what other products a customer had purchased. CRM enables the creation of long-term, profitable relationships with customers through increased customer intimacy and improved business process effectiveness.

CRM links back-office and front-office functions with all of a company's touch points (for example, call centers, the corporate Website) with the customer. CRM can refer to either a business strategy or software tools. Cherry Tree & Co defines CRM applications as front-end tools designed to facilitate the capture, consolidation, analysis, and enterprisewide dissemination of data from existing and potential customers. This process occurs throughout the marketing, sales, and service stages of the business (Cherry Tree & Co, 2000).

- CRM covers all customer touch points, like face-to-face, Internet (the corporate web site), or phone (call centers). It supports employee- and customer-facing roles and allows all individuals, whether employees, business partners or consumers, to collaborate.
- Further, e-CRM—the customer-centric e-Business—produces sustainable competitive advantage by enabling firms to:

1. Offer improved levels of service and support.
2. Build rich customer databases.
3. Personalize customer interactions.
4. Continuously capture valuable information about customer preferences, which improves product development efforts.

- CRM provides an integrated view of a company's customers to everyone in the organization and, thus, ensures that everyone in the enterprise is focused on the customer.
- CRM provides seamless integration between all applications and flexible deployment of solutions, merging front-office and back-office into one office that focuses on increased customer satisfaction.
- CRM ensures that the same information is readily available to all in the company so customers don't have to explain their problem or issue over and over to a different company representative. For example, if marketing runs an outbound campaign, all the information about the customers and the program should be retained for:
  (a) the sales people to follow up on,
  (b) the customer service people to answer any queries, and
  (c) the technical support to provide any field support (Khera, 2000).
- Another related term is *Enterprise Relationship Management (ERM)*, which puts more emphasis on extending CRM to include other relationships, such as those with business partners and suppliers.

At the end of the day, excellent customer relationship management is not only or even mainly about improving profits, but about long-term competitive survival. If the expectations of today's customers are not met, they will simply defect. Further, disgruntled customers have much more power to broadcast their grievances in a networked world, as experienced by many companies on the web. And also with the Internet, the customer knows the product variety, quality, and prices offered by any supplier anywhere in the world. As a countermeasure, companies are ratcheting up their marketing and sales efforts, leading to an escalating battle for customer attention and allegiance.

12.2 CRM OBJECTIVES

1. Enable the Company to Quickly Identify, Contact, Attract, and Acquire New Customers

This allows the company to focus its limited marketing resources on the most promising target markets with the highest potential value. This is typically done using the information generated by CRM applications, which automatically generate customer and market profiles, identify target markets with high revenue and profitability potential, generate leads, track marketing campaigns across a variety of media, select appropriate contact media (that is, mail, phone, print, etc.), plan promotions and incentives, and manage the quote/proposal process through negotiation to close.

2. Obtain a Better Understanding of the Customers—Their Wants and Needs

CRM applications, often used in combination with data warehousing, e-commerce applications, and call centers, allow companies to gather and access information about customers' buying histories, preferences, complaints, and other data so they can better anticipate what customers will want in terms of the products and/or services provided by the company. This information is used in the planing and execution of marketing campaigns. It also enables customers to seek products and reveal their preferences in an interactive manner.

3. Define the Appropriate Product and Service Offering and Match it to the Customer's Unique Needs

CRM provides customization and personalization capabilities that give customers the power to view the enterprise in a way that they can relate to, thereby making it easier for them to do business with it. This includes configuration, pricing, quotation, catalog, and proposal generation capabilities that harness the power of Internet while ensuring the flexibility to respond quickly to changing technical and business conditions.

4. Manage and Optimize a Company's Sales Cycle

The productivity of the sales process is increased by accelerating the contracting process and improving revenue velocity. This is accomplished through capabilities such as, online order entry, credit card processing, tax calculations, auctions, billing, order status and payment processing. CRM solutions also include tools (such as Internet Fulfillment Serve, IFS), which provide the ability to communicate important information from supply chain modules to the customer interface in real time. These tools can help in determining feasibility, profitability, and delivery dates, while understanding the constraints of the entire supply, production, and logistics chain across multiple channels and enterprises.

5. Identify Cross-Selling and Up-Selling Opportunities

CRM can help in identifying opportunities for cross-selling and up-selling of higher value-added services to existing customers based on their past purchasing behavior.

6. Increase Retention of Existing Customers through Improved After-sales Service and Support

CRM applications document all post-close service- and support-related interactions with the customers, track customer requests and feedback from a variety of communications channels, and use this information to anticipate the demand for service and technical assistance, and maximize customer satisfaction and retention while minimizing customer service staff. The goal is to instill greater customer loyalty. CRM provides capabilities for providing online support information, online product registration to an electronic help desk, self-service support logging and tracking, and integration with call centers. Sophisticated CRM solutions also provide decision support tools for spare part planning, field service, and scheduled maintenance, repair, and overhaul (MRO).

12.3 NEED FOR CRM

Companies have to Increasingly Pursue a Customer-Centric Competitive Strategy rather than Product-Centric One

Two trends have brought CRM to the forefront. First, with increased global competition and easy access to the latest technology, products have become harder to differentiate, hence companies are increasingly pursuing a customer-centric competitive strategy rather than product-centric one. Second, now it's possible to put customer information from all over the enterprise into a single system rather than in a number of different systems and access it from anywhere in the world through the Internet (Mills, 1999).

CRM was the Logical Next Step to ERP

Since traditional ERP applications did not include a customer management aspect, CRM was the logical next step.

E-Customers Demand Constant Access, Immediate Response, and a Personalized Touch

The customers in the e-business age expect constant access to a company, through e-mails, call centers, faxes and websites. They demand immediate response and a personalized touch. Meeting their needs places new demands on the enterprise, which CRM alone can satisfy.

Focus is Shifting from Supply Chain to Demand Chain effectiveness

With the product quality at all time highs, manufacturers find it increasingly difficult to gain competitive advantage based solely on product attributes. Therefore the focus now is on channel operations and customer relationships, where there is great potential for adding value and differentiating the offerings. Increasingly, companies are applying sophisticated information technology to identify, acquire, and retain the most profitable customers. Their ultimate goal is to deliver greater value to customers by continuously improving the quality of the customer experience, thus, creating and sustaining the highest levels of customer satisfaction. Thus, in the era of e-business, the key source of competitive advantage is shifting, from optimizing the supply chain to taking control of the demand chain. Although an efficient supply chain is still necessary to stay competitive, it no longer provides long-term competitive advantage on its own. Hence, forward-thinking firms are revising their strategy to focus on demand chain effectiveness that is, on continuously improving their ability to identify, acquire, and retain profitable customers.

Better Understanding and Intelligent Management of Customer Relationships is Essential for Survival

The effect of increased commoditization of products and production processes is that customers now have more and more choices. Hence, giving them better, more personalized product offerings and services is the only real way to make a difference. Winning the battle for share of mind and share of wallet in this new economy depends on understanding and intelligently managing relationships with customers. It is no longer the privileged domain of particularly successful companies; it has become the make-or-break challenge for every company.

12.4 EFFECT OF IT AND INTERNET ON SUPPLIER-CUSTOMER RELATIONSHIP

IT Facilitates Better and Faster Understanding of Customer Needs Data collection and interpretation tools provide insights into what customers want, and how they can be served in the most profitable way.

Internet has Opened a New Interaction Channel Enabling the Company to Access Each and Every e-Customer in the Cyber Space Direct, real-time interaction with customers via the Internet is dramatically changing the way companies manage and drive customer relationships. It has rapidly become an important new interaction channel for financial institutions, retailers and hightech companies. A company can now access (or be accessed by) virtually every customer on the web. As the Internet is becoming increasingly more important for business transactions, companies now have to collaborate over the web in order to meet customer demands—both with consumers (in order to better know their preferences) and with business partners (in order to offer even better choices and delivery quality to customers).

Internet has also Opened the Floodgates of Global Competition making Customer Retention Extremely Difficult The range of choice and convenience of the Internet make it much more difficult to retain customers—other e-competitors' offerings are *always* just one click away.

Whether through the click of the mouse, a toll-free call, or a visit to a store down the street, today's customers can switch to a competitor with unprecedented ease. Falling transaction costs have increased the international division of labor. Many enterprises in the retail and automotive industry are now competing on a global scale. Transaction costs close to zero, via the Internet are now increasing competition even more, since companies can reach out for global markets. New technology (such as, hand-held computing devices such as smart phones or palm pilots) have extended the Internet's reach even further.

Today's e-Customers Expect Interaction with the Companies Serving them at Internet Speed.

12.4.1 What is Expected of a Comprehensive CRM Solution?

Today's CRM Solutions have to be Web-enabled, must be Integrated with the Existing Systems, and Synchronized with other Processes In the networked economy, companies will only be truly customer-centric when they are able to take their CRM solution to the Internet. As customers expect consistent treatment across all touch points, CRM processes and applications need to be closely synchronized. As customer relationships are linked to almost every process in a company, integration with existing systems and synchronization with other processes are crucial requirements. For example, excellence in customer relationships and higher customer satisfaction levels cannot be attained without seamless and excellent fulfillment processes in operations. Synchronization of data with all other systems is also necessary to achieve a consistent view of the customer throughout the organization and provide personalized experience for customers.

CRM Solutions must be 'open'—Having Flexibility for Incremental Implementation and Improvement Most companies grow customer-centric along a migration path. This means changing processes, creating new ones, and building the corresponding capabilities across the organization. CRM solutions must therefore be open, so that they can be adapted to specific business needs and implementation speeds. And companies are concerned about how today's decisions may affect their application landscape in the future. Can new CRM applications be easily added to those being installed today? What happens to the CRM applications installed when a new standard in enterprise applications emerges? Reliable, long-term partnership with the CRM vendor and flexibility for incremental implementation and improvement can protect CRM investments over time.

12.5 WHAT CRM CAN DO FOR VARIOUS USERS?

(a) Internet Customer

When a customer wants to buy a product on-line, he can simply log onto a company's web site, where he is greeted by name and receives recommendations of new products based on his past buying behavior. He can select the desired product from a catalog, configure it according to his wants and needs, add it to his shopping basket and place the order there and then.

(b) Field Sales Representative

A field sales rep, who is frequently on the road, can have a mobile sales component installed on a laptop or a hand-held device to keep him up to date with the latest information on products, customers, prices and competitors. He can also organize his daily activities, his calendar and his sales leads.

(c) Production Manager

When a production manager has a problem with part of the machinery, he can log onto the manufacturer's portal. The system will recognize which customer is accessing the site, and immediately displays information on his installed base, including the model type. A problem description is entered and the system comes back with a suggestion.

(d) Contact (Call) Center Agent

When contact center agent needs information on the customer, the equipment, and the service contract, he can have it at his fingertips. When a customer rings the call center, the system identifies the customer's number and routes the call immediately to the agent in charge. This person can see the customer's warranty coverage, repair and billing history, and other documentation related to the customer's equipment.

12.6 THE CRM PROCESS

A fully integrated CRM system as shown in Fig. 12.1 collects data about customers from many different customer touch points (for example, through telephone, fax, e-mail, Internet, or face-to-face contacts) and stores the information in a centralized database. Once collected, stored, and organized, this information can be analyzed and accessed in a number of ways by various users within the enterprise. A comprehensive CRM system has the ability to seamlessly integrate customer information from across the enterprise into a user-friendly, real time format, and transform the entire enterprise into a single integrated sales network. This model requires sufficient cohesion such that at every instance of contact with a customer or prospect, enterprise personnel, have access to a comprehensive overview of the marketing/sales/service relationship. Such an overview would include, but would not be limited to, the customer's or prospect's initial contact with the enterprise, their current status within the sales cycle, account status, post-sale interac-

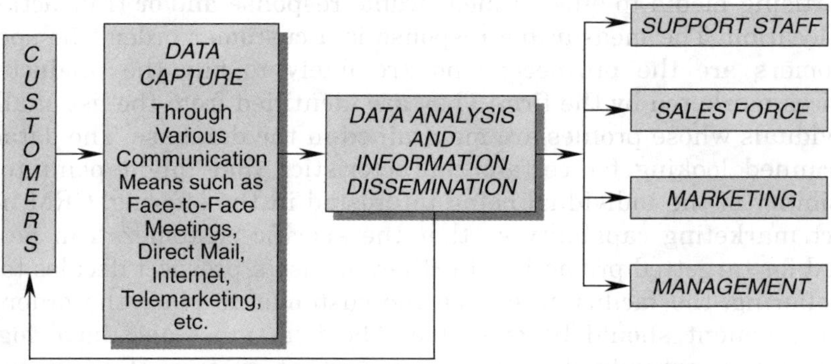

Fig. 12.1 *The CRM Process.*

tions with customer service and support personnel, and suggestions for additional sales opportunities based upon previous purchases and inquiries (Cherry Tree & Co, 2000).

12.7 CRM PROCESS COMPETENCIES

1. Cross-Selling and Up-Selling

Cross-sell and up-sell software typically include the capability to qualify prospects, track contacts, and refer them to salespersons when appropriate. *Event-driven marketing* is one aspect of cross-selling, for example, in a bank an event would be a large deposit, which would then trigger a salesperson to call the customer and ask if he or she would be interested in investment options. Cross-selling and up-selling can also be based on identifying the life-path needs of the prospects, for example, banks could recommend assets such as money markets, bonds, and annuities to those of their customers who are approaching retirement. If customers with young children could be identified, then banks could cross-sell education savings plans. Cross-sell and up-sell software may be used to schedule sales calls, keep detailed records about sales activities, and check on the status of customer orders. This software may also be integrated with inventory software (to see what products are in stock) or field servicc/external customer support (to learn how the product is working for the customer).

2. Direct Marketing and Fulfillment

Direct marketing uses mail, telephone, fax, e-mail, and/or Internet to communicate directly with or solicit a direct response from specific customers and prospects. It has its roots in direct mail and catalog marketing. The Direct Marketing Association (DMA) defines direct marketing as an interactive marketing system that uses one or more advertising media to effect a measurable response and/or transaction at any location. The measurable response is a customer order. The specific customers are the prospects who are likely to buy the product and services marketed by the firm. They are identified from the list of all the individuals whose profiles are maintained in the database. The database is scanned looking for certain characteristics that might point to the possibility of the individual being interested in the offering. CRM needs direct marketing capability so that the specific customers can be contacted for targetted promotion. Further, in case a prospect decides to buy the offering, the facility to enable the customer to place the order and make payment should be available. The firm must also have logistic capability to actually deliver the product at customer's place in the shortest possible time, and thus fulfill the customer requirements.

3. Customer Service and Support

The applications include support for service request management, account management, contact and activity management, customer surveys, return material authorizations, and detailed service agreements. These discrete applications work together to ensure that customer service representatives can quickly assign, create, and manage service requests, as well as look up detailed information about customer service contracts, contacts, and activities. Customer support capabilities are used to manage customers who are having problems with a product or service and to resolve those problems. Help-desk software automates the management and resolution of support calls. These applications typically include capabilities to verify customer status (for example, what level of support they are entitled to), open trouble tickets, track specific tasks needed to resolve problems across multiple workgroups, monitor service-level agreements, maintain permanent incident histories, and capture support costs for charge backs (Kalakota and Robinson, 2000).

4. Field Service Operations

Field service is the hands-on extension of external customer support, activated when a problem cannot be solved over the phone and requires sending a repair person to the customer site to perform maintenance or repair services. Field service and dispatch applications have become mission-critical tools that affect a company's ability to deliver effective customer service and contain costs. Field service software provides service organizations with features for scheduling and dispatching repair personnel, managing inventory and logistics, and handling contracts and accounting (Kalakota and Robinson, 2000).

5. Retention Management

Effective CRM requires that we differentiate between customers based on account and transaction histories and spend the valuable company resources wisely on the customers who count. Detailed knowledge about customers allows companies to treat all customers individually and, in many cases, disengage from or 'fire' customers who are high maintenance, low margin prospects (Kalakota and Robinson, 2000).

12.8 THREE KEY ENABLERS FOR IMPLEMENTING A CRM STRATEGY

1. Qualified Professional People

It is critical to hire professional people who know how to use the technology and then extensively train them on dealing with customers in a professional manner.

2. Well-Designed Processes

Without well-designed processes, any discipline will fail to meet its objectives. CRM is no exception. Companies should define their business requirements and objectives and develop CRM processes to meet these requirements.

3. Leading-Edge Technology

Finally technology should include at least the following elements:

(a) Business Rules for Efficient Transaction Processing Business rules are needed to ensure that any transaction with the customer is processed in an efficient manner. For example, if a company wants the most profitable and high volume customers to be serviced by experts, the business rules should clearly define what that criteria is. Based on the complexity of transactions, an organization may need hundreds of business rules.

(b) Data Warehousing Managing relationships with the customers depends on customer information, which is usually in various disparate databases. Consolidating the relevant information in one place and making sure that the information interrelates is not an easy task. Once done however, data warehousing augments a company's revenue potential and customer service. For example, a company can segment the types of customer it has in the data warehouse and launch a marketing campaign geared toward specific types of customers. Similarly, good data warehousing can help in presenting the information based on certain business rules to help in cross-selling and up-selling to customers calling for other reasons.

(c) Web to Facilitate Self-Service, EBPP, and Instant Messaging

The most important use of the web from the CRM perspective is self-service so customers can make inquiries about their accounts any time from any where. The web should also be used for Electronic Bill Presentment and Payment (EBPP), so customers can see what they owe and pay online if appropriate. For revenue-enhancement, companies can also provide instant messages to be used for cross-selling and up-selling services based on the profiles of customers using their Web site.

(d) Interactive Voice Response (IVR) System for Self-Service through Telephone

It is required for customers to do self-service inquiries via the phone instead of the web.

(e) Reporting Tools

Good reporting tools are necessary for both customer reports as well as internal reports.

(f) Call Center Technology

A call center is a group of agents and Voice Response Units (VRUs) that assist customers with support, inquiry, and transaction functions. Some type of call center technology with PBX or VoIP (Voice over Internet Protocol) integrated with intelligent call-routing is an absolute must for an interface with the live customer service representatives.

(g) Integration Framework

A technology framework that allows all the applications and databases that have customer information to be integrated can make a big difference in implementation.

12.9 CRM BENEFITS

- Identifying and targeting the most profitable customers and having a deeper knowledge of customers.
- Getting more marketing or cross-selling opportunities.
- Ability to manage marketing campaigns with clear objectives.
- Doing more one-to-one marketing enabling the formation of individualized relationships with customers.
- Obtaining information that can be shared with the company's business partners to effectively build relationships between the company, its distribution partners, and the customer base (Mills, 1999).

CRM solution increases profitability and boosts market share through:

A. Increased Revenues and Reduced Costs

- Increased market share due to superior customer relationships.
- Decreased marketing and promotion expenses.

- Improved channel efficiency.
- Increased parts and service revenues.
- Improved demand-creating collaboration.
- Profitable cross-selling and up-selling opportunities to existing customers.
- Increased repeat business revenues from customized, personalized, customer online experience—maximize lifetime profit per customer.
- Reduced service expenses with online self-service.

B. Increased Return on Service/Support Investments

- Reduced service parts inventory.
- Reduced physical storage space of service parts.
- Improved service personnel productivity.
- Increased equipment uptime, increased first-time fix rates.
- Improved inventory forecasting with service parts suppliers.

C. Increased Customer Satisfaction and Responsiveness

- Faster response to customer inquiries.
- Reduced lost sales and stockouts due to poor planning.
- Increased critical parts availability.
- Increased efficiency through automation.
- Receiving customer feedback that leads to new and improved products or services.
- Increased desire fulfillment and service response times.
- Better management of wider product assortments that meet segmented customer demands while managing profitability.

12.10 CRM MARKET LEADERS

The top vendors of CRM software include Siebel, Vantive (now part of PeopleSoft), and Clarify along with ERP/SCM vendors SAP, Baan (Baan FrontOffice), i2, and Oracle Corp. In the growing segment of CRM professional services, market leaders include Andersen Consulting, Cambridge Technology Partners, CSC, Deloitte Consulting, EDS/Centrobe, eLoyalty, Ernst & Young, IBM Global Services, KPMG, and PriceWaterhouseCoopers (Mills, 1999).

12.11 HOW CRM CAN SUPPORTS VARIOUS USER ROLES?

(adapted from www.sap.com)

SAP is now offering an open Customer Relationship Management solution as a key component of its mySAP.com environment, to be accessed like all other mySAP.com components through role-based portals. For example, Customer Relationship Management with mySAP.com is delivered through user roles as described below.

1. Internet Customer (e-Customer)

To the Internet customer, the Internet customer portal can provide with a new, unassisted, personalized sales channel offering interactive sales dialogues based on:

(a) customer profiles,
(b) multi-media product catalog searching,
(c) online configuration and pricing capabilities,
(d) availability checks, and
(e) shopping basket management.

- The Internet customer portal can be used as a marketing instrument to direct product information at specific target groups so that each e-Customer receives personalized product, service and support information from the company.
- Through the Internet customer portal, the Internet customers can access effective self-service functions, such as advanced decision support for problem determination and resolution, enabling them to dramatically reduce service costs.

In addition to offering these services in a business-to-consumer context, it also offers specific role-based portals for business partners and resellers, e.g. for purchasing agents. These business-to-business and business-to-reseller scenarios create direct links and enable companies to streamline their collaboration processes.

2. Webshop Manager

The CRM user role for the Webshop Manager allows him to create and update product catalogs by automatically or manually assigning products, variants, multimedia objects such as text, picture, sound, etc. and other attributes. It enables him to set up attractive webshops that provide unique and user-friendly shopping experiences for customers and business partners.

3. Sales Manager

The CRM user role for the sales manager enables him to plan, organize and implement sales strategies with efficiency. It provides him with a consolidated view of the sales pipeline, sales performance and sales budget. In addition, the sales manager can monitor the sales force and distributors, review sales activities and track opportunities. With this integrated view of sales activities, the sales manager is empowered in his decision-making, resulting in more effective execution and more precise monitoring.

4. *Sales Representative*

The CRM provides the sales representative all the functions and tools he needs to plan, execute and monitor sales activities. For example, CRM can be used to prepare and generate quotations, take orders, and provide sales and product-related information to the customers.

5. *Field Sales Representative*

The field sales representative can access the latest facts and figures on customers, products and competitors through his laptop or hand-held device. He can manage his calendar, his sales activities, and his opportunities. Functions are provided for integrated order management, pricing and configuration. Sales representatives are better informed and no longer cut off from their organizations. Information about changes in product offerings or moves by the competition is continually made available to them. This results in higher productivity of the mobile sales force, shorter sales cycles, and greater customer satisfaction.

6. *Customer Service Manager (Sales Focus)*

The customer service manager oversees the activities of customer service representatives and/or order processing staff. Through CRM system, he can monitor the global sales administration process, including quotations, order management, delivery and billing. He can track customer problems and provide solutions, ensuring overall customer satisfaction. Thus with CRM, customer service manager has all functions, information and services he needs to ensure consistent customer service and satisfaction.

7. *Customer Service Representative*

The customer service representative has to respond to customer requests, questions, complaints, and inquiries about product and account information. He can use CRM system to access data on customers, orders, deliveries and invoices data, as well as service and contract details. He also processes sales documents such as quotations, orders, contracts, and scheduling agreements. He has all the information he needs at his fingertips to immediately and precisely respond to customer inquiries.

8. *Business Sales Analyst*

The main tasks of the business sales analyst include supporting managers and executives in monitoring global sales profitability and implementing sales strategies. Via CRM system he can analyze past and future product performance. He also plays a supporting role in the formation of pricing strategy and price list analysis. He creates and processes reports covering profitability analysis, price lists, the profitability of promotions, and much more. Using the Internet he can track information about the competition. With these comprehensive analyses, taking into account market as well as customer information from all touch points and the entire supply chain, he can provide management with effective support for decision-making.

9. Contact (Call) Center Manager

The contact center manager is responsible for the primary point of interaction with customers. Customers contact the center via telephone, e-mail, fax and mail to make inquiries, submit orders, report problems and gain access to every kind of customer service. The contact center manager's responsibility is to assure their wants and needs are satisfied. Using his CRM system, he analyzes calls and e-mails as well as warranty and other contractual agreements. He handles complex or critical cases himself, such as complaints from key customers. As the contact center agents' team leader, he maintains the organizational structure, organizes the agents' schedules and assesses their skills and performance, assuring efficiency and the highest possible levels of customer satisfaction.

10. Contact Center Agent

The contact center agent is a company's first line of customer interaction: he answers customer calls and e-mails, handles requests and complaints, and takes orders. CRM system enables him to resolve customers' complex problems quickly, using the interactive intelligent agent, an adaptive learning engine that leverages the knowledge gained in past interactions. If necessary, he can also forward requests to service engineers. Interactive and reminder scripting support the contact center agent in marketing and sales-related situations as well, enabling companies to leverage the interaction center for cross- and up-selling opportunities.

11. Knowledge Engineer

As a product specialist, the knowledge engineer has to help customers and contact center agents resolve complex problems. He is also directly involved in working on, resolving and closing service requests, and forwarding problems to field service engineers. He is responsible for training Contact Center Agents to deal with technical questions. Besides maintaining customer and contact master data, the knowledge engineer uses CRM system to maintain data on equipment specifications, symptoms of problems and their solutions. He also analyzes the technical root causes of problems and complaints and informs Quality Management and R&D accordingly. He ensures that employees directly interacting with customers have all the technical knowledge at their disposal to handle requests promptly and provide optimum solutions.

12. Resource Planner for Service Centers

Using CRM system, the resource planner for service centers plans and schedules the activities of field service engineers and keeps them informed about service orders based on request priority using fax, e-mail, or WAP. He keeps track of the skills requirements of field service engineers and maintains profiles of their current qualifications. With CRM system, he monitors order fulfillment, analyzes service requests and service orders as well as information on warranties and contracts. He also checks field service activities and plans future resources. He ensures that the right resources are assigned to requests, so that customer needs can be met speedily and to their complete satisfaction.

13. Customer Service Manager (Service Focus)

The customer service manger initiates and coordinates *retention management strategies.* Using CRM system, he analyzes levels of satisfaction with products and services. A primary task is to develop key performance indicators for service processes, and to define services and standard contracts to meet demands of customers in a competitive market. He monitors revenues and profitability of services, and communicates findings to R&D, production, as well as sales and marketing. He also establishes and maintains relationships with external service providers and partners. With CRM system, he has access to all processes and activities within customer service, enabling him to leverage information provided by employees, customers, surveys and third-party organizations. Graphical analysis and reports based on key performance indicators (both qualitative and quantitative) allow him to monitor operational service processes, including complaints, faults and defects, returns and exchanges, escalations, problem resolution, costs, and revenues.

14. Field Service Engineer

The Field Service Engineer uses his laptop or handheld device to review his daily schedule, to prepare his work and to report on work done. He performs onsite service activities, fulfilling service orders and confirming time and material spent on the job. Supported by the power of mobile computing, the field service engineer is more effective, cutting costs and improving the quality of the work performed, for example, in repair and maintenance operations.

15. Contract Administrator

The contract administrator advises customers and creates, maintains, and monitors quotes and contracts for them. He uses CRM system to manage billing processes, including assessing credit worthiness and payment history. He maintains information on the installed base, and analyzes warranty details. Other functions include analyzing requests and order history of contracts, monitoring receivables, maintaining pricing, and dealing with complaints. The Contract Administrator is also in charge of creating reports on revenues, and aggregated reports on customers and contracts.

16. Marketing Manager

The marketing manager develops the corporate marketing strategy. CRM system allows him to define and monitor marketing objectives and the corresponding budget, and ensure the effectiveness of the marketing organization. He establishes partnerships with key customers and suppliers, and monitors the market, including the competition. The marketing manager uses his CRM system to access information on all marketing operations and findings of research and analysis. He benefits from powerful functionality for flexible marketing and campaign planning, and much more.

17. Campaign Manager

The CRM user role for the campaign manager offers integrated and synchronized capabilities. It features campaign development, including planning and target group optimization, campaign execution functionality across all touch points and monitoring and performance analysis. It is possible to initiate and coordinate a variety of activities, achieving proactive management of customer interactions and improved marketing accountability and effectiveness.

18. Marketing Analyst

Using CRM system, the marketing analyst can view up-to-the-minute information on product lines, key customer accounts, competitors, etc. He has the analytical tools to leverage internal and external sources that allow him to determine, for example, customer lifetime value based on transaction data. He can also generate and refine customer profiles, generate competitor analyses, and determine market potential. He can use the findings for improved decision-making, and to build effective marketing campaigns, such as for customer retention.

12.12 THE EXTENDED ENTERPRISE

Traditional ERP is now being viewed not as a comprehensive solution but as the backbone to which customer- and supplier-facing applications may be linked as a means of extending the functionality of enterprise software. At the core of the extended enterprise sits the core ERP backbone, which resides *within* the enterprise and can be described as being primarily *inward* facing applications that track *internal* flow of information. An enterprise starts to become 'extended' when its information systems face *outward* by enabling connectivity with customers, suppliers, and distributors as shown in Fig. 12.2. SCM and CRM are the examples of enterprise applications that enable the connectivity between all the participants in the business process. A company completes its evolution and becomes a truly extended enterprise when its connectivity with its business partners becomes fully integrated into its ERP backbone (Cherry Tree & Co., 2000).

Despite the value that extended enterprise applications such as SCM

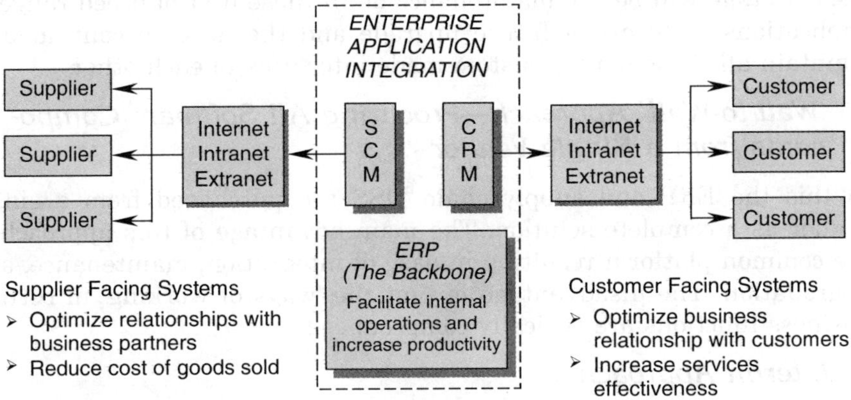

Fig. 12.2 *The Extended Enterprise* (Adapted from Cherry Tree & Co. 2000).

and CRM bring to the table, their full potential can never be realized without the technological 'glue' that integrates them into the core of the enterprise. A diverse set of tools and technologies have evolved out of the general category of middleware that are loosely grouped together under the fabric of enterprise application integration (EAI) to serve as the integrating environment for the enterprise applications.

12.13 ENTERPRISE IT SOLUTIONS: 'BEST-OF-BREED' VERSUS 'WALL-TO-WALL'

Enterprise IT solutions consist of many pieces that need to be assembled in order to achieve competitive edge. They include infrastructure (ERP) and various decision support systems (DSS), such as SCM and CRM. Two extreme approaches can be taken:

1. Build a 'Best-of-Breed' Solution Using Several Best-Fit Products from Different Vendors

In this case, the best-fit solution in each category is procured from a different vendor, thus producing a system that better fits each function in the company. The main advantages of this approach are that it allows an organization to choose the best fit application for it, so that the company is assured of the leading IT solution in the chosen area, thereby having access to the best knowledge and expertise, it provides greater long-term flexibility and makes the company less dependent on any specific vendor. The main disadvantages are in the areas of interfaces, with other systems and also suppliers, high cost, and longer implementation period and difficult maintenance due to the complexities involved in the variety of infrastructure, database, and software. The greatest disadvantage will be the maintenance of all these best-of-breed range of applications in terms of future upgrade and the need to continuously maintain all these mostly custom-made interfaces to each other.

2. 'Wall-to-Wall' Approach—Procuring All Software Components from a Single Vendor

In this the ERP and supply chain DSS are purchased from a single vendor as a complete solution. The main advantage of this approach is the common platform resulting in ease of integration, maintenance, and upgradation. The disadvantage is that the 'ways of working' in certain business functions are typically compromised.

3. Interim Approach

Many companies choose an interim approach which has a dominant ERP provider; the functionality that cannot be provided by the vendor, or does not suit the company, is provided through best-of-breed or in-house system.

4. In-House Proprietary Software Development

There are companies, for example Wal-Mart, that still prefer in-house, proprietary software development. This probably makes sense for extremely large companies with expert IT departments and systems that already serve the company well. Table 12.1 gives a comparison between the three kinds of IT Enterprise Solutions.

Table 12.1

Comparison of Best-of-Breed, Single Vendor, and Proprietary Solutions

Implementation Issues	*Best of Breed*	*Single Vendor (Wall-to-Wall)*	*Proprietary (In-House)*
Length	2–4 Years	Up to 24 months	Not known
Cost	Higher	Lower	Depends on expertise
Flexibility	Higher	Lower	Highest
Complexity	Higher	Lower	Highest
Quality of Solution	Higher	Lower	Uncertain
Fit to the enterprise	Higher	Lower	Highest
Staff training	Longer	Shorter	Shortest

References

Cherry Tree & Co, *Extended Enterprise Applications: Spotlight Report,* January 2000. A white paper prepared by Cherry Tree & Co available at the ITtoolbox Portal for SCM at the site http://scm.ittoolbox.com

Kalakota, Ravi and Marcia Robinson, *e-Business: Roadmap for Success.* Reading, Addison-Wesley, Massachusetts, 2000.

Khera, Mandeep, *Customer Relationship Management: Beyond the 'Buzz.'* Available at the ITtoolbox Portal for CRM at the website http://www.crmassist.com, 2000.

Mills, Josh, *CRM Overview.* Available at the *ITtoolbox Portal for CRM* at the web site http://www.crmassist.com, 1999.

Appendix 1

Related Websites

A.1 Forecasting Related Sites

1. Delphus	www.delphus.com
2. SFT Inc	www.MRP3.com
3. Institute for Business Forecasting	ibforecast.com

A.2 Scheduling Related Sites

1. Waterloo Software, Inc	www.waterloo.software.com
2. Scheduling Technology Group	www.stg.co.uk

A.3 Inventory-Control-Related Site

www.effectiveinventory.com

A.4 MRP-Related Sites

1. Alliance Manufacturing, Inc	www.alliancemfg.com
2. Cimpal Inc	www.cimpal.com
3. Expandable Software, Inc	www.expandable.com
4. Expert Buying Systems, Inc	www.choosesmart.com
5. Hampton-Tilley Associates, Inc	www.isenet.com
6. Info Tech Systems, Inc	www.itsystems.com
7. Logistics Planning Associates, LLC	www.psiplanner.com
8. Manufacturing Action Group, Inc	www.mrpii.net
9. Obvious MRP	www.obvious.com
10. SFT Inc	www.MRP3.com
11. User Solutions, Inc	www.usersol.com

A.5 ERP-Related Sites

1. Baan Fan Club	www.baanfans.com
2. Cyber India On Line	www.ciol.com
3. ERP Fan Club	www.erpfans.com

4. Midrange ERP	www.midrangeERP.com
5. Oracle Fan Club	www.oraclefans.com
6. PeopleSoft Fan Club	www.peoplesoftfans.com
7. SAP Fan Club	www.sapfans.com

A.6 ERP Vendors' Websites

1. Abacus	www.abacus.ch
2. BaaN	www.baan.com
3. CINCOM	www.cincom.com
4. Computer Associates	www.cai.com
5. Dataworks	www.dataworks.com
6. Dun & Bradstreet Software	www.dnb.com
7. ERPlite	www.erplite.com
8. IFS	www.ifs.com
9. J D Edwards	www.jdedwards.com
10. Marcam	www.marcam.com
11. Navision	www.navision.com
12. Oracle	www.oracle.com
13. PeopleSoft	www.peoplesoft.com
14. QAD Inc	www.qad.com
15. Ramco System	www.ramco.com
16. Ross System	www.rossInccom
17. SAP AG	www.sap.com
18. SSA	www.ssax.com

A.7 SCM/CRM Related Websites

1. Baan Company	www.baan.com
2. Clarify eBusiness Applications	www.clarify.com
3. Ernst & Young:	www.ey.com/mc/sc
4. Fed Ex Learning Laboratory:	www.fedex.com/logistics
5. i2	www.i2.com
6. Manugistics	www.manugistics.com
7. Oracle	www.oracle.com
8. PeopleSoft	www.peoplesoft.com
9. SAP	www.sap.com
10. Siebel	www.siebel.com
11. Strategic Supply Chain Management Program at the Consortium for Advanced Manufacturing International (CAMI)	http://biz.onramp.net

Index

Accelerator (bullwhip) Effect 539–40
Action buckets 123
Action notices 130–33
  number of 148
  report 133
Adaptive mass customizers 232
Add-ons (additive choices) 30
Aggregate planning 255
Allocated stock 117
Allocations
  time-phased 139
  uses of 108–9
APICS Certification (CPIM, CFPIM, and CIRM) 89–90
Assemble to order (ATO) 29–41
  farm tractor family example 33–4
Assumptions and vulnerability analysis 265
Attachments 30
Available-to-promise (ATP) 312–4, 357–63
  backward method of computing 359–60
  cumulative with lookahead 360
  cumulative without lookahead 358
  discrete with lookahead 358–60
  discrete without lookahead 357
  logic with modular BOM architecture 38

BaanERP solution 509–12
Baan IV ERP solution 503–8
  menu browser 504
Backflushing 121–3
  single-level 123
Backward integration 222
Basic industries
  classification of 5
  examples of 12
Basic producers 13
Batch continuous production 7
Bill of material
  accuracy, methods of checking 417
  computerized 388
  costed 412–4
  definition 387
  errors, types of 413
  exploding through 93
  explosion process 97
  file(s) 116, 388
  functions of 412
  generic 34–6, 401–2
  indented 394–6
  items to be included in 389
  kit, K-bill, kit number 403
  maintenance of 416–7
  manufacturing, or M-bills 404–5
  master 389
  matrix (cross classification chart) 411
  modular 398
  multilevel or indented 394–6
  nomenclature 393
  phantom, blowthrough, or transient 405–10
  planning 34–6, 396 404
    variations of 398
  processor 389
  restructuring 391
  single-level 393–4
    advantages of using 394
  single-stage and multiple-stage 392–3
  structurc 388
  structuring 390
    principles of 391

summarized (part list) 410
super, S-, top-level planning, or master 400
super company 400
super family 401
super modular 34–6, 401–2
tree structure 390
types 393
where-used 414–6
Bill of resources 82, 441
Blanket ordering 183–4
Book date and quantity 310
Bottom-up replanning 145–7
Bucketless scheduling 339
Bucketless/vertical/list MRP I display 136–8
Buckets, time 123
Buffering against demand uncertainties 286
Bullwhip (accelerator) effect 539–40
Business planning 69–242
Business requirements planning (BRP) 175
Buyer (purchasing agent) 184

Calendar, manufacturing/shop 108–10, 449–50
Capacity
bills 82
control 425
decisions 246
definition 424
demonstrated 177, 435–6
expansion strategies (lead, lag, and average) 425–6
long-range to short-range balancing 428
management 424
measuring 432
rated 177
unit of measure 432
theoretical 433
Capacity available 432–6
Capacity-driven master scheduling 379–80
Capacity planner, role of 454
Capacity planning 425
in repetitive manufacturing 460
planning levels 429
principles as applied to repetitive and intermittent manufacturing 460–1
process 428
Capacity planning using overall factors (CPOF) 82, 441
Capacity requirements planning (CRP) 175–9, 430–2, 442–3
advantages and limitations 459
effects of not doing 459
inputs to 443
input/output model 443
outputs of 452
principles as applied to repetitive and intermittent manufacturing 460–1
CFPIM 89
CIRM 90
Closed-loop
approach 188
MRP 174
Collaborative mass customizers 232
Computer planned orders (CPOs) 124, 346
Configurator, product 36, 73
Construction and manufacturing industries 2
Contingency plans 226
Continuous replenishment 290
Converters 14
Core competence 219
Corporate decisions, framework for reflecting manufacturing issues in 240
Corporate strategy 218–9
Corporate strategy debate, necessity of involving manufacturing in 237
Cosmetic mass customizers 232
Cost leadership 227–8
CPIM 89
Critical resources 81
Cross classification chart (matrix BOM) 411
Cross docking 548–9
Cross-selling 559
Customer order
decoupling point (CODP) 41
deviations from the forecast 343
methods for analyzing 314
pieces of information required for 312

Customer relationship management,
benefits of 562
introduction 552
key enablers for implementing 561
market leaders 563
need for 555
objectives of 553
process competencies 559
process of 558–9
what it can do for various users 557, 563–8
Customer Service Operations 560
Customer-supplier linking 186

Data-omissions and errors 148
Decision-making levels, characteristics of 210
Decisions
capacity 246
facilities 246
make-or-buy 245
operational 208
process and technology 248
strategic 208
tactical 208
Deduct list 122
Deduct point 122
Deexpediting 181
Demand
abnormal/extra 314
comparison of dependent and independent 78
dependent 77
filter 302
independent 77
interfaces 308
Demand management 71
an information centered activity 278
definition 277
functions 281
fundamental requirements 280
interactions with MPS 283
interfaces 278
linkage with production planning 283
objectives of 72, 287
Demand time fence (DTF) 284, 369–70
Dependency
horizontal 81
vertical 81
Dependent Demand 77
Design parameters, hidden 31
Design rules, visible 31
Direct Marketing and Fulfillment 559–60
Discrete-item manufacturing industries 8
Dispatch list 477
Dispatcher 470
Dispatching 181
rules 478
Distinctive competence 224
Distribution 3

Econometrics 294
Effectivity dates 84
Efficiency 434
Engineering change control 83
through effectivity dates 84
through serial numbers 84
Engineering changes
discretionary, optional, or deferrable 85
factors to be considered in determining
the timing of introduction of 86
Mandatory or immediate 85
Engineer to order (ETO) 43
Enterprise IT solutions 569–70
Enterprise resource planning (ERP)
benefits of 521
definition 494
evaluation criteria 513–7
ease of implementation 515
future road map 517
product evaluation 513
future of 523
implementation 517–21
major difficulties in 521
methodology 518–20
teams 518–20
need for 497
selecting the right package 512
ERP and SCM systems, comparison of 543
ERP implementation case studies 524–35
Calcom Vision, New Delhi, India 527–9
Glaxo India Ltd. Mumbai 531–2

Godrej Telecom Ltd., Mumbai and Hyderabad 533–5
IBM 529–30
Mahindra and Mahindra, Mumbai, India 524–7
Raychem RPG Limited, Mumbai, India 532–3
Exception messages (MPS) 366–8
Exception messages/report (MRP I) 133–4
Expediting 181
Exploding through BOM 93
Explode to deduct (backflushing) 122
Extended enterprise 568–9
Extraction industries 2

Fabricators and assemblers 14
Facilities decisions 246–8
Factory coordination 180, 465–9
control task within 468–9
building blocks 468
production environment design 466–7
Faxban 187
Field service operations 560
Final assembly scheduling 76, 380
benefits of 384
comparision with MPS 382–3
mechanics of 383
Finite loading 180, 479
Firm/frozen MPS period 285
Firm planned orders (FPOs) 125, 347
First-to-market 229
Flexible manufacturing systems (FMS) 249
Flexible/slushy MPS period 285
Flow
of costs 59
of information 58
of materials 58
Flows supporting effective manufacturing 57
Focus forecasting 309
Focused organization 230
areas of focus 231
Forecast
accuracy, the principles of 309–10
aggregating and disaggregating of 307–9
and MPS 342
control 301
customer intelligence 46
disaggregating through super family BOM 308
rolling the 318
sales and production 341
updating using manager's views 305–6
uses of 289
Forecast consumption 315–26
example 324–6
parameter
backward 324
forward 324
procedure 322
Forecasting
associative 294
business 70
focus 309
informal or intuitive 292, 296
judgmental 293, 302–5
pyramid approach 309
qualitative 293, 302–5
seasonal adjustments 299
strategic role in SCM and TQM 289–91
subjective 293
system 288
basic functions of 306

Forecasting error measurements 299–300
bias or average error 300
cumulative error 300
mean absolute deviation (MAD) 299
mean absolute percent deviation 300
standard deviation 299
tracking ratio/signal 301
Forecasting methods
autoprojection 293
causal 294
classification of 291
Delphi 304
exponential smoothing 297
adaptive 301
factors influencing the choice of 295
historical analogy 304
life-cycle analysis 305
market survey 303
panel consensus 303

personal insight 303
projective 293, 295
naive or intuitive 296
simple moving average 295
quantitative or statistical 293
regression analysis of demand against a market indicator 294
testing before using (focus) 309
weighted moving average 296
Formalization of informal MRP Process 169
Forward integration 222
Free (unrestricted/liquid) MPS period 286
Free stock 117
Functional strategies 232
Funnel (product) structure 335

Game planning 255
Gross requirements 95, 97, 99, 138–9

Hedge or overplanning 352–3
product mix 353
volume 353
Hewlett Packard's manufacturing strategy development process 244
Hollow (or hollowed out) firm 220
Horizontal integration 222
Hourglass (product) structure 336
Human resources decisions 253

Independent demand 77
Infinite scheduling 176
Information
qualitative or symbolic 209
quantitative or numeric 210
Input/output control 181, 431, 482–8
Internet Fulfillment Serve (IFS) 554
Inventory
pipeline/transportation 72
level projections (forecasts) 134
turnover ratio 147
Intermittent (batch) production 18
Item 121
Identification (part numbers) 417–8
inactivity report 148
obsolescence report 148
Item master file 115–6, 471–2
basic or main 116
dynamic or subordinate 116

Job shop production 15
advantages of 17
disadvantages of 17

Kanban 489
supplier 187
Kit bills 403

Labor ticket 128
Late-to-market 229
Layout
functional or process 15
fixed-position 15
Lead time 121
offsetting 110
purchasing 182
safety 119
Life of a product, ways of extending 54
Load(s)
activity 449
planned 427
profiles 81
types of 427
unplanned 427
Loading 180
finite 180, 451, 479
horizontal 481
infinite 451
vertical 481
Logistics Network 531
Lot size 121
Lot sizing 100–8
adjustments 105
commonly-used policies 102–3
costs associated with 101
discrete 101
dynamic 104
nondiscrete 104
significance of 104
static 104
Low-level coding 94, 121

Machine loading 177, 456
Make-or-buy decisions 245
Make to order (MTO) 41
Make to stock (MTO) 27
Managerial Resource Planning (MRP

III) 175
Management levers 245
Manufacturing 1
calender 108
cycle 449
environments 26–46
evolution of 48
first law of 59
infrastructure decisions 252
infrastructure to support production 243
lead time elements 448
principles of effective 60
strategy 233–5
supervisor, role of 455
tactics 235
Manufacturing processes
difficulties in categorizing 8
Manufacturing resource planning (MRP II)
and JIT 197–200
and the realities of manufacturing: the mismatches 192–3
benefits of and prospects for 190–2
definition 63
differences in functionalities as applied to
discrete and process industries 202
framework 64–5
from MRP I to 174
implementation problems 190–2
information flow in 66
problems associated with 189–93
role in process manufacturing 201–202
Manufacturing strategy 233–9
development of 235
development process developed at Hewlett Packard 244
Japanese approach to 244
Market segmentation 230
Mass customization 231
categories of 232
Master planning 376
Master production scheduling (MPS) 75
activities 373–4
an iterative process 333
and the management 331
capacity-driven 379
communication link between strategic and tactical planning 332
computer assistance in 378–9
decisions 368, 371–3
definition 328
factors that go into developing 341
family totals per month 364
file 115
functions 330
in MTO environments 364
in process industries 379
inputs to 341
branch warehouse demands 343
customer orders 343
interplant orders 343
management decisions 343
vendor, material and capacity limitations 344
items to be included in 334
measures of performance 377
mixed-model 365
multi-level 337
need for 329
objectives 330
other items likely to be included in 337
providing stability through time fences 369
report (master schedule) 353
revision frequency 340
symptoms of mismanagement 377
time fence guidelines 286, 371
time fences 284–6, 369–71
Master Schedule (MPS report) 353–63
an example of 361
Master Scheduler
characteristics of 376
finite scheduling algorithm to assist 379
role of 374

Material requirements planning (MRP I)
bucketless/vertical/list display 136–7
comparison with horizontal display 138
changes that affect plans 154
daily transactions and planner's response 154–60
definition 87

development of 88
fields used in spreadsheet 138–42
inputs 115
nervousness
causes of 167–8
methods of reducing 168–9
net change (transaction driven) 164
characteristics of 166
disadvantages of 166
objectives and applicability 90–3
operating philosophy of 91
outputs 115, 124
program health monitors 148
regeneration frequency 163
regenerative 161
replanning 161
comparison of different method, 167
requirement alteration 164
sample daily transactions 154–60
sample exercise 150–3
schedule regeneration 161
spreadsheet 135
showing time-phased allocations 140
to MRP II 174
versus informal system 170
versus order point 170–4
Material shortage problem, possible solutions 145–7
Mini plant 25
Mission and vision statements, examples of 215
Mission statement 213
Mixed mode assembly lines 250
Modular design 30
advantages 31
developing 31
drawbacks 32
Monitor (expediter) 468, 471
Move ticket 128
Mover (material handler) 471
Move time 448
matrix 448
MPS family totals per month 364
MPS time fences 284–5, 369–71
for different manufacturing environments 286, 371
MPS periods
firm/frozen 285
flexible/slushy 285
free/unrestricted/liquid 286
MRO supplies 79
MRP III 175
MTO and ETO, common characteristics 43
MTS, ATO, M/ETO environments, comparison of 45
Multiple-source arrangement 186–7

Nervousness in MRP I 125, 167–9
Net requirements 141
Netting 100
New product introduction 253
Non-issues 237

Objectives
corporate 215, 241
examples of 218
financial performance 216
strategic performance 217
On-hand inventory balance 116
Open orders 129
file 130, 443, 473
Open requisitions
priority dispatching of 187
Operations
alternate 446
overlapping 479
sequencing 482
Oracles of Delphi 304
Order
blanket 183
closing 130
computer-planned (CPO) 124, 346
disposition 491
entry system 311
firm-planned (FPO) 125, 347
launching 126
management 345
modifiers 91
policy code 121
preparation 473–4
qualifiers 236
released 129–30, 348
releasing 126
winners 236
winning criteria 243
Outsourcing 220
Overplanning 352

Part list (summarized BOM) 410
Part numbering 417–8
  prerequisites and principles 41--22
Part numbers 417
  nonsignificant 418
  semisignificant 418
  significance 418
  significant 418
Past due 136
Pegging 142–5
  full 143
  report(s) 144–5
  single-level 143
  versus where-used 144
Performance measurements 147–8
  capacity 147
  customer service 147
  inventory 147
  productivity 147
  supplier 148
Performance reporting 147
PEST Analysis 224
Phantom assemblies, need for 406–8
Phantom, blowthrough, or transient BOM 405–10
  ensuring equivalence of actual and simplified BOMs 409
  use to phase-out old items and phase-in new items while implementing engineering changes 410
Physical distribution 3
Picking list 128
Pipeline inventory 72
Planned factory order releases 124
Planned orders 124–30
Planned order receipts 141
Planned order releases 142, 444
Planned purchase order releases 124
Planner
  basic responsibilities of 148–50
  code or number 121, 150
  response to sample daily MRP I transactions 155–60
Planning
  factors 123
  horizon 123, 338
  time fence (PTF) 285, 370
Plant scheduling 179, 488–9
Plant structures (VAT analysis) 335–7
  combinations 336
  funnel, inverted, or V 335
  hourglass, modular, or T 336
  pyramid, standard, tree, or A 335
Post-deduct inventory transaction processing (backflushing) 122
Pre-deduct inventory transaction processing 122
Pre-S&OP process 264–5
Primary industries 2
Primary (MRP I) Report 124
Priority planning 87
Process and discrete-item industries, the differences between 9
Process optimization 549
Process and technology decisions 248–52
  based on level of customization 249
  based on volume of product 249
Process choice 243
  relationship with product, process technology, and plant layout 252
Process industries 6
  job shop 7
Processes, comparison of various types 251
Producer (machine operator) 470–1
Product configuration, farm tractor family example 32–4
Product configurator 36, 73
Product differentiation 228
Product families, definition 258
Product life cycle
  and product positioning strategy 49
  characteristics 51
  comparison of various stages 52
Product structure 390
  file 472
  planning approach to 397
  record 390
  traditional approach to 397
Product variants (exclusive choices) 30
Production 1
  cellular 25
    advantages 25
    disadvantages 26
  classification based on production volume 14
  continuous 23

flow 22
characteristics of 122
intermittent (batch) 18
job shop 15
mass 19
modes of 2
quantity 23
repetitive 20–21
mixed model 21
advantages of 24
disadvantages of 24
reporting and status 490
smoothing, policies of 178, 456–9
Production activity control,
data requirements 462
overview of 462–3, 469–71
Production environment design 466–7
Production industries 6
classification of 13
Production planning 255
Production smoothing, policies of 178, 456–59
Productivity factor 434
Project execution 47
Project industries 11
Projected available balance (PAB) 140, 355–7
analysis of 356
formulas for 355
Projected on-hand 118, 139
Purchase commitment report 135
Purchase routing 187–8
Purchasing agent (buyer) 184
Purchasing lead time 182
Pyramid (product) structure 335

Queue time 447–9

Rate generator 379
Record errors 148
Release-to-stock-to-dock tracking 187
Released orders 129–30, 348
Replanning
bottom-up 145
frequency 124
Requisition tracking 187
Rescheduling messages 130–3
Resource profiles 82
Resource requirements planning (RRP) 70, 430, 436–7
Retention management 560
Rolling schedule 136
Rolling the plan 340
Rough-cut capacity planning (RCCP) 81, 430, 437–42
check 133
Routing
alternate 445
file 444, 472
organization of 446
primary 445
purchase 187–8

Safety, various ways of achieving 353
Safety capacity 352
Safety stock 119, 350–1
mechanics of using 351
a word of caution about 63
Safety lead time 119, 351–2
Sales and operations planning (S&OP) 73, 255–8
and MPS 262
as an input-output process 259
benefits 275
definition 255
factors that affect 259
integration issues 273
issues to be addressed 261
key management considerations 263
measurements 274
meeting 265, 271
activities performed in 266
agenda 260
attendees 270
decisions 266
need for management's direct involvement in 269–70
objectives of 258
process 263–4
as a monthly cycle 271
spreadsheet 266–9
strategic and tactical nature of 262
summary 275
SAP R/3 ERP solution
applications 500–3
basis 499–500
Schedule, controlling 478
Scheduled receipts 129–30, 139, 348–9
Scheduler 129, 139, 470
supplier (vendor) 182

Scheduling
  backward 450, 475–6
  bucketless or daily transaction 339
  code (order policy) 121
  detailed 181
  forward 450, 476
  finite 451, 479
  infinite 451, 477
  plant 179
  rate-based,
  strategies 450
  supplier (vendor) 182
Scrap 111, 114
  factor or rate 112
  fixed 111
  two methods of applying 112
Seasonal adjustments 299
Second-to-market 229
Secondary industries 2
Sequencing 181, 431
Service and other industries 12
Service industries 2, 12
Service levels 278
Services and goods, the differences 4
Setup—internal and external 449
Shipment date and quantity 310
Shop floor control
  activities performed in 179
  control 464
  control files 473
  definition 462
  implementation 464
  importance of 492
  measurement for control 490–1
  planning 463
  planning files 471–3
Shrinkage 112, 114
Smoothing constant 298
Spare parts issue 287
Speculative items 46
Staging 47
Stock
  allocated 117
  free 117
  physical 117
Strategic
  integration 239
  management, five tasks of 211–2
  mission 213
  planning 69, 205
  vision 213
Strategies, models for linking corporate, marketing, and manufacturing 241
Strategy
  adaptive 206
  and tactics: the differences 208
  business 218
  business process 232
  characteristics 205
  competitive 227
    categorization based on new product introduction timing 229
  corporate 218–9
  definition 205
  intended 206
  manufacturing 233–4, 244
    Japanese approach 244
  marketing 242
  need for devising 207
  organization 218
  planned 206
  unplanned 206
Strategy levels, primary responsibility and areas of focus at different 211
Strategy-making as a group task 213
Strategy-making process,
  characteristics of 212
Superflush 123
Supplier 186-7
  capacity planning 431
  invoices, matching of 188
  scheduling 182
  scheduler 185
Supplier-customer relationship, effect of IT and Internet on 556
Supplier kanban 187
Supplier scheduling and control (SSC) system 187
Supply chain,
  characteristics of 539
  definition 538
  performance measures 546
  objectives 541
  virtual 539
Supply chain application modules 549
Supply chain execution applications 550
Supply chain infrastructure, improving

the 548
Supply chain integration, difficulties and issues in 544
Supply chain planning applications 549
Supply chain structure, improving the 547
Supply chain management 540
and ERP 543
and logistics management 542
functions of IT systems in 545
objectives of 541
Supply chain relationships 253
SWOT analysis 224–225
Synchroflushing 122

Tertiary industries 2
Ticket
labor 128
move 128
Time-based competition 236
Time buckets 123, 338
Time fences 284–287, 369–71
demand time fence (DTF) 284, 369–70
fallacy of 349–50
planning time fence (PTF) 285, 370
providing stability through 284
Tracking
release-to-dock-to-stock 187
requisition 187
Transaction errors 148
Transaction messages 134
Transformation chain 13
Transparent mass customizers 232
Transportation inventory 72

Universal manufacturing equation 67
Utilization 433

Value analysis and engineering 185
Variant 30
VAT analysis 335
Vendor requirements planning (VRP) 181
Vertical integration 222, 233, 245
Virtual Company 539
Virtual Supply Chain 539
Visible design rules 31
Vision statement 215
examples of 215

Wait time 448
Web sites 572–3
What-if analysis 425
Wheelam and Hunger's model of corporate strategy formulation 226
Where-used 144–5, 414–6
indented 415–6
single-level 415
summarized 416
Work center
cumulative load profile 453
data elements 447
file 446, 472
load profile 452
load report 452–4

Yield 112